T0313732

ALLIS-CHALMERS

SHOP MANUAL AC-202

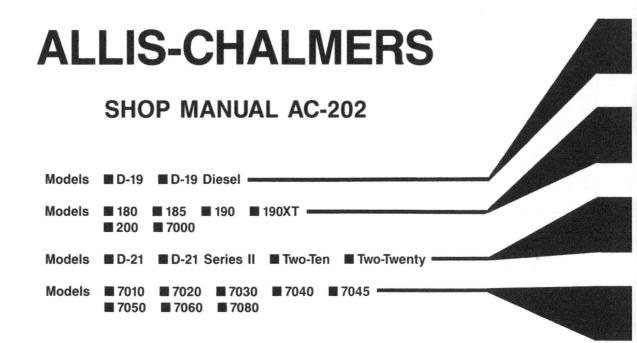

Models ■ D-19 ■ D-19 Diesel

Models ■ 180 ■ 185 ■ 190 ■ 190XT
■ 200 ■ 7000

Models ■ D-21 ■ D-21 Series II ■ Two-Ten ■ Two-Twenty

Models ■ 7010 ■ 7020 ■ 7030 ■ 7040 ■ 7045
■ 7050 ■ 7060 ■ 7080

Information and Instructions

This shop manual contains several sections each covering a specific group of wheel type tractors. The Tab Index on the preceding page can be used to locate the section pertaining to each group of tractors. Each section contains the necessary specifications and the brief but terse procedural data needed by a mechanic when repairing a tractor on which he has had no previous actual experience.

Within each section, the material is arranged in a systematic order beginning with an index which is followed immediately by a Table of Condensed Service Specifications. These specifications include dimensions, fits, clearances and timing instructions. Next in order of arrangement is the procedures paragraphs.

In the procedures paragraphs, the order of presentation starts with the front axle system and steering and proceeding toward the rear axle. The last paragraphs are devoted to the power take-off and power lift systems. Interspersed where needed are additional tabular specifications pertaining to wear limits, torquing, etc.

HOW TO USE THE INDEX

Suppose you want to know the procedure for R&R (remove and reinstall) of the engine camshaft. Your first step is to look in the index under the main heading of ENGINE until you find the entry "Camshaft." Now read to the right where under the column covering the tractor you are repairing, you will find a number which indicates the beginning paragraph pertaining to the camshaft. To locate this wanted paragraph in the manual, turn the pages until the running index appearing on the top outside corner of each page contains the number you are seeking. In this paragraph you will find the information concerning the removal of the camshaft.

More information available at Clymer.com
Phone: 805-498-6703

Haynes Publishing Group
Sparkford Nr Yeovil
Somerset BA22 7JJ England

Haynes North America, Inc
859 Lawrence Drive
Newbury Park
California 91320 USA

ISBN-10: 0-87288-364-7
ISBN-13: 978-0-87288-364-2

ALLIS-CHALMERS

Models ■ D-19 ■ D-19 Diesel

Previously contained in I&T Shop Service Manual No. AC-23

SHOP MANUAL
ALLIS-CHALMERS

MODELS D-19   D-19 DIESEL

Tractor serial number is stamped on the left front of torque tube. Engine serial number is stamped on the center left side of the engine block. Transmission serial number is stamped on the lower right hand corner of rear face of transmission case.

The D-19 tractor is available with an LP-Gas, gasoline or a turbo-charged diesel engine in dual wheel tricycle, single wheel tricycle or adjustable axle versions.

INDEX (By Starting Paragraph)

CONDENSED SERVICE DATA

GENERAL	NON-DIESEL	DIESEL
Engine Make	Own	Own
Engine Model	G-262	D-262T
Cylinders	6	6
Bore—Inches	3.5625	3.5625
Stroke—Inches	4.375	4.375
Displacement—Cubic Inches	262	262
Compression Ratio, Except LPG	8:1	14:1
Compression Ratio, LPG	9.65:1	
Pistons Removed From	Above	Above
Main Bearings, Number of	7	7
Main Bearings Adjustable?	No	No
Rob Bearings Adjustable?	No	No
Cylinder Sleeves	Wet	Wet
Forward Speeds	8	8
Reverse Speeds	2	2
Generator & Starter Make	D-R	D-R
Tightening Torques		
General Recommendations	See End of Shop Manual	

TUNE-UP	NON-DIESEL	DIESEL
Firing Order	——1-5-3-6-2-4——	
Valve Tappet Gap (Hot)		
Intake	0.015	Refer to
Exhaust	0.020	Paragraph 40A
Valve Seat & Face Angle		
Intake	45°	30°
Exhaust	45°	45°
Ignition Distributor Make	D-R	......
Ignition Distributor Model	1112615	......
Breaker Gap	0.022	......
Retarded Timing	TDC	......
Full Advanced Timing Degrees	25° BTDC	......
Mark Indicating:		
Retarded Timing	"TDC"	......
Full Advanced Timing	25°	......
Mark Location	Crankshaft Pulley	......
Spark Plugs	See Paragraph 101	......
Carburetor Make, LP-Gas	Ensign	......
Model, LP-Gas	XG	......

TUNE-UP (Cont.)	NON-DIESEL	DIESEL
Carburetor make, Gasoline	Marvel-Schebler	......
Model, Gasoline	TSX-848	......
Float Setting (Inches)	1/4	......
Injection Pump Make	......	Roosa-Master
Injection Pump Model & Timing	......	See Paragraph 86
Engine Low Idle RPM	375-425	650-700
Engine High Idle RPM	2275-2300	2275-2300
Engine Loaded RPM	2000	2000

SIZES—CAPACITIES—CLEARANCES	NON-DIESEL	DIESEL
Crankshaft Journals Diameter	2.4975	2.4975
Crankpin Diameter	1.998	1.998
Camshaft Journal Diameter		
Rear	1.2485	1.2485
All Except Rear	1.9985	1.9985
Piston Pin Diameter	0.9996	0.9996
Valve Stem Diameter		
Intake	0.3426	0.3095
Exhaust	0.3412	0.3095
Piston Ring End Gap (Minimum)		
Top Ring	0.007	0.022
2nd & 3rd Rings	0.014	0.014
4th Ring	0.014	0.014
5th Ring	......	0.007
Main Bearing Diam. Clearance	——0.0013-0.004——	
Rod Bearing Diam. Clearance	——0.0011-0.0036——	
Piston Skirt Clearance	0.0023-0.0048	0.004-0.0065
Crankshaft End Play	——0.003-0.009——	
Camshaft Bearing Clearance	——0.002-0.0046——	
Camshaft End Play	——0.003-0.008——	
Cooling System—Quarts	16	17
Crankcase Oil (With Filter)—Quarts	7	7
"Power-Director" & Hydraulic Sump, Quarts	22	22
Transmission & Differential, Quarts	32	32
Final Drives, Each—Quarts	8	8

FRONT SYSTEM

SINGLE WHEEL TRICYCLE

1. WHEEL ASSEMBLY. The single front wheel assembly may be removed after raising front of tractor and removing bolt (3—Fig. AC1) at each end of wheel spindle (1).

To renew bearings and/or seals, first remove wheel assembly; then, unbolt and remove bearing retainer (10—Fig. AC2), seal (4), seal retainer (5) and shims (9). Drive or press on opposite end of spindle to remove spindle (8), bearing cones (7) and bearing cup from retainer side of hub. Then drive remaining seal and bearing cup out of hub. Remove bearing cones from spindle.

Soak new felt seals in oil prior to installing seals and seal retainers. Drive bearing cup into hub until cup is firmly seated. Drive bearing cones tightly against shoulders on spindle. Pack bearings with No. 2 wheel bearing grease. Install spindle and bearings in hub and drive remaining bearing cup in against cone. When installing bearing retainer, vary number of shims (9—Fig. AC2) to give a free rolling fit of bearings with no end play.

Fig. AC2 — Exploded view of single front wheel assembly.

1. Side rings (2)	4. Seals (2)	8. Spindle
2. Tire	5. Seal retainers (2)	9. Shims
3. Wheel hub	6. Bearing cups (2)	10. Bearing retainer
	7. Bearing cones (2)	

Front wheel bearings should be repacked with No. 2 wheel bearing grease after each 500 hours of use.

If necessary to renew wheel hub or repair tire, deflate tire before unbolting and removing tire retaining rings (1—Fig. AC2).

2. R&R SINGLE WHEEL FORK. Raise front end of tractor, remove bolts (3—Fig. AC1) from each end of wheel spindle and remove wheel assembly from fork. Unbolt and remove fork (2) from steering gear sector shaft (14—Fig. AC8 or Fig. AC17).

Make sure that steering gear is centered and reinstall fork with caster to rear. Tighten cap screws that retain fork to sector gear shaft to a torque of 130-140 Ft.-Lbs.

DUAL WHEEL TRICYCLE

3. WHEEL ASSEMBLY. Front wheel and bearing construction on dual wheel tricycle models is of conventional design. Stamped steel wheel disc is reversible on hub. Bearing adjustment is made by tightening retaining nut on spindle until bearings are firmly seated, backing nut off one castellation and installing cotter pin. Bearings should be repacked with No. 2 wheel bearing grease after each 500 hours of use.

Dual wheel pedestal spindles are equipped with bearing spacers (15—

Fig. AC3). Place spacer on spindle with flange of spacer against shoulder on spindle. Install seal over spacer with crimped edge of seal shell towards spacer (lettered side of shell towards bearing).

4. PEDESTAL. Dual wheel pedestal (10—Fig. AC3) can be unbolted and removed from the sector shaft bearing retainer (62—Fig. AC6 or Fig. AC16) after raising front end of tractor.

To disassemble pedestal, remove cap screw (4—Fig. AC3), lock washer (5), flat washer (6), shims (7) and sleeve (9) with internal snap ring (8). Spindle (14), seal (13) and bearing cone (12) can then be removed from bottom end of pedestal. Drive bearing cup (11) from bottom of pedestal and bearing cone from spindle.

To reassemble, drive bearing cup into bottom of pedestal until cup is firmly seated against shoulder. Pack bearing cone with No. 2 wheel bearing grease and insert cone in cup. Seal should be soaked in oil prior to installation. Apply sealer to outer rim of seal and install seal with lip towards bearing. Carefully insert spindle through seal and bearing making sure that shoulder on spindle is seated firmly against bearing cone. Place sleeve with end nearest internal snap ring (8) on spindle shaft. Vary number of shims (7) to remove all end play from spindle shaft without creating any binding tendency.

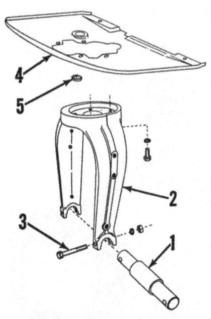

Fig. AC1 — Exploded view of the single front wheel fork and associated parts.

1. Spindle	4. Mud shield
2. Fork	5. Plug
3. Bolts (2)	

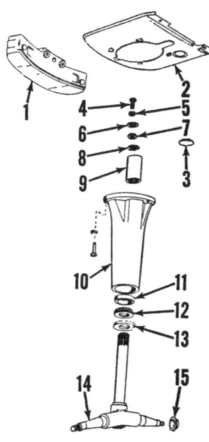

Fig. AC3 — Exploded view of dual wheel pedestal and associated parts.

1. Front weight
2. Mud shield
3. Plug
4. Cap screw
5. Lock washer
6. Washer
7. Shims
8. Snap ring
9. Coupling sleeve
10. Pedestal
11. Bearing cup
12. Bearing cone
13. Seal
14. Spindle
15. Bearing spacers (2)

Make sure that hole in bottom of sector gear shaft (57—Fig. AC6 or Fig. AC16) is exactly crosswise with tractor and install pedestal with wheels in straight ahead position (caster to rear of tractor). Tighten cap screws retaining pedestal to bearing retainer to a torque of 70-75 Ft.-Lbs.

ADJUSTABLE FRONT AXLE

5. **WHEEL ASSEMBLY.** Front wheel and bearing construction on adjustable front axle models is of conventional design. Stamped steel wheel disc is reversible on hub. Bearing adjustment is made by tightening retaining nut on spindle until bearings are firmly seated, backing nut off one castellation and installing cotter pin. Bearings should be repacked with No. 2 wheel bearing grease after each 500 hours of use.

Crimped edge of seal shell should be towards shoulder on spindle when renewing seal (lettered side of seal shell towards bearing).

6. **ADJUSTMENTS.** The adjustable front axle provides wheel tread widths of 60-84 inches or 65-89 inches depending on whether front wheels are mounted with dish in or out.

Toe-in should be correct at each adjustment position when mark in notch of tie rod is in line with mark on tie rod shaft. (See Fig. AC5). However, it may be advisable to measure front wheel toe-in and adjust to 1/16-1/8 inch if necessary. Tighten bolts in tie rod clamps securely.

7. **RENEW AXLE CENTER (MAIN) MEMBER AND RADIUS ROD ASSEMBLY.** Support front end of tractor and remove bolts through center member and axle extensions (spindle supports). Loosen bolts in tie rod clamps and slide axle extensions with front wheels and spindles out of main member and tie rod shafts out of tie rod tubes. Place floor jack under center member and unbolt rear pivot bracket from engine rear adapter plate. Lower the rear end of center member until pivot bracket clears the torque housing, roll to rear until front pivot pin clears front axle support and then pull center member out from under tractor. Remove rear pivot pin and bracket.

Reverse removal procedure to install new center member.

8. **RENEW AXLE PIVOT PINS AND FRONT PIVOT PIN BUSHING.** To renew axle rear pivot pin (19—Fig. AC4), support front of tractor, unbolt pivot pin bracket (20) from engine rear adapter plate and lower bracket until nut and pivot pin can be removed. Install new pin through radius rod and pivot bracket. Tighten nut securely. Install cotter pin, then bolt pivot bracket back to engine rear adapter plate. No rear bushing is provided.

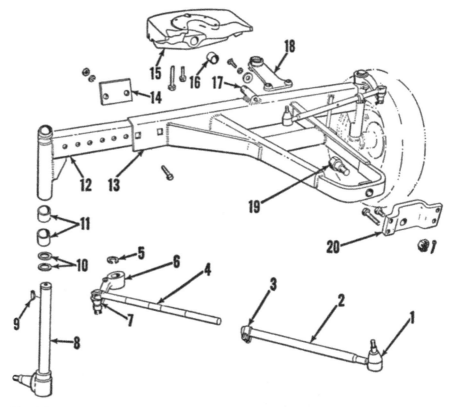

Fig. AC4 — Exploded view of adjustable front axle assembly and associated parts. Center (main) member and radius rod (13) is a welded assembly.

1. Dust cover
2. Socket assy., inner
3. Tie rod clamp
4. Socket assy., outer
5. Snap ring
6. Spindle arm
7. Dust cover
8. Spindle
9. Woodruff key
10. Thrust washers
11. Bushings
12. Spindle support
13. Axle main member
14. Adjusting plate
15. Axle support
16. Bushing
17. Pivot pin, front
18. Steering arm
19. Pivot pin, rear
20. Pivot plate

Fig. AC5 — Toe-in adjustment marks on adjustable front axle tie rod. Mark is provided on tie rod outer (rod) end for each adjustment position of front axle. Toe-in should measure 1/16 to 1/8-inch when marks on both tie rod assemblies are aligned.

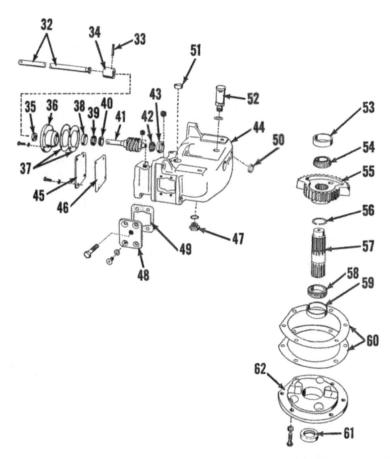

Fig. AC6 — Exploded view of manual steering front support and associated parts. Shims (37) are available in 0.005 vellum or steel; shims (60) are available in 0.005 vellum or 0.010 steel. Vellum and steel shims should be used alternately for proper sealing.

32. Steering shaft	39. Bearing	47. Drain plug	55. Steering gear
33. Roll pin	40. Bearing cone	48. Cover	56. Snap ring
34. Splined coupling	41. Steering worm	49. Gasket	57. Steering shaft
35. Oil seal	42. Bearing cone	50. Plug	58. Bearing cone
36. Bearing retainer	43. Bearing cup	51. Plug	59. Bearing cup
37. Shims (0.005)	44. Front support	52. Breather	60. Shims (0.005 & 0.010)
38. Bearing cup	45. Cover	53. Bearing cup	61. Oil seal
	46. Gasket	54. Bearing cone	62. Shaft retainer

To renew axle front pivot pin (17), support front end of tractor, disconnect tie rods from center steering arm (18), unbolt rear pivot bracket from engine rear adapter plate and lower rear end of radius rod until bracket will clear torque housing. Then roll axle assembly to rear until front pivot pin will clear front axle support, raise tractor and roll front axle assembly forward. Remove pivot pin from axle main (center) member with cutting torch. Carefully position new pivot pin on axle and weld in place using electric welder. Bushing (16) in front axle support may also be renewed at this time.

9. RENEW SPINDLE BUSHINGS. Support front end of tractor and remove front wheels. Remove snap ring (5—Fig. AC4), spindle arm (6) and woodruff key (9). Spindle can then be removed from bottom of axle extension (spindle support). Drive bushings (11) from top and bottom of spindle bore and remove thrust washers (10) from spindles.

New spindle bushings are pre-sized and if care is taken in pressing or driving bushings into support flush with ends of spindle bore, reaming should not be necessary. Install two new thrust washers on spindle and reassemble.

10. RENEW TIE ROD SOCKETS. Tie rod sockets are non-adjustable automotive type and are renewable only as either a complete tie rod assembly or one end of tie rod as required. Dust covers (1 & 7—Fig. AC4) are renewable.

MANUAL STEERING SYSTEM

The worm and sector type manual steering unit is contained in the front support casting (44—Fig. AC6). Recommended steering gear lubricant is SAE 80 EP gear lube. Capacity is approximately 3¾ quarts. Oil level should be maintained at top of steering (sector) gear (55).

11. ADJUSTMENTS. The gear unit is provided with two adjustments as follows:

WORMSHAFT BEARINGS. To adjust the wormshaft bearings, it is first necessary to remove the front support as outlined in paragraph 12. (On tricycle models, remove front wheel fork or pedestal and wheel unit with front support.)

Unbolt and remove bearing retainer (36—Fig. AC6) and vary the number of shims (37) to remove all shaft end play without causing any binding tendency. Alternate paper and steel shims for proper sealing. Paper and steel shims are 0.005 thick.

STEERING SHAFT BEARINGS. Support front end of tractor. On single wheel tricycle models, unbolt and remove fork (2—Fig. AC1) and wheel assembly from steering sector shaft (14—Fig. AC8). On dual wheel tricycle models, unbolt pedestal from steering shaft bearing retainer. On adjustable front axle models, unbolt front axle support from front support (steering gear unit); then raise front of tractor so that front axle support can be removed. Drain oil from front support on all models.

On single wheel tricycle models, refer to Fig. AC8 and proceed as follows: Unbolt retainer (9) from bottom of front support and remove retainer, shaft and gear unit from front support. Remove cap screw (1), lock washer (2), flat washer (3) and vary the number of shims (4) to remove all end play of bearings without causing any binding tendency. Reinstall unit using two gaskets (8) and tighten cap screws to a torque of 75 Ft.-Lbs. Timing of sector and worm gears is not necessary. Reinstall front wheel and fork assembly and tighten fork retaining cap screws to a torque of 130-140 Ft.-Lbs.

On dual wheel tricycle and adjustable front axle models, steering shaft end play is adjusted by varying the number of shims (60—Fig. AC6) between bearing retainer (62) and front support casting (44). Unbolt and remove retainer, shaft and gear assembly and vary the number of shims to remove all bearing end play without causing any binding tendency. Alternate paper and steel shims for proper sealing. Tighten cap screws retaining bearing retainer to front support to a torque of 70-75 Ft.-Lbs. On dual wheel tricycle models, tighten pedestal retaining cap screws to a torque of 70-75 Ft.-Lbs. Same torque value also applies to adjustable front axle support retaining cap screws.

12. R&R FRONT SUPPORT. Remove the front support from tractor as follows: Remove grille and drain radiator. Remove both hood side panels and unbolt hood center channel from radiator shell. Disconnect both radiator hoses and unbolt radiator shell from side rails and radiator from front support. Remove front support breather and lift radiator and radiator shell from tractor as a unit. Support front end of tractor. Attach hoist to front support. On adjustable front axle models, disconnect tie rods, unbolt front support from side rails and lift front support and weight unit from front axle pivot pin. On other models, unbolt front support from side rails and remove front wheel unit with front support. NOTE: If front support is to be disassembled, proceed as follows prior to removal

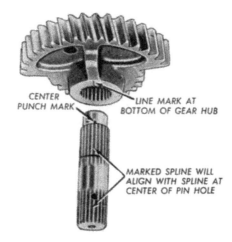

Fig. AC7 — Align punch mark on steering shaft with line at bottom of steering gear hub as shown.

CENTER PUNCH MARK

LINE MARK AT BOTTOM OF GEAR HUB

MARKED SPLINE WILL ALIGN WITH SPLINE AT CENTER OF PIN HOLE

of front support from tractor: On single wheel tricycle models, unbolt and remove fork and wheel assembly from steering gear sector shaft. On dual wheel tricycle models, unbolt and remove pedestal and front wheel assembly from front support. On adjustable axle models, unbolt front axle support (15—Fig. AC4) from front support casting, raise front of tractor slightly and remove axle support with a floor jack. Then, drive pin from steering arm (18) and remove steering arm from sector shaft. Drain steering gear lubricant from front support on all models.

Reverse removal procedures to reinstall front support. Tighten adjustable front axle support to front support retaining cap screws to a torque of 70-75 Ft.-Lbs. Tighten single front wheel fork to sector shaft retaining cap screws to a torque of 130-140 Ft.-Lbs. Tighten dual wheel pedestal to front support retaining cap screws to a torque of 70-75 Ft.-Lbs.

13. OVERHAUL GEAR UNIT. Unbolt and remove steering shaft bearing retainer, shaft, bearings and sector gear from bottom of casting. Unbolt and remove wormshaft bearing retainer, wormshaft and bearings from rear of casting. Drive expansion plug (50—Fig. AC6) from front of casting; then, remove front bearing cup (43). Use bearing cup puller to remove cup (53) on dual wheel tricycle and adjustable front axle models.

14. SINGLE FRONT WHEEL SECTOR SHAFT. On single front wheel tricycle models, refer to Fig. AC8; then, overhaul the removed sector gear, shaft and retainer assembly as follows: Remove capscrew (1), lock washer, flat washer and shims; then,

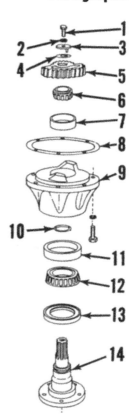

Fig. AC8 — On single front wheel tractors, above parts are used in manual steering gear instead of items 53 through 62 shown in Fig. AC6. Mark on sector gear (5) should be aligned with punch mark on top of sector shaft (14).

1. Cap screw	8. Gaskets (2)
2. Lock washer	9. Bearing support
3. Washer	10. "O" ring
4. Shims	11. Bearing cup
5. Sector gear	12. Bearing cone
6. Bearing cone	13. Seal
7. Bearing cup	14. Sector shaft

drive the shaft (14) out of sector gear, bearings and retainer. Procedure for further disassembly is evident.

Renew any questionable parts. Reassemble using new "O" ring (10) and seal (13) as follows: Drive bearing cups (7 & 11) into retainer (9) making sure that they are firmly seated. Pack lower bearing cone with No. 2 wheel bearing grease and place cone in lower cup. Soak new seal (13) in oil prior to installation. Apply gasket sealer to outer rim of seal and install with lip towards bearing.

Install new "O" ring in groove of shaft and insert shaft through seal and bearing cone. Make sure that shoulder on shaft is firmly seated against lower bearing cone. Install upper bearing cone on shaft. Install sector gear on shaft with line mark on hub of gear down and aligned with marked spline on shaft (mark located on top end of shaft). Install proper number of shims (4) to provide free rolling fit of bearings without end play when cap screw (1) is securely tightened. Install unit in front support using two gaskets (8) and tighten retaining cap screws to a torque of 70-75 Ft.-Lbs.

15. DUAL WHEEL & WIDE FRONT AXLE SECTOR SHAFT. On dual wheel tricycle and adjustable front axle models, disassembly of sector shaft unit is evident from exploded view in Fig. AC6. To reassemble, drive bearing cups into front support and bearing retainer making sure that the cups are firmly seated. Soak new seal in oil prior to installation. Apply gasket sealer to outer rim of seal and install seal in retainer with lip towards bearing. Drive lower bearing cone firmly against snap ring (56) on shaft. Install sector gear (See Fig. AC7) with line mark on bottom

of gear hub aligned with marked spline on shaft. Install upper bearing cone making sure that sector gear is seated against snap ring and upper cone is tight against sector gear. Insert shaft with bearings and gear into front support casting and install bearing retainer (62—Fig. AC6) with proper number of shims (60) to provide free rolling fit of bearings without end play. Alternate paper and steel shims for proper sealing. Tighten retaining cap screws to a torque of 70-75 Ft.-Lbs.

16. WORMSHAFT UNIT — ALL MODELS. Refer to Fig. AC6 for disassembly of wormshaft unit. Rear wormshaft bearing is in three pieces: cup (38), roller assembly (39) and cone (40). Drive front bearing cup (43) into front support until cup is firmly seated against shoulder in bore.

Apply gasket sealer to rim of expansion plug (50) and drive plug into front support casting only far enough to seal hole. Drive rear bearing cup into retainer (36) until cup is firmly seated. Soak new seal (35) in oil, apply gasket sealer to outer rim of seal and install with lip towards front of retainer. Drive bearing cones onto wormshaft and make sure that they are firmly seated against shoulders on shaft. Insert shaft into front support casting; then, install rear bearing assembly and retainer. Use proper number of shims (37) between retainer and front support casting to provide free rolling fit of bearings without end play. Alternate paper and steel shims (each are 0.005 thick) for proper sealing. No timing of worm gear is necessary. Fill front support with SAE 80 EP lubricant to top of segment gear. (Capacity is approximately 3¾ quarts.)

POWER STEERING SYSTEM

NOTE: The maintenance of absolute cleanliness of all parts is of utmost importance in the operation and servicing of the hydraulic power steering system. Of equal importance is the avoidance of nicks or burrs on any of the working parts of the pump or control valve unit.

LUBRICATION AND BLEEDING

17. The fluid reservoir is located in the front support casting. Fluid level should be maintained at ⅝-inch above the top of the steering (sector) gear. Whenever the power steering oil lines have been disconnected, reconnect the lines, fill the reservoir, start tractor engine and cycle the system several times to bleed out any trapped air in the pump or lines. Recheck fluid level and add fluid as necessary. Fluid capacity is 5 quarts of type "A" automatic transmission fluid.

SYSTEM OPERATING PRESSURE AND RELIEF VALVE

18. A pressure test of the hydraulic circuit will disclose whether the pump, relief valve or some other unit

in the system is malfunctioning. To make such a test, proceed as follows: Connect a pressure gage in series with the pump discharge tube (3—Fig. AC9), run the tractor engine until power steering fluid is warm, then turn the steering wheel to either extreme right or extreme left position and observe the pressure gage reading. CAUTION: The steering wheel should be held in the extreme position only long enough to observe the gage reading. Pump may be damaged if wheel is held in this extreme position for an excessive length of time. If gage reading is 1000-1200 psi, the pump and relief valve are O.K. and any trouble is located in the control valve, power steering cylinder (ram) and/or connections.

If the pump output pressure is less than 1000 psi, either the relief valve is improperly adjusted, defective or the pump is in need of repair. If the pump output pressure is more than 1200 psi, the relief valve is either improperly adjusted or is stuck in the closed position. In any event, the first step in eliminating trouble is to adjust the relief valve and recheck pressure.

Remove the relief valve plug (1—Fig. AC9) and vary the number of shims (12—Fig. AC11) as required to obtain the correct opening pressure of 1000-1200 psi, or prove that the pump is in need of repair by being unable to raise the pressure to a minimum of 1000 psi.

Fig. AC9—Installed view of power steering pump and lines. Pump relief valve is located under hex plug (1). Oil by-passed by relief valve flows back to sump through line (4). Oil enters pump through suction (intake) line (2). Line (3) is pressure (outlet) line to control valve.

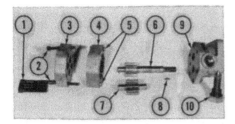

Fig. AC10 — Exploded view of gear-type power steering pump.

1. ⅜-in. cap screws (4)
2. ¼-in. cap screws (2)
3. Cover assembly
4. Gear plate
5. Hollow dowels (2)
6. Drive gear & shaft
7. Driven gear & shaft
8. Woodruff key
9. Body assembly
10. Relief valve assembly

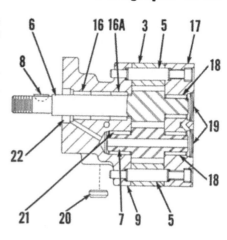

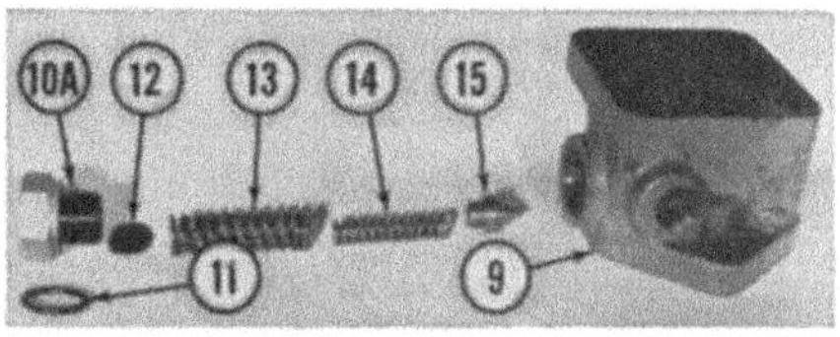

Fig. AC12 — Cross-section view of power steering pump showing bearing and seal location.

3. Gear plate
5. Hollow dowels (2)
6. Drive gear & shaft
7. Drive gear & shaft
9. Pump body
16. Needle bearing
16A. Needle bearing
17. Pump cover
18. Needle bearings
19. Expansion plugs
20. Tube seat
21. Needle bearing
22. Shaft seal

Fig. AC11 — Exploded view of power steering pump relief valve assembly.

9. Pump body
10A. Hex plug
11. "O" ring
12. Shims
13. Outer spring
14. Inner spring
15. Relief valve

PUMP

19. REMOVE AND REINSTALL. The power steering pump is mounted on a plate which is attached to front of engine. (See Fig. AC9). To remove pump, disconnect the suction (intake) tube (2), pressure tube (3) and by-pass tube (4) from pump. Then, loosen cap screws retaining pump to mounting plate to relieve drive belt tension. Unbolt mounting plate from engine and remove pump.

Reverse removal procedures to reinstall pump. Adjust drive belt tension to ¼-inch deflection at 10 pounds pull, then tighten cap screws retaining pump to mounting plate. After connecting oil lines to pump, bleed the system as outlined in paragraph 17.

20. OVERHAUL PUMP. To disassemble the power steering pump, first remove nut, pulley, woodruff key and mounting plate from drive end of pump. Scribe a line across cover, gear plate and pump body to aid in reassembly. Remove pressure relief valve plug (10A—Fig. AC11), "O" ring (11), shims (12), outer spring (13), inner spring (14) and relief valve (15) from body of pump. Be

careful not to lose any of the shims (12). After removing the six socket head cap screws (1 and 2—Fig. AC10) from cover (rear) end of the pump, carefully separate the cover (3), gear plate (4) and body of pump to avoid damage to the mating surfaces. No gaskets are used and machined surfaces are depended upon for sealing. The hollow dowel pins (5) are a tight fit in cover, gear plate and body.

Inspect all parts for wear, scoring or other damage and renew as necessary. New pump body includes bearings, seal, by-pass tube seat and relief valve assembly. New pump cover includes bearings, expansion plugs, discharge (pressure) tube seat and suction (inlet) tube seat. However, all other pump parts (including those in the body and cover assemblies) are available separately.

For method of removal and installation of the pump shaft bearings, refer to Fig. AC12. Bearings (18) in cover may be driven out towards rear of pump after removing the expansion plugs (19). Driven shaft bearing (21) in pump body must be pulled from blind hole. Drive shaft bearings (16 and 16A) can be driven out front end of body after removing seal (22).

New bearings should be pressed into place. Press on lettered end of bearing cage only as opposite end is soft and is easily distorted. Press bearing cages into cover 0.020 below flush with surface toward gear plate. Press driven shaft bearing into body 0.020 below surface. Press rear drive shaft bearing (16A) into body in against shoulder in bearing bore. (NOTE: Do not use excessive force on this bearing as pressure against shoulder in bore may distort bearing shell.) Press front drive shaft bearing (16) into body flush with counterbore. Press new double-lip seal (22) in flush with mounting surface of pump body with heaviest sealing lip inward.

To install drive shaft and gear in pump body, use a seal protector on stepped end of shaft or use a suitable smooth pointed tool to work inner lip of seal over shoulder on drive shaft. Install idler gear, gear plate and cover making sure that the previously scribed mark across cover, gear plate and body is realigned. (The ⅛-inch hole in the gear plate must align with

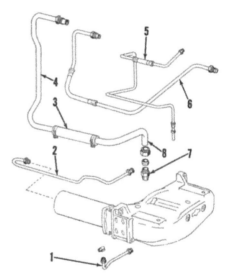

Fig. AC13 — Exploded view showing general layout of power steering tubes.

1. Control valve to ram (front) tube
2. Control valve to ram (rear) tube
3. Suction tube connector hose
4. Upper suction tube
5. By-pass tube
6. Pump to control valve (pressure) tube
7. Suction tube fitting
8. Lower suction tube

⅛-inch hole in rear cover.) Install the two ¼-inch cap screws through the holes with the hollow dowel pins and the 5/16-inch cap screws through the remaining holes. Tighten the ¼-inch cap screws to a torque of 95-105 inch-pounds and the 5/16-inch cap screws to a torque of 190-210 inch-pounds. Install mounting plate loosely on pump body; then, install woodruff key, pulley and retaining nut.

CONTROL VALVE AND RAM
(CYLINDER)

The power steering control valve and ram units are a part of the front support assembly as shown in Fig. AC16. Service work on these units requires R&R of the front support assembly as outlined in paragraph 21. As the lubricating oil in the front support is also used for power steering fluid, cleanliness when working on any part or component of the front support assembly is of utmost importance.

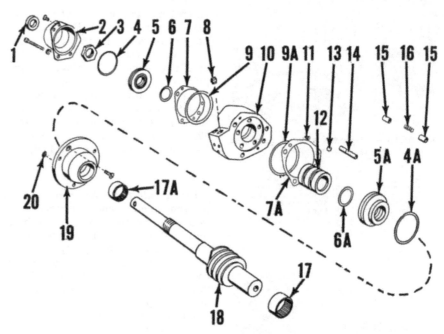

Fig. AC14 — Exploded view of power steering control valve unit.

1. Seal	6. "O" ring	10. Control valve body	15. Active plungers
2. Rear cover	6A. "O" ring	11. Check valve	16. Spring
3. Adjusting nut	7. Shim	12. Control valve spool	17. Needle bearing
4. "O" ring	7A. Shim		17A. Needle bearing
4A. "O" ring	8. Tube seat	13. Plug	18. Wormshaft
5. Thrust bearing	9. "O" ring	14. Inactive plunger	19. Front cover
5A. Thrust bearing	9A. "O" ring		20. "O" ring

21. R&R FRONT SUPPORT ASSEMBLY. Remove grille and drain radiator. Remove both hood side panels and unbolt hood center channel from radiator shell. Disconnect both radiator hoses. Unbolt radiator shell from side rails and radiator from front support. Remove front support breather, then lift radiator and radiator shell from front support as a unit. Support front end of tractor. Disconnect pressure and by-pass tubes from power steering pump and suction tube from front support as shown in Fig. AC18. Attach hoist to front support and unbolt front support from side rails. On adjustable front axle models disconnect tie rods from the center steering arm and lift front support, with front axle support attached, from front axle pivot pin. On tricycle models, roll front support and front wheel assembly forward while supported with hoist.

Reverse removal procedures to reinstall. After installation is completed, and reservoir in front support refilled if necessary, start tractor engine and cycle the power steering system several times to bleed air from system. Then check fluid level and add type "A" automatic transmission fluid as necessary.

22. R&R CONTROL VALVE ASSEMBLY. Remove front support assembly as outlined in paragraph 21. Drain fluid from front support and disconnect power steering tubes from control valve. Unbolt control valve and wormshaft from front support and withdraw the assembly.

Wormshaft front needle bearing (17—Fig. AC14) may be renewed in front support casting at this time if necessary. Drive plug (50—Fig. AC16) forward out of casting, then drive needle bearing out to rear. Drive new needle bearing cage just below flush with the counterbore in casting. (Do not drive needle bearing in against shoulder as this may distort soft end of cage. Drive or press on lettered end of bearing cage only.) Apply sealer to rim of plug and drive plug into casting only far enough to seal the hole.

Reinstall control valve and wormshaft unit in front support casting using new gasket (37—Fig. AC16). Tighten retaining cap screws to a torque of 24 Ft.-Lbs. NOTE: Wormshaft must have some end play in order to actuate control valve spool.

23. OVERHAUL CONTROL VALVE. Prior to disassembly, scribe a mark across the rear cover (2—Fig. AC14), body (10) and front cover (19) of control valve assembly to aid in reassembly of unit. Unbolt and remove the rear cover (2), shim (7) and "O" ring (9). Unstake and remove the bearing adjusting nut (3) and lift out the thrust bearing assembly (5).

Pull body (10) and front cover (19) from the wormshaft as an assembly. Remove the two screws retaining the front cover to the body and remove front cover (19), thrust bearing (5A), shim (7A) and "O" rings from the valve body.

The inactive plungers (14) need not be removed if they are tight in their bores and the ends of the plungers are flush with the ends of the valve body. The inactive plungers are steel rods serving no purpose other than filling the extra holes in the valve body in which centering plungers are not used. If for some reason, the inactive plungers have been removed, they should be reinstalled with the stake mark on the plungers towards the outside of the valve body to prevent distortion of the valve spool bore. NOTE: Later production may not have the extra holes and inactive plungers in the valve body.

The two active plungers (15) and centering spring (16) should be removed from their bore in the valve body. Check the plungers for free fit in the bore and the spring for any signs of distortion, wear or other defect and renew as necessary.

Carefully clean the control valve parts in fuel oil or other solvent and be sure that the restrictor passageway is open and clean as well as other passages in the valve body. The restrictor bore has a 0.031 I.D. orifice pressed into the center of the passageway. This orifice may be checked and cleaned with a No. 68 wire size drill. Do not remove the orifice or enlarge it above the 0.031 dimension.

As the control valve body and spool are a matched assembly, they are not available as separate parts. Diametrical clearance between spool and body should be 0.00025-0.00045. Renew all questionable parts, all "O" rings, seal and adjusting nut (3) when reassembling control valve. Renew needle bearing (17) in front support or (17A) in control valve front cover if loose or damaged.

To renew needle bearing (17) in front support (44—Fig. AC16), proceed as follows. Drive plug (50) out to front; then, drive bearing out to rear. Install new bearing from rear by driving or pressing on lettered end of bearing cage only. Be sure to leave approximately 1/16-inch clearance between front end of bearing cage and counterbore (shoulder). Apply sealer to outer rim of plug (50) and drive plug into casting only far enough to seal the hole.

To facilitate reassembly of control valve, clamp wormshaft with rear end up in a vise. Be careful not to damage gear or bearing surfaces. Clean and lubricate all parts prior to reassembly. Drive or press needle bearing (17A—Fig. AC14) into cover (19) until front end of cage is 1/32-inch below front end of cover. Install cover on wormshaft. Install new "O" ring (4A) on rear race of thrust bearing (5A) and install thrust bearing assembly on wormshaft with rear face of bearing flush with rear face of cover. Install new "O" ring (20) in groove in control valve body, new "O" ring (6A) in groove of thrust bearing and place shim (11) on cover. Install spool (12) with identifying groove in I.D. of spool towards front cover. Place new "O" ring (9A) in front face of valve body and install valve body (with inactive plungers) making sure that previously scribed marks are aligned. Attach front cover to valve body with two flat-head screws. Place one active plunger in bore, drop in centering spring and install second active plunger. Place new "O" ring in groove of thrust bearing (5) face and install

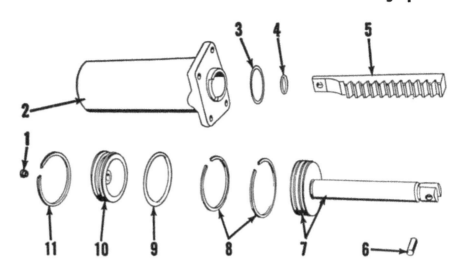

Fig. AC15 — Exploded view of power steering ram (cylinder) and ram rack.

1. Tube seat	4. "O" ring	7. Piston & rod unit	10. End cap
2. Cylinder	5. Ram rack	8. Piston rings	11. Snap ring
3. "O" ring	6. Rack pin	9. "O" ring	

thrust bearing assembly with "O" ring towards valve spool. Install new nut (3) and tighten same to a torque of 5 Ft.-Lbs. (60 inch-pounds). Back nut off ⅓-turn (two flats) and stake nut to wormshaft. Install new "O" ring (4) on thrust bearing and place new "O" ring (9) in groove and shim (7) on control valve body. Drive new seal (1) with lip facing inside into rear cover (2); then, install cover to control valve body with previously scribed marks aligned and secure cover with two flat-head screws. Control valve and wormshaft unit is then ready to reinstall in front support.

24. R&R AND OVERHAUL STEERING RAM (CYLINDER). First remove the front support as outlined in paragraph 21. Then, proceed as follows: Remove the rack adjusting block (48—Fig. AC16) taking care not to lose shims (49). Disconnect oil lines from the ram, remove the retaining cap screws and, while holding rack away from idler gear (64), withdraw the ram assembly from the front support casting.

To disassemble the removed unit, refer to Fig. AC15 and proceed as follows: Remove the pin (6) attaching rack (5) to piston rod (7). Then, extract snap ring (11) retaining rear cap (10) in cylinder (2), remove cap and withdraw piston unit.

Examine all parts and renew any that are scored, scuffed or show excessive wear. Lubricate all parts prior to assembly and use all new "O" rings.

Prior to attaching the rack (5—Fig. AC15) to the piston rod, insert the rack into the steering gear housing in mesh with the idler gear, reinstall rack adjusting block and shims, and check for backlash between rack and idler gear. The rack should move freely without backlash. If not, vary the number of shims between the rack adjusting block and housing to provide this condiiton. Paper shims (0.005 thick) and steel shims (0.003 thick) should be alternately placed for proper sealing. When proper adjustment is obtained, remove the rack adjusting block, taking care not to lose or damage the shims, and remove rack from housing. Attach rack to piston rod with pin and rivet pin securely taking care not to draw ears of piston rod together.

To reinstall ram, first rotate the steering shaft (57—Fig. AC16) to full right (counter-clockwise as viewed from lower end of shaft) and pull the ram rack to fully extended position. Engage rack and idler gear teeth and install rack adjusting block (48) with the previously selected shims (49) for proper rack to idler gear backlash.

POWER STEERING GEAR UNIT

The worm and sector type gear unit is contained in the front support casting (44—Fig. AC16). Lubricating oil for the gear unit is also used as power steering fluid. Oil level should be maintained at ⅝-inch above the sector (steering) gear with type "A" automatic transmission fluid.

25. ADJUSTMENTS. The gear unit is provided with two adjustments: Rack mesh position is adjusted by the number of shims between the front support casting and the rack adjusting block (48—Fig. AC16). Steering shaft bearing end play is also adjustable by varying the number of shims (4—Fig. AC17) between the sector gear (5) and flat washer (3) on single front wheel models or by varying the number of shims (60—Fig. AC16) between the bearing retainer (62) and the front support casting (44) on dual wheel tricycle and adjustable front axle models. However, these adjustments are more in the nature of assembly procedure when overhauling front support assembly than routine adjustment to provide better power steering operation. Therefore, adjustments will be discussed under reassembly of gear unit in front support. Refer to paragraph 27.

26. R&R FRONT SUPPORT (GEAR UNIT). Support front end of tractor. Unbolt and remove single front wheel fork and wheel assembly, dual wheel tricycle pedestal and wheel assembly or adjustable front axle support casting. On adjustable front axle models, drive pin from center steering arm and remove steering arm from shaft. Drain power steering fluid from front support on all models. Then remove front support assembly as outlined in paragraph 21.

Reverse removal procedures to reinstall front support. Refill to ⅝-inch above sector gear with type "A" automatic transmission fluid. Start engine and cycle system several times after assembly is completed to bleed any trapped air from system. Check fluid level and add fluid as necessary.

27. OVERHAUL FRONT SUPPORT. Remove power steering tubes from ram cylinder and power steering control valve. Unbolt and remove ram cylinder as outlined in paragraph 24.

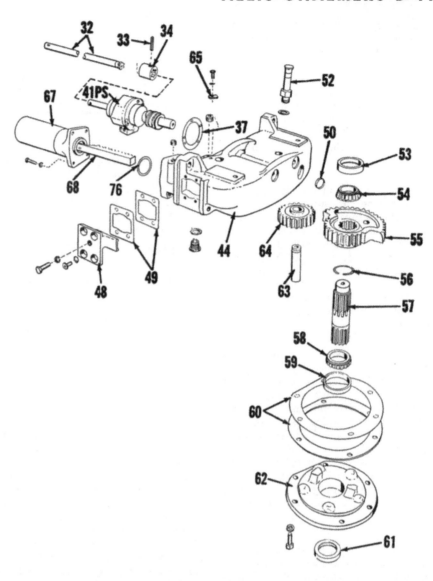

Fig. AC16 — Partially exploded view of the power steering front support. For exploded views of the control valve unit (41PS) and ram (67), refer to Fig. AC14 and Fig. AC15 respectively.

32. Lower steering shaft	48. Rack adjusting block	54. Bearing cone	61. Oil seal
33. Roll pin	49. Rack adjusting shims (0.005 & 0.003)	55. Steering gear	62. Shaft retainer
34. Splined coupling		56. Snap ring	63. Idler gear shaft
37. Gasket		57. Steering shaft	64. Idler gear
41PS. Steering control valve and wormshaft unit	50. Plug	58. Bearing cone	65. Lock plate
	52. Breather	59. Bearing cup	67. Steering ram
	53. Bearing cup	60. Shims (0.005 & 0.010)	68. Rack
44. Front support			76. "O" ring

Unbolt and remove control valve and wormshaft unit as outlined in paragraph 22.

Unbolt bearing retainer (9—Fig. AC17) on single front wheel models or (62—Fig. AC16) on other models and remove shaft and sector gear assembly from front support. Be careful not to lose or damage shims (60—Fig. AC16) on dual wheel tricycle and adjustable axle models.

Remove idler shaft lock (65—Fig. AC16), then pull idler shaft from front support. (Top end of shaft has threaded hole to facilitate pulling shaft from casting.) Withdraw idler gear through bottom opening in casting.

Thoroughly clean front support casting prior to reassembly of steering gear unit. As front support is used for power steering fluid reservoir, cleanliness is of utmost importance. Following procedure should be observed in inspecting and renewing parts and in reassembly of unit:

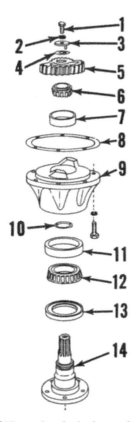

Fig. AC17 — On single front wheel tractors, above parts are used in power steering front support unit instead of items 53 through 61 as shown in Fig. AC16. Mark on sector gear (5) should be aligned with punch mark on top of sector shaft (14).

1. Cap screw
2. Lock washer
3. Washer
4. Shims
5. Sector gear
6. Bearing cone
7. Bearing cup
8. Gaskets (2)
9. Bearing support
10. "O" ring
11. Bearing cup
12. Bearing cone
13. Seal
14. Sector shaft

Fig. AC18 — View showing power steering tubes disconnected and arranged to facilitate removal of front support unit.

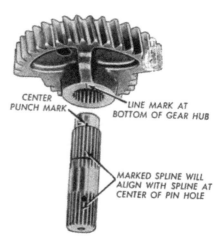

Fig. AC19—Align punch marks on steering shaft with line at bottom of steering gear hub as shown.

SINGLE WHEEL TRICYCLE STEERING SHAFT ASSEMBLY. Refer to Fig. AC17. Remove cap screw (1), lock washer (2), flat washer (3) and shims (4). Be careful not to lose or damage shims. Drive or press shaft (14) out of sector gear, bearings and retainer. Further disassembly is evident from an inspection of the unit and reference Fig. AC17.

Check teeth of sector gear and splines in gear and on steering shaft for wear. Any excessive play between gears or looseness of sector gear on shaft may cause shimmy of front wheels. Renew any questionable parts. Inspect bearings for damage or wear and renew if necessary.

To reassemble, install new "O" ring (10) in groove on steering shaft. Drive both bearing cups into retainer until they are firmly seated and pack

lower bearing cone with No. 2 wheel bearing grease. Place lower bearing cone in cup. Soak new seal (13) in oil, apply gasket sealer to rim of seal and install seal in bearing retainer with lip towards bearing. Insert steering shaft through seal and lower bearing cone making sure that shaft is seated firmly against cone. Install upper bearing cone on shaft. Install sector gear on shaft with line mark on bottom of gear hub aligned with marked spline on shaft (mark is located on top end of shaft). Install cap screw, lock washer and flat washer with proper amount of shims (4) to provide a slight preload on bearings.

DUAL WHEEL TRICYCLE OR ADJUSTABLE AXLE STEERING SHAFT ASSEMBLY. Refer to Fig. AC16 for disassembly of sector shaft unit (items 53 through 62). Check teeth of sector gear and splines in gear and on steering shaft for wear. Any excessive play (backlash) between gears or looseness of sector gear on shaft may cause shimmy of front wheels. Renew any questionable parts. Inspect bearings for damage or wear and renew if necessary.

Drive lower bearing cup into bearing retainer (62) and upper bearing cup (53) into front support casting (44) until cups are firmly seated. Soak new seal (61) in oil, apply gasket sealer to outer rim of seal and install seal in bearing retainer (62) with lip towards bearing.

Install snap ring (56) in groove on steering shaft. Drive lower bearing cone tightly against lower side of snap ring. Refer to Fig. AC19 and install sector gear on shaft in proper alignment. Be sure that hub of gear is tight against upper side of snap ring (56—Fig. AC16). Install upper bearing cone tightly against upper hub of sector gear.

IDLER GEAR AND SHAFT. Check idler gear (64—Fig. AC16) for any wear or damage of gear teeth or looseness on shaft (63) and renew gear and/or shaft as necessary. Place gear in front support casting through bottom opening and drive the shaft into place from top. Install shaft lock (65).

ASSEMBLY AND ADJUSTMENT. Install the previously assembled single front wheel steering shaft assembly (Fig. AC17) using two new gaskets (8); or, on other models, install steering shaft assembly (items 53 through 62—Fig. AC16) using proper number of shims (60) to give a slight preload to bearings. Use paper and steel shims alternately for proper sealing. Paper (vellum) shims are 0.005 thick; steel shims are 0.010 thick. Tighten retaining cap screws to a torque of 75 Ft.-Lbs. Check backlash of idler gear to sector gear. (It may be possible to remove a small amout of backlash by repositioning the idler gear to the sector gear; however, if much backlash is present, either the sector gear, the idler gear, or both gears must be renewed.) Remove sector shaft assembly after being sure that no noticeable backlash is present between sector gear and idler gear.

Install rack adjusting block (48— Fig. AC16) with proper number of shims (49) so that rack (68) will mesh with idler gear throughout the length of the rack with no backlash and without any binding tendency. (Tighten retaining cap screws to a torque of 30-35 Ft.-Lbs. while checking rack mesh adjustment and on final assembly.) Alternate paper and steel shims for proper sealing. Paper (vellum) shims are 0.005 thick and steel shims are 0.003 thick. When mesh of rack

and idler gear is properly adjusted, withdraw rack and install rack on ram piston. Rivet pin securely, but do not draw ears of piston rod together.

Extend ram piston to fully extended position and install ram assembly using new "O" ring (76—Fig. AC16). Be sure that forward end of rack contacts internal stop in front support casting. Install sector shaft assembly with shaft turned to full right (counter-clockwise as viewed from bottom end of shaft) against stop on bearing retainer. Tighten retainer cap screws to a torque of 90-100 Ft.-Lbs. Check to see that sector (steering

shaft) can be turned equal distance each way from centered position.

Install power steering control valve and wormshaft unit using new gasket (37—Fig. AC16) and tighten cap screws to a torque of 24 Ft.-Lbs. No timing of the wormshaft gear to the sector gear is necessary. NOTE: The wormshaft is mounted on straight needle bearings to allow end play in the shaft which is necessary to actuate the power steering control valve spool.

Connect power steering oil lines to ram and control valve. Reinstall the front support assembly as outlined in paragraph 26.

ENGINE AND COMPONENTS

PRODUCTION CHANGES

Diesel

27A. At engine serial number D-D1499, different cylinder head, valves, injector nozzles, camshaft, pistons and associated parts were used during production, than were used on earlier diesel engines. All or some of the later parts may be installed on early engines, but some combinations of early and late parts will prevent the engine from running, other combinations of early and late parts are not desirable.

Early pistons have a recess for valves, later pistons are flat. The late flat top pistons cannot be used in conjunction with the early (high lift) camshaft. Early valves are larger in diameter than later valves. The cylinder head used for early production has larger volume in the combustion chamber than later models. Because of casting differences, the late cylinder head is less likely to crack than the early cylinder head. The late cylinder head assembly, flat top pistons and late camshaft will result in easier starting. The early injector tips are larger than the type used for late cylinder head. The camshaft used in early production engines has higher lift (opens valves further) than the later camshaft. The late camshaft can be used with recessed pistons and early cylinder head, but it is suggested that later cylinder head and flat pistons also be installed.

R&R ENGINE WITH CLUTCH

Non-Diesel

28. Remove grille and drain coolant from radiator and engine block. If engine is to be disassembled, drain oil

from pan. Remove both hood side panels, disconnect hoses from air cleaner, unbolt center hood channel from radiator shell and fuel tank bracket; then, remove hood center channel and air cleaner as a unit. Disconnect both radiator hoses. Unbolt radiator shell from side rails and radiator from front support. Remove front support breather and lift off radiator and radiator shell as a unit.

On power steering models, disconnect pressure and by-pass tubes from power steering pump and suction tube from front support. Arrange tubes as shown in Fig. AC18.

Disconnect battery ground strap, wiring from generator and ignition coil and temperature gage bulb from water manifold. Disconnect starter wire, unbolt and remove starter.

Remove the left side cover from below fuel tank. Disconnect and remove the governor control rod. Disconnect oil gage line at rear of engine and remove fuel line from filter to fuel pump. Disconnect choke wire from carburetor and tachometer cable from distributor.

Support front end of tractor under torque housing and attach hoist to front support. On adjustable front axle models, disconnect tie rods, unbolt front support from side rails and lift front assembly from front axle; then, unbolt rear pivot bracket and roll front axle assembly from tractor. On other models, unbolt front support from side rails and, using hoist, remove front support with pedestal and wheels or fork and wheel attached. Disconnect lower (horizontal) steering shaft from universal joint. Unbolt

and remove side rails from tractor. Attach hoist to the two extended head bolts with lifting chain or fixture. Unbolt engine adapter plate from torque housing, pull engine forward until clear of torque housing and move engine to stand or bench.

To reinstall engine and clutch assembly, reverse removal procedures. Following torque values in Ft.-Lbs. will apply:

Engine to torque housing.......70-75
Side rail to torque housing...130-140
Front support to side rails....130-140
Front engine plate to side rails..70-75
Axle pivot bracket to engine
 rear adapter plate..........90-100

On power steering models, after assembly is complete, start engine and cycle the power steering system several times to bleed any trapped air from system. Check fluid level and add type "A" automatic transmission fluid as necessary. Refer to paragraph 17.

Diesel

29. Remove grille and drain coolant from radiator and engine block. If engine is to be disassembled, drain oil pan. Remove both hood side panels. Disconnect battery ground strap, manifold heater wire, and wires to generator and voltage regulator. Disconnect hoses from air cleaner. Unbolt hood center channel from radiator shell and fuel tank bracket; then, remove channel, air cleaner and voltage regulator as a unit. Disconnect both radiator hoses. Unbolt radiator shell from side rails and radiator from front support. Remove front support breather and lift radiator and shell from tractor as a unit.

Remove the side cover and fuel shut-off control rod from below the left hand side of fuel tank. Remove the fuel line from tank to primary fuel filter. Remove the throttle control rod from bell crank to fuel injection pump. Disconnect the fuel return line and the oil pressure gage line at rear of engine, the tachometer cable from drive unit and temperature gage bulb from water manifold. Disconnect starter wire, unbolt and remove starter.

On power steering models, disconnect pressure and by-pass lines from the power steering pump and the suction line from front support. Arrange tubes as shown in Fig. AC18.

Support front end of tractor under torque housing and attach hoist to front support. On adjustable front axle models, disconnect tie rods, unbolt front support from side rails and lift front assembly from front axle. Then, unbolt rear pivot bracket and roll front axle assembly from tractor. On other models, unbolt front support from side rails, and, using hoist, remove front support with wheels attached. Disconnect lower (horizontal) steering shaft from U-joint. Unbolt and remove side rails from tractor. Attach lifting chain or fixture to the two extended head bolts and hook hoist to chain or fixture. Unbolt engine adaptor plate from torque housing, pull engine forward to clear housing and move engine to stand or bench.

To reinstall engine and clutch assembly, reverse removal procedures. Following torque values in Ft.-Lbs. will apply:

Engine to torque housing......70-75
Side rails to torque housing..130-140
Front support to side rails....130-140
Front engine plate to side rails..70-75
Axle pivot bracket to engine
    rear adapter plate.........90-100

On power steering models, after assembly is complete, start tractor engine and cycle the power steering system several times to bleed any trapped air from system. Check fluid level and add type "A" automatic transmission fluid as necessary. Refer to paragraph 17.

CYLINDER HEAD

Non-Diesel

30. **REMOVE AND REINSTALL.** Remove grille and drain complete cooling system. Remove both hood side panels. Disconnect air cleaner hoses and remove air cleaner from center channel. Unbolt carburetor from intake manifold and leave carburetor suspended on fuel line and

linkage. Unbolt and remove manifold and muffler assembly from engine. Loosen water tube flare nuts at water pump and water manifold. Disconnect upper radiator hose from water manifold. Unbolt water manifold and, leaving temperature gage bulb attached, lay unit back on battery tray. Disconnect spark plug wires. Remove rocker arm cover, disconnect oil line to rocker arm shaft and remove rocker arm assembly and push rods. Disconnect oil line from cylinder block to cylinder head. Remove the cylinder head cap screws and lift cylinder head from engine.

Cylinder head gasket has individual "fire rings" for each cylinder. Head gaskets for gasoline and LP-Gas models are not interchangeable. Clean cylinder head and block surfaces. Place gasket on cylinder block with imprint "THIS SIDE DOWN" against block. Hold gasket in position with guide studs and place "fire rings" in gasket with rounded side of ring up. Set cylinder head down over guide studs taking care not to disturb gasket placement. Place water manifold and steel water tube in position and install long cap screws through water manifold loosely. Remove guide studs and install remainder of cap screws loosely. Install two extended cap screws on manifold side of head in same holes from which removed. Partially tighten all cap screws; then, working from center of cylinder head, tighten all cap screws to a torque of 110-120 Ft.-Lbs. Install push rods and rocker arms and adjust valve tappet gap to 0.017 cold on intake valves and 0.022 cold on exhaust valves.

Manifold gasket consists of individual asbestos gaskets and metal shields for each port. To facilitate reinstalling manifold, stick asbestos gaskets and metal shields to manifold ports with a small amount of gasket sealer.

After completing reassembly of tractor, start engine and operate until normal operating temperature is reached. Recheck cylinder head cap screw torque and readjust valves to 0.015 hot on intake valves and 0.020 hot on exhaust valves.

Diesel

31. **REMOVE AND REINSTALL.** Refer to paragraph 27A. Remove grille and drain coolant from radiator and cylinder block. Remove both hood side panels. Disconnect battery ground cable and wiring from voltage regulator and manifold heater. Disconnect hoses from air cleaner; then, unbolt and remove hood center channel with air cleaner and voltage regulator as a unit.

Remove air tube from turbo-charger to intake manifold. Remove oil drain tube from turbo-charger and disconnect oil supply tube at turbo-charger. Remove manifold retaining nuts, washers and clamps. Remove intake manifold with manifold heater attached. Remove exhaust manifold with turbo-charger attached. Loosen nuts on steel water tube between water manifold and water pump. Disconnect upper radiator hose. Unbolt water manifold from cylinder head and, leaving temperature gage bulb installed, lay unit back over batteries making sure insulation is placed between batteries and unit.

Disconnect fuel return line at rear of cylinder head and remove fuel return line between fuel injection pump and injector leak-off line. Disconnect fuel injector pressure lines at injectors. Remove oil pressure line from cylinder block to cylinder head. Remove rocker arm cover, rocker arm assembly and push rods. Unbolt and remove cylinder head from engine.

Cylinder head gasket has individual "fire rings" for each cylinder. After cleaning head and block surfaces, place gasket on block with imprint "THIS SIDE DOWN" against block. Hold gasket in place with guide studs and place fire ring in each cylinder opening of gasket with rounded side of ring up. Set cylinder head down over guide studs taking care not to disturb placement of cylinder head gasket. Position water manifold on cylinder head and install steel water tube between manifold and water pump. Install long cap screws through water manifold and remove guide studs. Install extended cap screws on manifold side of head in same holes from which removed. (Longest extended cap screw goes through water opening cover at rear.) Install remainder of cap screws and partially tighten. Starting in center of cylinder head, tighten the cap screws to a torque of 110-120 Ft.-Lbs. Install push rods and rocker arms; then, adjust valve tappet gap as outlined in paragraph 40A.

Complete the reassembly by reversing the disassembly procedures. NOTE: Oil return line from Turbo-Charger should be left disconnected until it is assured that oil is flowing. After starting engine, reduce engine speed immediately to low idle until oil is flowing through the turbo-charger return line; then, reconnect line. Operate engine at 900-1000 RPM until normal operating temperature is reached. Recheck cylinder head cap screw torque. Refer to paragraph 40A and readjust valve tappet gap.

VALVES, VALVE SEALS, SEATS AND ROTATORS

Non-Diesel

32. Intake and exhaust valves have a face and seat angle of 45 degrees. The seat width can be narrowed by using 30 and 60 degree cutters to obtain the desired seat width of 5/64 to 3/32-inch. Intake valves seat directly in the cylinder head. Exhaust valves seat in renewable ring type inserts which are available in standard size and 0.005 oversize. Inserts should have 0.002-0.004 interference fit. Standard bore diameter for inserts is 1.4345-1.4355. Machine bore out to 1.4395-1.4405 for 0.005 oversize insert.

Intake valve stem diameter is 0.3423-0.3430 and exhaust valve stem diameter is 0.3407-0.3417.

Intake valve stem seals should be renewed whenever intake valves are removed for service. Remove old seals from valve guides. When reinstalling valves, refer to Fig. AC20 and install new intake valve seals as follows: Install intake valve in guide and place plastic sleeve (contained in seal kit) over stem as shown in view A. If sleeve extends over 1/16-inch below lower groove of valve stem, cut off the excess length of sleeve. Lubricate the sleeve and, while holding against head of valve, push seal assembly down over sleeve and valve stem as shown in views B and C. Rubber sleeve of seal should be pushed down over intake valve guide with two screwdrivers as shown in view D. Make sure that seal is tight against top of valve guide. Remove plastic sleeve from valve stem and when installing spring retainer, compress valve spring only far enough to install keepers. Compressing spring too far may result in damage to valve seal.

The positive type valve rotators used on the non-diesel exhaust valves require no maintenance, but should be observed when engine is running to make sure that each exhaust valve rotates clockwise slightly when opening. Renew the rotator of any exhaust valve that fails to turn.

Tappet gap should be set hot to 0.015 on intake valves and 0.020 on exhaust valves. Preliminary cold settings is 0.017 on intake valves and 0.022 on exhaust valves.

Diesel

33. Intake valves have a face and seat angle of 30 degrees. The seat width can be narrowed by using 10

and 70 degree stones to obtain the desired seat width of 5/64 to 3/32-inch. The intake valves seat directly in the cylinder head. Intake valves should be recessed 0.016 to 0.043 into the cylinder head. This can be measured by placing a straight edge across cylinder head surface and measuring clearance between intake valve heads and straight edge with a feeler gage. Reseat intake valves deeper into cylinder head if clearance measures less than 0.016.

Intake valve stem seals should be renewed whenever the intake valves are removed for service. Remove old seals from intake valve guides. When reinstalling valves, refer to Fig. AC20 and install new intake valve seals as follows: Install intake valve in guide and place plastic sleeve (contained in seal kit) over stem as shown in view A. If sleeve extends over 1/16-inch below groove in valve stem, cut off excess length of sleeve. Lubricate the sleeve and, while holding against head of valve, push seal assembly down over sleeve and valve stem as shown in view A and view B. Rubber sleeve of seal should be pushed down over intake valve guide with two screwdrivers as shown in view D making sure that seal is tight against top of valve guide. Remove plastic sleeve from valve stem and when installing valve rotator, compress spring only far enough to install keepers. Compressing spring too far may result in damage to seal.

Intake valves are equipped with positive type valve rotators. No maintenance of the rotators is required, but rotators should be observed while engine is running to be sure that each intake valve is rotating slightly in a counter-clockwise direction. Renew the rotator of any intake valve that fails to turn.

The exhaust valves have a face and seat angle of 45 degrees. The seat width can be narrowed by using 30 and 70 degree stones to obtain the desired seat width of 3/64 to 1/16-inch. The exhaust valves seat in renewable ring type inserts which are available in standard size and 0.005 oversize. Inserts should have 0.002-0.004 interference fit in standard bore diameter of 1.468-1.469. Machine bore out to 1.473-1.474 to install oversize insert.

Intake and exhaust valve stem diameter is 0.3090-0.3100.

Refer to paragraph 40A before adjusting valve tappet gap.

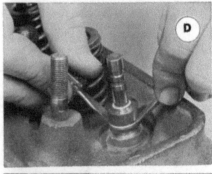

Photos courtesy of Perfect Circle Corporation

Fig. AC20 — Views A through E illustrate procedure for installing intake valve stem seals on both non-diesel and diesel engines. Refer to text.

VALVE GUIDES
Non-Diesel

34. Valve guides should be renewed if clearance between stems and guides exceeds 0.006. Intake and exhaust valve guides are not interchangeable. New valve guides should be pressed into cylinder head until tops of guides are 1/8-inch below machined rocker arm cover gasket surface of the cylinder head. New guides should be reamed to 0.344-0.345 to provide 0.001-0.0027 clearance between guides and intake valve stems and 0.0023-0.0043 clearance between the guides and exhaust valve stems. NOTE: Upper ends of intake valve guides are machined for valve seals.

Diesel

35. Valve guides should be renewed if clearance between valve stems and guides exceeds 0.006. Intake and exhaust valve guides are not interchangeable. New valve guides should be pressed into cylinder head until tops of guides use 5/16-inch above the machined rocker cover gasket surface of the cylinder head. New guides should be reamed to 0.3125-0.3135 to provide 0.0025-0.0045 clearance between the guides and valve stems. NOTE: Upper ends of intake valve guides are machined for valve seals.

VALVE SPRINGS
Non-Diesel

36. The interchangeable inlet and exhaust valve springs should be renewed if they are rusted, distorted, or fail to meet the following test specifications:

Pounds pressure @ $1\frac{22}{32}$ in. . . . 40.5-46.5
Pounds pressure @ $1\frac{15}{32}$ in. . . . 89.5-99.5
Spring free length. . . . . $2\frac{9}{32}$ in.

Diesel

37. Intake and exhaust valve springs are not interchangeable. Install stamped steel valve spring dampeners with flange between spring and cylinder head. Renew springs if they are rusted, distorted or fail to meet the following test specifications:

INTAKE VALVE SPRINGS
Pounds pressure @ 1.584 in. 40-45
Pounds pressure @ 1.240 in. 86-92
Spring free length. . . . . $1\frac{22}{32}$ in.
EXHAUST VALVE SPRINGS
Pounds pressure @ 1.756 in. 40-45
Pounds pressure @ 1.412 in. 86-92
Spring free length. . . . . $2\frac{3}{32}$ in.

CAM FOLLOWERS
All Engines

38. The 0.560-0.5605 diameter mushroom type cam followers (valve lifters) ride directly in unbushed cylinder block bores and can be removed after removing the camshaft. Cam followers are available in standard size only and should be renewed if either end is chipped or worn or if they are loose in cylinder block bores. Desired follower to bore diametrical clearance is 0.001-0.0025.

ROCKER ARMS
Non-Diesel

39. **R&R AND OVERHAUL.** Rocker arms and shaft assembly can be removed after removing the left hood side panel, air cleaner assembly, rocker arm cover and oiling tube; then removing the retaining cap screws and stud nuts.

The hollow rocker arm shaft is drilled for lubrication to each rocker arm bushing. Lubricating oil to the oil tube and drilled cylinder head passage is supplied by an external oil line which is connected to the main oil gallery at the left side of the engine.

To disassemble the rocker arms and shaft assembly, remove the cotter pin and washer from end of shaft; then, slide rocker arms, shaft supports, springs and shaft oiling collar from shaft. The valve contacting surface of the rocker arms is not ground on a radius and rocker arm should be renewed if contact surface is worn excessively. Desired clearance between rocker arms and shaft is 0.0025-0.0055. If clearance is excessive, renew the rocker arms and/or shaft. Rocker arm bushings are not available separately. The rocker arms are offset to right or left, but are not identified as intake or exhaust valve rocker arms. When reinstalling rocker arms on shaft, be sure that offset at adjusting screw end of rocker arm is away from the nearest rocker arm shaft support. Install rocker arm shaft with oiling holes toward the cylinder head. Renew corks in each end of shaft if they are loose.

39A. **VALVE TAPPET GAP.** Valve tappet gap for non-diesel models should be adjusted after engine reaches operating temperature. Clearance should be 0.015 hot for inlet valves, 0.020 hot for exhaust valves. Preliminary setting with engine cold is 0.017 for inlet; 0.022 for exhaust.

Two-position adjustment of all valves is possible as shown in Figs. AC22 & AC22A. To make the adjustment, turn the crankshaft until No. 1 cylinder is at TDC on compression stroke. Timing mark on crankshaft pulley (Fig. AC79) indicates TDC and if clearances are nearly correct, both front rocker arms will be loose and both rear rocker arms will be tight with No. 1 cylinder on compression

Fig. AC21 — Diesel engine rocker arm assembly is lubricated by oil passing through slot (S) in stud from drilled passage in cylinder head.

stroke. Adjust the six valves indicated in Fig. AC22 to the correct clearances. Turn the crankshaft one complete revolution until TDC marks are again aligned. This will position No. 6 cylinder at TDC on compression stroke. Adjust the remaining six valves shown in Fig. AC22A.

Diesel

40. **R&R AND OVERHAUL.** Rocker arms and shaft assembly can be removed after removing the left hood side panel, air cleaner assembly, disconnecting breather tube from rocker arm cover, removing the rocker arm cover; then removing the retaining cap screws and stud nuts.

The hollow rocker arm shaft is drilled for lubrication to each rocker arm bushing. Lubricating oil to the drilled cylinder head passage and slotted oil stud (S—Fig. AC21) is supplied by an external oil line which is connected to the main oil gallery on left side of engine. If the slotted stud is tight in the cylinder head and the end of the stud is above the drilled passageway, it is not necessary that the slot be in line with the passageway. However, this should be checked and if the end of the stud is lower than the passageway, be sure that the slot in the stud is in line with the drilled passage. If oil does not flow from the hole in the top of each rocker arm, check for foreign material in the external oil line or in the cylinder head passage.

The procedure for disassembling and reassembling the rocker arms and shaft is evident. Check the rocker arm shaft and bushing in each rocker arm for excessive wear. Desirable clearance between the shaft and bushings is 0.001 - 0.0035. Renew rocker arms and/or shaft if clearance is excessive. Rocker arm bushings are not available separately.

Inspect the valve stem contact button in the end of each rocker arm for being mutilated or excessively loose. If either condition is found, renew the contact button. **Extract the button retaining clip and remove the button and oil wick. Install new oil wick and button and test the button for a free fit in the rocker arm socket.** NOTE: If a new contact button has any binding tendency in the rocker arm socket, use a fine lapping compound and hand lap the mating surfaces.

40A. VALVE TAPPET GAP. Two different camshafts have been used in D-19 diesel engines and before valve tappet gap can be set, it must be determined which camshaft is used.

Turn the crankshaft until No. 1 cylinder is at TDC on compression stroke. (Timing marks are located on crankshaft pulley.) Set tappet gap for No. 1 cylinder inlet valve at 0.015. Leave feeler gage between No. 1 cylinder rocker arm and inlet valve stem, then rotate crankshaft slowly until feeler gage becomes tight. If feeler gage becomes tight (inlet valve begins to open) at 30 degrees BTDC, the early type camshaft (4513674) is used. If feeler gage does not become tight until approximately 10 degrees BTDC, the late type camshaft (4513204) is installed.

Tappet gap for engines equipped with early camshaft (4513674) should be set at 0.015 with engine at operating temperature for both inlet and exhaust valves.

Tappet gap for engines equipped with the late camshaft (4513204) should be set at 0.010 hot for inlet valves; 0.019 hot for exhaust valves. NOTE: The late camshaft (4513204) was used after engine Serial No. D-01498; however, earlier models may have the later camshaft installed.

Two-position adjustment of all valves is possible as shown in Figs. AC22 & AC22A. To make the adjustment, turn the crankshaft until No. 1 cylinder is at TDC on compression stroke. Timing mark on crankshaft pulley (Fig. AC47) indicates TDC and if clearances are nearly correct, both front rocker arms will be loose and both rear rocker arms will be tight with No. 1 cylinder on compression

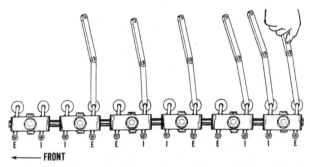

Fig. AC22A—With number 6 piston at TDC on compression stroke, valve clearances (tappet gap) can be set for the six valves indicated. Non-diesel engines use coil springs to position rocker arms.

stroke. Adjust the six valves indicated in Fig. AC22 to the correct clearances. Turn the crankshaft one complete revolution until TDC marks are again aligned. This will position No. 6 cylinder at TDC on compression stroke. Adjust the remaining six valves shown in Fig. AC22A.

TIMING GEAR COVER AND CRANKSHAFT FRONT OIL SEAL
All Engines

41. REMOVE AND REINSTALL. To remove the timing gear cover, it is first necessary to remove the front support as outlined in paragraph 12 or 21.

Remove generator brace and fan belt. If equipped with power steering, loosen power steering pump on bracket and remove pump drive belt. Remove the crankshaft pulley retaining nut and withdraw pulley from crankshaft. The timing gear cover can now be unbolted and removed. Cover is located with two dowel pins. NOTE: On diesel engines, a spring loaded pin is located in the fuel injection pump drive shaft and pin contacts the timing gear cover. On non-diesel engines, the front end of the governor shaft rides in a bushing in the timing gear cover. Therefore, care should be taken when removing timing gear cover from either engine.

The crankshaft front oil seal is installed with lip facing inside. On non-diesel engines, the governor shaft front bushing is renewable at the timing gear cover. On diesel engines, the fuel injection pump drive gear and shaft can be removed and the seals renewed while timing gear cover is removed. See paragraph 46.

Reinstall cover by reversing removal procedures. On diesel engines, be sure that injection pump drive gear, shaft and spring loaded thrust pin are in place and timing marks are aligned before installing timing gear cover.

TIMING GEARS

All Engines

42. VALVE TIMING. Valves are properly timed when punch marked tooth of the crankshaft gear is in register with the punch mark between two teeth on the camshaft gear as shown in Fig. AC25. Valve opening and closing specifications are as follows:

NON-DIESEL
Intake valve opens @...27° B.T.D.C.
closes @...57° A.B.D.C.
Exhaust valve opens @...59° B.B.D.C.
closes @...25° A.T.D.C.
DIESEL (Prior to Eng. Serial No. D-01499)
Intake valve opens @...30° B.T.D.C.
closes @...40° A.B.D.C.
Exhaust valve opens @...50° B.B.D.C.
closes @...70° A.T.D.C.
DIESEL (After Eng. Serial No. D-01498)
Intake valve opens @ .10° 30' B.T.D.C.
Closes @....38° A.B.D.C.
Exhaust valve opens @...40° B.B.D.C.
closes @ .14° 30' A.T.D.C.

43. CAMSHAFT GEAR AND CRANKSHAFT GEAR BACKLASH. Desired backlash between camshaft gear and crankshaft gear is 0.001-0.005. Camshaft gear and/or crankshaft gear should be renewed if backlash exceeds 0.008. Gears are available in standard size only. NOTE: While checking gear backlash, be sure to hold all end play out of camshaft.

44. R&R CRANKSHAFT GEAR. The crankshaft gear is keyed and press fitted to the crankshaft. The gear can be removed by using a suitable puller after first removing the timing gear cover as outlined in paragraph 41.

New gear can be installed by heating it in oil for fifteen minutes prior to installation and drifting it on

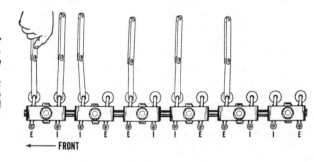

Fig. AC22—With number 1 piston at TDC on compression stroke, valve clearances (tappet gap) can be set for the six valves indicated. Refer to text for recommended clearances.

Fig. AC25 — Diesel engine timing gears with timing marks aligned. Timing marks on non-diesel engine crankshaft and camshaft gears are similar.

Install camshaft gear and thrust plate on camshaft and reinstall assembly by reversing removal procedures. NOTE: Both camshaft thrust plate retaining cap screws are drilled. Never substitute solid cap screws for this installation as lubricating oil for timing gears must pass through hollow cap screw in lower position. Be sure timing marks are aligned as shown in Fig. AC25 before bolting camshaft thrust plate to block.

DIESEL INJECTION PUMP DRIVE AND DRIVEN GEARS

46. REMOVE AND REINSTALL. The diesel fuel injection pump drive and driven gears can be removed and reinstalled after the timing gear cover is removed as outlined in paragraph 41. The pump driven gear and shaft assembly is removed by pulling it from the fuel injection pump. NOTE: Fuel in the pump will drain out through the shaft opening unless the fuel has been shut off and the pump drained by removing timing inspection cover on pump prior to removal of gear and shaft. The pump drive gear is retained to the camshaft gear by three wired cap screws and one dowel pin.

crankshaft or by pressing gear on shaft using crankshaft pulley retaining nut and suitable washers and spacers. Be sure timing marks are aligned as described in paragraph 42.

45. R&R CAMSHAFT GEAR. It is recommended that camshaft be removed from engine to remove and reinstall timing gear. After timing gear cover is removed as outlined in paragraph 41, proceed as follows: Remove oil pan and engine oil pump. Remove the air cleaner assembly, rocker arm cover and rocker arm assembly. Remove all push rods and drive wood dowel pins of suitable size into the hollow cam followers (valve lifters). Pull each cam follower up against cylinder block with the dowel pins and hold in that position with pincher-type clothes pins. On diesel engines, pull injection pump driven gear and shaft from pump; then, unbolt and remove the injection pump drive gear from front of camshaft gear. On gasoline engines, remove the fuel pump. Working through holes in camshaft gear, remove the two cap screws retaining camshaft thrust plate to the cylinder block. Camshaft and gear assembly can then be withdrawn from front of engine.

Camshaft gear is retained by a snap ring in front of gear and is keyed and press fitted to the camshaft.

Camshaft end play is controlled by the thrust plate that retains the camshaft assembly in the cylinder block. End play should be 0.003-0.008 and can be measured with a dial indicator, or when camshaft is removed, with a feeler gage as shown in Fig. AC26. If end play exceds 0.014, worn thrust plate should be renewed or end play reduced by filing off the rear face of the camshaft gear as shown in Fig. AC27.

Fig. AC26 — To check the camshaft end play on both diesel and non-diesel engines, insert feeler gage as shown between shaft journal and thrust plate. Camshaft end play is equal to clearance between thrust plate and journal.

Fig. AC27 — Camshaft end play on both diesel and non-diesel engines may be reduced by filing metal from rear face of camshaft gear hub as shown. Refer to text.

The injection pump driven gear can be removed from the pump shaft by removing the retaining nut and pressing shaft from gear. The two lip seals on the shaft should be renewed whenever the shaft is removed from the pump. Lip of each seal should be towards end of shaft (opposed).

To install pump driven gear and shaft unit, lubricate seals with Lubriplate or similar lubricant and insert shaft in pump with off-center hole on drive tang of shaft and off-center hole in pump rotor slot aligned. Carefully work shaft into pump to avoid rolling lip of rear seal back. NOTE: If lip of seal has been rolled back during installation of shaft in pump, pull shaft back out and renew seal before proceeding further. The seal will have been damaged and early failure of seal will occur. Be sure drive tang of shaft enters drive slot in pump rotor. Place spring and plunger in front end of shaft.

To install pump drive gear, turn engine until timing marks on camshaft gear and crankshaft gear are aligned and secure pump drive gear to camshaft gear with dowel pin and cap screws. Wire cap screws as shown in Fig. AC25. Be sure that punch-marked tooth of injection pump driven gear is in register with punch-mark between two teeth of the drive gear as in Fig. AC25.

CAMSHAFT AND BUSHINGS

All Engines

47. **REMOVE AND REINSTALL.** To remove the camshaft, follow the procedure outlined in paragraph 45. Refer to paragraph 27A for explanation of later camshaft used in diesel models.

Dimensions of camshaft journals (new) are as follows: Front three camshaft journals should measure 1.998-1.999; rear journal should measure 1.248-1.249. Desired diametrical clearance of camshaft in bushings is 0.002-0.0046. Camshaft bushings should be renewed if clearance exceeds 0.0065. Cam follower lift is 0.312.

Dimensions listed above apply to all camshafts. Refer to paragraph 42 for timing specifications and paragraphs 39A and 40A for correct valve tappet gap.

48. **CAMSHAFT BUSHINGS.** The camshaft is supported in four precision steel-backed babbit lined bushings. The camshaft journals should have 0.002 - 0.0046 clearance in the bushings. Renew bushings if clearance exceeds 0.0065. Renew camshaft if clearance exceeds 0.0065 in new bushings. New bushing I.D. is 1.2510-1.2526 for rear journal and 2.001-2.0026 for the three front journals.

To renew camshaft bushings, the engine must be removed from the tractor, and the flywheel, engine rear plate and soft plug behind the rear bushing must be removed. Remove camshaft as outlined in paragraph 45. Remove and reinstall bushings using Kent-Moore No. J-6582 bushing remover and replacer kit or equivalent.

New rear bushing should have 0.001-0.003 interference fit in bore of block and front three bushings should have 0.002-0.004 interference fit in bores. Although front bushing has same diametrical dimensions as the two intermediate bushings, it is wider and oil holes are different. Be sure that oil holes in bushings line up with oil passageways in cylinder block. Bearings are pre-sized, but should be checked after installation to be sure there are no localized high spots.

Use Permatex or other suitable sealer when installing plug at rear of camshaft bore in cylinder block.

ROD AND PISTON UNITS

All Engines

49. Piston and connecting rod units are removed from above after removing cylinder head, oil pan and rod bearing caps.

Cylinder numbers are stamped on the connecting rod and cap. When reinstalling rod and piston units, make certain that the cylinder identifying numbers are in register and face away from the camshaft side of the engine. (Both bearing insert tangs must be towards same side of rod assembly.)

Tighten the connecting rod nuts to a torque of 40-50 Ft.-Lbs. and install new cotter pins.

PISTONS, LINERS AND RINGS

Non-Diesel

50. The cam ground aluminum pistons are fitted with three compression rings and one three-piece oil control ring. Pistons and rings are available in standard size only.

Install compression rings with side of ring marked "TOP" to top of piston. To install the three-piece oil ring, place expander in groove with ends of expander butted together over piston pin end. Install top steel rail over expander with end gap 90 degrees from expander gap. Install bottom steel rail over expander with end gap 180 degrees away from end gap of top rail. Be sure that ends of expander are butted together and not overlapped.

After removing piston and rod unit, use suitable puller to remove the wet type cylinder liners (sleeves). Clean and lubricate mating surfaces of block and liner prior to installing liner. Install seal rings in grooves in cylinder block, lubricate with soap or thinned white lead and push liner into place. Top of liner should be from 0.002 below to 0.002 above top surface of block. If liner is more than 0.002 below top of block, counterbore in block may be bored deeper and liners with 0.020 thicker flange installed.

Check pistons, rings and sleeves against the following specifications:
Piston Ring Side Clearance:

    Top ring, desired......0.0015-0.003
      Max. allowable...........0.008

2nd and 3rd, desired..0.0015-0.0035
4th (oil control), desired...0-0.0055
Ring End Gap (Minimum):
    Top compression 0.007
    2nd and 3rd compression.......0.014
    4th oil (side rails only).......0.014
Piston Skirt to Sleeve Clearance:
    Desired 0.0023-0.0048

Renew cylinder sleeve if wear at top of ring travel (taper) exceeds 0.007. Inside diameter of new sleeve is 3.5623-3.5638.

Diesel

51. The cam ground aluminum pistons are fitted with three compression rings, one three-piece oil control ring above the piston pin and one one-piece scraper type oil control ring below the piston pin. Pistons and rings are available in standard size only.

Refer to paragraph 27A for explanation of later (flat top) pistons.

Install compression rings with side of ring marked "TOP" to top of piston. To install the three piece oil ring, place expander in oil ring groove with ends of expander butted together over either end of piston pin. Install top steel rail over expander with gap 90 degrees from expander gap. Install bottom steel rail over expander with gap 180 degrees away from end gap of top rail. Be sure that ends of expander are butted together and not overlapped. Install the scraper type oil ring in groove below piston pin with scraper edge down.

With the piston and connecting rod removed from the block, use a suitable puller to remove the wet type cylinder sleeve. Clean and lubricate all sealing and mating surfaces of block and liner and renew sealing "O" ring before new liner is installed. Use soap or thinned white lead as a lubricant. Top of liner should be from 0.002 below to 0.002 above the top of block when installed. If top of liner is more than 0.002 below the top of block, sleeves with flange 0.020 thicker than standard sleeve flange are available for service. To install these sleeves, the counterbore in block must be milled out to the correct depth to give proper sleeve standout of −0.002 to +0.002.

Check pistons, rings and sleeves against the following values:
Piston Ring Side Clearance:

    Top ring, desired......0.003-0.0045
      maximum allowable........0.008
    2nd and 3rd, desired.....0.002-0.004
    4th, desired 0-0.0055
    5th, desired 0.0015-0.0035

Ring End Gap (Minimum):
Top compression 0.022
2nd and 3rd compression......0.014
4th oil (side rails only)......0.014
5th oil 0.007
Piston Skirt to Sleeve Clearance:
Desired 0.004-0.0065

Renew cylinder sleeve if wear at top of ring travel (taper) exceeds 0.007. Inside diameter of new sleeve is 3.5623-3.5638.

PISTON PINS

All Engines

51A. The full floating type piston pins are retained in the piston pin bosses by snap rings and are available in standard size only. Check the piston pin fit against the following specifications:

NON-DIESEL
Piston pin bore in
piston 0.99985-1.00005
I.D. of connecting rod
bushing 1.0001-1.0006
Piston pin diameter...0.99955-0.99975
Desired clearance between
pin and bore in piston
at 70 degrees F. 0.0001-0.0005
Desired clearance between
pin and connecting
rod bushing 0.00035-0.00105
Renew pin and/or bushing if
clearance exceeds 0.002
DIESEL
Piston pin bore in
piston 0.99985-1.00005
I.D. of connecting rod
bushing 0.9999-1.0004
Piston pin diameter...0.99955-0.99975
Desired clearance between
pin and bore in piston
at 70 degrees F. 0.0001-0.0005
Desired clearance between
pin and connecting
rod bushing 0.00015-0.00085
Renew pin and/or bushing if
clearance exceeds 0.002

jections (tangs) engage the milled slot in connecting rod and bearing cap and that the cylinder numbers on the rod and cap are in register and face away from the camshaft side of the engine. Bearing inserts are available in standard size and undersizes of 0.002, 0.010, 0.020 and 0.040. Check crankshaft crankpin journals and connecting rod bearings against the following values:

Crankpin diameter
(Std.) 1.9975-1.9985
Rod bearing clearance (Non-Diesel):
Desired clearance 0.0011-0.004
Maximum allowable 0.006
Rod bearing clearance (Diesel):
Desired clearance 0.0011-0.0036
Maximum allowable 0.006
Rod side play:
Desired 0.003-0.009
Maximum 0.015
Rod bolt torque (Ft.-Lbs.).....40-50

CRANKSHAFT AND MAIN BEARINGS

All Engines

53. The crankshaft is supported in seven slip in type main bearings. Main bearing inserts (liners) can be renewed after removing oil pan, oil pump, oil tube and main bearing caps. Desired crankshaft end play is 0.003-0.009. Maximum allowable end play is 0.015. Crankshaft end play is controlled by the flanges on the center main bearing insert. Center main bearing cap is doweled to cylinder block.

To remove the crankshaft, first remove the engine as outlined in paragraph 28 or 29. Then remove clutch, flywheel, engine rear adapter plate, crankshaft pulley, timing gear cover, oil pan, oil pump, oil tube, camshaft, engine front plate and rod and main bearing caps. Lift crankshaft from en-

gine. Remove crankshaft gear with suitable puller.

Check crankshaft and main bearing inserts against the following specifications:
Crankpin diameter
(Std.) 1.9975-1.9985
Main journal diameter
(Std.) 2.4970-2.4980
Main bearing clearance:
Desired clearance 0.0013-0.004
Maximum clearance 0.007
Main bearing bolt torque
(Ft.-Lbs.) 120-130

Main bearing inserts are available in standard size and undersizes of 0.002, 0.010, 0.020 and 0.040.

CRANKSHAFT OIL SEALS

All Engines

54. **FRONT SEAL.** The crankshaft front oil seal is located in the timing gear cover and can be renewed as outlined in paragraph 41.

55. **REAR SEAL.** The crankshaft rear oil seal is installed in the adapter plate at rear of engine. The seal consists of two parts: A steel seal retainer with an integral inner seal (1—Fig. AC28) and an outer seal ring (2). The seal retainer is pressed into the engine rear adapter plate from the front side. The outer seal ring fits around the seal retainer and forms the rear seal for the oil pan.

To remove the retainer and integral inner seal, first remove the flywheel and oil pan (sump); then, remove adapter plate from rear of engine and drive seal retainer out of plate. Before installing new seal, apply gasket sealer to outer rim of retainer that contacts adapter plate. Apply gasket sealer to exposed outer rim and install outer sealing ring. Reverse disassembly procedure to reassemble tractor.

CONNECTING RODS AND BEARINGS

All Engines

52. Connecting rod bearings are of the renewable slip in type. The bearings can be renewed after removing oil pan and bearing caps. When installing new bearing inserts (liners), make sure that the bearing liner pro-

1. Oil seal
2. Outer seal ring
3. Engine rear plate
4. Flywheel
5. Flywheel ring gear
6. Cap screws
7. Dowel pins (2)

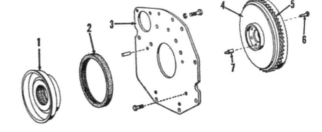

Fig. AC28 — Exploded view of both diesel and non-diesel flywheel and crankshaft rear oil seal installation.

Fig. AC29 — D-19 diesel and non-diesel flywheel is retained to crankshaft by four cap screws and two dowel pins. Installation of flywheel is possible in one position only.

FLYWHEEL

All Engines

56. **REMOVE AND REINSTALL.** To remove flywheel, first remove engine clutch as outlined in paragraph 109. The flywheel is retained to the engine crankshaft with four unequally spaced cap screws and two dowel pins. See Figs. AC28 and AC29. To renew flywheel ring gear, flywheel must be removed from crankshaft. Inspect clutch friction surface and crankshaft rear oil seal surface of flywheel. Drive flywheel ring gear from flywheel. Heat new ring gear evenly until it will fit over flywheel and install ring gear with tooth bevel to front. Clutch pilot bearing in flywheel is a sealed unit.

When reinstalling flywheel, tighten retaining cap screws to a torque of 95-105 Ft.-Lbs. Complete reassembly by reversing disassembly procedure.

OIL PAN (SUMP)

All Engines

57. **REMOVE AND REINSTALL.** To remove the oil pan, it is first necessary to remove the front support as outlined in paragraph 12 or 21.

After removing front support, remove the four cap screws retaining lower portion of timing gear cover to pan, then remove cap screws holding pan to cylinder block. Pan generally will be tightly sealed to block and care should be taken in prying pan loose in order not to distort rear arch and gasket sealing surface of pan.

As gasket between engine front plate and cylinder block will probably be damaged when pan is re-

moved and renewal of complete gasket would require removal of engine front plate, cut lower portion of new gasket to fit end of pan and then apply this cut portion of the gasket to the front plate with gasket cement. Use a heavy gasket sealer where gasket meets cylinder block. Apply gasket sealer to rubber sealing ring around crankshaft rear oil seal retainer and to both sides of pan gasket. Stick gasket to cylinder block and carefully lift pan in place after being sure any distorted part of pan sealing surface has been straightened. Install cap screws retaining pan to cylinder block finger tight; then, apply gasket sealer to the four cap screws retaining pan to front cover and install these cap screws tightly. Starting at front of pan, tighten all cap screws to a torque of 18-21 Ft.-Lbs. Reinstall front support by reversing removal procedures. On power steering models, start engine and cycle power steering system several times to bleed any trapped air from system. Check power steering fluid level and add type "A" automatic transmission fluid as necessary.

OIL PUMP AND RELIEF VALVES

All Engines

58. **R&R AND OVERHAUL PUMP.** Removal procedure will be self-evident after removal of oil pan as outlined in paragraph 57.

To disassemble the removed pump, refer to Fig. AC30 and proceed as follows: Remove screen (13) and cover (12). Extract pin (2), then press shaft (9) out of gear (1) and body (4). To remove either pump gear (11), press shaft out of gear.

Renew any parts which are excessively worn, scored or are in any way questionable. Pump gears (11) should have not more than 0.020 back lash or more than 0.006 end play. Pump body and/or shafts should be renewed if shaft to body clearance exceeds 0.004.

When reinstalling oil pump on diesel engines, be sure that slot in oil pump gear drive engages drive pin on operation (hour) meter drive shaft. Tighten the pump retaining cap screw to a torque of 18-21 Ft.-Lbs. NOTE: Flange on pump does not fit against cylinder block.

On non-diesel engines, turn crankshaft until piston for number one cylinder is at TDC on compression stroke (Refer to Fig. AC79). Turn the dis-

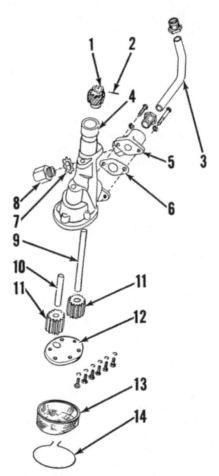

Fig. AC30 — Exploded view of D-19 diesel and non-diesel oil pump assembly. Relief valve (8) is non-adjustable and is used to prevent surge pressures only. Normal oil pressure is controlled by relief valve shown in Fig. AC31.

1. Drive gear	8. Relief valve
2. Pin	9. Drive shaft
3. Discharge tube	10. Driven shaft
4. Pump body	11. Pump gears
5. Adapter flange	12. Cover plate
6. Gasket	13. Screen
7. Lock washer	14. Retainer

tributor rotor until it points toward the number one cylinder terminal on distributor cap and breaker points are just open. Install pump with slot in pump drive gear engaging drive tang on distributor drive shaft. NOTE: The driving slot and tang are off-set from center line of shafts. Tighten pump retaining cap screw to 18-21 Ft.-Lbs. torque. Check ignition timing as outlined in paragraph 103.

59. **RELIEF VALVES.** Two oil pressure relief valves are used in both diesel and non-diesel engines. The relief valve (8—Fig. AC30) on the oil pump discharge tube adapter is non-adjustable and should by-pass oil directly into the sump at approximately 80 psi. As this is well above normal oil pressure of 25-40 psi, valve (8) opens only due to surge pressure when oil is cold.

The relief valve (See Fig. AC31) located on the left front side of the engine crankcase is adjustable to control normal engine oil pressure to 25-40 psi. Oil by-passing this relief valve flows through the hollow (lower) cap screw retaining the camshaft thrust plate to the front face of the cylinder block. Oil passing through this hollow cap screw lubricates the timing gears and fuel injection pump or governor gears.

To adjust the relief valve (Fig. AC31) loosen lock nut (25) and turn adjusting screw (26) in or out until oil pressure is within 25-40 psi range. The regular oil pressure gage does not indicate pounds pressure; therefore, a master gage should be installed in the oil pressure line when adjusting the relief valve. Adjust relief valve only when engine is running at normal operating temperature and speed. Do not attempt to adjust relief valve to compensate for worn oil pump or engine parts.

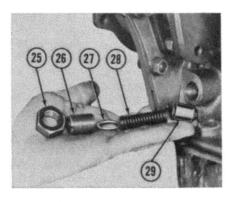

Fig. AC31 — Diesel and non-diesel oil pressure relief valve is located in left front side of cylinder block.

25. Nut	28. Spring
26. Adjusting screw	29. Regulator piston
27. Gasket	

CARBURETOR

Gasoline

60. Gasoline models are equipped with a Marvel-Schebler TSX-848 carburetor. Calibration data follows:

Float setting	¼-inch
Basic repair kit	286-1390
Gasket set	16-594
Float valve and seat	233-543
Idle jet	49-101L
Main adjusting needle seat	36-297
Main adjusting needle	43-631
Idle adjusting needle	43-58
Nozzle	47-A71
Venturi	46-A1
Power jet	49-A10

Fuel Pump

61A. Gasoline models are equipped with an automotive type diaphragm fuel pump mounted on the right hand side of the engine and actuated by a cam on the engine camshaft. Refer to Fig. AC32 for exploded view of fuel pump.

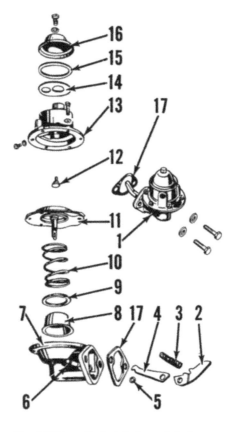

Fig. AC32 — Exploded view of fuel pump used on gasoline D-19 tractor.

1. Fuel pump		10. Spring
2. Arm		11. Diaphragm assy.
3. Spring		12. Valve
4. Link		13. Cover assy.
5. Washer		14. Screen
6. Pin		15. Gasket
7. Body		16. Plate
8. Seal		17. Gasket
9. Washer		

LP-GAS SYSTEM

The D-19 tractor is available with an LP-Gas system manufactured by the Ensign Products Section, American Bosch Arma Corporation. Like other LP-Gas systems, this system is designed to operate with the fuel tank not more than 80% filled.

The American Bosch Series CBX (Ensign model XG) carburetor and Series RDG (Ensign model W) regulator have three points of mixture adjustment, plus an idle stop screw.

Repair data follows:
Carburetor part number:
(American Bosch) .CBX 125A5477A
Carburetor repair kit part number:
(American Bosch) ....KT 46115
Regulator part number:
(American
Bosch) ....RDG 100A5466A
Regulator repair kit part number:
(American Bosch) ....KT 4613
(Ensign) 2115

ADJUSTMENTS

61. **INITIAL ADJUSTMENTS.** After overhauling or installing new carburetor or regulator, make following initial adjustments: Open idle screw on regulator 1½ turns. Open starting adjustment screw on carburetor 1¼ turns. Open load adjustment screw 4 turns. Close choke and open throttle ½ way to start engine.

62. **STARTING SCREW ADJUSTMENT.** As soon as the engine is started, fully open throttle while leaving choke closed. Adjust starting screw to give highest engine speed; then, open starting adjustment slightly further to give a slightly richer mixture. Return throttle to idle position and open choke.

63. **IDLE STOP SCREW.** Adjust stop screw on carburetor throttle to obtain an engine low idle speed of 375-425 RPM.

64. **IDLE MIXTURE SCREW.** Adjust idle mixture screw on regulator for best engine idle performance when engine is warm. Readjust idle stop screw if necessary.

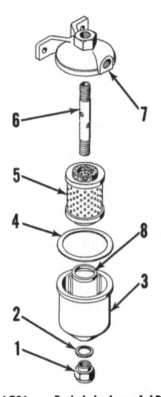

Fig. AC34 — Exploded view of LP-Gas filter.

1. Cap nut
2. Seal ring
3. Filter bowl
4. Sealing ring
5. Filter element
6. Stem
7. Filter body
8. Magnetic ring

65. LOAD SCREW ADJUSTMENT (WITH ANALYZER). Be sure to follow the gas analyzer operating instructions and set load screw to give reading of 12.8 on gasoline scale or 14.3 on LP-Gas scale with engine warm and running at high idle speed. Recheck idle adjustment as outlined in paragraph 64.

66. LOAD SCREW ADJUSTMENT (WITHOUT ANALYZER). With engine running at full throttle and at normal operating temperature, apply load until governor opens carburetor throttle wide open. Find adjustment point where engine speed begins to drop from mixture being too rich, then too lean. Set adjustment midway between these two points and tighten jam nut.

67. LOAD SCREW ADJUSTMENT (WITHOUT LOAD OR ANALYZER). Make idle adjustment carefully as outlined in paragraph 64. With engine running at high idle speed and at normal operating temperature, adjust load screw to give maximum engine RPM; then, slowly turn load screw in until engine speed begins to fall. Set load screw mid-way between these two positions.

FILTER

68. The Ensign filter (Fig. AC34) used with the LP-Gas system is equipped with a felt filtering element and a magnetic ring. When servicing the LP-system or on major engine overhauls, it is advisable to remove the lower part of the filter and clean or renew the filtering elements. CAUTION: Shut off both liquid and vapor valves at fuel tank and run engine until fuel is exhausted before attempting to remove filter bowl.

REGULATOR

The Ensign Model W regulator combines a heat exchanger to vaporize liquid LP-Gas with a two stage regulator to reduce fuel pressure to slightly below atmospheric. The primary regulator reduces the fuel from tank pressure to approximately 4 psi as it enters the final regulator. Heat for vaporizing the liquid fuel is obtained from coolant from the tractor cooling system. Refer to Fig. AC35 for cross-sectional views of regulator and principles of operation.

69. R&R REGULATOR. Shut off both liquid and vapor withdrawal valves at the fuel tank and run engine until fuel in system is exhausted. Drain engine cooling system. Remove

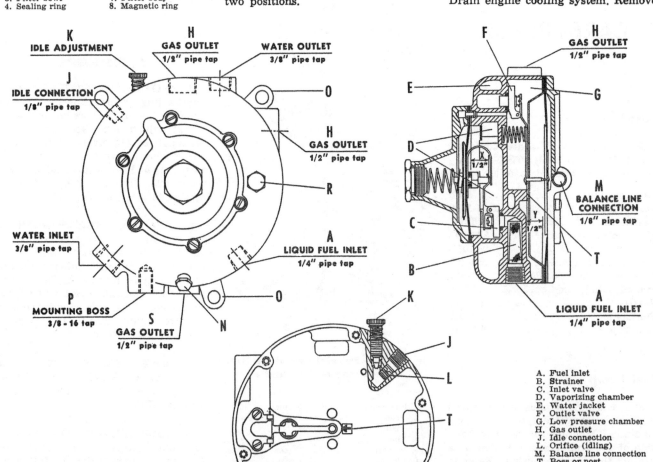

A. Fuel inlet
B. Strainer
C. Inlet valve
D. Vaporizing chamber
E. Water jacket
F. Outlet valve
G. Low pressure chamber
H. Gas outlet
J. Idle connection
L. Orifice (idling)
M. Balance line connection
T. Boss or post

Fig. AC35 — Series RDG American Bosch (Ensign Model W) regulator of similar construction to that used on the D-19 tractor. For exploded view, refer to Fig. AC35A.

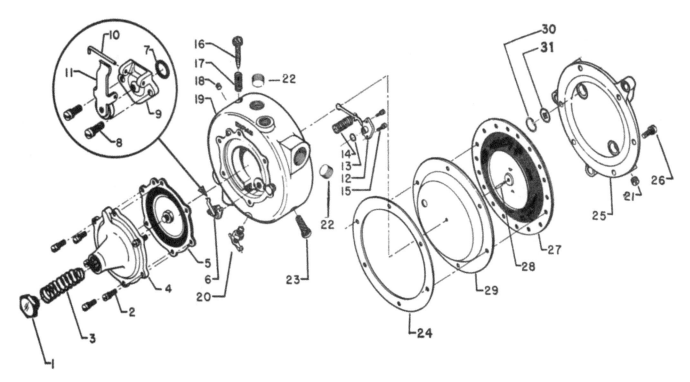

Fig. AC35A — Exploded view of American Bosch Series RDG (Ensign Model W) regulator.

1. Spring retainer	5. Inlet pressure diaphragm	11. Inlet diaphragm lever	16. Idle adjusting screw	20. Drain cock
3. Inlet diagphragm spring	6. Inlet valve assy.	12. Outlet valve assy.	17. Idle screw spring	23. Strainer
4. Regulator cover	7. "O" ring	13. "O" ring	18. Bleed screw	24. Partition plate
	9. Valve seat	14. Outlet diaphragm spring	19. Regulator body	25. Back cover plate
	10. Pivot pin			27. Outlet pressure diaphragm

28. Push pin	
29. Partition plate	
30. Retainer (snap) ring	
31. Compensator	

water connections to engine cooling system. Disconnect fuel and balance lines and remove regulator from tractor. Reverse removal procedures to reinstall.

70. OVERHAUL REGULATOR. Refer to exploded view of regulator in Fig. AC35A. Disassemble regulator and clean parts with solvent and dry with air hose. Inspect and renew valve seats, valves, diaphragms and springs as necessary. Parts are available individually or in a repair kit.

After inlet valve (primary regulator valve) assembly is installed, open and close valve several times; then, check distance from face of housing to bottom of groove in inlet valve lever (See Fig. AC36 and dimension "X" in Fig. AC35). Bend inlet valve lever if necessary so that this measurement is exactly ½-inch.

When installing outlet valve (final regulator valve) assembly, align center of valve lever with arrow on boss (See Fig. AC37) and, after operating valve several times, bend valve lever if necessary so that lever is flush with top of boss.

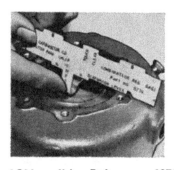

Fig. AC36 — Using Ensign gage 8276 to set the fuel inlet valve lever to the dimension as indicated at "X" in Fig. AC35.

71. TROUBLE SHOOTING REGULATOR PROBLEMS. Generally, difficulties encountered with regulator are due to leakage of gas past valves. Trouble will generally show up as excessive fuel consumption, decrease in power, inability to properly adjust fuel mixtures and/or loss of gas through carburetor when engine is not running. To test regulator, remove plug (R—Fig. AC35) and install a 0-10 psi gage in opening. If gage pressure gradually builds up after engine

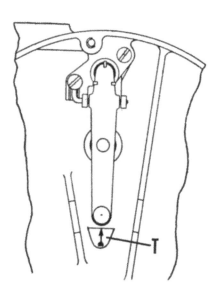

Fig. AC37 — Location of post or boss with stamped arrow for the purpose of setting the fuel inlet valve lever.

is stopped, inlet valve is leaking and same should be cleaned or renewed. Remove fuel hose from regulator to carburetor. Soap bubble should hold over fuel opening at regulator. If not, fuel outlet valve or low pressure diaphragm is leaking. Clean or renew valve and check diaphragm.

CARBURETOR

The Ensign model XG carburetor is equipped with starting and load adjustment screws and with an economizer unit which richens the fuel mixture under load conditions. See Fig. AC37A for exploded view of carburetor. Idle fuel mixture adjustment is provided on the regulator.

72. OVERHAUL CARBURETOR. Refer to Fig. AC37A. A repair kit is available for the carburetor, and parts are also available separately. Other than renewing throttle shaft and bushings, carburetor servicing generally concerns renewal of economizer diaphragm (24) and making sure that the starter valve (32) is working properly. Check to see that economizer diaphragm will hold vacuum. Inspect starting valve for proper position when choke is closed. Valve should fit tightly against carburetor body and completely cover main fuel opening in carburetor body.

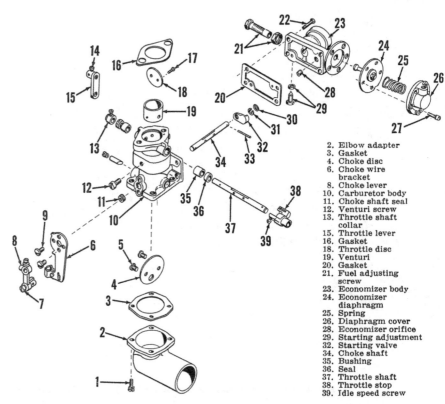

2. Elbow adapter
3. Gasket
4. Choke disc
6. Choke wire bracket
8. Choke lever
10. Carburetor body
11. Choke shaft seal
12. Venturi screw
13. Throttle shaft collar
15. Throttle lever
16. Gasket
18. Throttle disc
19. Venturi
20. Gasket
21. Fuel adjusting screw
23. Economizer body
24. Economizer diaphragm
25. Spring
26. Diaphragm cover
28. Economizer orifice
29. Starting adjustment
32. Starting valve
34. Choke shaft
35. Bushing
36. Seal
37. Throttle shaft
38. Throttle stop
39. Idle speed screw

Fig. AC37A—Exploded view of American Bosch Series CBX (Ensign XG) carburetor used on LP-Gas engines.

DIESEL FUEL SYSTEM

The diesel fuel system consists of three basic units; the fuel filter, injection pump and injection nozzles. When servicing any unit associated with the fuel system the maintenance of absolute cleanliness is of utmost importance.

Probably the most important precaution that servicing personnel can impart to owners of diesel powered tractors is to urge them to use an approved fuel that is absolutely clean and free from foreign material. Extra precaution should be taken to make certain that no water enters the fuel storage tanks. This last precaution is based on the fact that all diesel fuels contain some sulphur. When water is mixed with sulphur, sulphuric acid is formed and the acid will quickly erode the closely fitting parts of the injection pump and nozzles.

75. QUICK CHECKS — UNITS ON TRACTOR. If the diesel engine does not run properly, and the diesel fuel system is suspected as the source of trouble, refer to the Diesel System Trouble Shooting Chart and locate points which require further checking. Many of the chart items are self-explanatory; however, if the difficulty points to the fuel filters, injection pump and/or injection nozzles, refer to the appropriate paragraphs which follow.

FILTERS AND BLEEDING

The fuel filtering system consists of a fuel strainer and sediment bowl which incorporates the fuel shut-off valve, a first stage filter (of the replaceable element type) and a second stage filter (of the replaceable element type).

76. BLEEDING. Each time the filter elements are renewed or if fuel lines are disconnected, it will be necessary to bleed air from the system.

To bleed the fuel filters, remove the air bleed plug at the top of the filter head assembly (Refer to Fig. AC38) and open the fuel shut-off valve (incorporated in the fuel strainer and sediment bowl unit). As soon as all air has escaped and a solid flow of fuel is escaping from the air bleed hole, reinstall the plug.

Normally the injection pump is self bleeding; however, in some cases it may be necessary to proceed as follows:

Loosen the pump inlet line, turn the fuel on at the tank shut-off valve and allow fuel to flow from the connection until the stream is free from air bubbles; then, tighten the connection.

DIESEL SYSTEM TROUBLE SHOOTING CHART

	Sudden Stopping of Engine	Lack of Power	Engine Hard to Start	Irregular Engine Operation	Engine Knocks	Engine Smoking	Excessive Fuel Consumption
Lack of fuel	★	★	★	★			
Water or dirt in fuel	★	★	★	★			
Clogged fuel lines	★	★	★	★			
Inferior fuel	★	★	★	★			★
Faulty transfer pump	★	★	★	★			
Faulty injection pump timing		★	★	★	★	★	★
Air traps in system	★	★	★	★			
Clogged fuel filters		★	★	★			
Deteriorated fuel lines	★						★
Air leak in suction line	★						
Faulty nozzle		★	★	★	★	★	★
Sticking pump plunger		★	★	★			
Binding pump control rod				★			
Weak or broken governor spring				★			
Faulty governor and/or linkage adjustment	★	★		★	★	★	
Faulty distribution of fuel		★		★	★	★	

Loosen the high pressure fuel line connections at the injectors and crank engine with starting motor until fuel appears. Tighten the fuel line connections and start the engine.

77. FILTERS. The first and second stage fuel filtering elements should be renewed every 500 hours of operation. Poor fuel handling and storage facilities will decrease the effective life of the filter elements; conversely, clean fuel will increase the life of the filters. Filter elements should never remain in the fuel filtering system until a decrease in engine speed cr power is noticed because some dirt may enter the pump and/or nozzles and result in severe damage.

Fig. AC38—To bleed the fuel filters, remove the air bleed plug and open fuel shut-off valve. As soon as all air has escaped and a steady flow of fuel is flowing, reinstall air bleed plug.

INJECTION NOZZLES

The D-262-T engine is equipped with throttling pintle type fuel injection nozzles. In operation, some fuel is atomized to start the combustion process, but much of the fuel is emitted from the nozzle as a solid "core" which crosses the combustion chamber and enters the energy cell. As the power stroke continues, the fuel-air mixture is ejected from the energy cell into the combustion chamber where burning of the fuel is completed.

WARNING: Fuel leaves the injection nozzles with sufficient force to penetrate the skin which could cause blood poisoning. When testing nozzles, keep your person clear of the nozzle spray.

Different injector nozzle tips are used with late production cylinder head than with early. Refer to paragraph 27A.

78. LOCATING FAULTY NOZZLE. If rough or uneven engine operation, or misfiring, indicates a faulty injector,

the defective unit can usually be located as follows:

Fig. AC39 — Typical spray patterns of throttling pintle type injection nozzles. Left: Poor spray pattern. Right: Ideal spray pattern. NOTE: Very little or no atomization as shown in ideal pattern at right will be noted on injector tester due to slow operation on tester. Refer to text.

With the engine operating at low idle speed, loosen the high pressure connection at each injector in turn. As in checking spark plugs, the faulty unit is the one which, when its line is loosened, least affects the running of the engine.

If a faulty nozzle is found and considerable time has elapsed since the injectors have been serviced, it is recommended that all injectors be removed and serviced or that new or reconditioned units be installed.

79. REMOVE AND REINSTALL. Before loosening any lines, wash the nozzle holder and connections with clean diesel fuel or kerosene. After disconnecting the high pressure and leak-off lines, cover open ends of connections with composition caps to prevent the entrance of dirt or other foreign material. Remove the nozzle holder stud nuts and carefully withdraw the nozzles from cylinder head, being careful not to strike the tip end of the nozzle against any hard surface.

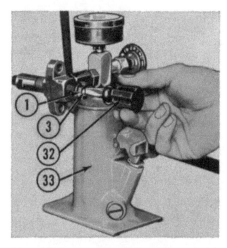

Fig. AC40 — Adjusting nozzle opening pressure on nozzle tester.

1. Nut	32. Screw driver
3. Adjusting screw	33. Nozzle tester

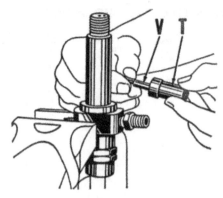

Fig. AC42 — Removing injection nozzle valve (V) from tip (T). If the valve is difficult to remove, soak the assembly in a suitable carbon solvent.

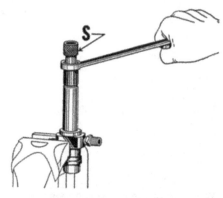

Fig. AC43 — Using Bosch tool (S) to center the nozzle tip while tightening the cap nut.

Thoroughly clean the nozzle recess in the cylinder head before reinstalling the nozzle and holder assembly. Use only wood or brass cleaning tools to avoid damage to seating surfaces and make sure the recess is free of all dirt and carbon. Even a small particle could cause the unit to be cocked and result in compression loss and improper cooling of the fuel injector.

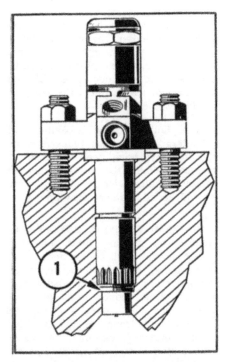

Fig. AC41 — Sectional view showing the injection nozzle installation. Whenever the nozzle has been removed, always renew the copper gasket (1).

When reinstalling the injector, always renew the copper gasket (1—Fig. AC41). Torque each of the two nozzle holder stud nuts in 2 Ft.-Lb. progressions until each reaches the final torque of 12-15 Ft.-Lbs. This method of tightening will prevent the holder from being cocked in the bore.

80. **NOZZLE TESTING.** A complete job of testing and adjusting the nozzle requires the use of a special tester such as the one shown in Fig. AC40. Use only clean, approved testing oil in the tester tank. Operate the tester lever until oil flows; then, attach injector to tester and make the following tests:

81. **OPENING PRESSURE.** Close gage valve and operate tester lever several times to clear all air from injector. Then open gage valve and observe pressure required to open nozzle valve while operating lever slowly. Opening pressure should be 2000 psi. NOTE: On new injectors, or where new spring has been installed in injector, opening pressure should be set to 2100 psi as a new spring will take a set and reduce the opening pressure about 100 psi after several hours operation. If the opening pressure is not as specified, remove cap from top of injector, loosen jam nut and turn adjusting screw in or out to obtain correct opening pressure. Tighten jam nut and reinstall cap before making final opening pressure check.

82. **SPRAY PATTERN.** Prior to testing for spray pattern, check opening pressure as outlined in paragraph 81. Close the valve to the tester gage; then, operate lever at about 100 strokes per minute while observing nozzle spray pattern. As the tester pump cannot duplicate the injection velocity necessary to obtain the operating spray pattern of throttling pintle nozzles, very little or no atomization may be noted. However, the solid core of fuel from the nozzle opening should be in a straight line with the injector body, with no branches, splits or dribbling. NOTE: Slow operation of tester pump may cause some dribble. Also, throttling pintle type nozzles do not usually "chatter" or make a popping sound when operated on a tester pump as some nozzles do.

Under operating velocities, the solid core of fuel from the nozzle will cross the combustion chamber and enter the energy cell. In addition, a fine conical mist surrounding the core will ignite in the combustion chamber area above the piston. The solid core cannot vary more than 7½ degrees in any direction and enter the energy cell. While the core is the only spray characteristic which can be observed on the tester, it is of utmost importance that the core be absent of any deviations.

83. **SEAT LEAKAGE.** Open valve to tester gage and slowly pump pressure up to 1700 psi. The nozzle valve seat should hold this pressure for several seconds without noticeable leakage.

84. **MINOR OVERHAUL OF NOZZLE VALVE AND BODY.** Hard or sharp tools, emery cloth, crocus cloth, grinding compounds or abrasives of any kind should **never** be used in the

Fig. AC44 — Exploded view of throttling pintle type nozzle.

1. Cap nut	4. Adjusting screw	7. Nozzle holder
2. Gasket	5. Spring	body
3. Lock nut	6. Spindle	8. Nozzle
		9. Cap nut
		10. Copper washer

Fig. AC45 — The injection pump timing marks can be viewed after removing the timing hole cover. To change timing, loosen the pump retaining nuts and turn pump until timing marks are aligned as in Fig. AC46.

cleaning of nozzles. A nozzle cleaning and maintenance kit is available through any diesel service agency.

Wipe all dirt and loose carbon from the nozzle and holder assembly with a clean, lint free cloth or paper wiper. Carefully clamp nozzle holder assembly in a soft jawed vise and remove the nozzle holder nut and spray nozzle. Reinstall the holder nut to protect the lapped end of the holder body. Normally, the nozzle valve (Fig. AC42) can be easily withdrawn from the nozzle body. If the valve cannot be easily withdrawn, soak the assembly in fuel oil, acetone, carbon tetrachloride or similar carbon solvent to facilitate removal. Be careful not to permit the valve or body to come in contact with any hard surface.

Clean the nozzle valve with mutton tallow used on a soft, lint free cloth or pad. The valve may be held by its stem in a revolving chuck during this cleaning operation. A piece of soft wood well soaked in oil will be helpful in removing carbon deposits from the valve.

The inside of the nozzle body (tip) can be cleaned by forming a piece of soft wood to a point which will correspond to the angle of the nozzle valve seat. The wood should be well soaked in oil. The orifice of the tip can be cleaned with wood splinter. The outer surfaces of the nozzle body should be cleaned with a brass wire brush and a soft, lint free cloth or wiper soaked in a suitable carbon solvent.

Thoroughly wash the nozzle valve and body in clean diesel fuel and clean the pintle and its seat as follows: Hold the valve at the stem end only and, using light oil as a lubricant, rotate the valve back and forth in the body. Some time may be required in removing the particles of dirt from the pintle valve; however, abrasive materials should never be used in the cleaning process.

Test the fit of the nozzle valve in the nozzle body as follows: Hold the body at a 45 degree angle and start the valve in the body. The valve should slide slowly into the body under its own weight. NOTE: Dirt particles, too small to be seen by the naked eye, will restrict the valve action. If the valve sticks, and it is known to be clean, free-up the valve by working the valve in the body with mutton tallow.

Before reassembling, thoroughly rinse all parts in clean diesel fuel and make certain that all carbon is removed from the nozzle holder nut. Install nozzle body and holder nut, making certain that the valve stem is located in the hole of the holder body. It is essential that the nozzle be perfectly centered in the holder nut. A centering sleeve is supplied in American Bosch kit TSE 7779 for this purpose. Slide the sleeve over the nozzle with the tapered end centering in the holder nut. Tighten the holder nut, making certain that the sleeve is free while tightening. Refer to Fig. AC43.

Adjust the injector opening pressure and test the nozzle for spray pattern and leakage as in paragraphs 81, 82 & 83. If the nozzle does not leak under 1700 psi, and if the spray pattern is symmetrical as shown in right hand view of Fig. AC39, the nozzle is ready for use. If the nozzle will not pass the leakage and spray pattern tests, renew the nozzle valve and seat or send the nozzle and holder assembly to an official diesel service station for complete overhaul which includes reseating the nozzle valve pintle and seat.

85. OVERHAUL OF NOZZLE HOLDER. Refer to Fig. AC44. Remove cap nut (1) and gasket. Loosen jam nut (3) and remove the adjusting screw (4). Remove the spring (5) and spindle (6). Thoroughly wash all parts in clean diesel fuel and examine the end of the spindle which contacts the nozzle valve stem for any irregularities. If the contact surface is pitted or rough, renew the spindle. Examine the spring seat and spindle for cracks or worn spots. Renew the spring seat and spindle unit if the condition is questionable. Renew any other questionable parts.

Reassemble the nozzle holder and leave the adjusting screw lock nut loose until after the nozzle opening pressure has been adjusted as outlined in paragraph 81.

INJECTION PUMP

86. TIMING. Refer to the following chart for correct injection pump static timing according to engine serial number and injection pump part number.

Pump Part Number	Eng. Serial No. Range	Degrees BTDC
4513709	Before 121791	16
4514017	121791 to D-01498	16
4514557	D-01499 to D-03150	20
4514756	After D-03150	20

Turn engine in normal direction of rotation until the number one piston is coming up on compression stroke. Continue to turn engine slowly until the correct timing mark located on timing strip attached to crankshaft pulley (See Fig. AC 47) is aligned with the pointer on timing gear cover. Shut off fuel supply and remove the timing hole cover (See Fig. AC45). If timing marks on cam and governor weight retainer are not aligned as shown in Fig. AC46, loosen pump mounting bolts and turn pump in either direction as required so that marks are aligned. Retighten pump mounting bolts.

Automatic speed advance should be 5 degrees for 4513709 and 4514017 pumps with engine at fuel load 1600 engine RPM. Automatic speed advance should be 3 degrees for 4514557 and 4514756 pumps with engine at full load 1600 RPM. High idle engine speed is 2300 RPM for all models.

Fig. AC46 — The injection timing marks as seen when timing hole cover is removed from pump.

Fig. AC48 — View of injection pump showing the governed speed adjusting screws. Screw (1) adjusts the high idle speed. Screw (2) adjusts low idle speed.

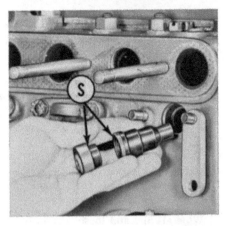

Fig. AC49 — Installing the energy cell. If the surfaces (S) are rough or pitted, they can be reconditioned by lapping. Refer to text.

87. REMOVE AND REINSTALL. To remove the complete injection pump unit, first shut off the fuel supply and thoroughly clean the pump, fuel lines and connections. Turn crankshaft in normal direction of rotation until the number one piston is coming up on compression stroke. Continue turning engine slowly until the correct timing mark on the crankshaft pulley (Refer to paragraph 86 and Fig. AC47) is in line with the pointer on the engine front cover. Disconnect the fuel supply line, throttle rod and shut-off rod from pump. Disconnect high pressure (nozzle) lines at injectors and excess fuel line from pump. Remove the fuel filters assembly, then remove the two pump mounting nuts and slide the pump off the drive shaft and remove from engine.

Before reinstalling pump, remove the timing hole cover from outer side of injection pump and make certain the pump timing lines are aligned as shown in Fig. AC46. Install new lip seals on injection pump drive shaft and apply lubricant such as Lubriplate on seals. Carefully work pump on shaft over seals and mount pump on engine. CAUTION: Be sure that lip of rear seal on pump shaft is not rolled or bent back as pump is worked on to shaft. If this occurs, renew seal as the seal will have been damaged and early failure may result.

Connect the engine controls and fuel supply line. Before connecting the pressure lines, flush lines with clean fuel oil. Recheck the pump timing as in paragraph 86 and bleed the fuel system as in paragraph 76.

88. SPEED ADJUSTMENTS. To adjust the governed engine speeds, first start engine and bring to normal operating temperature. Move the speed control hand lever to the wide open position, loosen the jam nut and turn the high speed adjusting screw (1—Fig. AC48) either way as required to obtain an engine high idle no-load speed of 2275-2300 RPM; then, tighten the adjusting screw jam nut. NOTE: With the high idle speed properly adjusted, the full load engine speed will be 2000 RPM.

Move the speed control hand lever to the low idle speed position, loosen the jam nut and turn adjusting screw (2) either way as required to obtain an engine slow idle speed of 650-700 RPM, then tighten the adjusting screw jam nut.

Screws (3 and 4) are provided to set the limits of shut-off arm travel and should not normally require adjustment in the field. Screw (3) adjusts for maximum travel toward the "shut-off" position and screw (4) adjusts for "run" position. Adjustment of either screw requires removal of pump control cover and should be done only by experienced diesel service personnel. Screw (5) is for synchronizing travel of speed control hand lever to travel of control arm on fuel injection pump.

ENERGY CELLS

89. R&R AND CLEAN. The necessity for cleaning the energy cells is usually indicated by excessive exhaust smoking, or when fuel economy and power drops. To remove the number 2, 3, 4 or 5 energy cells, it is necessary to remove the intake manifold. Remove the energy cell clamp and tap the energy cell cap with a hammer to break loose any carbon deposits. Using a pair of pliers, remove the energy cell cap. (A ¼-inch tapped hole is also provided in the cap to facilitate removal.)

The cell body can be removed by first removing the respective nozzle, and using a brass drift inserted through the nozzle hole, bump the cell out of the cylinder head. On some early models the outer end of energy cell body is tapped to permit the use of a screw type puller without removing nozzle.

The removed parts can be cleaned in an approved carbon solvent. After

Fig. AC47 — View of crankshaft pulley showing timing strip on pulley and pointer on timing gear cover.

parts are cleaned, visually inspect them for cracks and other damage. Renew any damaged parts. Inspect the seating surfaces between the cell body and the cell cap for being rough or pitted. The surfaces (S—Fig. AC49) can be reconditioned by lapping with a fine valve grinding compound. Surfaces must form an air tight seal. Make certain that the energy cell seating surface in cylinder head is clean and free from carbon deposits.

When installing the energy cell, tighten the clamp nuts to a torque of 18-21 Ft.-Lbs.

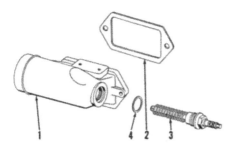

Fig. AC50 — Exploded view of D-19 diesel engine intake manifold pre-heater.

1. Adapter
2. Gasket
3. Glow plug
4. Gasket

Fig. AC51 — Cut-away view of diesel engine Turbo-Charger system.

1 Air from air cleaner
2 "Turbo-Charged" air to intake manifold
3. Exhaust gasses to Turbo-Charger
4. Exhaust gasses from Turbo-Charger

DIESEL TURBO-CHARGER

The diesel engine is equipped with a turbo-charger that is driven by the exhaust from the engine. See Fig. AC51 for operational details. When servicing the turbo-charger, extreme care must be taken to avoid damaging any of the moving parts as renewal of the turbine wheel and shaft or compressor rotor involves considerable expense.

90. R&R TURBO-CHARGER. Remove both hood side panels, disconnect hoses from air cleaner, disconnect battery ground cable and wiring to voltage regulator; then, unbolt and remove center hood channel with air cleaner and voltage regulator as a unit. Remove air intake hose, air hose to intake manifold, exhaust pipe and oil return line from turbo-charger. Disconnect oil pressure line at turbocharger. Unbolt and remove turbocharger assembly from exhaust manifold.

To reinstall, reverse removal procedure. Leave oil return line (7) disconnected and crank engine with starter until oil begins to flow from return port.

NOTE: Do not start engine until it is certain that turbocharger is receiving lubricating oil. Tighten stud nuts attaching turbocharger to exhaust manifold to 18-21 ft.-lbs. torque. Exhaust elbow mounting screws should be tightened to 28-33 ft.-lbs. torque.

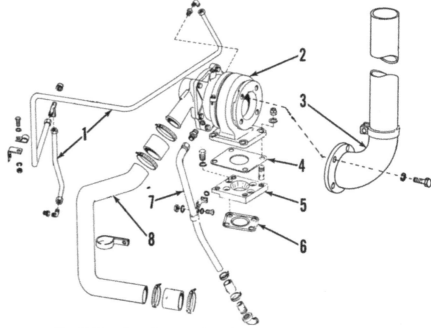

Fig. AC52 — Exploded view of diesel engine Turbo-Charger system.

1. Lubricating pressure tube
2. Turbo-Charger unit
3. Exhaust stack
4. Gasket
5. Adapter plate
6. Gasket
7. Oil return tube
8. Air tube to intake manifold

91. OVERHAUL. When servicing turbocharger as originally installed on early tractors, it is advisable to use special tools available from Kent-Moore Organization, Inc. Some of the early turbochargers have been changed by installing later type parts and the special tools are not required. Attention will be called to difference in service procedure.

Remove turbocharger unit as outlined in paragraph 90. Scribe a mark across turbine housing (22—Fig. AC53), clamp (20), bearing housing (13) and compressor housing (1) to aid reassembly and proceed as follows: Remove nut (19) and clamp (20); then, remove turbine housing (22). Remove cap screws attaching bearing housing (13) to compressor housing (1) and remove compressor housing. Hold blades of turbine wheel (18) with shop towel; remove nut (3) and washer (4) from turbine shaft (L.H. threads). On early models, impeller (5E) is a tight fit on turbine shaft (18) and shaft must be pressed out of impeller as shown in Fig. AC54. On later models, impeller (5—Fig. AC53) will slide easily off shaft.

CAUTION: On all models, do not allow turbine wheel and shaft to drop when impeller is removed.

Remove snap ring (6), then remove oil retainer plate (7 or 7E). On early type, retainer plate (7E) must be pressed out of bore using the Kent-Moore special tool. On later type (7), a ¾-inch driver can be used as shown in Fig. AC55 to remove seal and retainer plate (7—Fig. AC53), mating piece (10) and bearing (12).

It is recommended that late type seal (7) and mating ring (10) be renewed each time turbocharger is disassembled for service. The early type turbocharger can be changed to later type by installing new type parts (5, 7, 10 and 11).

Clean all parts except late type oil seal (7) by washing in kerosene or diesel fuel. A nylon bristle brush may be used to clean carbon from parts.

CAUTION: Do not use wire brush, caustic cleaners, etc., on turbocharger parts. Inspect all parts for burring, eroding, nicks, breaks, scoring, excessive carbon build-up or other defects and renew all questionable parts.

On early type turbocharger, compressor impeller (5E) should be press fit on turbine shaft (18). Thrust surfaces of spacer (11E) must not be rough. Inspect grooves in spacer (11E) for excessive wear and make certain

carbon is removed from bottom of grooves. Check O.D. of retainer plate (7E) for nicks that would damage "O" ring (9). Inspect bore in retainer plate for grooving caused by seal rings (10E) sticking and turning with shaft. Clean the bore chamfer to ease installation of seals (10E).

On late type turbocharger, inspect seal (7) carefully if the seal is to be reinstalled. The carbon face insert should move freely in and out and must not be scored or excessively worn. Mating ring (10) should not show evidence of wear or scuffing on either side. Remove any carbon from seal contact side. Examine sleeve (11) for burrs, scoring and wear. The sleeve is precision ground and any defect will distort the mating ring and cause seal leakage.

On all models, pay particular attention to blades of turbine wheel (18) and compressor impeller (5 or 5E) and to bushing (12). If bushing (12) and/or turbine shaft is worn excessively, the blades of turbine wheel and compressor impeller may have rubbed against housings. Shaft clearance in bushing should be measured at compressor impeller end of shaft using a dial indicator with unit assembled. If indicated clearance exceeds 0.022, bushing (12) and/or shaft (18) should be renewed. Inspect bore in bearing housing (13) for evidence of stuck seal ring (17). If seal ring was sticking in shaft groove, the bore in housing will be grooved and should be renewed. Make certain that seal bore chamfer is clean and smooth to ease installation of shaft and seal ring.

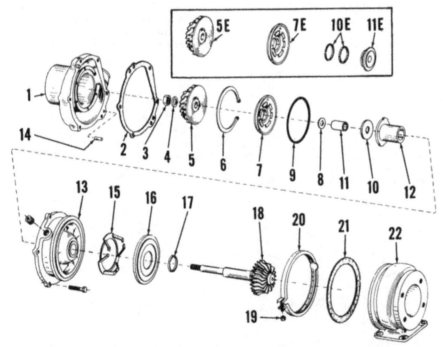

Fig. AC53—Exploded view of Thompson turbocharger unit used on diesel engines. Inset shows parts that are different on earlier unit.

1. Compressor housing	7E. Oil retaining plate	11E. Turbine shaft spacer	16. Turbine wheel
2. Gasket	8. Shims	12. Bearing	17. Turbine shaft seal
5 & 5E. Impeller	9. "O" ring	13. Bearing housing	18. Turbine wheel & shaft
6. Snap ring	10. Mating ring	14. Groove pins	20. Clamp
7. Oil seal & retaining plate	10E. Turbine shaft seals	15. Spring ring	21. Gasket
	11. Shaft sleeve		22. Turbine housing

Fig. AC54—View showing method of removing early type impeller using special Kent-Moore tool.

To reassemble turbocharger, proceed as follows: Assemble bearing (12), mating ring (10) and sleeve (11) or bearing (12), and spacer (11E) on turbine shaft (18). Hold sleeve and mating ring (or spacer) tight against shoulder of turbine shaft and check end play of bearing with feeler gage as shown in Fig. AC56. End play should be 0.004-0.006. If clearance exceeds 0.006, bearing (12—Fig. AC53) should be renewed. **Record amount of end play for use in later step in reassembly.** Remove parts from turbine shaft.

Lubricate bearing (12) with motor oil and position in bearing housing bore with tabs on bearing between lugs in housing. Lubricate "O" ring (9) with silicone lubricant (or light grease) and position in bottom of retainer bore in bearing housing (13).

On early type, lubricate grooves in spacer (11E) and install new seal rings (10E) with end gaps opposite each other. Carefully slide spacer (11E) with seal rings (10E) into bore of retainer (7E). Install retainer and seal assembly in bearing retainer bore with one lug centered over each of the two tabs on bearing (12). Pres retainer lightly into "O" ring (9) and install snap ring (6) with tapered side out.

On late type, lubricate mating ring (10) and position over hold in bearing (12). Install oil seal and retainer (7) in bearing retainer bore, press oil seal and retainer down lightly into "O" ring (9) and install snap ring (6) with

tapered side out. When correctly assembled, lugs on inside of retainer will hold mating ring (10) in position.

CAUTION: Do not damage mating ring, "O" ring, seal and retainer by attempting to force installation with parts misaligned.

Check clearance between bearing and retainer as follows: Insert turbine shaft (18—Fig. AC53) through bearing (12) and mating ring (10) or bearing and spacer (11E).

NOTE: Seal ring (17), spring ring (15) and shield (16) should not be installed for this check. On late type install sleeve (11) over end of shaft.

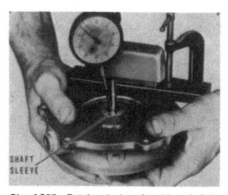

Fig. AC57—Total end play of turbine shaft in bearing and bearing housing can be measured as shown. Late type turbocharger is shown but early type is similarly measured. It is usually necessary to push sleeve or spacer down, then push up on turbine shaft when measuring.

On all models, attach a dial indicator as shown in Fig. AC57. Measure and record the total amount of end play measured at sleeve (11—Fig. AC53) or spacer (11E). If total end play is not within limits of 0.005-0.009, check for incorrect assembly. The end play of shaft in bearing (previously measured Fig. AC56) should be 0.004-0.006 and end play of bearing between retainer and housing should be 0.001-0.003. If end play measurement is again within limits, **record the total shaft and bearing end play for use when selecting shims (8—Fig. AC53).** Remove the turbine shaft (18) and proceed as follows:

Position impeller (5) over oil seal and retainer (7 or 7E). On late models, make certain that sleeve (11) is installed. On all models, install compressor housing (1) using a new gasket and three of the attaching screws torqued to 80 inch-pounds. Check end play of impeller in housing as shown in Fig. AC59. Subtract total shaft and bearing end play (found in preceding paragraph) from impeller end play. In final assembly, add shims (8—Fig. AC53) of total thickness of 0.015-0.020 less than this determined value. As an example, if impeller to housing clearance is 0.049 and total shaft and bearing end play is 0.006; subtract as follows:

Impeller end clearance		0.049
Shaft and bearing end play		−0.006
		0.043
Determined value	0.043	0.043
Desired impeller clearance	−0.015	−0.020
Shim thickness required	0.028	0.023

Shims (8) are available in thicknesses of 0.010 and 0.015. The addition of one

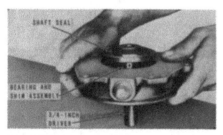

Fig. AC55—View showing method of removing oil seal and retainer from late type turbocharger. Parts should move out with only a small amount of pressure.

Fig. AC56—Measure turbine shaft end play with feeler gage. Early type is shown.

Fig. AC58 — Installing oil retaining plate using arbor press and Kent-Moore tool No. J-9498.

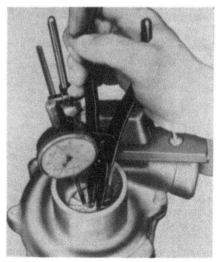

Fig. AC59—Measure impeller clearance in housing as shown. Turbine shaft is not installed when checking this clearance.

0.010 and one 0.015 thick shim will be within the range of the shim thickness required in the preceding example.

Remove compressor housing (1) and impeller (5). Lubricate groove in turbine shaft (18) with motor oil and install new seal ring (17) in groove. Place plastic seal compressor (furnished with seal ring) over seal ring. Install spring ring (15) in bearing housing and place shield (16) over spring with projections on shield against the flat sections of the spring. Use two small "C" clamps to compress spring and hold shield in place. Insert turbine shaft (18) through hole in shield and bore of bearing. Plastic band used to compress seal ring (17) will be pushed off of seal ring as shaft is inserted in bearing and plastic will disintegrate from heat as soon as engine is started. Rotate turbine shaft to be sure it is a free fit in bearing. Place shim pack (8) (of thickness determine in previous step) on turbine shaft; then, install impeller onto turbine shaft. On early type, impeller must be pressed onto turbine shaft. On all models, install washer (4) and nut (3) on turbine shaft (L.H.

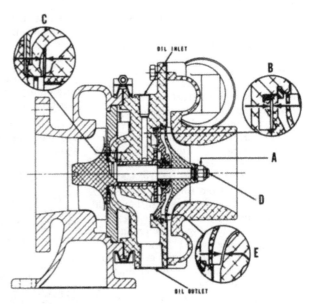

Fig. AC60—Cross-section of Thompson turbocharger. Shaft side play in bearing, measured at (A) should not exceed 0.022. Shaft end play in bearing (C) is measured as shown in Fig. AC56. Total shaft end play (D) is measured as shown in Fig. AC57. End clearance of bearing in housing (B) is difference of end play (D & C). Impeller to compressor housing clearance (E) is measured as shown in Fig. AC59.

threads) and tighten nut to a torque of 80-100 inch-pounds while holding turbine with shop towel. Check turbine shaft assembly for free rotation.

Reinstall compressor housing (1) and gasket (2) to bearing housing and tighten screws to 80-100 inch-pounds.

Install turbine housing and gasket (21). Tighten clamp screw nut to 15-20 inch-pounds. Remove plug from bearing housing and fill reservoir with same type oil as used in engine. Protect all openings of turbocharger until unit is installed on tractor.

NON-DIESEL GOVERNOR

95. **SPEED ADJUSTMENT.** Before attempting to adjust the governed speed of engine, refer to Fig. AC71 and check the length of carburetor link as follows: Unhook link rod from carburetor throttle shaft. With carburetor throttle and governor arm both in wide open position, link should be $\frac{1}{32}$-inch too short to reattach to carburetor arm. If not, adjust length of link at points "A" and "B". Be sure adjustment lock nuts are securely tightened and re-attach link to carburetor throttle arm.

Adjust low idle stop screw (1—Fig. AC70) so that governor spring just clears governor shaft at "C" when controls are in idle position. Back out the anti-surge adjusting screw (3), start engine and adjust idle speed to 375-425 RPM with idle stop screw on throttle shaft of carburetor. Then, adjust the high idle stop screw (2) on governor to give a high idle speed of 2275 rpm. Turn anti-surge screw (3) in until high idle speed picks up to 2300 RPM. Be sure that all lock nuts on adjustment points are secure.

Fig. AC70 — When adjusting D-19 non-diesel governor, adjust low idle stop screw (1) so that governor spring just clears governor shaft at (C). High idle speed is adjusted by screw (2). Screw (3) is anti-surge adjustment.

NOTE: It is possible that travel of control lever may not be synchronized with travel of governor arm and proper adjustment of engine speed will require readjustment of control linkage (S—Fig. AC72). Also, readjust tension spring (T—Fig. AC73) if control lever is difficult to move or if lever will not stay in place.

Fig. AC71 — View showing carburetor link length adjustment points (A and B).

Fig. AC72 — Throttle control lever travel may be synchronized by adjusting linkage at (S).

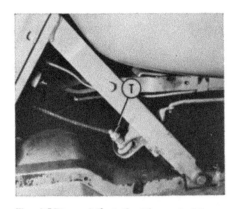

Fig. AC73 — Adjust throttle control lever tension at (T) so that lever is not difficult to move, yet will retain position at full throttle.

96. R&R OVERHAUL GOVERNOR. To remove the governor, first remove the ignition coil and the oil line to governor housing. Disconnect governor linkage. Then, unbolt and remove governor from engine. NOTE: Governor shaft and weight unit is not retained in governor housing. Withdraw the shaft and weight assembly from timing gear cover if not removed with governor housing. The governor shaft

front bushing in the timing gear cover is renewable after removing governor assembly.

Procedure for further disassembly of governor is evident from inspection of unit and reference to the cross-sectional views in Fig. AC74. Inspect all parts for wear or other damage and renew all questionable parts. Governor drive gear (23), spacer (24)

and shaft (25) are available as an assembly only. Any binding tendency should be removed from governor parts during assembly. Desired backlash of governor drive gear to engine camshaft gear is 0.004-0.005. No timing of drive gears is necessary when reinstalling governor. Tighten housing retaining cap screws to a torque of 18-21 Ft.-Lbs.

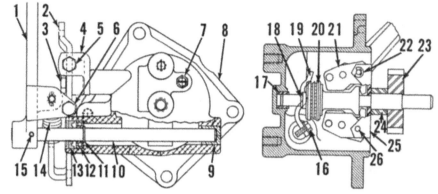

Fig. AC74 — Cross-section views of D-19 non-diesel governor unit.

1. Governor lever	8. Governor body	14. Governor spring	21. Weight assembly
2. Speed change lever	9. Bushing	15. Roll pin	22. Retaining ring
3. Shoulder screw	10. Rocker shaft	16. Fastener	23. Drive gear
4. Stop screw bracket	11. Snap ring	17. Bushing	24. Gear spacer
5. High idle stop screw	12. Bearing	18. Thrust bearing	25. Shaft assembly
6. Low idle stop screw	13. Seal	19. Governor yoke	26. Weight pin
7. Anti-surge screw		20. Thrust bearing	

COOLING SYSTEM

RADIATOR

All Models

97. REMOVE AND REINSTALL. To remove the radiator, remove grille and drain cooling system. Remove both hood side panels and unbolt hood center channel from radiator shell. Disconnect both radiator hoses. Unbolt radiator shell from side rails and radiator from front support. Remove front support breather and lift off radiator and radiator shell as a unit. Remove radiator from radiator shell. Reverse removal procedures to install radiator.

WATER PUMP

All Engines

98. REMOVE AND REINSTALL. Although it is possible to remove the water pump without removing the

radiator, the following procedure is recommended: Remove radiator and radiator shell as outlined in paragraph 97. Remove fan blades, fan belt and pulley. Loosen fittings on steel

tube between water pump and water manifold. Unbolt and remove water pump from cylinder block. Reverse removal procedures to reinstall water pump.

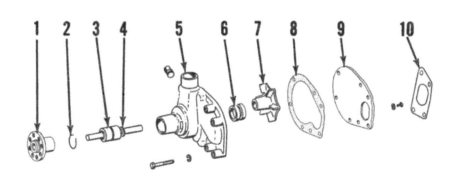

Fig. AC75 — Exploded view of water pump used on diesel engine. Water pump used on non-diesel engine is basically similar.

1. Pulley hub	3. Shaft & bearing assy.	5. Pump body	8. Cover gasket
2. Snap ring	4. Water slinger	6. Seal assembly	9. Cover plate
		7. Impeller	10. Mounting gasket

99. OVERHAUL WATER PUMP. Refer to Fig. AC75. Using suitable puller, remove pulley hub (1) from shaft. (See Fig. AC76). Extract snap ring (2—Fig. AC75) from pump body (5). Remove rear cover (9). Press shaft and bearing unit out of the impeller and pump body as shown in Fig. AC76A). Remove seal (6 — Fig. AC75) from pump body.

Renew pump shaft and bearing assembly if bearing is dry or rough. Shaft and bearing assembly is available only in water pump repair kit which also includes the following parts: Snap ring (2), seal assembly (6), impeller assembly (7) and gaskets (8) and (10).

To reassemble pump, apply sealer to outer diameter of new seal assembly and press seal into body being careful not to damage seal face. Press shaft and bearing assembly into body just far enough to allow installation of snap ring; then press shaft and bearing forward against snap ring. Press impeller onto rear end of shaft until rear face of impeller is 0.030-0.040 below gasket surface of pump body. CAUTION: Be careful not to cock impeller on shaft while installing as this may crack the ceramic sealing surface of the impeller and cause the pump to leak.

Support the rear end of shaft and press pulley hub onto shaft so that front edge of hub extends about ½-inch from end of shaft. Install rear cover and gasket.

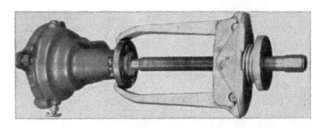

Fig. AC76 — Removing pulley hub from water pump.

Fig. AC76A — Pressing water pump drive shaft and bearing unit out of impeller and housing.

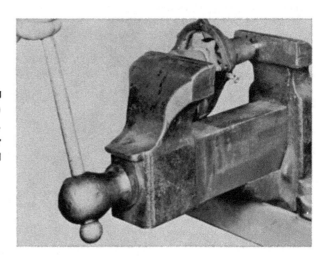

Fig. AC76B — Pressing water pump impeller on the pump drive shaft. Rear face of impeller should be flush with end of shaft.

IGNITION AND ELECTRICAL SYSTEM

Non-Diesel

101. SPARK PLUGS. A-C 45XL, Autolite AG5A, or Champion N-8 spark plugs should be used in gasoline engines. For LP-Gas engines, use Champion N-3 spark plugs. Set electrode gap to 0.025 for all applications.

102. DISTRIBUTOR. Delco-Remy 1112615 ignition distributor is used. Specification data follows:

Breaker contact gap............0.022

Breaker arm spring pressure (measured at center of contact) ....17-21 oz.

Cam angle ....31-37 degrees

Advance data in distributor degrees and distributor RPM:
Start advance ......0°-2° @ 275 RPM
Intermediate
   advance ....5°-7° @ 400 RPM,
            11°-13° @ 900 RPM
Maximum
   advance ......17°-19° @ 1400 RPM

103. IGNITION TIMING. Turn engine in normal direction of rotation until number one piston is coming up on compression stroke, then continue turning engine slowly until the TDC mark on crankshaft pulley timing strip (See Fig. AC79) is in line with the pointer on engine timing gear cover. Set contact gap at 0.022 and position rotor so that it points to the number one spark plug terminal posi-

tion. Place distributor in engine block engaging driving tang with notch in oil pump drive gear. (The tang and notch are approximately 0.030 off-center with center-line of shafts.) Loosen clamp nut (C—Fig. AC80) and tighten bolt (B) with minor timing indicator at 0 on scale. Turn distributor body in opposite direction from distributor shaft rotation until points just start to open. Tighten clamp nut (C) with distributor body in this position.

With the distributor installed as previously outlined, the static timing is approximately correct; it is important, however that the advanced timing be checked with a timing light with engine running at exactly 2000

1. Cable from battery negative terminal to starter solenoid.
2. Cable from starter solenoid to starter.
3. Blue wire from starter solenoid to ammeter positive terminal.
4. White wire from starter solenoid small terminal to ignition switch "SOL" terminal.
5. Red wire from ammeter negative terminal to voltage regulator "BAT" terminal.
6. Green wire from voltage regulator "F" terminal to generator "F" terminal.
7. Brown wire from voltage regulator "G" terminal to generator "A" terminal.
8. Black wire from voltage regulator "L" terminal to ignition switch "BAT" terminal.
9. Yellow wire from ignition switch "IGN" terminal to ignition coil negative terminal.
10. Wire from ignition coil positive terminal to distributor primary lead terminal.
11. Green wire from ignition switch "BAT" terminal to light fuse holder.
12. Purple wire from light fuse holder to light switch.
13. Wire from dash lamp to light switch terminal with wire adapter.
14. Orange wire from light switch adapter terminal to head lamps, tail lamp and auxiliary outlet socket.
15. Wire from head lamps to head light terminal connector.
16. Green wire from ignition switch "IGN" terminal to "Power-Director" oil pressure indicator light and pressure switch.
17. Battery ground strap from battery positive terminal to ground.

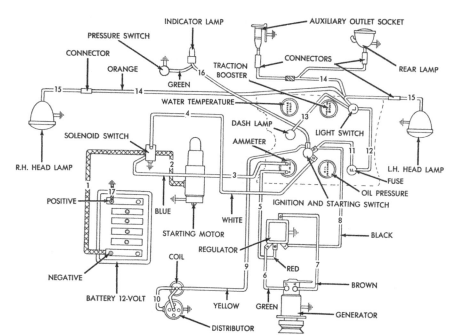

Fig. AC77—Wiring diagram for non-diesel Model D-19 Allis-Chalmers tractor. Late tractors have push button starter switch and a special ignition coil which must be used only with an external resistor.

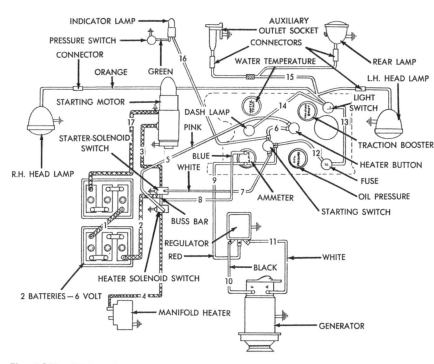

Fig. AC78—Wiring diagram for Allis-Chalmers Model D-19 diesel tractors. Late production tractors have a push button starter switch.

1. Cable from L.H. battery negative terminal to R.H. battery positive terminal.
2. Cable from R.H. battery negative terminal to starter solenoid.
3. Cable from starter solenoid to starter.
4. Cable from manifold heater solenoid to manifold heater.
5. Pink wire from manifold heater solenoid small terminal to manifold heater switch.
6. Pink wire from manifold heater switch to starting switch "IGN" terminal.
7. White wire from starting switch "SOL" terminal to starter solenoid small terminal.
8. Blue wire from starter solenoid to ammeter positive terminal.
9. Red wire from ammeter negative terminal to voltage regulator "BAT" terminal.

10. Black wire from voltage regulator "F" terminal to generator "F" terminal.
11. White wire from voltage regulator "L" terminal to generator "ARM" terminal.
12. Black wire from starter switch "BAT" terminal to fuse holder.
13. Black wire from fuse holder to light switch.
14. Wire from dash lamp to wire adapter on light switch.
15. Orange wire from light switch wire adapter to head lamps, rear lamp and auxiliary outlet socket.
16. Green wire from starter switch "IGN" terminal to "Power-Director" oil pressure indicator lamp and pressure switch.
17. Ground strap from L.H. battery positive terminal to ground.

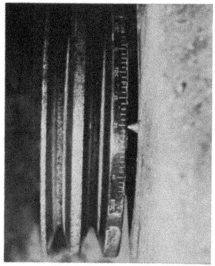

Fig. AC79 — View of crankshaft pulley showing timing strip on pulley and pointer on timing gear cover.

Fig. AC80 — View of ignition distributor showing minor (B) and major (C) timing adjustment points.

RPM. At this speed, timing flash should occur when 25° BTDC mark on timing strip (See Fig. AC79) is exactly in line with pointer on engine front cover. If 25° BTDC mark is not in line with pointer at 2000 engine RPM, loosen clamp nut (C—Fig. AC80) and adjust distributor body position so that running timing is correct. Adjustment at minor timing indicator may be made to compensate for minor fuel and engine variations to achive best engine performance.

104. GENERATOR. Delco-Remy 1100305 generator and a Delco-Remy 1118993 voltage regulator are used. Specification data for each unit follows:

Generator—1100305
Brush spring tension........28 oz.
Field draw:
   Volts 12.0
   Amperes 1.5-1.67
Output (cold):
   Max. amperes 20.0
   Volts 14.0
   Max. RPM 2300

Regulator—1118993
Cut-out relay:
   Air gap 0.020
   Point gap 0.020
Closing voltage:
   Range 11.8-14.0
   Adjust to 12.8
Voltage regulator:
   Air gap 0.020
   Voltage range 13.6-14.5
   Adjust to 14.0
   Ground polarity ....Positive

105. STARTING MOTOR. A Delco-Remy 1107235 starting motor is used. Specification data follows:
Brush spring tension (Min.)....35 oz.
No load test:
   Volts 10.3
   Amperes 75
   RPM (Min.) 6900

Diesel

106. GENERATOR. Delco-Remy 1100426 generator and a Delco-Remy 1119191 voltage regulator are used. Specification data for each unit follows:

Generator—1100426
Brush spring tension........28 oz.
Field draw:
   Volts 12
   Amperes 1.5-1.62
Output (cold):
   Max. amperes 25
   Volts 14.0
   Max. RPM 2710

Regulator—1119191
Cut-out relay:
   Air gap 0.020
   Point gap 0.020
Closing voltage:
   Range 11.8-14.0
   Adjust to 12.8
Voltage regulator:
   Air gap 0.075
   Voltage range 13.6-14.5
   Adjust to 14.0
   Ground polarity ....Positive

107. STARTING MOTOR. Specification data for the Delco-Remy 1113152 starting motor are as follows:
Brush spring tension ....80 oz.
No load test:
   Volts 11.5
   Amperes 37-50
   RPM 5000-7400
Lock test:
   Volts (approx.) 3.4
   Amperes 500
   Torque, Ft.-Lbs. (Min.).........22

ENGINE CLUTCH

All models are equipped with a 11-inch single plate, dry disc, spring loaded clutch assembly. Ferro-metalic lining is bonded to clutch disc.

108. PEDAL ADJUSTMENT. Pedal height may be adjusted to suit the operator by turning the pedal link rod in or out of the throw-out bearing fork. This adjustment does not affect the release bearing clearance, but merely positions the clutch pedal position.

109. R&R ENGINE CLUTCH. The engine clutch may be removed after splitting the tractor between the engine and torque housing as follows: Remove both hood side panels. Disconnect battery ground cable and unbolt and remove starter. Disconnect air hoses from air cleaner (on diesel models, disconnect wiring from voltage regulator), unbolt and remove hood center channel with air cleaner and diesel voltage regulator as a unit. Drain coolant down below water manifold and disconnect temperature

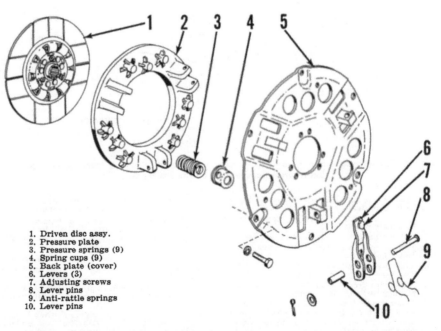

1. Driven disc assy.
2. Pressure plate
3. Pressure springs (9)
4. Spring cups (9)
5. Back plate (cover)
6. Levers (3)
7. Adjusting screws
8. Lever pins
9. Anti-rattle springs
10. Lever pins

Fig. AC81 — Exploded view of engine clutch assembly. Driven disc (1) has bonded ferro-metallic lining.

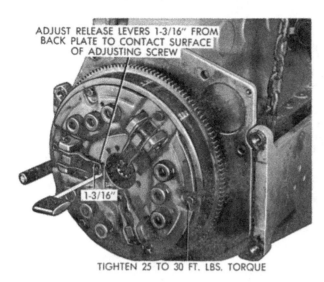

ADJUST RELEASE LEVERS 1-3/16" FROM
BACK PLATE TO CONTACT SURFACE
OF ADJUSTING SCREW

1-3/16"

TIGHTEN 25 TO 30 FT. LBS. TORQUE

Fig. AC81A—View showing method of adjusting clutch release lever contact screws. Adjustment must be made using new clutch friction disc.

Clutch cover to flywheel.......25-30
Torque housing to engine.......70-75
Torque housing to side rails...130-140

110. OVERHAUL CLUTCH UNIT. Clutch disc has non-renewable ferrometalic lining (facing) bonded to disc, and must be renewed as an assembly. Renew clutch disc assembly if disc is warped, hub assembly is worn or damaged, or if lining is burned or excessively worn. New disc thickness is 0.215-0.225.

Overhaul of clutch cover assembly is conventional. Refer to Fig. AC81 for exploded view of unit. Following specifications will apply:

Pressure spring free length....$2\frac{5}{32}$ in.
Lbs. pressure @ $1\frac{7}{16}$ in. 140

Reassemble the clutch and install same on flywheel using a NEW lined disc. Do not attempt to adjust the release lever adjusting screws with a used lined disc. Distance from surface of back plate (5—Fig. AC81) to release bearing contacting surface of release lever adjusting screws is $1\frac{3}{16}$ inches. Refer to Fig. AC81A. After release lever adjusting screws are properly positioned, the cover assembly can be removed and a used but still serviceable lined disc can be installed if so desired.

111. RELEASE BEARING. After engine is detached from torque housing as outlined in paragraph 109, remove the bolt (B—Fig. AC82) joining the two halves of the shifter yoke (20), spread yoke and withdraw release bearing (22) and bearing shifter (21). The shifter may be pressed out of the bearing.

112. ENGINE CLUTCH SHAFT. To remove the engine clutch shaft, the torque housing must first be removed from the tractor as outlined in paragraph 113. The clutch shaft can then

gage bulb from manifold. Unhook tachometer cable at engine. Remove L. H. side sheet from below fuel tank (requires prior removal of diesel fuel shut-off knob). Disconnect oil pressure gage line at rear of engine. Disconnect front steering shaft from U-joint at side of torque housing.

On diesel models, remove control rods from bell-crank to fuel injection pump and from joint in shut-off rod to fuel injection pump. Disconnect fuel return line at rear of engine and remove fuel supply line from fuel tank to filter assembly. Disconnect wire to intake manifold heater.

On gasoline models, remove governor control rod and disconnect choke wire and housing from carburetor. Remove line from fuel tank to fuel pump. Disconnect wires to ignition coil and generator.

On adjustable front axle models, place wedges between front axle and front axle support. Support tractor under torque housing and attach hoist to the two extended cylinder head cap screws. Unbolt engine rear adapter plate and side rails from torque housing. (Leave the two upper bolts retaining the front axle rear pivot bracket to engine rear adapter plate.) Roll engine and front end assembly away from torque housing.

On tricycle models, exact method for splitting tractor will depend upon shop equipment available. Some mechanics may prefer to adequately

block and brace front end of tractor, and use floor jack or hoist to roll rear assembly away from engine after unbolting engine and side rails from torque housing. In other instances, suitable equipment may be available for supporting front assembly safely and procedure as outlined for adjustable front axle models may be followed. In any case, caution should be taken with tricycle models.

After splitting tractor, unbolt and remove clutch cover assembly and clutch disc from engine flywheel. The pilot bearing in the flywheel may be renewed at this time.

Reverse removal procedures to reinstall clutch assembly. Use suitable pilot in clutch disc and clutch pilot bearing to aid in reassembly of tractor. Following torque values in Ft.-Lbs. will apply:

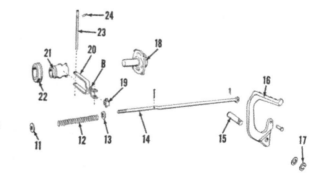

B. Bolt
11. Washer
12. Pedal return spring
13. Washer
14. Pedal rod
15. Pedal shaft
16. Pedal
17. Snap ring
18. Retainer
19. Shifter trunnion
20. Shifter lever
21. Clutch shifter
22. Throw-out (release)
    bearing
23. Shift lever pivot
24. Clip

Fig. AC82 — Exploded view of the clutch throw-out (release) bearing linkage and associated parts.

Fig. AC83 — View showing battery support, fuel tank support, battery, fuel tank, instrument panel and steering shaft removed as a unit from torque housing. To prevent scratching fuel tank, etc., lift unit from tractor with rope sling.

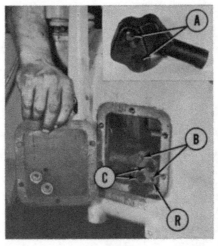

Fig. AC84 — View showing "Power-Director" clutch cover removed from torque housing. Inset shows reverse side of relief valve assembly (R).

A. "O" rings   C. "O" rings
B. Cap screw holes   R. Relief valve assembly

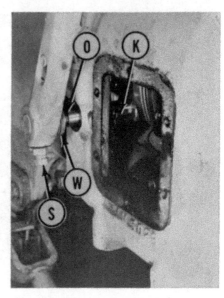

Fig. AC85 — View showing "Power Director" lever and shaft partially removed.

K. Woodruff keys (2)   S. Setscrew
O. "O" ring   W. Washers

be removed from the torque housing as outlined in paragraph 114.

113. R&R TORQUE HOUSING ASSEMBLY. To remove the torque housing assembly, first split the tractor between the engine and torque housing as outlined in paragraph 109; then, proceed as follows: Disconnect "Power-Director" oil tubes from cover on R.H. side of torque housing, unbolt line support bracket from torque tube, unbolt filter base from battery support frame and remove filter assembly and lines as a unit. Remove R.H. side sheet from below fuel tank. Disconnect U-joint coupling on "Traction-Booster" control shaft, "Traction-Booster" gage line at gage and light wire from slip connection behind instrument panel. Place a rope sling around the fuel tank, battery and support frame as shown in Fig. AC83 and attach hoist to sling. Unbolt fuel tank and battery support frame from torque housing, tie instrument panel up to steering wheel shaft with cord; then lift complete unit from torque tube and place unit on blocks out of way. Refer to Fig. AC83.

Support tractor under transmission and drain oil from hydraulic and "Power-Director" compartments. Disconnect "snap-coupler" release rod, then remove both step plates from tractor. Disconnect "make-up" control cable from control shaft. Unbolt and remove linkage guard from hydraulic pump and "snap-coupler" spring housing; then, remove all hydraulic lines from pump to control valve (including "Traction-Booster" gage line). NOTE: Two different types of line connections are used at hydraulic pump cover. .Early type has extra-long flare

nut fittings holding flared end of lines directly to hydraulic pump, and packing nuts were used to seal the outer circumference of the flare nuts. To disconnect this early type connection, loosen packing nut; then unscrew flare nut fitting and pull line and fitting out of packing nut. Late type connectors use a short flare nut connection outside of hydraulic pump cover.

Disconnect brake rods from actuating arms. Unbolt "snap-coupler" (drawbar) spring housing from torque tube and lower the unit to floor.

Remove the two hex-head cap screws from "Power-Director" clutch cover; then unbolt and remove cover from torque housing. Pull the relief valve body from the oil tubes. Remove snap ring and washer from "Power-Director" clutch shaft at left side of torque housing. Push shaft far enough to right to remove the two woodruff keys from the shaft. (See Fig. AC85). Then remove set screw (S), lever, washers (W) and "O" ring (O) from shaft. Shaft can then be removed out left side of torque housing and yoke can be removed from opening in side of housing. Attach hoist to torque housing; then unbolt and remove torque housing from transmission.

Reinstall torque housing assembly by reversing removal procedures. Use all new gaskets and "O" rings. Apply sealer to both sides of gaskets.

114. R&R ENGINE CLUTCH SHAFT. After torque housing is removed from tractor as outlined in paragraph 113, remove engine clutch shaft as follows: Unbolt and remove hydraulic pump cover from bottom of torque housing. Compress the linkage loading spring (S—Fig. AC86) and install a nail in hole of shaft at

Fig. AC86 — View showing torque housing with hydraulic pump cover removed. Late style outlet fittings (F) are shown. Note that spring (S) is compressed and nail (N) is placed through hole in guide to facilitate removal.

B. Pump mounting   F. Late style pump
    capscrews   fittings
C. Cotter pin   N. Nail (See text)
                       S. Spring

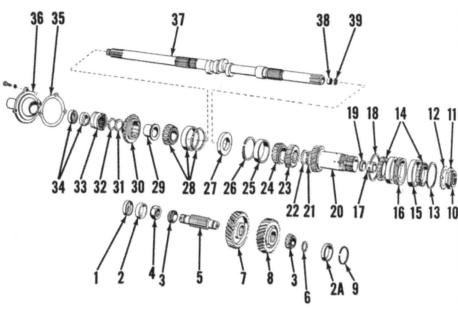

Fig. AC87 — Exploded view of engine clutch shaft and related parts. Lip seals (34) face to rear.

1. Plug	11. Nut	21. Snap ring (selective)	30. Pinion gear
2. Bearing cup	12. Lockwasher	22. Thrust washer	31. Thrust washer
3. Bearing cone	13. Snap ring (selective)	23. Drive gear	32. Snap ring
4. Nut	14. Bearing cone	24. Bearing cone	33. Collar
5. Intermediate shaft	15. Bearing cup	25. Bearing cup	34. Oil seals
6. Snap ring	16. Bearing spacer	26. Snap ring	35. Gasket
7. Intermediate driven gear	17. Bushing	27. Separator ring	36. Retainer
8. Intermediate drive gear	18. Snap ring	28. Bearing assy. (includes snap ring)	37. Engine clutch shaft
9. Snap ring (selective)	19. Bushing	29. Bushing	38. Bearing
10. Snap ring	20. Hollow (outer) shaft		39. Sleeve

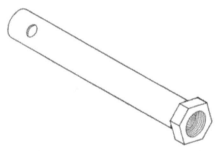

Fig. AC88 — Suggested tool for removing the "Power-Director" clutch outer (hollow) shaft can be made by welding a pipe to a nut (Allis-Chalmers part No. 229428). This tool can be screwed on the shaft in place of the standard nut and pipe can be bumped rearward withdrawing the bearing cones, cups, spacer and the outer shaft.

(N). Then remove cotter pin (C) and withdraw shaft and spring assembly. Remove the three cap screws (B) and withdraw hydraulic pump assembly from torque housing.

Refer to Fig. AC87. Remove snap ring (10), nut (11), retaining washer (12) and large snap ring (13). Using a puller similar to that shown in Fig. AC88, pull hollow shaft and bearings from rear of torque housing. Remove snap ring (21), thrust washer (22) and gear (23). Remove belt pulley assembly or belt pulley opening cover from

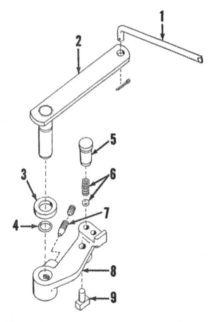

Fig. AC89 — Exploded view of the belt pulley shifter (8) and associated parts.

1. Shifter rod	6. Detent spring and ball
2. Shifter lever	7. Set screw
3. Spacer	8. Shifter
4. "O" ring	9. Insert
5. Spring retainer	

torque housing. Working through belt pulley opening, remove the lock screw and set screw (7—Fig. AC89) with an Allen wrench. Remove shifter lever (2) out of top of housing; then, withdraw shifter arm (8), spring and ball (6) and fork insert (9).

Remove front retainer (36 — Fig. AC87) and bump shaft (37) forward out of torque housing. Bearing cone (24) will be free when shaft is removed. Remove belt pulley shifter collar (33) from shaft; then remove snap ring (32), spacer (31), bevel gear (30) and drive the bushing (29) and front bearing cone from shaft. Rear end of clutch shaft contains transmission input shaft pilot bearing (38) and sleeve (39). Inspect all parts for wear or damage and renew as necessary. Front bearing cup (28) may be renewed at this time. To renew rear bearing cup (25), torque housing must be completely stripped of all other shafts and gears as bearing cup will not clear the intermediate shaft gear (7). Refer to paragraphs 120, 146, 147 and 148 for procedure to remove pto shafts and gears and "Power-Director" intermediate shaft and gears.

To install new engine clutch shaft, proceed as follows: Press front bearing cone (28—Fig. AC87) and bushing (29) far enough onto shaft so that bevel gear (30), spacer (31) and snap ring (32) may be installed. Then press bearing cone, bushing and spacer forward on shaft against snap ring so that there is no end play of the bushing between the spacer and bearing cone. Check to see that bevel gear turns freely. Install shaft in housing and drive rear bearing cone (24) onto rear of shaft until it is tight against rear bearing cup (25); then, install gear (23), spacer (22) and snap ring (21). Snap ring should be of sufficient thickness to limit shaft end play from zero to 0.004. Snap rings are available in thicknesses of 0.093 to 0.137 in steps of 0.004 inch.

Prior to reinstalling the belt pulley shifter fork, drive the detent retainer (5—Fig. AC89) far enough out of the torque housing to allow the detent ball (6) to be installed after shifter assembly is installed. Then drive retainer in until it is flush with top of torque housing.

Complete remainder of reassembly by reversing disassembly procedures. Refer to paragraph 119 for reinstalling hollow shaft at rear of engine clutch shaft.

"POWER DIRECTOR"

The "Power-Director" (See Fig. AC90) consists of two multiple disc wet type clutch packs contained in a common housing that is splined to the transmission input shaft. A reduction gear drive in front of the clutch housing turns the discs of the front clutch pack through a hollow shaft. The engine clutch shaft passes through the hollow shaft and turns the discs of the rear clutch pack at engine speed. Both clutch packs are controlled by a single lever and over-center type linkage. With the control lever in the forward position, the rear clutch pack is engaged and the clutch housing and transmission input shaft are turned at engine speed. When the lever is in the rear position, the front clutch pack is engaged and the clutch housing and transmission input shaft are turned at a reduced speed. Placing the control lever in center position disengages both clutch packs.

The "Power-Director" clutch packs are lubricated by oil from a pump in the bottom of the torque housing. The pump is driven by the front power take-off shaft. Oil from the pump passes through lines to an external filter and then to the clutch packs and reduction gear drive for lubrication of these units. The "Power-Director" and hydraulic oil sumps are connected. Sump capacity is approximately 22 quarts of SAE 20-20W non-detergent service "ML" motor oil. Oil level is checked with the hydraulic lift arms lowered and any remote cylinders retracted. Run engine at high idle speed for at least three minutes, shut engine off and immediately check oil level on dipstick.

Fig. AC90 — View showing "Power-Director" clutch unit on transmission input shaft.

New clutch assemblies are provided with three 0.090 stacks of shims (51A & 51B) between each housing and center plate and three 0.025 stacks of shims (53) between the two center plates. Thus, the total height of the shim packs is 0.205 and this total height must be maintained when adjusting the clutch to avoid changing clutch housing dimensions. For any thickness of shims added or removed from the shim stacks between the housing and center plate of either clutch pack, a like thickness must be removed or added to the shim stacks between the two center plates.

To gain access to the clutch packs to check clutch adjustment, first disconnect the filter lines (5 & 6—Fig. AC92) from the elbow fittings on the clutch cover (7); then remove filter and line assembly as a unit. Remove the two cap screws (8) and the quadrant (4) from clutch cover; then remove cover from clutch housing. Pull relief valve body (See Fig. AC93) from oil lines.

"POWER-DIRECTOR" CLUTCH

115. **CLUTCH ADJUSTMENT.** Refer to Fig. AC91. Clutch plate pressure is applied through a spring (Belleville) washer (46) that is located between the pre-load plate (45) and pressure plate (47) of each clutch pack. The spring washer must be compressed 0.042 to 0.048 inch when clutch pack is engaged. If compression is less, slippage of clutch will result. If compression of spring washer is greater than 0.048, clutch pack will not release properly. Adjustment is provided with shim packs (51A, 51B and 53) placed between the clutch housings and adjoining center plates and between the center plates.

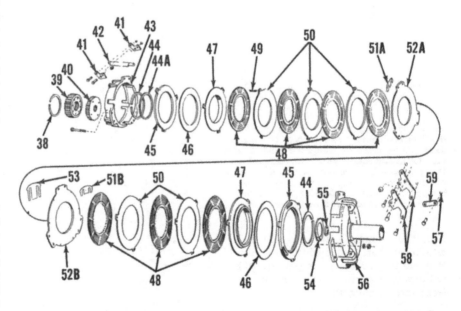

Fig. AC91 — Exploded view of "Power-Director" clutch assembly. Unit is equipped with more drive and driven plates than shown. Refer to text.

38. Snap ring	44. Snap ring	49. Clutch releasing spring	54. Thrust washer
39. Front hub	44A. Pressure plate spacer	50. Clutch plate	55. Snap ring
40. Rear hub	45. Pre-load plate	51A & 51B. Outer shims (0.010 & 0.015)	56. Rear housing
41. Release lever (front)	46. Pressure washer	52A & 52B. Center plates	57. Snap ring
42. Clutch link	47. Pressure plate	53. Center shims (0.010 & 0.015)	58. Release lever (rear)
43. Front housing	48. Clutch splined disc		59. Release lever link

Fig. AC92 — View of right hand side of torque housing showing "Power-Director" lever, clutch cover and related parts.

Fig. AC94 — View of "Power-Director" clutch unit.

A. Adjustment dimension	1. Shims
B. Adjustment dimension	2. Shims
C. Clutch collar	3. Shims
R. Snap ring	4. Bolts
S. Spacer	

Fig. AC95 — View of "Power-Director" clutch showing drive hubs removed. Remove snap ring (SR) to remove unit from transmission input shaft. Be sure tang of thrust washer enters hole (H) when reinstalling.

Using an inside hole gage of 2/10 to 3/10-inch capacity and a micrometer, measure clearance between the pre-load plate and the pressure plate (A & B—Fig. AC94) of each clutch pack; first with the clutch pack engaged, then in the disengaged position. Measurements should be made at each of the three openings around the clutch housings and an average of these dimensions used. Subtract the average engaged dimension from the average disengaged dimension. If the difference between the two average dimensions of a clutch pack is be-

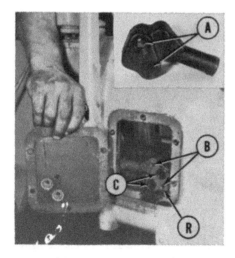

Fig. AC93 — View showing "Power-Director" clutch cover removed from torque housing. Inset shows reverse side of relief valve assembly (R).

A. "O" rings   C. "O" rings
B. Cap screw holes   R. Relief valve assembly

tween 0.042 and 0.048, no adjustment of that clutch pack is necessary. If the difference is less than 0.042, remove sufficient shim thickness from between clutch pack housing and center plate (1 or 3—Fig. AC94) to increase spring compression to 0.042-0.048 and add this same thickness between the center plates (2). For example, if the difference between the average engaged and disengaged dimension (A) of the front clutch pack was 0.035, removing 0.010 thickness of shims at (1) and adding 0.010 thickness of shims at (2) would increase the compression of the spring washer to 0.045.

If the difference between the averaged engaged and disengaged dimensions is more than 0.048, add sufficient shim thickness between clutch pack housing and center plate and remove this same thickness from between the two center plates. For example, if the difference between the average engaged and disengaged dimension (B) of the rear clutch pack was 0.050, adding 0.005 thickness of shims at (3) and removing 0.005 thickness of shims at (2) would decrease the compression of the spring washer to 0.045. NOTE: As 0.005 shims are not provided for use between the clutch housing and center plate (1 and 3) add a 0.010 shim and remove a 0.015 shim at each of the three shim stacks to reduce the shim stack height by 0.005; or add a 0.015 shim and remove a 0.010 shim at each of the three shim

stacks to add 0.005 to the shim height. Shims of 0.005 thickness are provided for service use between the two center plates; however, no 0.005 shims are used in the original assembly of the clutch.

116. **LEVER ADJUSTMENT.** To adjust lever quadrant (4—Fig. AC92), place lever in center detent position (1) and loosen the two quadrant retaining stud nuts (3). Start the engine, shift transmission into gear and release the engine clutch. Move lever to position where tractor has least tendency to creep and tighten quadrant stud nuts.

After the lever neutral position is located, adjust the forward stop position by varying the number of washers (2—Fig. AC92) until the control lever strikes the head of the stop bolt just as clutch links snap over center.

117. **R&R "POWER-DIRECTOR" CLUTCH.** To remove the "Power-Director" clutch assembly, it is necessary to split the torque tube from the transmission as outlined in paragraph 123; then, proceed as follows: Remove snap ring (R—Fig. AC94) and spacer (S). Then withdraw the two clutch hubs (39 & 40—Fig. AC91) and thrust washer (54). Remove the snap ring (SR—Fig. AC95) and withdraw the clutch assembly from the transmission input shaft.

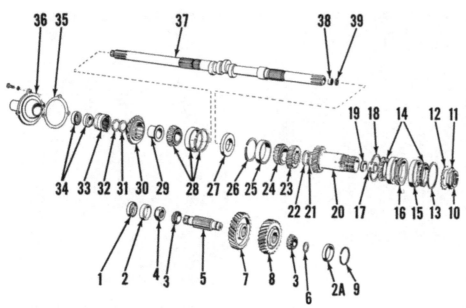

Fig. AC96—Exploded view of engine clutch shaft and related parts. Refer to Fig. AC87 for legend.

Fig. AC98 — View showing rear end of torque housing after unit is split from transmission.

To reinstall, reverse the removal procedures making sure that the tab on the thrust washer (54—Fig. AC91) engages the hole in the clutch housing (H—Fig. AC95).

118. OVERHAUL CLUTCH ASSEMBLY. The clutch assembly is a balanced unit; therefore, the front and rear housings should be marked prior to disassembly in order to maintain the balance when clutch is reassembled.

Refer to Fig. AC91 and proceed as follows: Disconnect the three clutch links (42) from the release levers, loosen the six bolts attaching the housings (43 and 56) together and remove the three shim stacks at each of the three bolting points. Unbolt and separate the clutch housings, discs and center plates. Compress the pre-load plate (45), spring washer (46) and pressure plate (47) assembly to remove snap rings (44).

Inspect the clutch discs and renew any which are excessively worn or have damaged notches for the clutch hub splines. Inspect all other parts and renew any which are questionable. The free height of the spring washer in each clutch pack should be 0.270-0.302 inch. Clutch plates with internal notches for clutch hub splines should measure 0.117-0.123 in thickness. Steel plates should be renewed if scored or showing signs of being over-heated. All plates should be flat within 0.009.

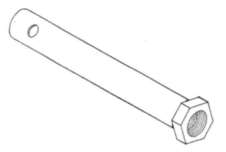

Fig. AC97 — Suggested tool for removing the "Power-Director" clutch outer (hollow) shaft can be made by welding a pipe to a nut (Allis-Chalmers part No. 229428). This tool can be screwed on the shaft in place of the standard nut and pipe can be bumped rearward withdrawing the bearing cones, cups, spacer and outer shaft.

To reassemble pre-load plate, spring washer and pressure plate, place parts in clutch housing to keep drive tangs aligned, compress spring and install snap ring. Reassemble clutch packs and install 0.090 shim stacks at each of the three positions (1 & 3—Fig. AC94) between the clutch housings and center plates and 0.025 shim stacks (2) between the two center plates. All pins should be installed with heads in direction of clutch rotation to prevent failure of snap rings that retain pins in linkage.

Clutch can be adjusted as outlined in paragraph 115 prior to or after attaching torque housing to transmission. Engage and disengage clutch with pry-bar against clutch collar (C—Fig. AC94) if adjusting prior to reassembly of tractor.

"POWER-DIRECTOR" CLUTCH OUTER (HOLLOW) SHAFT AND INTERMEDIATE GEARS

119. OUTER (HOLLOW) SHAFT. To remove the outer (hollow) clutch shaft, it is necessary to first split the torque tube from the transmission as outlined in paragraph 123; then, proceed as follows: Remove snap rings (10 and 13—Fig. AC96), then bend tangs of lock-tab washer away from nut (11) and remove the nut and washer. Using a tool similar to that shown in Fig. AC97 screwed onto the outer shaft (20—Fig. AC96), bump the shaft, bearings (14 & 15) and spacer (16) out of the torque tube. After snap ring (21) is removed, thrust washer (22) and drive gear (23) can be removed.

Bushings (17 & 19) are pressed into the outer shaft and in some cases may need honing to provide a free fit on the bearing surface of the inner clutch shaft. In renewing bushings, press ½-inch wide front bushing (19) into hollow shaft bore 3²¹⁄₃₂ inches from rear end of shaft. Press 1-inch wide rear bushing (17) into hollow shaft $\frac{21}{32}$-inch from rear end of shaft. Ream or hone new bushings to give 0.001-0.003 clearance between bushings and engine clutch shaft bearing surface.

To reassemble, install snap ring (18 —Fig. AC96) in groove in torque housing. Assemble front bearing cone (14), cup (15), spacer (16), rear bearing cup (15), rear bearing cone (14), lock-tab washer (12) and nut (11) on outer shaft. Do not tighten nut at this time. Install the assembly in bore of housing using tool that was used to pull shaft. Install as thick a snap ring as possible in rear groove in torque housing. Snap rings (13) are available in thicknesses of 0.094 to 0.109 in steps of 0.003. Adjust nut (11) to give 0.0005-0.004 end play of outer shaft in bearings and bend four tangs of lock-tab washer over opposite flats of nut. Install snap ring (10).

120. INTERMEDIATE SHAFT AND GEARS. To remove the intermediate shaft and gears, it is first necessary to remove the hydraulic pump as outlined in paragraph 157 and the power take-off driven gear and shaft as outlined in paragraph 147; then, proceed as follows: Remove the plug (1—Fig. AC96) and carefully unstake nut (4) from intermediate shaft (5). Hold shaft from turning by placing soft wedge between intermediate shaft gear and pto idler gear and remove nut with off-set box end wrench.

Remove snap ring (9) and thread slide hammer adapter into threaded hole in rear of intermediate shaft. Bump shaft out to rear of torque housing. If necessary to renew front bearing cup (2), drive cup out of rear of housing. Rear bearing cup (2A) will be driven out as shaft is pulled from housing. To remove rear bearing cone from shaft, first remove snap ring (6).

To reinstall, first drive front bearing cup (2) in tight against shoulder in bore of housing. Drive rear bearing cone on rear end of intermediate shaft and install snap ring (6). Place front bearing cone in cup and intermediate gears in housing. Insert shaft from rear into splines of gears and hold shaft against front bearing cone. Thread slide hammer adapter into rear end of shaft and bump shaft

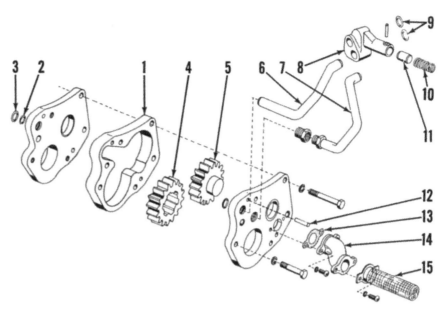

Fig. AC99 — Exploded view of D-19 "Power-Director" lubricating oil pump and related parts.

1. Gear plate	5. Driven gear	9. "O" rings	13. Gasket
2. "O" ring	6. Oil return tube	10. Spring	14. Manifold
3. "O" ring	7. Oil pressure tube	11. Relief valve	15. Inlet screen
4. Drive gear	8. Relief valve body	12. Dowel pin	

into front bearing cone. Install rear bearing cup in bore. Install new nut (4) on front end of shaft and tighten to a torque of 50-60 Ft.-Lbs. Check to see if snap ring (6) is tight against rear bearing cone and that outer rims of gears are pulled tightly together; then, stake nut securely at milled slot in shaft. Install snap ring (9) in groove behind rear bearing cup and seat cup against snap ring with slide hammer. Remove slide hammer and measure shaft end play with dial indicator. If end play is not within limits of 0.0005-0.0045, remove snap ring and install another snap ring of proper thickness to bring shaft end play within prescribed limits. Snap rings are available in thicknesses of 0.069 to 0.109 in steps of 0.004 inch.

"POWER-DIRECTOR" OIL PUMP

The "Power-Director" oil pump (See (Fig. AC98) provides lubricating oil for the wet type multiple disc clutch packs and for the reduction gears and bearings. Oil flow from the pump passes through lines to filter located outside of torque housing, then back

into torque housing to clutch packs, bearings and gears. A pressure relief valve (11—Fig. AC99) in the line to the filter by-passes oil to the return line if oil pressure exceeds 50-60 psi. Normal oil pressure is about 20 psi at filter base. Oil pressure warning light on instrument panel lights if pressure drops below 4-6 psi.

121. R&R AND OVERHAUL PUMP. To remove the pump, it is necessary to split the torque housing from the transmission as outlined in paragraph 123 and then remove the pto shifter assembly. The pump can then be unbolted and removed from the torque housing. NOTE: Remove the three cap screws with plain heads to remove pump. Cap screws with "L" on face of head retain pump cover to pump housing.

An exploded view of the pump and related parts is shown in Fig. AC99. Disassemble the pump and renew any parts which are excessively worn or deeply scored. As pump operates at a relatively low pressure, some wear can be tolerated. Renew "O" rings and inlet adapter gasket when reassembling pump. Reverse removal procedures to reinstall.

TRANSMISSION

SHIFTER ASSEMBLY

122. R&R AND OVERHAUL. The transmission shifter assembly is removed with the transmission cover by shifting gear selector lever to neutral position; then, unbolting and removing the cover and shifter assembly. (See Fig. AC100.)

NOTE: Prior to disassembly of shifter mechanism, location of overshift limiting washers (W—Fig. AC100), if any, should be noted so that they can be reinstalled in same position. Washers may be used as required to reduce excessive overshift if necessary after renewing shifter mechanism parts.

To remove the gear shift lever, remove snap ring (29—Fig. AC101) and oil shield (28); then, remove dust cover (16), snap ring (20) and pivot washer (19) and pull lever from cover. The two lever pivot pins (17) in cover are renewable after removing lever. To reinstall lever, reverse removal procedure.

To remove the reverse shifter fork (8), remove lock screw, rotate shift rail ¼-turn, and catch detent ball (25) and spring (24) while sliding rail forward out of cover. Withdraw rail and fork. With reverse shift rail removed, long interlock plunger (15— Fig. AC102) and interlock pin (30) may be removed from cover. Remove lock screw from first and second gear shifter fork (4—Fig. AC101), rotate rail ¼-turn and catch detent ball and spring at rear of cover as rail is removed out to front of cover. Rotate third and fourth gear shift rail (3) and fork (2) ¼-turn and slide the assembly forward out of cover while catching the detent ball and spring at rear end of cover. Be sure not to lose the short interlock plunger (26 — Fig. AC102). Renew bent, sprung or worn shifter forks or shift rails.

Wear on ends of interlock plungers and interlock pin (See Fig. AC102) will allow more than one shift rail to be moved at a time. Length of pins and plungers (new) are as follows: Interlock pin (30) should measure 0.551; short interlock plunger (26)

Fig. AC100 — Transmission cover and shifter assembly. Washers (W) are used as required on shift rails to limit overshift. Be sure to reinstall washers, if present, in same location from which they were removed.

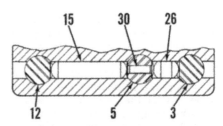

Fig. AC102 — Cut-away view of transmission cover showing location of interlock plungers (15 & 26) and interlock pin (30).

3. 3rd & 4th shift rail	26. Short interlock
5. 1st & 2nd shift rail	plunger (0.566)
12. Reverse shift rail	30. Interlock pin (0.551)
15. Long interlock plunger (1.691)	

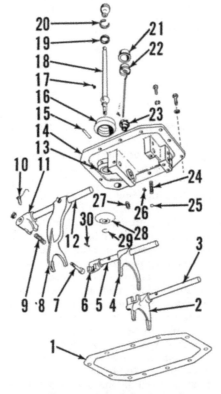

Fig. AC101 — Exploded view of transmission cover and shifter assembly.

1. Gasket	15. Long plunger
2. Shift fork (3rd & 4th)	16. Dust cover
3. Shift rail	17. Pivot pins (2)
4. Shift fork (1st & 2nd)	18. Shift lever
5. Shift rail	19. Pivot washer
6. Shift lug	20. Snap ring
7. Reverse latch plunger	21. Breather cap
8. Shift fork (reverse)	22. Cap & dipstick
9. Spring	23. Oil filler tube
10. Lock screw	24. Detent spring
11. Shift lug	25. Detent ball
12. Shift rail	26. Short plunger
13. Insert	27. Washer (use as required)
14. Transmission cover	28. Oil shield
	29. Snap ring
	30. Interlock pin

should measure 0.566; and long interlock plunger (15) should measure 1.691. Renew pin or plungers where measurement with micrometer indicates wear.

Face of reverse latch plunger (7— Fig. AC101) should be flush with reverse shift lug (11).

To reassemble cover, reverse disassembly procedures. Insert detent spring and ball in cover and depress ball with pin punch while sliding shift rails in place. Place shift lever and rails in neutral position, apply sealer to new gasket and reinstall cover and shifter assembly on transmission. Test shift for proper pattern before installing all cover retaining cap screws.

SPLIT TRANSMISSION FROM TORQUE HOUSING

123. Drain "Power-Director" and hydraulic oil sumps. If transmission is to be disassembled, drain transmission and final drive housings. Disconnect "Power-Director" oil filter lines at clutch cover and remove filter assembly and lines as a unit. Remove platforms and transmission heat shield. Disconnect brake rods.

Remove cotter pin from snap coupler pivot pin (2—Fig. AC129) and drive pin out with a special pin of same diameter and of length slightly less than inside width of clevis on drawbar spring rod (6). Center this short pin in the coupler bell (1) to hold the snap coupler latch (25) and spring (24) in place and allow snap coupler linkage to be separated at the pivot point. Remove cap screws (7A) and step plate bracket (7) from rear end of drawbar spring bracket (8).

On diesel models, remove diesel fuel shut-off knob. On all models, remove both side sheets from below fuel tank. Unhook hydraulic "make-up" cable from control shaft and bracket on torque housing. Disconnect light wire at instrument panel. Remove snap coupler linkage guard and disconnect link from pump control arm. Remove all hydraulic tubes to control valve.

Remove the two hex-head cap screws and lever quadrant from "Power-Director" clutch cover; then unbolt and remove cover from torque housing. Pull relief valve body from oil tubes.

Remove snap ring and washer from left end of "Power-Director" clutch shaft and push shaft to right far enough to allow removal of woodruff keys (K—Fig. AC85). Remove set-screw (S), lever, washers (W) and "O" ring (O) from shaft and push shaft out to left side of housing. Remove clutch fork out through clutch cover opening.

Support rear section of tractor with blocks both behind and in front of rear axle to prevent rear section from tipping. Attach hoist to front section at rear of torque housing (provide adequate additional support for tricycle models), unbolt torque housing from transmission and move front section forward.

TRANSMISSION OVERHAUL

Data on overhauling the various components of the transmission are outlined in the following paragraphs.

NOTE: Transmission bevel pinion shaft (1—Fig. AC103) is available for service only in a matched differential ring gear and bevel pinion set.

124. MAIN DRIVE (INPUT) SHAFT. To remove and overhaul this shaft, first split the transmission housing from the torque housing as outlined in paragraph 123; then, proceed as follows: Remove the large snap ring retaining the spacer and clutch hubs in the "Power-Director" clutch and

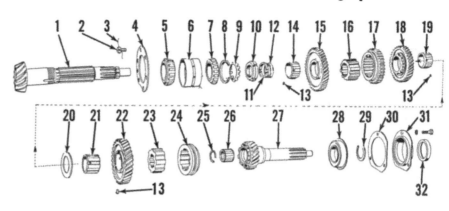

Fig. AC103 — Exploded view of the transmission main drive (input) shaft (27), bevel pinion shaft (1) and associated parts.

1. Bevel pinion	16. Collar (1.658)
2. Drilled cap screws	17. Gear (1st & 2nd) (34T)
3. Wire	18. Gear (2nd) (35T)
4. Retainer	19. Bushing (1.435)
5. Wide bearing cone	20. Thrust washer
6. Double bearing cup	21. Bushing (1.252)
7. Narrow bearing cone	22. Gear (3rd) (31T)
8. Snap ring (selective)	23. Collar (1.094)
9. Locking washer	24. Coupling
10. Lock nut	25. Snap ring
11. Split collar	26. Pilot bearing
12. Retainer	27. Main (input) shaft
13. Pins	28. Bearing assembly
14. Bushing (1.252)	29. Snap ring
15. Gear (1st) (40T)	30. Gasket
	31. Retainer
	32. Seal

withdraw the spacer, hubs and thrust washer. Remove the small snap ring retaining the clutch to transmission input shaft and pull clutch assembly from shaft. Shift transmission lever to neutral; then, unbolt and remove transmission cover and shifter assembly. Remove the cap screws attaching the front bearing retainer (31—Fig. AC103) to the transmission housing. Working through top cover opening, bump input shaft and retainer forward out of transmission.

Transmission bevel pinion pilot bearing (26) is contained in the rear end of the transmission input shaft and can be renewed at this time. Transmission input shaft pilot bearing (38 & 39—Fig. AC96) is located in rear end of engine clutch shaft and can be renewed at this time.

125. BEVEL PINION SHAFT. To remove the transmission bevel pinion shaft, first remove differential and bevel ring gear unit as outlined in paragraph 129; then, separate transmission housing from torque housing following general procedures outlined in paragraph 123. However, instead of blocking up the rear end of the tractor and moving the front section away, block up front section of tractor and lift transmission housing away from torque housing with hoist.

After removing transmission input shaft as outlined in paragraph 124, remove bevel pinion bearing retainer (4—Fig. AC103) and snap ring (25). Bump or push bevel pinion shaft, together with bearing cup and cones, rearward and remove the gears, collars and washers through top opening of transmission housing. Mark rear end of bearing cup (6) and unlock and remove nut (10); then, remove bearing cups and cones.

NOTE: The first, second and third speed gears rotate on hardened steel bushings (14, 19 & 21). Each bushing is equipped with two small removable steel pins (13) which engage the splines on the bevel pinion shaft.

The mesh position (center distance) of the bevel gears is controlled by the thickness of a snap ring (8) which is available in thicknesses of 0.177 to 0.191 in steps of 0.002. If necessary to renew the snap ring only, be sure to use replacement of exactly the same thickness as original. However, if the bevel pinion shaft and differential ring gear, bearings or transmission housing is renewed, it will be necessary to follow the procedure outlined in paragraph 125A to select the proper thickness of the snap ring (8).

If the bevel pinion shaft, gears and/or bushings are renewed, it will

Fig. AC104 — View showing degree of disassembly necessary to overhaul the complete transmission.

be advisable to first make a trial assembly on the bench of all parts located between and including the front snap ring (25) and collar (11) to check for free rotation of gears (15, 18 and 22). The front snap ring is available in thicknesses of 0.085 to 0.109 in graduations of 0.006. Select a snap ring which will eliminate all end play of the assembled parts when all are installed on the pinion shaft. If any gear fails to rotate freely when bevel pinion shaft is held stationary, it is an indication that the gear hub is wider than the bushing on which it rotates. To provide gear side clearance for free rotation, renew bushing or place a fine grade of emery cloth on a flat surface and lap the hub of the gear down to proper dimension to allow free rotation.

Reverse disassembly procedures to reassemble bevel pinion shaft and gears in transmission. Assemble pinion shaft and bearings on the bench. Install the wider roller bearing cone (5) next to the bevel pinion. Bearing cup (6) is reversible; however, it should be reinstalled in same position it was when removed. Adjust bearings by means of the nut (10) so that 5-8 inch-pounds of torque is required to rotate shaft in bearings and lock the adjusting nut with one of the tabs of the washer (9). NOTE: When bevel pinion shaft is installed in transmission housing, torque required to rotate shaft should increase to 8-35 inch-pounds. Figure of 5-8 inch-pounds is for use during bench check only. To facilitate reinstallation of pinion assembly in transmission housing, chill the assembly in freezer or dry ice for at least 30 minutes prior to installation.

Enter pinion shaft and bearing assembly in transmission housing and install pinion shaft components in the following order: Split collar (11) is installed in groove at front of adjusting nut (10) and retained with recessed ring (12). Install first gear

bushing (14) (1.252 wide) with pins forward. (NOTE: Heads of lock pins are placed to inside of bushing. Peen or rivet pins into bushings to hold them in place during reassembly.) Install first gear (15) (40 teeth) with clutch jaws forward. Install splined collar (16) and sliding gear (17) (34 teeth) with shift fork groove to rear. Install second gear bushing (19) (1.435 wide) with lock pins forward. Install second gear (18) (35 teeth) with clutch jaws to rear. Install thrust washer (20). Install third gear bushing (21) (1.252 wide) with lock pins to rear. Install third gear (22) (31 teeth) with clutch jaws forward. At this time, bevel pinion shaft bearing should be tight against snap ring in transmission housing. Install splined collar (23) ($1\frac{3}{32}$ inches wide); then, install snap ring (25) selected during trial assembly on bench or thickest snap ring that can be installed in groove. Install shifter coupling (24) over splined collar with beveled edge of coupling to front. Install bevel pinion shaft bearing retainer (4), tighten capscrews to a torque of 35 Ft.-Lbs. and wire the drilled heads of capscrews with soft wire.

Reinstall main drive (input) shaft and retainer assembly. Reassemble tractor in reverse of disassembly procedure.

125A. BEVEL PINION BEARING SNAP RING SELECTION. The mesh position of the bevel gears is controlled by the thickness of snap ring (8—Fig. AC103) located in front of the bearing cup (6) in bearing bore of transmission housing. If bevel pinion (1) and ring gear set, bearing set (5, 6 & 7) and/or transmission housing is renewed, the following procedure must be observed in selecting a new snap ring.

Assemble bearing cones and cup on the bevel pinion shaft as shown in Fig. AC105 and adjust bearing preload so that 5 to 8 inch-pounds of torque is required to rotate pinion shaft in bearings and lock nut in this position. (Widest bearing cone must be placed to pinion gear, bearing cup is reversible.) Then measure distance (See Fig. AC105) from rear face of rear bearing cone inner race to front edge of bearing cup. Add to this measurement the figure etched on the rear face of the pinion shaft (See Fig. AC106). Subtract this sum from the figure stamped on the lower right

Fig. AC105 — View showing bearing assembled on transmission bevel pinion shaft. To determine proper thickness of snap ring (8—Fig. AC103) required, measure distance from rear face of rear bearing cone inner race to front edge of bearing cup (dimension "D") and refer to text.

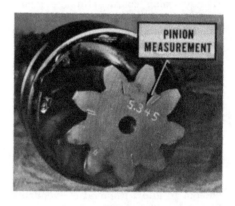

Fig. AC106 — Cone measurement is etched on rear face of pinion gear and is used in determining thickness of snap ring (8—Fig. AC103) for controlling pinion mesh position.

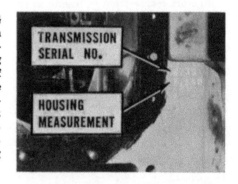

Fig. AC107 — Transmission housing measurement used in determining thickness of snap ring (8—Fig. AC103) is stamped on rear face of housing below transmission serial number.

hand corner of the rear face of transmission housing (See Fig. AC107). The result should be the size (thickness) of snap ring (8—Fig. AC103) to be used in reassembly of transmission. Snap rings are available in thicknesses of 0.177-0.191 in steps of 0.002.

As an example of above procedure,

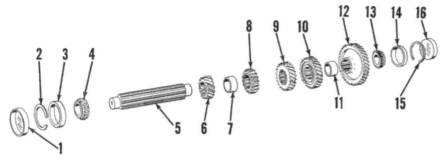

Fig. AC108 — Exploded view of transmission counter-shaft assembly.

1. Plug	5. Counter shaft	9. Gear (2nd) (22T)
2. Snap ring	6. Gear (1st) (16T)	10. Gear (3rd) (26T)
3. Bearing cup	7. Spacer (0.868)	11. Spacer (0.966)
4. Bearing cone	8. Gear (R) (18T)	12. Gear, driven (37T)

13. Bearing cone
14. Bearing cup
15. Snap ring (selective)
16. Plug

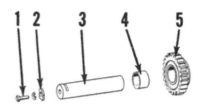

Fig. AC109 — Exploded view of transmission reverse idler gear and shaft. Bushing (4) is renewable. Front end of shaft (3) is 0.002 larger in diameter than rear.

1. Cap screw	4. Bushing
2. Lock plate	5. Idler gear
3. Shaft	

let measurement "D" (See Fig. AC-105) be 2.310. Add this to etched figure (See Fig. AC106) on rear of pinion gear, 5.345. This gives a sum of 7.655. Subtract this amount from the figure stamped on the transmission case (See Fig. AC107) which is, in this example, 7.840. Thus, 0.185 is the thickness of the snap ring that should be used for this particular set of bearings, bevel pinion and ring gear set, and transmission housing. Use closest thickness of snap ring to determined value.

126. R&R COUNTERSHAFT AND GEARS. To remove countershaft and gears, the bevel pinion shaft must first be removed as outlined in paragraph 125. Then, remove plug (16—Fig. AC108) and snap ring (15) from front face of transmission housing. Thread slide hammer adapter into front end of countershaft and bump shaft forward out of housing. Gears (6, 8, 9, 10 & 12), spacers (7 & 11) and rear bearing cone (4) can be removed out top opening. Front bearing cone (13) and cup (14) will be pulled with shaft. If necessary to renew rear bearing cup (3), drive plug (1) out to rear; then drive cup out to front of transmission.

End play of countershaft in bearings is controlled by varying the thickness of the snap ring (15) at front end of shaft. This snap ring is available in thicknesses of 0.069 to 0.109 in steps of 0.004. If use of the thickest snap ring (0.109) is not sufficient to control end play, the rear snap ring (1) may be removed and a 0.109 snap ring installed in that position; then readjust end play with front snap ring. As the bearing cones fit against the gear and spacer stack

instead of against shoulders on countershaft, end play can be checked only when the shaft is installed with all gears and spacers in place and all end play removed from gears and spacers on the shaft.

To reinstall countershaft, proceed as follows: Drive front bearing cone onto threaded end of countershaft. Install rear snap ring (normally 0.093 thick) and drive rear bearing cup in tightly against snap ring. Insert shaft through front bearing bore and install gears, spacers and rear bearing cone in following order: Install driven gear (12) (37 teeth) with long hub to rear; spacer (11) (0.966 wide); third gear (10) (26 teeth) with long hub to front; second gear (9) (22 teeth) with long hub to rear; reverse gear (8) (18 teeth) with beveled edge of teeth to rear; spacer (7) (0.868 wide); place rear bearing cone in cup and then install first gear (6) (16 teeth). NOTE: First gear is reversible but should be reinstalled in same position as it was removed due to developed wear pattern.

Drive or push the countershaft into the rear bearing cone. Install the front bearing cup and snap ring. Seat front bearing cup against snap ring with slide hammer threaded into countershaft; then remove slide hammer and remove all end play from countershaft gears and spacers by driving against inner race of front bearing cone with a hollow driver. Note: Tightness of bearing cones on shaft will retain gears and spacers in this "no end play" position. Then, check end play of the complete shaft, gears and spacers assembly with a dial indicator while moving the assembly back and forth between the front and

rear bearing cups. If end play is not within 0.0005-0.0045, remove snap ring (15) and install snap ring of proper thickness to bring end play within limits.

Apply sealer to rims of plugs (1 and 16) and drive plugs into bearing bores (flat side in) until rims of plugs are flush with transmission housing. Reassemble tractor in reverse of disassembly procedure.

127. REVERSE IDLER GEAR AND SHAFT. The reverse idler gear and shaft can be removed after the tractor is split between transmission and torque housing and both transmission housing top covers are removed; however, in most instances it will be removed when bevel pinion shaft and countershaft assemblies are being overhauled as outlined in paragraphs 125 and 126.

To remove reverse idler shaft, proceed as follows: Remove expansion plug located in line with idler shaft in front face of transmission housing. Remove cap screw (1—Fig. AC109) and lockplate (2) from differential compartment. Thread slide hammer adapter into front end of shaft, pull shaft from housing and remove gear.

Reverse idler rotates on renewable bronze bushing (4). When renewing bushing, press new bushing in flush with flat side of gear and hone bushing to 1.2375-1.2385. Service shaft (3) has lock notch at each end. Renew shaft if clearance between shaft and new bushing is excessive. Clearance should be 0.002-0.004. NOTE: Threaded end of shaft is 0.002 larger diameter than other end.

Install shaft with end having threaded hole to front and with shifter groove of reverse gear to front. Install lockplate, cap screw and lockwasher at rear end of shaft. Apply gasket sealer to edge of expansion plug and drive plug into housing.

DIFFERENTIAL, FINAL DRIVE AND REAR AXLE

DIFFERENTIAL

128. BEARING ADJUSTMENT. Adjustment of the differential bearings is controlled by shims (3—Fig. AC110) located between each bearing carrier (1) and the transmission housing. Transferring shims from under one bearing carrier to under the opposite bearing carrier will adjust the backlash of the bevel ring gear and bevel pinion gear. It is important that there is some backlash between the bevel pinion and ring gear before and after a bearing adjustment is made.

To adjust the differential carrier bearings, remove the bull gears as outlined in paragraph 133; then, proceed as follows: Vary the number of shims (3—Fig. AC110) between the bearing carriers and transmission housing to remove all end play of the differential unit, but permitting the differential to turn without binding. Then transfer shims from under one bearing carrier to under the opposite bearing carrier to adjust bevel gear backlash to 0.007-0.012. Check backlash at several points to be sure that point of minimum backlash is 0.007 or more and that point of maximum backlash is 0.012 or less. Apply No. 3 Permatex or similar gasket sealer to threads of retaining cap screws and to surface of bearing carriers that contacts transmission. (Do not apply sealer to shim contact surfaces.) NOTE: Adjustment of differential bearing end play and bevel gear backlash will affect bull pinion shaft end play. Be sure to check bull pinion shaft end play when reassembling tractor. See paragraph 132.

129. R&R AND OVERHAUL. To remove the differential assembly from tractor, first remove both bull gears and pinions as outlined in paragraph 132 and 133; then, proceed as follows: Unbolt and remove left differential bearing carrier. Unbolt right differential bearing carrier and support differential assembly while removing bearing carrier. Then remove differential assembly out rear opening of transmission housing.

To disassemble the differential, drive out the lock pin (14—Fig. AC-110), then drive pinion shaft (13) from differential housing. Differential pinions (4), side gears (6) and thrust

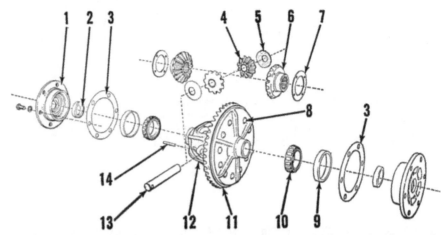

Fig. AC110 — Exploded view of differential assembly. Bull pinion inner bearing cup is located in differential bearing carrier (1). Shims (3) are available in thicknesses of 0.005, 0.007 and 0.012.

1. Bearing carrier	5. Thrust washers	8. Rivet or bolt	12. Differential case
2. Seal	6. Side gears	9. Bearing cup	13. Pinion shaft
3. Shims	7. Thrust washers	10. Bearing cone	14. Lock pin
4. Differential pinions		11. Ring gear	

washers (5 & 7) can then be removed from differential housing. Renew all questionable parts.

If backlash between teeth of side gears and pinion gears is excessive, renew side gear thrust washers (7) and/or pinion gear thrust washers (5). If backlash is still excessive after renewing thrust washers, it may be necessary to renew side gears and differential pinions. Pinions and side gears are available in matched sets only.

For renewal of bevel ring gear, refer to paragraph 131. To reinstall differential unit, reverse removal procedures. Install new oil seals in differential bearing carriers with lips of

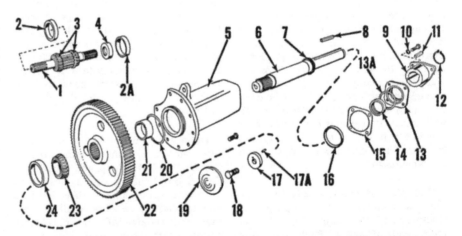

Fig. AC111 — Exploded view of final drive gears and rear axle assembly. Bearing cup (2) is located in differential carrier; (2A) is located in brake housing; and (24) is located in inner wall of bull gear compartment in transmission housing.

1. Bull pinion	7. Bearing cone	13A. "O" ring	19. Plug
2. Bearing cup	8. Key	14. Seal	20. "O" ring
2A. Bearing cup	9. Hub	15. Shims	21. Spacer
3. Bearing cones	10. Flat washer	16. Bearing cup	22. Bull gear
4. Seal	11. Key	17. Washer	23. Bearing cone
5. Axle housing	12. Snap ring	17A. Roll pin	24. Bearing cup
6. Axle shaft	13. Retainer	18. Cap screw	

seals facing differential. Check differential bearing end play, bevel gear gear backlash and bull pinion shaft end play when reassembling tractor. The following torque values should be observed during reassembly: (All values in Ft.-Lbs.)

Bearing carrier to transmission . . 45-50

Brake housing to
　　transmission 100-110

Axle housing to
　　transmission 130-140

Bull gear (axle inner end) 160-170

MAIN DRIVE BEVEL GEARS

130. **MESH AND BACKLASH ADJUSTMENT.** The mesh position of the bevel pinion gear is controlled by the thickness of snap ring (8—Fig. AC103) which is fitted to a particular assembly of a transmission housing, bevel gear set and bevel pinion bearings. Refer to paragraph 125A.

To adjust the backlash between bevel pinion and bevel ring gear, refer to paragraph 128.

131. **RENEW BEVEL GEARS.** As the bevel pinion is an integral part of the transmission output shaft, refer to paragraph 125 for service procedures.

NOTE: Main drive bevel pinion and ring gear are available only in a matched set.

The bevel ring gear is renewable when differential assembly is removed as outlined in paragraph 129. On factory assembled differential units, the ring gear is riveted to the differential housing. Special bolts and nuts are available for service installation of ring gear. Drill rivets from ring gear side to remove the original ring gear from housing. Install bolts with heads on ring gear side of assembly and tighten nuts securely. Install cotter pins through castellated nuts and drilled bolts.

FINAL DRIVE GEARS

132. **BULL PINION SHAFTS.** Bull pinion gear shafts are carried in tapered roller bearings located in differential bearing carriers and in brake housings. Number of shims used between differential bearing carriers and transmission housing to adjust differential bearing end play and bevel gear backlash will affect bull pinion shaft end play; shims are used between brake housings and transmission housing to control bull pinion shaft end play.

To remove bull pinion shafts, proceed as follows: Remove both rear axles as outlined in paragraph 135. Unbolt and remove both brake housing covers. Pull brake disc and drum

Fig. AC112 — Removing tapered, split hub from rear wheel. Special bolts (See Fig. AC113) are provided with each tractor to press wheel off of tapered hub.

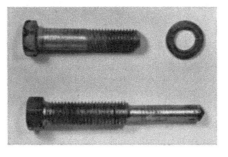

Fig. AC113 — Standard rear wheel hub cap screw and washer (top) compared with special cap screw (bottom) used for removing hub. Three of the special cap screws are provided with each tractor.

assemblies from bull pinion shafts. Unscrew brake adjusting cap screws and remove contracting brake bands from housings. Unbolt and remove brake housings from transmission housing. Pull pinion shafts from differential.

Reassemble by reversing disassembly procedures. Use all new "O" rings and gaskets during reassembly. Apply sealer to both sides of lift shaft housing gasket. Use proper number of shims between brake housings and transmission housing to allow 0.001-0.003 end play of bull pinion shafts. Shims are available in thicknesses of 0.003, 0.010 and 0.020.

Fig. AC114—Hydraulic jack in position to press rear axle shaft out of bull gear and inner bearing cone. Refer to text for caution during this operation.

133. **BULL GEARS.** To remove bull gears, first remove bull pinion shafts as outlined in paragraph 132; then, lift bull gears from compartments in transmission housing. Reassemble as in paragraph 132 after first placing bull gears in transmission housing.

AXLE SHAFTS AND BEARINGS

134. **BEARING ADJUSTMENT.** Bearing adjustment of each rear axle is controlled by shims (15—Fig. AC111) between axle housing (5) and outer bearing and seal retainer (13). To adjust bearings, support tractor securely and remove rear wheel and hub (See Fig. AC112) with special bolts (See Fig. AC113) that are provided with each tractor. Remove bearing retainer (13—Fig. AC111) and vary number of shims (15) to allow free turning of rear axle with no end play in bearings. The final installation of retainer should be made using new "O" ring (13A) and seal (14). Tighten retainer cap screws securely. Reinstall rear wheel and tighten hub retaining cap screws to a torque of 160-170 Ft.-Lbs. Retighten cap screws after tractor has been operated for several hours.

135. **RENEW REAR AXLE SHAFT AND/OR BEARINGS.** To remove either rear axle shaft, proceed as follows: Drain lubricant from both bull gear compartments, "Power-Director" and hydraulic sump, transmission and differential compartments. Support tractor solidly and remove rear wheel and hub. Special bolts (See Fig. AC113) are provided with each trac-

tor for pushing rear wheel from tapered hub. Disconnect lines from hydraulic control valve; then, unbolt and remove valve and manifold as a unit from transmission rear top cover. Unbolt and remove PTO output shaft and bearing retainer from rear face of lift shaft housing. Remove pins from lift pistons and hydraulic lift shaft pivot points. Unbolt and remove lift shaft housing from rear face of transmission housing. Remove seat assembly. Unbolt and remove transmission rear top cover.

Remove cap at inner end of rear axle and remove cap screw and pinned washer from inner end of rear axle. Remove rear axle outer bearing retainer. Remove final drive compartment magnetic plug to prevent damage to plug. Support weight of bull gear and buck up between gear and outer wall of bull gear compartment. Place porta-power ram or hydraulic jack between inner end of rear axle to be removed and opposite wall of differential compartment and press axle out of inner bearing cone and bull gear. Outer bearing cup will be pushed out with axle. CAUTION: When pressing rear axle out, be sure jack or ram is supported by opposite wall of differential compartment and not by inner end of opposite rear axle or cap covering inner end of axle.

Outer bearing cone may now be removed from axle shaft and outer seal from bearing retainer. To remove inner bearing cone, lift bull gear as far as possible and work cone out between bull gear and inner wall of bull gear compartment. (Bull gear cannot be removed from compartment unless bull pinion shaft is removed). Drive inner bearing cup out towards bull gear and work cup out between bull gear and inner wall of bull gear compartment.

Reverse removal procedures to reinstall rear axle and bearings. Use all new "O" rings and gaskets during reassembly. Apply sealer to both sides of lift shaft housing gasket. To pull axle through the inner bearing cone, remove roll pin (17A—Fig. AC111) from washer (17) and use 2½ inch cap screw with washer to pull axle until the long cap screw bottoms in axle; then, use original cap screw (18) to complete the operation. Remove cap screw, drive the pin (17A) back into washer (17) and reinstall washer with pin inserted into off-center hole in inner end of axle shaft. Install cap screw (18) and tighten

to a torque of 160-170 Ft.-Lbs. Apply No. 3 Permatex or similar sealer to rim of cap (19) before installing cap at inner end of axle shaft.

136. R&R REAR AXLE HOUSING. To remove a rear axle housing, it is first necessary to remove the rear axle from housing as outlined in para-

graph 135 and hydraulic ram (cylinder) from axle housing. Then, unbolt and remove rear axle housing from transmission housing.

Renew "O" ring (20—Fig. AC111) and tighten retaining cap screws to a torque of 130-140 Ft.-Lbs. when reinstalling axle housing. Reinstall rear axle as outlined in paragraph 135.

BRAKES

Bendix Band/Disc brakes are used. Refer to Fig. AC116.

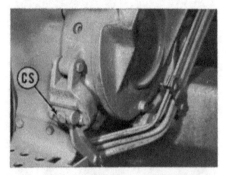

Fig. AC115 — Adjust cap screw (CS) so that brake pedals apply equally and have approximately two inches of free travel. Cap screws are self-locking.

137. ADJUSTMENT. To adjust either brake, turn cap screws (CS— Fig. AC115) in or out so that brake pedal has 2 inches of free travel. Cap screw at each brake assembly is self-locking. Be sure that brakes are adjusted so that both pedals are aligned when brakes are applied.

138. R&R AND OVERHAUL. To remove brake assemblies, first unbolt and remove rear fenders from brake housing covers (1—Fig. AC116). Unbolt and remove brake housing cover. Drum and disc unit (Fig. AC117) can then be removed from bull pinion shaft. To remove contracting band,

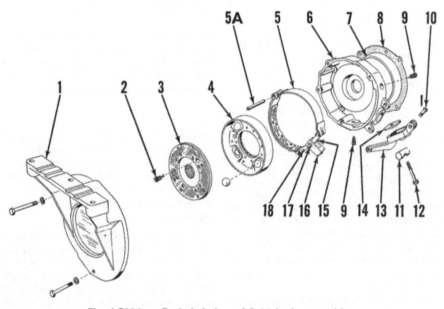

Fig. AC116 — Exploded view of D-19 brake assembly.

1. Brake housing cover	5. Band assembly	9. Springs	14. Seal
2. Spring	5A. Band pivot pin	10. Pin	15. Link (outer)
3. Disc assembly	6. Brake housing	11. Bar	16. Yoke
4. Drum assembly	7. "O" ring	12. Adjusting screw	17. Pin
	8. Shims	13. Lever	18. Link (inner)

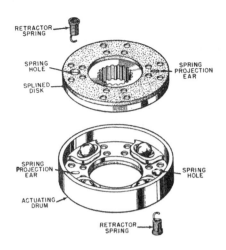

Fig. AC117 — View of brake drum and disc assembly showing component parts.

BELT PULLEY AND PTO

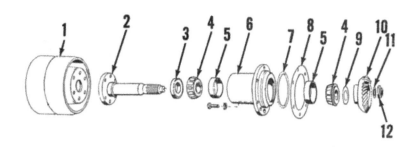

Fig. AC120—Exploded view of the belt pulley assembly. The pulley drive gear and shifter collar are shown in Fig. AC122.

1. Pulley	4. Bearing cone	7. "O" ring	10. Pulley gear
2. Pulley shaft	5. Bearing cup	8. Shims	11. Washers
3. Oil seal	6. Pulley housing		12. Nut

use slotted screw driver tip to disengage springs (9—Fig. AC116) from band (5), unscrew adjusting screw (12), and remove band from pivot pins (5A).

To disassemble drum and disc unit (Fig. AC117), insert slotted screwdriver tip through open end of springs and stretch springs only far enough to unhook. Remove disc and steel balls from drum.

Linings are not available separately from disc, drum or band. If necessary to renew brake linings, renew complete disc, drum or band assembly. Inspect friction surfaces of brake housing (6—Fig. AC116) and cover (1) for scoring or scuffing and renew either if friction surface is not suitable for further use. If brake housing is renewed, install new housing with proper number of shims (8) to limit bull pinion (differential) shaft end play to 0.001-0.003. Tighten the housing retaining cap screws to a torque of 100-110 Ft.-Lbs.

Condition of return springs (2 & 9—Fig. AC116) is of utmost importance when servicing band/disc brakes. Renew any spring if the coils of the spring do not fit tightly together and be careful not to stretch the springs any farther than necessary when reassembling brakes. Insufficient spring tension will allow the brakes to drag as they will fail to return the bands and discs to proper position when brakes are released.

Reverse disassembly procedures to reassemble brakes. Adjust brakes as outlined in paragraph 137.

BELT PULLEY UNIT

140. GEAR ADJUSTMENT. To adjust the backlash of the belt pulley gears (10—Fig. AC120 and 7—Fig. AC122) first remove the pulley as outlined in paragraph 141. Remove shims (8—Fig. AC120), then reinstall unit with the threaded hole in housing toward the bottom. Install a cap screw in the most forward position and another in the rear-most position; tighten these two cap screws finger tight. Install a cap screw in the threaded hole and tighten same until the flange of the housing (6) is evenly spaced from the surface of the torque tube. Insert as many shims (8) between the housing and torque tube as possible; then, remove the pulley assembly and reinstall same with these shims plus 0.030 thickness of shims to provide the proper gear backlash.

141. REMOVE AND REINSTALL. To remove the pulley assembly, drain hydraulic oil to below belt pulley opening, remove the six attaching cap screws; and withdraw the unit. Make sure that shims (8—Fig. AC120) are not lost or damaged.

When reinstalling, if there is any reason to suspect improper gear backlash adjustment, refer to paragraph 140 for information concerning backlash adjustment.

142. OVERHAUL. To overhaul the removed belt pulley unit, refer to Fig. AC120 and proceed as follows: Unstake and remove nut (12); then bump shaft from gear and inner bearing. The need and procedure for further disassembly will be evident after examination of the unit.

Inspect all parts for visible damage. Renew seal (3) and bearings (4 & 5) if questionable.

Reassemble in reverse of disassembly procedure. Tighten nut (12) to provide bearings with 4-10 inch pounds preload and stake nut to shaft.

PULLEY DRIVE GEAR

143. The belt pulley drive gear (7—Fig. AC122) located on the clutch

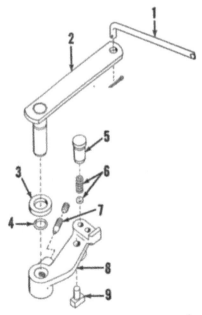

Fig. AC121 — Exploded view of the belt pulley shifter (8) and associated parts.

1. Shifter rod	6. Detent spring and
2. Shifter lever	ball
3. Spacer	7. Setscrew
4. "O" ring	8. Shifter
5. Spring retainer	9. Insert

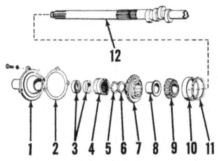

Fig. AC122 — Exploded view of the parts located on front part of engine clutch shaft (12) including the shifter collar (4).

1. Seal retainer	8. Bushing
2. Gasket	9. Bearing cone
3. Seals	10. Bearing cup
4. Shifter collar	11. Snap ring
5. Snap ring	12. Engine clutch
6. Thrust washer	shaft
7. Bevel gear	

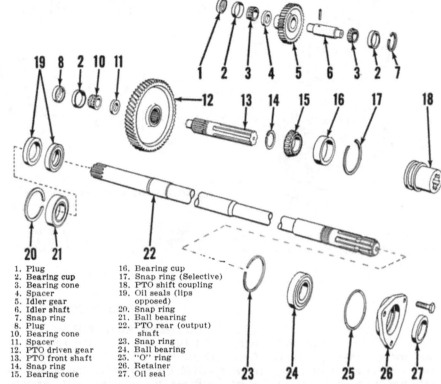

Fig. AC123 — Exploded view of power take-off assembly. Adjustment of bearings (10 & 15) is controlled by thickness of snap ring (17); adjustment of bearings (3) is controlled by thickness of snap ring (7).

1. Plug	16. Bearing cup
2. Bearing cup	17. Snap ring (Selective)
3. Bearing cone	18. PTO shift coupling
4. Spacer	19. Oil seals (lips
5. Idler gear	opposed)
6. Idler shaft	20. Snap ring
7. Snap ring	21. Ball bearing
8. Plug	22. PTO rear (output)
10. Bearing cone	shaft
11. Spacer	23. Snap ring
12. PTO driven gear	24. Ball bearing
13. PTO front shaft	25. "O" ring
14. Snap ring	26. Retainer
15. Bearing cone	27. Oil seal

shaft can be removed as follows: Split the engine from torque housing as outlined in paragraph 109, remove release bearing as outlined in paragraph 111 and belt pulley unit as outlined in paragraph 141. Remove the belt pulley shifter fork as follows: Working through the belt pulley opening, remove the lock screw and set screw (7—Fig. AC121) with a socket type Allen wrench. Remove the shifter lever (2) out top of torque housing; then remove shifter (8), spring and ball (6) and fork insert (9).

Remove front retainer (1—Fig. AC122) and slide belt pulley shifter collar (4) from front end of engine clutch shaft. After removing snap ring (5) and spacer (6) remove belt pulley drive gear (7) from clutch shaft.

To reinstall, reverse removal procedures. Drive retainer (5—Fig. AC-121) up far enough to allow installation of ball (6) after shifter fork is installed. Then drive retainer down flush with top of torque housing.

POWER TAKE-OFF UNIT

145. **REMOVE AND REINSTALL.** To remove the PTO output shaft (22—Fig. AC123), first drain the "Power-Director" and hydraulic sump, transmission and differential compartments. Then remove the cap screws attaching the PTO shaft bearing retainer (26) to the rear face of the lift shaft housing and pull the shaft assembly from the tractor.

When reinstalling the PTO shaft, take care not to damage seals (19) and/or bearing (21) which are located in the front wall of the transmission housing.

146. **R&R PTO SHIFTER COUPLING.** To remove the PTO shift coup-

ling (18—Fig. AC123) and/or shifter arm (2—Fig. AC124), it is necessary to split the transmission from the torque housing as outlined in paragraph 123. Further procedure will be evident after an examination of the unit and reference to Figs. AC123 and AC124.

Be sure to wire the pin (6—Fig. AC124) in place before reattaching transmission to torque housing.

147. **R&R PTO DRIVEN GEAR.** To remove the PTO driven gear (12—Fig. AC123), it is necessary to split the

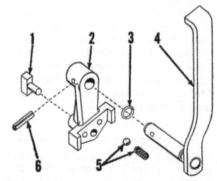

Fig. AC124 — Exploded view of the PTO shifter assembly. Shifter collar is shown at 18 in Fig. AC123.

1 Insert	4. Shifter lever
2. Shifter arm	5. Detent spring & ball
3. "O" ring	6. Roll pin

transmission from the torque housing as outlined in paragraph 123, remove the "Power-Director" pump as outlined in paragraph 121, the cover on bottom of torque housing and the snap ring (17). Then thread slide hammer adapter into rear end of shaft (13) and bump shaft out to rear of torque housing. Remove PTO driven gear out through bottom opening of torque housing.

Reverse removal procedure to reinstall PTO driven gear. Select proper thickness of snap ring (17) to give 0.0005-0.0045 end play of shaft (13). Snap rings are available in thicknesses of 0.061 to 0.105 in steps of 0.004.

148. **R&R PTO IDLER GEAR.** To remove the PTO idler gear (5—Fig. AC123), first remove the PTO driven gear as outlined in paragraph 147. Then remove snap ring (7), thread slide hammer adapter into rear end of shaft (6) and bump shaft out towards rear of torque housing. Remove gear (5), spacer (4) and front bearing cone (3) out bottom opening of housing.

Reverse removal procedure to reinstall PTO idler gear. Select proper thickness of snap ring (7) to give 0.0005-0.0045 end play of shaft (6). Snap rings are available in thicknesses of 0.069 to 0.109 in steps of 0.004.

HYDRAULIC POWER LIFT SYSTEM

149. The hydraulic lift system consists of a four-plunger constant stroke pump actuated by cams on the engine clutch shaft, either a single or three-spool control valve mounted on the transmission rear top cover, a "transport" valve mounted on the control valve manifold and a lift shaft actuated by two externally mounted hydraulic cylinders (rams).

Three of the pump plungers are ¾-inch diameter. Oil pumped by these plungers is directed to the inlet of the control valve (See Fig. AC125) located on the transmission rear top cover. The control valve, when all spools are in neutral position, returns the oil to the hydraulic sump. Relief valve for the three ¾-inch plungers is located in the control valve body.

The fourth pump plunger is of $\frac{7}{16}$-inch diameter. Oil pumped by this plunger is for "Traction - Booster" action (See paragraph 150), or for "make up" oil (See paragraph 156) to keep lift arms in raised position when transporting mounted implements. Pressure relief valve and control valve for $\frac{7}{16}$-inch pump plunger circuit are located in hydraulic pump body.

When tractor is equipped with a three-spool control valve (Fig. AC-125), spool next to operator is used to control tractor lift system or single acting remote cylinder attached to "transport" valve. The other two spools may be used to operate double acting remote cylinders. When equipped with single-spool valve, valve is used to control lift arms or single acting remote cylinder attached to "transport" valve only.

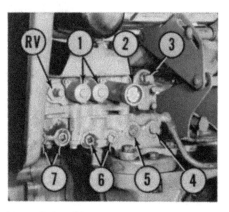

Fig. AC125—Three-spool hydraulic control valve installation. One-spool control valve (optional with three-spool) is mounted in same location.

RV. Relief valve
1. Remote control spools
2. Lift circuit spool
3. Transport valve outlet port
4. Pressure test port
5. Pressure test port
6. Remote cylinder ports
7. Remote cylinder ports

The "Traction-Booster" circuit is connected to the lift circuit when the control valve spool (Fig. AC126) is moved to lowering position, then allowed to return to detent position. The control valve in the hydraulic pump body then controls implement working depth and continually maintains a pressure on the tractor lift cylinders with oil from the $\frac{7}{16}$-inch pump plunger.

In operation, after implement working depth is adjusted, lift control lever is placed in "Traction-Booster" detent position and "Traction-Booster" lever is moved upwards on quadrant which, along with linkage movement from compression of "snap-coupler" spring by implement draft, partially closes the control valve in hydraulic pump. Pressure is then built up in the lift circuit. Quadrant lever should be so adjusted that "Traction-Booster" gage indicator is in first half of gage dial.

"Traction-Booster" system may be blocked out if desired by placing drawbar stop (See Fig. AC127) between "snap-coupler" spring housing and stop nut on spring rod.

PRESSURE TESTS AND ADJUSTMENTS

151. **LIFT CIRCUIT OPERATING PRESSURE.** Operate tractor until hydraulic oil is warm. Install high pressure gage in valve manifold port (5—Fig. AC125) and hold one of the control valve spools in lift position until gage reading can be observed. Gage reading should be 2000-2050 psi with engine running at 2000 RPM. If not within these limits, remove cap nut from pressure relief valve adjusting screw (RV), loosen jam nut and

"TRACTION-BOOSTER" SYSTEM

150. The "Traction-Booster" system provides draft control of implements attached to the spring loaded "snap coupler" hitch and also, with certain implements, increases traction by carrying part of the implements weight on the hydraulic lift arms.

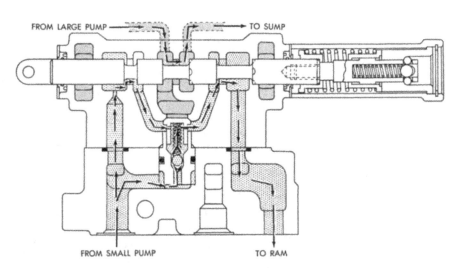

Fig. AC126 — Schematic diagram of oil flow during "Traction-Booster" action operation. Refer to text for principles of operation.

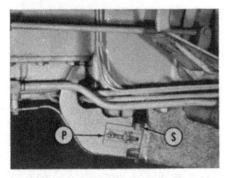

Fig. AC127 — View showing drawbar stop (S) installed and retained by plate (P). Stop is installed when "Traction-Booster" action is not desired.

1. Tubing clamp
2. Packing nut
3. Flare fitting
4. Linkage guard
5. Adjusting nut
6. Pressure tube
7. Linkage rod
8. Linkage joint
9. Stop nut
10. Drawbar stop
11. Spring housing

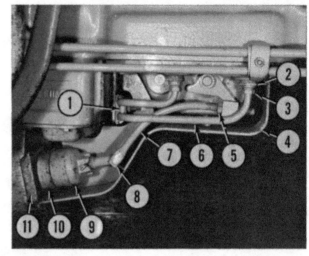

Fig. AC128 — View showing "Traction-Booster" linkage and associated hydraulic components.

turn the screw either way as required to give the proper relief pressure. If specified pressure cannot be obtained, remove and overhaul hydraulic pump.

152. "TRACTION-BOOSTER" AND "MAKE-UP" CIRCUIT PRESSURE. Operate tractor until hydraulic oil is warm. Install high pressure gage in port (4—Fig. AC125) of valve manifold. Engage "make-up" cable to tractor lift arm as shown in Fig. AC130A. With engine running at 1650 RPM, raise lift arms to top position with lift control valve; then, return control valve to hold position.

For early tractors without diverter valve (shown in Fig. AC133), gage reading should be steady at 2100-2350 psi. On later tractors (after tractor Serial No. D19-12461), gage reading should be steady at 1900-2250 psi. If not within these limits, record pressure so that proper number of shims (28—Fig. AC132 or AC133) may be added or removed. The pump must be removed as in paragraph 157 to remove or install shims. Refer to paragraph 161 when varying the number of shims.

153. TRANSPORT VALVE OPERATING PRESSURE. Operate tractor until hydraulic oil is warm. Install high pressure gage in port (5—Fig. AC125) of control valve manifold. Attach single acting remote ram to transport valve port (3) with ram piston retracted. Engage "make-up" cable to tractor lift arm as shown in Fig. AC130A. Turn transport valve knob all the way out (counter-clockwise). With engine running at slow idle speed, raise lift arms slowly with lift control valve until "make-up" control operates pump control valve, then return lift control spool to hold position. Read pressure gage while

$\frac{7}{16}$-inch plunger output is extending remote ram. Gage reading should be approximately 1225 psi. Transport valve operating pressure may be adjusted by varying the number of shims (5—Fig. AC138) between the transport valve spring (4) and spool cup (6). Addition of one shim will increase operating pressure approximately 50 psi.

154. SNAP COUPLER SPRING ADJUSTMENT. To adjust snap coupler (drawbar) spring tension, proceed as follows: Remove linkage guard (4—Fig. AC128). If oil line fittings (2 and 3) are as shown, drain hydraulic sump, loosen packing nut (2) and screw fitting (3) out of hydraulic pump body. If tractor is equipped with later type external pump connections (as shown on pump body in Fig. AC131A), do not drain hydraulic sump; disconnect tube (6—Fig. AC128) from fitting. Remove clamp (1) and pull oil tube down out of way. Push "Traction-Booster" lever to top of quadrant. Back off nut (5) until linkage is free and disconnect linkage at pivot joint (8). Remove drawbar stop (10) if installed. Remove cotter pin and backoff the adjusting nut through spring housing. Remove cotter pin and back-off the adjusting nut (14—Fig. AC129) until rod (6) and spring (17) have some end play. Retighten nut (14) until all end play of rod and spring is just removed; then, turn nut $\frac{7}{8}$-turn farther and continue until next castellation of nut is aligned with hole in rod for cotter pin. Install cotter pin. This procedure should pre-load the spring (17) about $\frac{1}{16}$ to $\frac{3}{32}$-inch.

Install drawbar stop (10—Fig. AC-

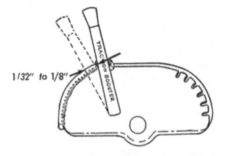

Fig. AC128A — "Traction-Booster" linkage should be adjusted so that "contact point" is felt within lever travel as indicated.

128). Install stop nut (9) and tighten nut down against drawbar stop; then, back-off nut $\frac{1}{8}$-turn and install cotter pin. This should give a dimension of approximately 61/64-inch between stop nut (9) and spring housing (11). Reinstall all parts removed to gain access to adjustment points. During reassembly, the "Traction-Booster" linkage should be adjusted as outlined in paragraph 155.

155. "TRACTION-BOOSTER" LINKAGE ADJUSTMENT. To adjust "Traction-Booster" linkage, first adjust snap coupler (drawbar) spring and stop nut as outlined in paragraph 154. Then, proceed as follows: With lift arms fully lowered and engine not running, move "Traction-Booster" lever towards top of quadrant. At a point $\frac{1}{32}$ to $\frac{1}{8}$-inch from top quadrant position (Fig. AC128A), increased tension on lever should occur due to contact of linkage with "Traction-Booster" ($\frac{7}{16}$-inch plunger) control valve. If contact point is not within the dimension given, readjust nut (5—Fig. AC128) on linkage rod (7) until

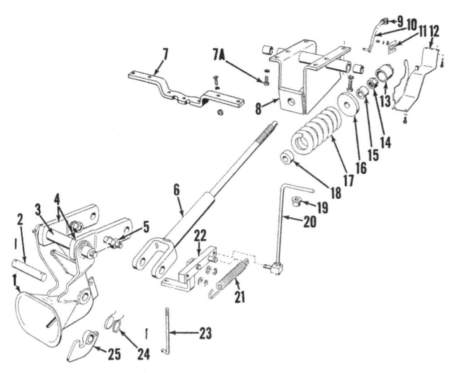

Fig. AC129 — Exploded view of "Snap Coupler" and "Traction-Booster" linkage.

1. Snap coupler bell	7A. Special cap screw
2. Pivot pin	8. Spring housing
3. Pivot shaft	9. Adjusting nuts
4. Brackets	10. Link rod
5. Pivot screw	11. Stop retaining plate
6. Drawbar spring rod	12. Linkage guard
7. Bracket	

13. Stop nut	20. Release lever rod
14. Pre-load nut	21. Spring
15. Spacer	22. Release lever
16. Washer	23. Release link
17. Drawbar spring	24. Spring
18. Pilot	25. Coupler hook
19. Guide	

Fig. AC130 — View showing left side sheet removed.

1. "Make-up" cable
2. "Make-up" control shaft
3. "Traction-Booster" control U-joint
4. "Traction-Booster" gage tube
5. Head light wire

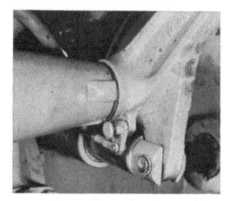

Fig. AC130A — View showing "make-up" control engaged by screwing lock pin into detent in left arm.

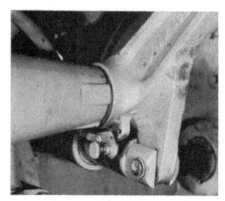

Fig. AC130B — View showing "make-up" control disengaged by screwing lock pin out as far as possible.

contact point is within proper location on "Traction-Booster" quadrant.

With engine running at low idle speed and lift arm control lever in "Traction-Booster" detent position, the lift arms should not raise with the "Traction-Booster" lever in top notch of quadrant. If lift arms raise under these conditions, readjust linkage.

156. "MAKE-UP" CONTROL LINKAGE. Due to normal leakage through transport valve to sump return line, some means must be provided to hold implements in raised position during transport. A cable attached to the control shaft for the $\frac{7}{16}$-inch plunger control valve runs back to the hydraulic lift arm shaft. When implements are to be transported some distance, the cable is hooked up to the right-hand lift arm by a lock-screw (See Figs. AC130A & AC130B) turned into a detent in the arm. Thus, when implement is raised, the control valve for the $\frac{7}{16}$-plunger is actuated and plunger will continually supply oil to the lift circuit to "make-up" for oil being lost through the transport valve. Excess oil is by-passed through

the $\frac{7}{16}$-inch plunger relief valve in pump body.

To check the "make-up" control adjustment, turn lock-screw (Fig. AC-130A) into lift arm detent. With engine at slow idle, raise lift arms about ⅔-way to top; then, continue to raise lift arms in small increments. During the last 10 degrees of arm travel, the "make-up" linkage should actuate the $\frac{7}{16}$-inch plunger control valve and lift arms should continue raising to top position automatically. If not, refer to Fig. AC130 and adjust linkage by repositioning stop on "make-up" cable (1).

PUMP

157. R&R HYDRAULIC PUMP. Drain the hydraulic reservoir (integral with "Power-Director" reservoir). Remove linkage guard (4—Fig. AC-128) and oil tubes between hydraulic pump and control valve. Push "Traction-Booster" lever to top of quadrant. Back off adjusting nut (5) and disconnect linkage at pivot joint (8). Unbolt and remove pump linkage.

Fig. AC131 — Exploded view of "Traction-Booster" and "make-up" control linkage.

1. Plug	14. U-joint
2. Bushing	15. Shaft
3. Pin	16. Lever
4. Linkage housing	17. Retaining ring
(pump cover)	18. Guide
5. Screen & gasket	19. Spring
assembly	20. Bearing
6. Lever	21. Retaining ring
7. Roller	22. Cam
8. Lever	23. Gasket
9. Bushing	24. Drain plug
10. Bushing	25. Pin
11. Seal	26. Seal
12. Lever	27. Lever
13. Pin	28. Adjusting block

Fig. AC131A—View showing torque housing with pump cover removed. Late style outlet fittings (F) are shown. Note that spring (S) is compressed and nail (N) is placed through hole in guide to facilitate removal.

B. Pump mounting	F. Late style pump
capscrews	fittings
C. Cotter pin	N. Nail (See text)
	S. Spring

158. OVERHAUL LINKAGE HOUSING (PUMP COVER) AND CONTROL LINKAGE. Refer to Fig. AC131 and proceed as follows: Drill hole through plug (1) and pry plug out. Remove lever pin (3) with bolt threaded into tapped hole in pin. Pivot pin (25) can be removed after driving roll pin out of pivot pin and cam (22). Control shafts (8 and 15) can be removed from below after roll pins attaching "make-up" lever (12) and U-joint (13) to shafts are driven out.

Renew all seals. Renew shafts and bushings where wear is evident. Reverse disassembly procedure to reassemble.

159. OVERHAUL HYDRAULIC PUMP. To disassemble and overhaul hydraulic pump, proceed as follows:

160. CAM FOLLOWERS AND PUMP PLUNGERS. Refer to Fig. AC-132. Drive out pin (1) to release cam followers (2). A washer (3) is located on the follower pin between each pair of followers. Renew the complete follower cam if rollers are pitted, worn or loose or if the cam is damaged.

After removing cam followers, withdraw the plungers (4 and 5) and the three ¾-inch plunger check balls (6), retainers (7) and return springs (9). Remove plug (20) to remove and inspect the $\frac{7}{16}$-inch plunger return spring (17) and guide (18). Renew plunger return springs if broken, rusted, corroded or if free length does not compare with that of new springs. Renew check balls if rusted or pitted and renew plungers if check ball seats are not perfect. Also renew plungers if scored, excessively worn or if they stick in bores after plungers and bores are cleaned. Minor defects can be corrected on plungers and in bores by use of crocus cloth. Wash parts thoroughly after cleaning with crocus cloth.

housing (pump cover) (4—Fig. AC-131). Compress spring (S—Fig. AC-131A) and insert nail in hole (N) in spring guide as shown. Remove cotter pin (C). Remove spring and guide assembly from linkage and plate on pump. Remove the three cap screws (B) and lower the pump from torque housing.

To reinstall pump, reverse removal procedure. After reinstalling spring and guide assembly, remove nail previously placed in guide to hold spring compressed. When reinstalling pump linkage housing, be sure that roller (7—Fig. AC131) is in relief of spring guide (18). If roller is not in this position when housing cap screws are tightened, spring guide will be damaged. After reservoir is refilled, operate system several times to bleed any trapped air out of system and refill reservoir to proper level with SAE 20-20W non-detergent motor oil.

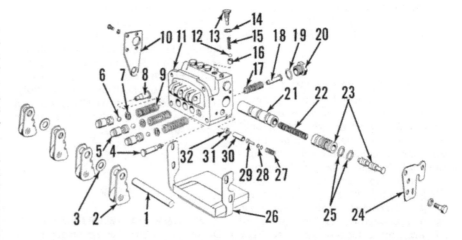

Fig. AC132—Exploded view of D-19 hydraulic pump assembly. Refer also to Fig. AC133.

1. Pin	9. Spring	17. Spring	25. "O" ring
2. Follower arm	10. End cover	18. Spring insert	26. Intake manifold
3. Washer	11. Pump body	19. "O" ring	27. Spring
4. 7/16-inch plunger	12. Check ball	20. Plug	28. Shim
5. ¾-inch plungers	13. Plug	21. Sleeve	29. Plunger
6. Check ball	14. Gasket	22. Spring	30. Sleeve
7. Retainer	15. Spring	23. Valve & sleeve	31. Relief valve
8. Spring insert	16. Seat	24. End cover	32. Seat

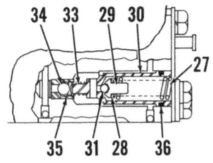

Fig. AC133—Cross sectional view of "Traction-Booster" relief valve and diverter valve used on late production pumps. Unit is installed in same location of pump body as early type (27 thru 32—Fig. AC132).

27. Relief spring	33. Diverter spring
28. Shim	34. Diverter valve ball
29. Relief plunger	35. Diverter seat
30. Sleeve	36. "O" ring
31. Relief valve ball	

Fig. AC134 — View of assembled pump showing manifold installed.

1. "Traction-Booster" pressure line
2. Lift circuit pressure line
3. Sump return line

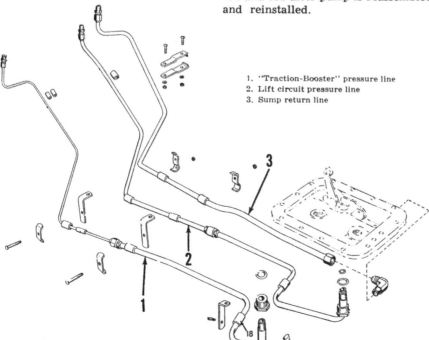

Fig. AC135—D-19 hydraulic lines. Old style flare fittings are shown on pump end of pressure lines.

161. **"TRACTION-BOOSTER" CONTROL AND RELIEF VALVES.** After removing end covers (10 and 24)—Fig. AC132), the control valve and bushing (23) and relief valve assembly (items 27 through 32) for the "Traction-Booster" circuit may be removed from pump body. Also, control valve return spring (22) and retainer (21) may be removed at this time.

Add shims (28—Fig. AC132 or AC-133) to increase "Traction-Booster" circuit pressure if pressure tested low prior to removing pump. (See paragraph 152.) On early type (prior to pump serial number 20001), addition of one shim (28—Fig. AS132) will raise pressure approximately 86 psi. No more than four shims should be installed on early type pump. On late production pumps, addition of one shim (28—Fig. AC133) will change pressure approximately 215 psi. No more than eight shims should be installed on late type pump.

If relief valve spring (27—Fig. AC-132 or AC133) is renewed, no shims should be added at this time. Free length of spring should be 1⅛ inches. When compressed to ⅞-inch, the spring should exert 13.3 to 14.7 pounds pressure. If spring does not meet these specifications, it should be renewed. If spring tests O.K. and circuit relief valve operating pressure tested low, install additional shims as required between spring and plunger (29). Recheck pressure per paragraphs 151 and 152 after pump is reassembled and reinstalled.

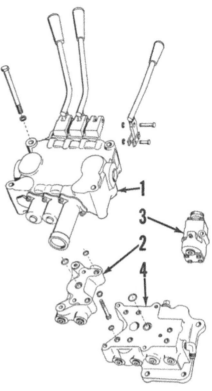

Fig. AC136—Three-spool control valve and associated parts.

1. Control valve
2. Manifold (used with 3-spool valve only)
3. Transport valve
4. Manifold (for one or three-spool valve)

162. **PUMP DISCHARGE VALVES.** The check (discharge) valves can be removed after removing the plugs (13—Fig. AC132). Seats (16) for the valves should not be removed unless known to be defective as they cannot be reused after being removed. A leaky valve can often be corrected by seating a new ball to the seat by tapping the ball with a soft drift and light hammer. Valve seat inserts are removed by threading the I.D. of the insert to permit use of a screw puller.

CONTROL VALVE

The D-19 tractor may be equipped with a single-spool or a three-spool Husco control valve. Either valve will be mounted on a manifold that is bolted to the rear top transmission cover. Cross-section view of the three-spool valve is shown in Fig. AC137. The one-spool valve is similar but has only the special spool (4).

163. **R&R CONTROL VALVE.** Be sure lift arms are fully lowered and that no pressure is exerted on any attached remote cylinder. On single-spool valve, unhook pump pressure

TRANSPORT VALVE

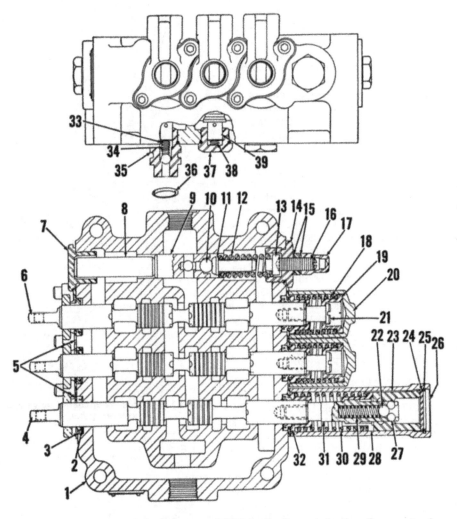

Fig. AC137—Cut-away view of three-spool D-19 hydraulic control valve. One-spool valve has special "Traction-Booster" action spool (4) only.

1. Valve housing	11. Plunger	21. Cap screw	31. Pin
2. Seal support rings	12. Spring	22. Steel ball, 11/32	32. Spring seat
3. Seals	13. Cap	23. Sleeve	33. Spring
4. Special spool	14. Plug	24. Plug	34. Pin
5. Seal plates	15. Copper washer	25. Snap ring	35. Fitting
6. Remote spools	16. Set screw	26. Cap	36. "O" ring
7. Plug	17. Nut	27. Steel ball, ¼	37. Plug
8. Spacer	18. Spring	28. Spring seat	38. Spring
9. Seat	19. Spring seat	29. Spring	39. Poppet
10. Ball, relief valve	20. Cap	30. Spring	

line from valve and unbolt and remove valve from manifold. On three-spool valve, unbolt and remove valve from manifolds. (The tail light bracket is retained by the two rear valve retaining capscrews.) To remove valve manifolds, disconnect all hydraulic lines; then, unbolt and remove manifold from transmission rear top cover.

164. OVERHAUL CONTROL VALVE. Valve spool front seals may be renewed after removing lever and lever bracket from valve body. Rear seals may be renewed after removing caps from rear of valve. Install seals with cup to inside of valve body. It is not necessary to remove

valve spools to renew either front or rear seals. Valve spool centering springs and/or detent mechanism may be inspected or renewed when rear spool caps are removed.

Valve spools and body are not serviced except as a complete valve assembly. As all renewable parts may be removed and reinstalled with valve spools remaining in valve body, it is recommended that spools not be removed unless necessary. Be sure not to mix spools in valve body bores. Unless valve body is cracked or broken, valve assembly may be returned to manufacturer (See nameplate on valve) for repairs.

165. VALVE OPERATION. Refer to Fig. AC138. When transport valve control knob (13) is turned all the way out, spring (4) moves valve spool (7) to close off remote ram outlet of transport valve. When pressure in hydraulic circuit is sufficient to move the valve spool against spring pressure, spool will move back and open remote ram outlet port to hydraulic circuit. Valve spool should operate when hydraulic pressure reaches approximately 1225 psi. Transport valve spool operating pressure is adjusted by shims (5). When valve control knob is turned all the way in, outlet port of valve is opened to lift circuit at all times (valve spool is held in open position).

Transport valve operating pressure may be checked as outlined in paragraph 153.

The transport valve is used to permit "Traction-Booster" action when using certain specially designed pull-type implements with remote cylinder operated transport wheels. The remote cylinder hose is attached to the outlet port of the transport valve and the valve control knob is turned all the way out (knob should be turned all the way in for all other applications). After the implement working depth has been established while leaving the "Traction-Booster" lever in its lowest position on quadrant, lever is then raised on quadrant until "Traction-Booster" gage registers in first half of dial.

As long as pressure in hydraulic lift circuit remains below 1225 psi, the remote outlet of the transport valve is closed off and pressure on lift arm rams will transfer implement weight to tractor. When lift arm control valve is placed in lift position, hydraulic pressure will build up above 1225 psi; spool in transport valve will move to open the lift circuit to outlet port and oil will flow to remote ram and raise implement on transport wheels.

166. R&R AND OVERHAUL TRANSPORT VALVE. To remove transport valve, first disconnect any remote cylinder line from outlet port of valve and be sure tractor lift arms are fully lowered. Then unbolt and remove valve from hydraulic manifold.

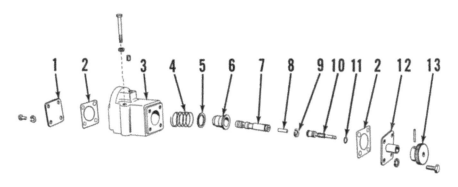

Fig. AC138—Exploded view of transport valve.

1. Front cover	5. Shims	8. Spool plunger	11. "O" ring
2. Gaskets	6. Spring cup	9. Snap ring	12. Rear cover
3. Body	7. Valve spool	10. Adjusting screw	13. Adjusting knob
4. Spring			

HYDRAULIC RAMS

167. **LIFT ARM (ROCKSHAFT) RAMS.** To remove one or both rams, proceed as follows: Be sure that lift arms are fully lowered to expel all oil from rams. Disconnect hoses to rams. Remove pins retaining ram pistons to lift arms and ram cylinders to tractor axle housings. Then, remove rams from tractor.

To disassemble rams, remove plug (5—Fig. AC139) and position ram so that the 24 steel balls (4) will roll out of hole (3). Ram must be fully retracted so that groove (G) of piston is aligned with hole. Remove scraper (8), "O" ring (7) and leather back-up rings (6) from cylinder and discard same.

Install two new leather back-up rings (6) with the "O" ring (7) between them. Clean all rust and burrs from outer end of piston, lubricate piston and install piston in cylinder. Install scraper (8) with sharp edge out after piston is installed. Push piston fully into cylinder and insert the 24 steel balls through hole in cylinder into groove in piston; then, install plug in hole using new gasket washer.

Reinstall rams on tractor in reverse of removal procedure.

An exploded view of the transport valve is shown in Fig. AC138. Disassembly procedures are evident from inspection of unit and reference to exploded view. Valve spool is a close tolerance fit to bore in valve housing and plunger is also a close tolerance fit to bore in valve spool; however, the parts should work freely without binding or sticking. Valve housing, spool and plunger are not available as separate items; therefore, if any of these parts are not serviceable, complete valve assembly must be renewed. If valve housing, spool and plunger appear serviceable, renew "O" ring seals on plunger and adjusting stem. Install extra shims if needed to bring valve operating pressure up to 1225 psi. (Addition of one shim will raise operating pressure approximately 50 psi.) Reassemble valve using new gasket and reinstall valve on tractor hydraulic manifold. Recheck the valve operating pressure as outlined in paragraph 153.

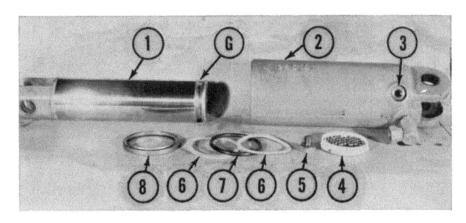

Fig. AC139—Exploded view of D-19 lift ram (cylinder).

G. Piston groove	3. Plugged hole	6. Leather rings (2)
1. Piston	4. Retaining balls (24)	7. "O" ring
2. Cylinder	5. Plug	8. Scraper

NOTES

ALLIS-CHALMERS

Models ■ 180 ■ 185 ■ 190 ■ 190XT
■ 200 ■ 7000

Previously contained in I&T Shop Service Manual No. AC-33

SHOP MANUAL

ALLIS-CHALMERS

MODELS 180, 185, 190, 190XT, 200, 7000

INDEX (By Starting Paragraph)

CONDENSED SERVICE DATA

	180, 185 & 190		One-Ninety XT			200,
	Gasoline	Diesel	Gasoline	LP-Gas	Diesel	7000
GENERAL						
Engine Make	Own	Own	Own	Own	Own	Own
Engine Model	G2500	2800	G2800XT	G2800LPXT	2900XT	2900
Number of Cylinders	6	6	6	6	6	6
Bore—Inches	3¾	3-7/8	3-7/8	3-7/8	3-7/8	3-7/8
Stroke—Inches	4	4¼	4¼	4¼	4¼	4¼
Displacement—Cubic Inches	265	301	301	301	301	301
Main Bearings, Number of	7	7	7	7	7	7
Cylinder Sleeves	Wet	Wet	Wet	Wet	Wet	Wet
Generator and Starter Make	Delco-Remy	Delco-Remy	Delco-Remy	Delco-Remy	Delco-Remy	Delco-Remy#

#Some models are equipped with a Lucas starter.

	180, 185 & 190 Gasoline	Diesel	One-Ninety XT Gasoline	LP-Gas	Diesel	200, 7000
TUNE-UP						
Firing Order	1-5-3-6-2-4	1-5-3-6-2-4	1-5-3-6-2-4	1-5-3-6-2-4	1-5-3-6-2-4	1-5-3-6-2-4
Valve Tappet Gap (Hot) Intake	.020	.015	.020	.020	.015	.015
Valve Tappet Gap (Hot) Exhaust	.025	.015	.025	.025	.015	.015
Inlet Valve Seat Angle (Degrees)	30	30	30	30	30	30
Exhaust Valve Seat Angle (Degrees)	45	45	45	45	45	45
Ignition Distributor Make	D-R	...	D-R	D-R	...	...
Breaker Gap	.016	...	.016	.016	...	...
Ignition Timing at 2200 rpm	26° BTDC	...	26° BTDC	26-28° BTDC	...	...
Timing Mark Location	Crankshaft Pulley		—————— Crankshaft Pulley ——————			
Spark Plug—						
A-C	45XL	...	45XL	42XL	...	...
Autolite	AG-3A	...	AG-3A	...	...	...
Champion	N-6	...	N-6	N-3	...	...
Electrode Gap	.025	...	.025	.020	...	...
Injection Timing						
190, 190XT and 200	...	28° BTDC	...	...	26° BTDC	24° BTDC##
180, 185 and 7000	...	26° BTDC	...	...	...	18° BTDC
Injection Pump Make	...	Roosa-Master	...	...	Roosa-Master	Roosa-Master
Carburetor Make	Holley	...	Holley	Ensign	...	...
Carburetor Model	1970	...	1970	CBX	...	...
Battery Terminal, Grounded	Neg.	Neg.	Neg.	Neg.	Neg.	Neg.
Engine Low Idle rpm	425†	675	425	425	675	775
Engine High Idle rpm, No Load	2450†	2400††	2450	2450	2400	2400
Engine Full Load rpm	2200†††	2200†††	2200	2200	2200	2200

†Refer to paragraph 76 for 180 non-diesel. ††Refer to paragraph 72 for 180 diesel.
†††Full load rpm should be 2000 for all 180 models. ##Model 200 after serial no. 2D-69515 same as 7000.

SIZES—CAPACITIES—CLEARANCES
(Clearances in Thousandths)

	180, 185 & 190 Gasoline	Diesel	One-Ninety XT Gasoline	LP-Gas	Diesel	200, 7000
Crankshaft Main Journal Diameter	——— 2.7475 ———		——————— 2.7475 ———————			
Crankpin Diameter	——— 2.373 ———		——————— 2.373 ———————			
Camshaft Journal Diameter, All	——— 2.1305 ———		——————— 2.1305 ———————			
Piston Pin Diameter	——— 1.2516 ———		——————— 1.2516 ———————			
Valve Stem Diameter, Inlet	——— 0.3717 ———		——————— 0.3717 ———————			
Exhaust	——— 0.3707 ———		——————— 0.3707 ———————			
Main Bearing Diametral Clearance	——— 1.6-4.3 ———		——————— 1.6-4.3 ———————			
Rod Bearing Diametral Clearance	——— 0.9-3.4 ———		——————— 0.9-3.4 ———————			
Piston Skirt Diametral Clearance	1.5-4.0	3.5-6.0	1.5-4.0	1.5-4.0	3.5-6.0	4.5-7.0
Camshaft End Play	——— 4-9 ———		——————— 4-9 ———————			
Camshaft Bearings Diametral Clearance	——— 2-6 ———		——————— 2-5 ———————			
Camshaft End Play	——— 3-9 ———		——————— 3-9 ———————			
Cooling System Capacity, Quarts	——— 18 ———		——————— 18* ———————			
Crankcase Oil, Quarts	——— 8-12** ———		——————— 8-12** ———————			
Power-Director and Hydraulic System, Quarts	——— 24-36** ———		——————— 24-36** ———————		26-28**	
Transmission, Quarts	——— 27-32** ———		——————— 27-32** ———————		32-38**	
Final Drive, Quarts (each)	——— 8*** ———		——————— 8 ———————			

*Capacity 22.5 qts. on 7000.
**Approximate capacity.
***Capacity 1 quart on 180 and 185 models.

CONDENSED SERVICE DATA CONT.

**TIGHTENING TORQUES—
FOOT-POUNDS**

General Recommendations	See End of Shop Manual			
Connecting Rod Screws	——40-45——	——40-45——		
Cylinder Head Screws	——150——	——150——		
Flywheel Screws	——68-73——	——68-73——		
Injection Nozzel Nuts	...	9-12	...	...	9-12	11-13
Main Bearing Screws	——130-140——	——130-140——		

FRONT SYSTEM

SINGLE WHEEL TRICYCLE

1. WHEEL ASSEMBLY. The single front wheel assembly may be removed after raising front of tractor and removing bolt (3—Fig. 1) at each end of wheel axle (1).

To renew bearings and/or seals, first remove wheel assembly; then unbolt and remove bearing retainer (10—Fig. 2), seal (4), seal retainer (5) and shims (9). Drive or press on opposite end to remove axle (8), bearing cones (7) and bearing cup from retainer side of hub. Drive the remaining seal and bearing cup out of hub. Remove bearing cones from spindle.

Soak new felt seals in oil prior to installing seals and seal retainers. Drive bearing cup into hub until cup is firmly seated. Drive bearing cones tightly against shoulders on axle. Pack bearings with No. 2 wheel bearing grease. Install axle and bearing cup in against cone. When installing bearing retainer, vary number of shims (9—Fig. 2) to give bearings a free rolling fit with no end play.

Front wheel bearings should be repacked with No. 2 wheel bearing grease after each 500 hours of use.

If necessary to renew wheel hub or repair tire, deflate tire before unbolting

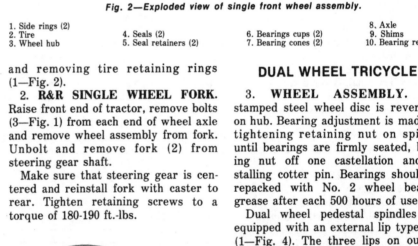

Fig. 2—Exploded view of single front wheel assembly.

1. Side rings (2)		8. Axle
2. Tire	4. Seals (2)	6. Bearings cups (2)	9. Shims
3. Wheel hub	5. Seal retainers (2)	7. Bearing cones (2)	10. Bearing retainer

and removing tire retaining rings (1—Fig. 2).

2. R&R SINGLE WHEEL FORK. Raise front end of tractor, remove bolts (3—Fig. 1) from each end of wheel axle and remove wheel assembly from fork. Unbolt and remove fork (2) from steering gear shaft.

Make sure that steering gear is centered and reinstall fork with caster to rear. Tighten retaining screws to a torque of 180-190 ft.-lbs.

DUAL WHEEL TRICYCLE

3. WHEEL ASSEMBLY. The stamped steel wheel disc is reversible on hub. Bearing adjustment is made by tightening retaining nut on spindle until bearings are firmly seated, backing nut off one castellation and installing cotter pin. Bearings should be repacked with No. 2 wheel bearing grease after each 500 hours of use.

Dual wheel pedestal spindles are equipped with an external lip type seal (1—Fig. 4). The three lips on outside diameter of seal contact a steel wear sleeve that is pressed into the front wheel hub. Drive wear sleeve into hub with crimped edge of wear sleeve

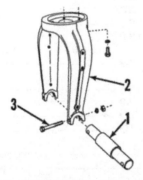

Fig. 1—Exploded view of the single front wheel fork and associated parts.

1. Axle
2. Fork
3. Bolts (2)

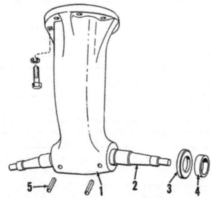

Fig. 3—Dual wheel tricycle spindles (2) can be removed after removing groove pins (5).

1. Pedestal
2. Spindle
3. Shield
4. Bearing spacer
5. Groove pin

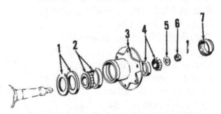

Fig. 4—Exploded view of wheel hub and bearing used on dual wheel tricycle and wide front axle models.

1. Wear sleeve and seal	4. Outer bearing
2. Inner bearing	5. Washer
3. Wheel hub	6. Nut
	7. Hub cap

towards bearing until sleeve is ¼ inch below flush with hub.

4. R&R PEDESTAL. Raise front of tractor, then remove cap screws retaining pedestal (1—Fig. 3) to steering gear shaft.

When reinstalling, make certain that the steering gear shaft is centered and install pedestal with wheels in straight ahead position and caster to rear. Tighten retaining screws to 130-140 ft.-lbs. torque.

ADJUSTABLE FRONT AXLE

5. WHEEL ASSEMBLY. Refer to Fig. 4. Stamped wheel disc is reversible on hub. Bearing adjustment is made by tightening retaining nut on spindle until bearings are firmly seated, backing nut off one castellation and installing cotter pin. Bearings should be repacked with No. 2 wheel bearing grease after each 500 hours of use.

Crimped edge of seal shell should be towards shoulder on spindle when renewing seal (lettered side of seal shell towards bearing).

6. AXLE ADJUSTMENTS. The adjustable front axle provides wheel tread widths of 62-86 inches on 7000 models and 60-84 inches on all other models. Front wheels may be mounted inside out on some models to increase tread width.

Tie rods are grooved or threaded to facilitate toe-in adjustment; however, it may be advisable to measure front wheel toe-in and adjust to 1/16 to 1/8 inch. Tighten bolts in tie rod clamps securely.

7. AXLE CENTER (MAIN) MEMBER AND RADIUS ROD ASSEMBLY. Support front end of tractor and remove bolts through center member and axle extensions (spindle supports). Loosen or remove bolts in tie rod clamps and slide axle extension (7—Fig. 5 or 5A) with front wheels and spindles out of main member and tie rod shafts out of tie rod tubes. Remove external power steering cylinder on models so equipped. On all models, place floor jack under center member and remove pivot pin (2—Fig. 5A) on 7000 models or rear pivot screw (4—Fig. 5) and front pivot pin (2) on all other models. Lower the axle until it clears the pivot brackets, then pull center member out from under tractor.

Bushing (3—Fig. 5) on all models except 7000 is pre-sized and if carefully installed should not need reaming. Pivot pin (2) should have 0.012-0.015 clearance in bushing (3) and 0.005-0.009 clearance in front support bore.

Reverse removal procedure to install new center member.

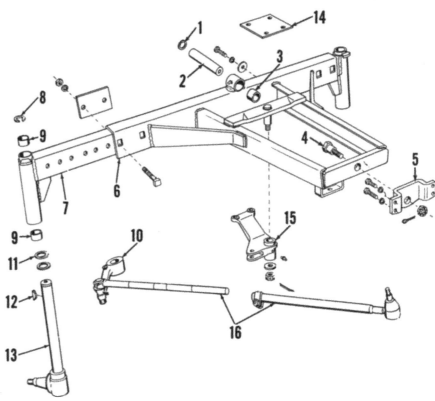

Fig. 5—Exploded view of adjustable front axle assembly used on all models except 7000. Steering arm (14) is used with internal cylinder type front support (Fig. 12). Steering arm (15) is used with external cylinder type steering (Fig. 11).

1. Snap ring	4. Rear pivot screw	6. Axle main member	10. Spindle arm
2. Axle front pivot pin	5. Rear pivot bracket	7. Axle extension	11. Thrust washers
3. Bushing		8. Snap rings	12. Key
		9. Spindle bushings	13. Spindle
			16. Tie rod

8. SPINDLE BUSHINGS. Support front end of tractor and remove front wheels. Remove snap ring (8—Fig. 5 or 5A), spindle arm (10) and Woodruff key (12). Spindle can then be removed from bottom of axle extension (spindle support). Drive bushings (9) from top and bottom of spindle bore and remove thrust washers (11) from spindles.

New spindle bushings are pre-sized and should not require reaming. Press or drive bushings into support flush with ends of spindle bore. Install two thrust washers on spindle and reassemble.

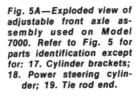

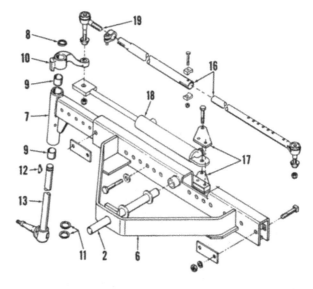

Fig. 5A—Exploded view of adjustable front axle assembly used on Model 7000. Refer to Fig. 5 for parts identification except for: 17. Cylinder brackets; 18. Power steering cylinder; 19. Tie rod end.

FRONT SUPPORT

Models With External Steering Cylinder

9. **REMOVE AND REINSTALL.** To remove front support casting from adjustable front axle models with an external steering cylinder, proceed as follows: Remove hood, grille and hood side panels. Drain cooling system and remove air cleaner and air cleaner tube. Disconnect hoses and remove radiator. Disconnect steering cylinder. Remove radius rod pivot pin. (4—Fig. 5) on all models except 7000 models. Support tractor under torque housing so there is no weight on front wheels. Attach a hoist to front support, remove cap screws attaching front support, to side rails and roll front support and axle away from tractor.

NOTE: If front support casting binds between the side rails, loosen the bolts attaching side rails to the engine front support plate. Front support casting can be separated from axle assembly after removing pivot pin.

When reinstalling, tighten the radius rod pivot screw (4—Fig. 5) on all models except 7000 models to at least 150 ft.-lbs. Tighten cap screws attaching side rails to the engine front support plate to 70-75 ft.-lbs. On tractors prior to Serial No. 190-2807, tighten the eight 5/8-inch screws attaching side rails to front support to 130-140 ft.-lbs torque. On tractors after Serial No. 190-2806, and all model 200 tractors, tighten side rail to front support ¾-inch cap screws to 310-320 ft.-lbs. torque.

Models With Internal (Dual) Steering Cylinders

10. **REMOVE AND REINSTALL.** The front support casting (5—Fig. 12) on all tricycle models and some adjustable axle models is used as the steering cylinder. To remove the front support, proceed as follows: Remove hood, grille and hood side panels. Drain cooling system and remove air cleaner and air cleaner tube. Disconnect hoses and remove radiator. Disconnect the steering hydraulic lines from "Tee" fittings on front support. On adjustable front axle models, remove radius rod pivot pin (4—Fig. 5). On all models, support tractor under torque housing so there is no weight on front wheels. Attach a hoist to the front support, remove cap screws attaching front support to side rails and roll the front support and wheels away from tractor.

NOTE: If front support casting binds between side rails, loosen bolts attaching side rails to engine front support plate.

When reinstalling, tighten cap screws attaching side rails to engine front support plate to a torque of 70-75 ft.-lbs. On tractors prior to Serial No. 190-2807, tighten the eight (5/8-inch) cap screws attaching side rails to front support to 130-140 ft.-lbs of torque. On tractors after Serial No. 190-2806, and all model 200 tractors, tighten side rail to front support (¾-inch) cap screws to 310-320 ft.-lbs. torque. On adjustable front axle models, radius rod pivot pin should be tightened to at least 150 ft.-lbs. torque.

STEERING SYSTEM

All models are equipped with hydrostatic steering system that has no mechanical linkage between the steering wheel and front axle. The control valve unit (Fig. 9) contains a rotary metering motor, a commutator feed valve sleeve and a selector valve spool. In the event of engine or hydraulic power failure, the metering motor becomes a rotary pump to actuate the power steering cylinder when the steering wheel is turned. A check valve in the control valve housing allows recirculation of fluid within the control valve and steering cylinder during manual operation.

Power for the steering system is supplied by a gear type pump that is driven from the engine camshaft gear. The power steering pump also supplies hydraulic pressure to "Power-Director" clutch and hydraulic pto clutch systems.

TROUBLE SHOOTING

11. Before attempting to adjust or repair the power steering system, the cause of any malfunction should be located. Refer to the following paragraphs for possible causes of power steering system malfunction:

Irregular or "Sticky" steering. If irregular or "sticky" feeling is noted when turning the steering wheel with forward motion of tractor stopped and with engine running at rated speed, or if steering wheel continues to rotate after being turned and released, foreign material in the power steering fluid is the probable cause of trouble. Clean the magnetic oil filter, renew throwaway type oil filter, drain "Power-Director" compartment (hydraulic sump) and refill with clean oil. If trouble is not corrected, the power steering valve assembly should be removed and serviced; refer to paragraph 15.

Steering Cylinder "Hesitates". If steering cylinder appears to pause in travel when steering wheel is being turned steadily, probable cause of trouble is air trapped in the power steering cylinder. Bleed the cylinder as outlined in paragraph 12.

Slow Steering. Slow steering may be caused by low oil flow from pump. Check time required for full stroke travel of power steering cylinder; first with tractor on the front wheels, then with front end of tractor supported by a jack. If time between the two checks varies considerably, overhaul the power steering pump as outlined in paragraph 186.

Loss of Power. Loss of steering power may be caused by system pressure too low. Check and adjust relief valve setting as outlined in paragraph 13, 13A or 13B.

LUBRICATION AND BLEEDING

12. The "Power-Director" compartment of the torque housing is utilized as a reservoir for the hydraulic systems including power steering. Capacity is approximately 28 quarts for 190, 190XT, 200 and 7000 models, 24 quarts for 180 and 185 models. Oil level on all models except 7000 should be maintained at mark on filler cap dipstick located to left and forward of gear shift. On 190, 190XT and 200 models, it is necessary to remove rear section of gear shift linkage cover for access to dipstick. Oil level on 7000 models should be maintained between two sight glass gages on right side of torque housing. Allis-Chalmers "Power Fluid" should be used.

The power steering system is usually self-bleeding. With engine running at high idle speed, cycle the system through several full strokes of the cylinder. In some cases, it may be necessary to loosen connections at cylinder to bleed trapped air.

TESTS AND ADJUSTMENTS

190 and 190XT Models

13. The "Power-Director" clutch and (on models so equipped) the pto clutch are actuated by pressure from the power steering system. Before checking the power steering relief pressure, refer to paragraphs 91 and 155 to check the "Power-Director" and pto pressures.

A pressure and volume (flow) test of the power steering system will disclose whether the pump or relief valve is malfunctioning. To make such a test, proceed as follows:

Remove hood right side panel and disconnect steering system return

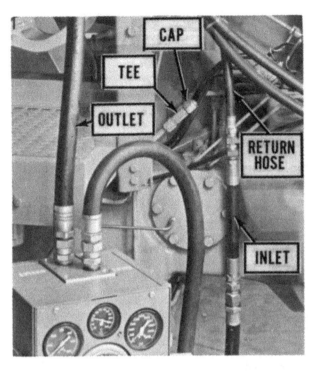

Fig. 6—View of OTC pressure-volume tester connected for checking power steering system on 190 and 190XT tractors. Open end of "Tee" fitting must be capped before engine is started.

To check volume, open the valve on tester until pressure is approximately 1200 psi and observe the volume of flow. With engine operating at 1600 rpm, volume should be 5.25 gpm. If volume of flow is incorrect, the flow control valve (4) or pump should be serviced as outlined in paragraph 186.

180 and 185 Models

13A. The "Power-Director" clutch and (on models so equipped) the pto clutch are actuated by pressure from the power steering system. Before checking the power steering relief pressure, refer to paragraphs 91 and 157 to check the "Power-Director" and pto pressures.

A pressure and volume (flow) test of the power steering system will disclose whether the pump or relief valve is malfunctioning. To make such a test, proceed as follows:

Remove cover from below instrument panel and disconnect steering system return tube from "tee" fitting shown in Fig. 7. Install a cap over open end of "Tee" fitting. Connect a pressure-volume tester inlet hose to steering system return tube as shown. Position outlet hose from tester in the torque housing filler opening so that oil will be returned to hydraulic reservoir. Start the engine and adjust speed to rated rpm (2000 rpm on 180 models; 2200 rpm on 185 models). Close valve on tester until pressure is 1200 psi and allow oil temperature to reach 120-135 degrees F.

Close the valve on tester completely and check pressure gage. If relief pressure is not 1500-1600 psi, remove and disassemble the relief valve (Fig. 7). Add or deduct shims (5—Fig. 7A) to set pressure. Addition of one shim

hose from "Tee" fitting shown in Fig. 6. Install a cap over open end of "tee"

fitting. Connect a pressure-volume tester inlet hose to steering system return hose as shown. Position outlet hose from tester in the "Power-Director" filler opening so that oil will be returned to the reservoir. Make certain that "Power-Director" control lever is in neutral and pto control is in neutral (not brake position). Start engine and and adjust speed to 1600 rpm. Close valve on tester until pressure is 1200 psi and allow oil temperature to reach 100 degrees F.

Close the valve on tester completely and check pressure gage. If relief pressure is not 1500 psi, remove plug (11—Fig. 6A) and add or deduct shims (9) to set pressure.

Fig. 6A—Exploded view of hydraulic pump manifold used on 190 models showing power steering flow control valve (4) and pressure relief valve (7).

1. Plug	
2. "O" ring	
3. Spring	7. Pressure relief
4. Flow control	valve
valve	8. Spring
5. Pump manifold	9. Shims
6. Bushing	10 Gasket
	11 Plug

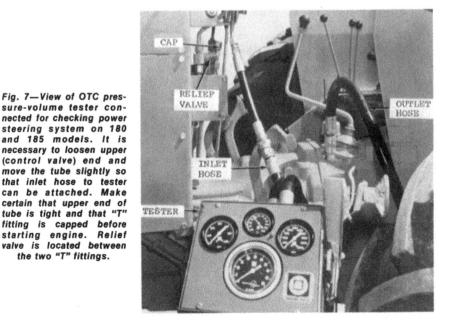

Fig. 7—View of OTC pressure-volume tester connected for checking power steering system on 180 and 185 models. It is necessary to loosen upper (control valve) end and move the tube slightly so that inlet hose to tester can be attached. Make certain that upper end of tube is tight and that "T" fitting is capped before starting engine. Relief valve is located between the two "T" fittings.

should raise pressure approximately 50 psi.

To check pump volume, open the valve on tester until pressure is approximately 200-300 psi below relief valve setting and observe volume of flow. With engine operating at rated speed (2000 rpm on 180 models ; 2200 rpm on 185 models), pump volume should be approximately 5.8 gpm. If volume is too low, the pump should be serviced as outlined in paragraph 189.

200 And 7000 Models

13B. The pto clutch (on models so equipped) is operated by return or by-pass oil from the power steering system. Pto clutch operating pressure must be tested as outlined in paragraphs 91 and 155 on model 200 or paragraphs 99 and 159 on Model 7000 before checking power steering relief pressure.

Power steering relief valve pressure can be checked by installing a high pressure (3000 psi) hydraulic gage in port (P—Fig. 8) after removing the plug. Turn the steering wheel briefly to one end of its travel and observe pressure. If tractor is **not** equipped with pto, power steering relief pressure should be 1575-1725 psi. If tractor is equipped with pto, relief pressure

should be 1400 psi on 7000 models and 1865-2015 psi on Model 200. Power steering relief pressure can be adjusted at screw (S) on early 200 models after removing cap nut (C). Relief pressure on 7000 and later 200 models is adjusted using shims.

The flow divider located in pump rear cover first supplies oil through the oil cooler to the power steering valve. Rate of flow is 4.25 gpm on Model 200 serial numbers prior to 200-4001, 6 gpm on Model 200 serial numbers 200-4001 and up and 8 gpm on Model 7000. A hydraulic flo-rater can be attached to port (P—Fig. 8) with outlet from flo-rater routed to sump filler opening. With engine operating at approximately 1600 rpm, first check relief pressure. Refer to preceding paragraph for recommended settings. Open flo-rater valve and turn steering wheel to full left or right and note gage reading. Volume should be 4.25 gpm for Model 200 serial numbers prior to 200-4001, 6 gpm for model 200 serial numbers 200-4001 and up and 8 gpm for Model 7000 with engine speed between 800 and 2200 rpm.

Caution: Use of flo-rater with steering in full left or right position disrupts oil circuit flow and should not be continued for extended periods.

The pump and/or flow divider valve should be serviced if volume of flow is incorrect.

POWER STEERING PUMP

14. A gear type pump driven from the engine camshaft gear is used as the power source for the steering system. The power steering pump is an integral part of the hydraulic system pump and the "Traction Booster" pump. For service information on pump, refer to paragraph 184.

CONTROL VALVE

15. **REMOVE AND REINSTALL.** On 190, 190XT, 200 and 7000 models, remove hood. On 180 and 185 models, remove cover from below instrument panel. On all models, disconnect hoses (or tubes) from control valve, then unbolt and remove control valve from the steering shaft housing.

Reinstall the control valve assembly by reversing the removal procedure. Refer to Fig. 10 for proper hose (or tube) locations. Install new "O" ring seals on fittings before connecting to control valve and tighten fittings securely. Bleed trapped air from the power steering cylinder as outlined in paragraph 12 after assembly is completed.

16. **OVERHAUL.** After removing control valve assembly, clean valve thoroughly and remove paint from points of separation with wire brush.

NOTE: Make certain that work area is clean and use only lint free paper shop towels for cleaning valve parts.

If oil leakage past seal (1—Fig. 9) is the only difficulty, a new seal can be installed after removing only the

Fig. 7A—Exploded view of the power steer-ing relief valve used on 180 and 185 models. Refer to Fig. 7 for location.

1. Relief valve
2. "O" ring
3. Ball
4. Spring
5. Shims

Fig. 8—Power steering pressure can be checked at port (P) on 200 and 7000 tractors. Relief valve is located on engine side of pump. Adjusting screw (S) and cap nut (C) are used on early 200 models.

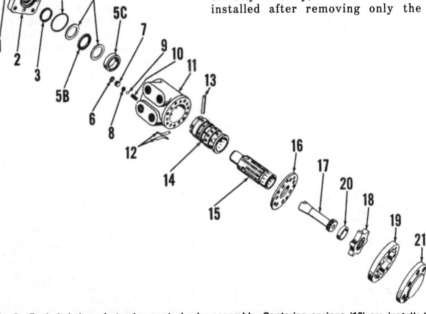

Fig. 9—Exploded view of steering control valve assembly. Centering springs (12) are installed in two groups of three springs with arch in each group back-to-back.

1. Oil seal
2. Mounting plate
3. Quad ring seal
4. "O"ring
5A. Bearing races
5B. Thrust bearing
5C. Locator bushing
6. "O" ring
7. Plug
8. Check seat
9. Check valve
10. Valve spring
11. Control valve body
12. Centering springs (6)
13. Centering pin
14. Sleeve
15. Valve spool
16. Plate
17. Drive shaft
18. Rotor
19. Ring
20. Spacer
21. Cover (cap)

mounting plate (2). Remove plug (7) from valve body using a bent wire inserted through out port (Fig. 10). Install new "O" ring (6—Fig. 9) on plug, lubricate plug and "O" ring with oil and reinstall in valve body. Install new quad seal (3) and "O" ring (4) in mounting plate and tighten retaining cap screws equally to a torque of 250 inch-pounds.

To completely disassemble and overhaul the control valve, refer to exploded view in Fig. 9 and proceed as follows: Clamp valve in vise with end cap (21) up and remove end cap retaining screws. Remove end cap (21), gerotor set (18 & 19), plate (16) and drive shaft (17) from valve body as a unit. Remove valve from vise, place a clean wood block in vise throat and set valve assembly on block with mounting plate (2) end up. Lightly clamp vise against port face of valve body and remove mounting plate retaining screws. Hold spool assembly down against wood block while removing mounting plate. Remove valve body from vise and place on work bench with port face down. Carefully remove spool and sleeve assembly from 14-hole end of valve body. Use bent wire inserted through outlet port to remove check valve plug (7) and use 3/16-inch Allen wrench to remove check valve seat (8). Then remove check valve ¼-inch steel ball (9) and spring (10) from valve body.

Remove centering pin (13) from spool and sleeve assembly, and push spool (splined end first) out of sleeve. Remove the six centering springs (12) from slot in spool.

Separate the end cap and plate from the gerotor set and remove the drive shaft, rotor and spacer.

Inspect all moving parts for scoring. Slightly scored parts can be cleaned by hand rubbing with 400 grit abrasive paper. To recondition gerotor section surfaces place a sheet of 600 grit paper on plate, glass, lapping plate or other absolutely flat surface. Stroke each surface of the gerotor section over the abrasive; any small bright areas indicate burrs that must be removed. Polish each part, rinse in clean solvent and air dry; keep these parts absolutely clean for reassembly.

Renew all parts that are excessively worn, scored or otherwise damaged. Install new seal kit when reassembling.

To reassemble valve, proceed as follows: Install check valve spring (10) with small end out. Drop check valve ball (9) on spring and install valve seat (8) with counterbored side towards valve ball. Tighten seat to a torque of 150 inch-pounds. Lubricate valve spool and carefully insert spool (15) in valve sleeve (14) using a twisting motion. Stand the spool and sleeve assembly on end with spring slots up and aligned. Assemble centering springs (12) in two groups with extended edges of springs down. Place the two groups of springs back-to-back (arched sections together) and install the springs into the spring slots in sleeve and spool in this position. Use a small screwdriver to guide springs through slots in opposite side of assembly. Center the springs with edges flush with upper surface of sleeve. Insert centering pin (13) through spool and sleeve assembly so that both ends of pin are below flush with outside of sleeve (14).

NOTE: On some models, a small nylon plug is located at each end of centering pin (13) to prevent contact with valve body.

Carefully insert spool and sleeve assembly, splined end first, in 14-hole end of valve body (11). Set body on clean surface with 14-hole end down. Install new "O" ring (6) in check valve plug (7), lubricate plug and insert in check valve bore. Install locator (5C) in valve bore with chamfered side up. Install thrust bearing assembly (5A and 5B)

over valve spool (15). Install new quad seal (3) and "O" ring (4) in mounting plate (2), lubricate seal and install mounting plate (2) over valve spool, thrust bearing and locator. Tighten the mounting plate retaining screws evenly to a torque of 250 inch-pounds. Clamp the mounting plate in a vise with 14-hole end of valve body up. Place plate (16) and gerotor outer ring (19) on valve body so that bolt holes align. Insert drive shaft (17) in gerotor inner rotor (18) so that slot in upper end of shaft is aligned with valleys in rotor and push shaft through rotor so that about ½ of length of the splines protrude. Holding the shaft and rotor in this position, insert them in valve housing so that notch in shaft engages centering pin in valve sleeve and spool. Install spacer (20) at end of drive shaft; if spacer does not drop down flush with rotor, shaft is not properly engaged with centering pin. Install end cap (21) and tighten the retaining screws equally to 150 in.-lbs. torque.

STEERING CYLINDER

External Cylinder

17. **REMOVE AND REINSTALL.** Thoroughly clean cylinder and hose fittings to avoid entry of dirt when lines are disconnected. Disconnect hose fittings at cylinder and remove cylinder from tractor. Work the cylinder piston each way to end of stroke to expel all excess oil.

When reinstalling cylinder, first attach hose fittings leaving the flare nuts loose. Install cylinder to crossbar and piston rod to steering arm, then tighten nuts and secure with cotter pins. Check to see that hoses are properly located, then tighten the flare nuts.

Bleed the power steering system as outlined in paragraph 12, check the "Power-Director" (hydraulic reservoir) oil level and add fluid as necessary to bring oil level to full mark on dipstick.

18. **OVERHAUL.** MODEL 7000. Remove steering cylinder as previously outlined and proceed as follows: Remove snap ring (11—Fig. 11). Push rod

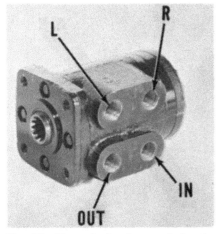

Fig. 10—View of steering control valve showing port locations. Pressure tube from pump connects to "IN" port; return tube to sump connects to "OUT" port; tube to rear end of cylinder connects to "L" (left turn) port; and tube to rod end of cylinder connects to "R" (right turn) port.

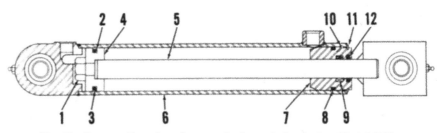

Fig. 11—Cross-section view of power steering cylinder used on Model 7000.

1. Nut	5. Piston rod
2. Back-up ring	6. Cylinder
3. "O" ring	7. Rod guide
4. Piston	

8. "O" ring & back-up ring	10. Snap ring
9. Seal	11. Snap ring
	12. Wiper seal

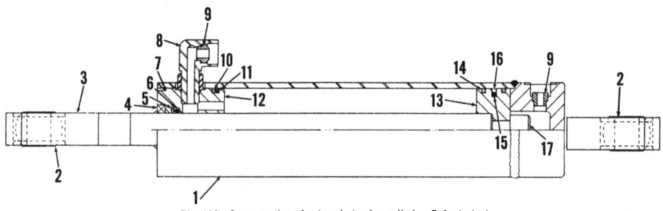

Fig. 11A—Cross section of external steering cylinder. Refer to text.

1. Cylinder	4. Wiper seal	8. Fitting	10. Back-up ring	13. Piston	16. Sealing ring
2. Bushings (2 used)	5. Back-up ring	9. Tube seats (2 used)	11. "O" ring	14. Wear rings (2 used)	17. Cap screw (or nut)
3. Piston rod	6. "O" ring		12. Rod guide bushing	15. "O" ring	
	7. Snap ring				

guide (7) into cylinder until snap ring (10) is accessible and remove snap ring. Pull rod assembly out of cylinder. Inspect components for excessive wear and damage. Tighten piston retaining nut to 100-115 ft.-lbs. Lubricate all seals and wipers prior to rod assembly installation in cylinder.

ALL OTHER MODELS. Remove the steering cylinder as outlined in paragraph 17 and proceed as follows: Remove fitting (8—Fig. 11A) and push the rod guide (12) into cylinder tube far enough to remove snap ring (7). Use piston and rod to bump guide (12) out of cylinder. Remove cap screw or nut (17) from end of rod and remove piston (13). Guide (12) can be withdrawn from rod after piston is removed.

Carefully inspect all parts for scoring or excessive wear. Install piston rod scraper (4) with sharp edge of seal out. Install back-up ring (5) with rough side toward inside of cylinder. Install "O" ring (6) in same groove on inner side of back-up ring. Install back-up ring (10) in outer groove of piston rod guide. Position "O" ring (11) in same groove between back-up ring and inner end of rod guide. Install "O" ring (15) in center groove of piston (13). Position sealing ring (16) in center groove over the "O" ring. On Model 200 install "O" ring (15) and back-up ring (16) side by side with "O" ring nearer piston nut. Install the two split wear rings (14) in the two outer grooves in piston. Model 200 has only one wear ring on piston.

Lubricate piston rod (3) and slide rod guide (with seals) over the rod. Install piston (13) and tighten lock nut or cap screw (17). Lubricate the cylinder bore, piston and rod guide. Slide piston and rod guide into cylinder and push rod guide into cylinder far enough to install snap ring (7).

NOTE: Port in rod guide should be in line with hole in cylinder. Bump rod guide out against snap ring with piston and install fitting (8).

Internal (Dual) Cylinders

19. R&R AND OVERHAUL. The front support casting on all tricycle models and some adjustable axle models is used as the steering cylinder. To remove front support, refer to paragraph 10. Remove adjustable axle, or tricycle pedestal from front support. Remove plug (11—Fig. 12), screw (10A), washer (10) and shims (9). Steering shaft (1) can be removed from the bottom. Remove cylinder plugs (18) and withdraw pistons (16).

When reassembling, renew all seals. Install one back-up ring (14) and "O" ring (15) in cylinder, then insert piston (16) from opposite side of cylinder. Install a back-up ring (14) and "O" ring (15) on the other side of cylinder, then center the piston.

NOTE: Make certain that cut-away (tooth) portion of piston is not pushed through new seals (15) or leakage may occur.

Install other piston and seals and align center tooth of both pistons (16) with centerline of hole (A). Install steering spindle (1) with index mark on flange centered at the rear. Steering shaft bearings (2 & 8) should be ad-

Fig. 12—Exploded view of dual internal (Zipper) steering front support. Front support is used as steering cylinder.

1. Steering shaft
2. Bearing cone
3. Bearing cap
4. Seal
5. Front support
6. Snap ring
7. Bearing cup
8. Bearing cone
9. Shims (0.003 & 0.010)
10. Washer
10A. Cap screw
11. Plug
12. Filler cap
13. Filler pipe
14. Back-up rings (4 used)
15. "O" rings (4 used)
16. Pistons (2 used)
17. "O" rings (4 used)
18. End plugs (4 used)

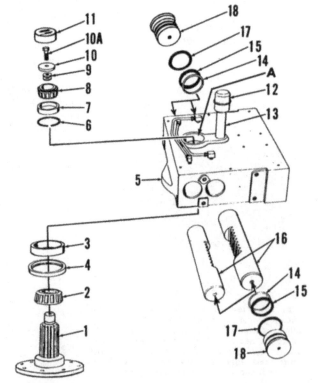

justed to a free rolling fit with no end play by varying number and thickness of shims (9). Cap screw (10A) should be tightened to 130-140 ft.-lbs. torque. Install "O" rings (17) on plugs (18).

Refer to paragraph 10 for installation procedure, the same type of oil (Allis-Chalmers "Power Fluid") as used in the hydraulic system should be used in front support. Oil level should be maintained at 2 inches from bottom as measured through breather tube (13). This oil is used to lubricate parts in the front support. Fluid for power steering is contained in 'Power-Director' compartment. Leakage past seals (15) will be noticed by an excessive amount of oil in front support housing.

ENGINE AND COMPONENTS

R&R ENGINE ASSEMBLY

All Models

20. To remove the engine, disconnect all battery ground straps. If engine is to be disassembled, drain oil pan. Remove front support casting as outlined in paragraph 9 or 10. Unbolt and remove both side rails.

NOTE: On models so equipped, detach hydraulic system filter from engine left side rail.

Disconnect wires and cables from engine electrical units. Disconnect hydraulic tubes from the pump, remove clamps and reposition tubes away from engine. Disconnect hourmeter cable, shut off fuel supply at tank and disconnect fuel supply line from gasoline carburetor, diesel injection pump or LP-Gas regulator. On diesel models, disconnect fuel return line, speed control cable and pump shut-off cable. On non-diesel models, disconnect governor control cable and choke cable.

NOTE: Depending on reason for engine removal, it may be easier to remove other components from engine at this time.

Attach hoist to lift hooks provided at front and rear end of cylinder head, then unbolt and remove engine from torque housing.

Reinstall engine and clutch assembly by reversing removal procedures. Bleed the diesel fuel system as outlined in paragraph 60 and, if necessary, bleed the power steering system as outlined in paragraph 12.

CYLINDER HEAD

All Models

21. **REMOVE AND REINSTALL.** Drain the cooling system and remove hood top panel and both hood rear side panels. On non-diesel models, shut off fuel at tank and remove inlet manifold, exhaust manifold and carburetor. On diesel models, remove the exhaust manifold. Disconnect fuel lines from injector nozzles and cap all openings as each line is disconnected.

NOTE: Although not required for removal of cylinder head, it is recommended that diesel fuel injector assemblies be removed at this time.

The injector nozzle tips protrude through the flat bottom surface of cylinder head and, if injectors are not removed, extreme care must be taken when removing and handling the removed cylinder head.

On all models, disconnect upper radiator hose from thermostat housing and remove inlet tube from air cleaner. Disconnect by-pass tube from water pump and thermostat housing, disconnect wire from temperature gage sending unit and remove thermostat housing. Disconnect bypass tube from water pump and thermostat housing, disconnect wire from temperature gage sending unit and remove thermostat housing. Disconnect breather tube and remove the rocker arm cover. Remove the rocker arm assembly and push rods. Remove cylinder head retaining cap screws and carefully pry cylinder head up off dowel pins located at front and rear of cylinder block. Lift cylinder head from tractor.

When reinstalling cylinder head, be sure that gasket surfaces of cylinder head and block are clean and free of burrs. Place new gasket with side marked "THIS SIDE DOWN" against cylinder block and be sure gasket is properly located on the two dowel pins.

NOTE: Do not apply grease or gasket sealer to gasket surfaces, block or cylinder head; new gasket is treated with phenolic sealer. Lubricate the retaining cap screws with motor oil before installing.

Install push rods and rocker arm assembly. Starting in center and working toward each end, tighten the cylinder head retaining cap screws first to 63-73 ft.-lbs. torque, then to final torque of 150 ft.-lbs. The small rocker arm bracket retaining cap screws on diesel models should be tightened to 28-33 ft.-lbs. torque. On non-diesel models, all rocker arm bracket retaining cap screws should be tightened to 18-20 ft.-lbs. Adjust valve tappet gap cold as outlined in paragraph 28.

When installing inlet and exhaust manifolds, observe the following: On diesel models, tighten inlet manifold retaining screws to 18-20 ft.-lbs. torque.

On late 190, late 190XT, 200, 7000 and all 180 and all 185 diesel models, the exhaust manifold retaining screws are self-locking type and should be tightened to 30 ft.-lbs. torque. On early 190 and 190XT diesel models, the retaining screws are locked with tab washers and should be tightened to 44-49 ft.-lbs. torque. On all non-diesel models, the two cap screws attaching each end of exhaust manifold should be tightened to 30-33 ft.-lbs. and all other inlet and exhaust manifold attaching screws should be tightened to 44-49 ft.-lbs.

Start and run engine for approximately 1 hour until it reaches normal operating temperature (preferably under load); then, retorque cylinder head, inlet manifold and exhaust manifold. Valve clearance should be adjusted to recommended clearance hot as outlined in paragraph 28.

VALVES AND SEATS

Diesel

22. Inlet valves have a face and seat angle of 30 degrees. A one degree interference angle may be used by machining the seat to 31 degrees. Seat width should be 3/32 inch and can be narrowed using 15 and 60 degree stones. Inlet valves seat directly on cylinder head of early models while a renewable valve seat insert is used on later models. Standard and 0.005-inch oversize seat inserts are available for service. When installing standard inserts, machine 1.6035-1.6045 diameter counterbore to a depth of 0.3825-0.3835 (measured from gasket surface of cylinder head). Insert should have 0.001-0.003 interference fit in counterbore. If standard size insert is loose, install 0.005-inch oversize insert. Surface of inlet valve head must be recessed at least 0.0345 from cylinder head surface to provide proper clearance between valve and piston. Inlet valve stem diameter is 0.3715-0.3720 inch.

Exhaust valves have a face and seat angle of 45 degrees on all models except 200 and 7000 which have a 30 degree seat. A one degree interference angle may be used by machining the seat to 46 degrees on all models except 200 and 7000 which should be machined to 31 degrees. Seat width should be 3/32 inch and can be narrowed using 30 and 60 degree stones on all models except models 200 and 7000 which can be narrowed using 15 and 60 degree stones. On early production 190 models and all 180 and 185 models, exhaust valves seat directly in the cylinder head. "One-Ninety" models after engine serial number (2D-03142) and all "XT" and all 200 and 7000 models are provided with renewable seat inserts. Counterbore for standard size seat insert is 1.4805-1.4815 diameter and depth is 0.4465-0.4485 (as measured from gasket surface of cylinder head). Inserts should have 0.001-0.003 interference fit. If new standard size insert is too loose, counterbore should be machined for 0.005 oversize insert. Stake the cylinder head at three points 120° apart near the exhaust seat insert after seat is installed.

NOTE: Do not stake cylinder head near the injector nozzle hole.

Surface of exhaust valve head should be recessed at least 0.0485 from cylinder head surface to provide proper clearance between valve and piston. Exhaust valve stem diameter is 0.3705-0.3710.

Non-Diesel

23. Inlet valves have a face and seat angle of 30 degrees. Seat width should be 3/32 inch and can be narrowed using 15 and 60 degree stones. Inlet valves seat directly on cylinder head, however seat inserts are available for service. When installing, service inserts on 180 and 190 models with G2500 engine (3¾-inch bore), counterbore should be machined 1.6085-1.6095 diameter and 0.3345-0.3365 deep (as measured from gasket surface of cylinder head). For 190XT models with 3-7/8-inch cylinder bore, counterbore for inlet valve insert should be 1.742-1.743 diameter and 0.399-0.401 deep (as measured from gasket surface of cylinder head). The insert should have 0.001-0.003 interference fit in counterbore on all models. Surface of inlet valve head should be recessed 0.0345 from cylinder head surface to provide proper clearance between valve and piston. Inlet valve stem diameter is 0.3715-0.3720.

Exhaust valves have a face and seat angle of 45 degrees. Seat width should be 3/32 inch and can be narrowed using 30 and 60 degree stones. Exhaust valves seat on renewable type seat inserts. Counterbore for standard size seat inserts on 180 and 190 models with 3¾-inch cylinder bore should be 1.4805-1.4815 diameter and 0.3835-0.3855 deep (as measured from gasket surface of cylinder head). For 190XT models with 3-7/8-inch cylinder bore, counterbore for seat insert should be 1.614-1.615 diameter and 0.4105-0.4125 deep (as measured from gasket surface of cylinder head). Inserts for all models should have 0.001-0.003 interference fit. If new standard size insert is too loose, counterbore should be machined for 0.005 oversize insert. Cylinder head should be punched at three equally spaced locations to stake the valve seat

in position after installation. Surface of exhaust valve head should be recessed 0.0485 from cylinder head surface to provide proper clearance between valve and piston. Exhaust valve stem diameter is 0.3705-0.3710.

VALVE GUIDES

All Models

24. Inlet valve stem diameter is 0.3715-0.3720 and exhaust valve stem diameter is 0.3705-0.3710. Desired stem to guide clearance if 0.001-0.0015 for inlet and 0.002-0.0025 for exhaust. Maximum allowable stem to guide clearance is 0.0035 for inlet valve and 0.0055 for exhaust.

Inlet and exhaust valve guides are interchangeable. When installing, press guides into cylinder head from bottom, with the chamfered end up. Distance from valve spring counterbore in head to top of valve guide (Fig. 13) is as follows:

```
Diesel—
    Inlet . . . . . . . . . . . . . . . .21/32 inch
    Exhaust . . . . . . . . . . . . . . .7/8 inch
Non-Diesel—
    Inlet . . . . . . . . . . . . . . . .15/16 inch
    Exhaust . . . . . . . . . . . . . .31/32 inch
```

New guides should be reamed to 0.373 diameter after installation to provide correct stem to guide clearance. Valve guides in factory service replacement cylinder head are reamed to correct size at factory. Valves should be reseated after installing new guides.

VALVE SPRINGS

All Models

25. The interchangeable intake and exhaust valve springs should be re-

Fig. 13—View showing method of measuring valve guide height above cylinder head counterbore when installing new valve guides. Refer to paragraph 24.

newed if rusted, distorted or if they vary more than 5% from following specifications:

Model 7000
```
Pounds pressure @
    2.23 inches . . . . . . . . . . . . . . . . . . .51-63
Pounds pressure @
    1.78 inches . . . . . . . . . . . . . . .121-133
Spring free length . . . . . . . . . . . . . .2.86
```
All Other Models
```
Pounds pressure @
    2.223 inches . . . . . . . . . . . . . . . . .41-45
Pounds pressure @
    1.836 inches . . . . . . . . . . . . . . . .73-81
Spring free length,
    approx. . . . . . . . . . . . . . .2-31/32 inches
```

CAM FOLLOWERS

All Models

26. The 0.748-0.7485 diameter mushroom type cam followers (valve lifters) ride directly in unbushed cylinder block bores and can be removed after removing the camshaft. Cam followers are available in standard size only and should be renewed if either end is chipped or worn or if they are loose in cylinder block bores. Desired follower to bore clearance is 0.001-0.0025.

ROCKER ARMS

All Models

27. **R&R AND OVERHAUL.** Rocker arms and shaft assembly can be removed after removing hood top and rear side panels, disconnecting breather tube and removing rocker arm cover and oiling tube. Loosen and remove all retaining cap screws equally, then lift rocker arms and shaft assembly from the cylinder head.

The hollow rocker arm shaft is drilled for lubrication to each rocker arm. Lubricating oil to the rocker arm oiling tube is supplied through a drilled passage in cylinder head and engine block.

Rocker arm shaft diameter is 0.999-1.000. Bore in rocker arm is 1.001-1.002 providing a clearance of 0.001-0.003 between shaft and rocker arms. If clearance is excessive, renew the shaft and/or rocker arms.

On non-diesel engines, rocker arms are offset toward shaft support bracket. On diesel engines, all rocker arms are alike.

Before reinstalling the rocker arm assembly, loosen each tappet adjusting screw about two turns to prevent interference between valve heads and pistons due to interchanged components. Tighten all of the rocker arm support retaining cap screws on non-diesel models to 18-20 ft.-lbs. On diesel models, tighten long rocker arm support (cylinder head) cap screws to 150 ft.-lbs. and the small cap screws to 28-33 ft.-lbs. Adjust valve clearance cold as outlined in paragraph 28, before engine is started. Complete reassembly of tractor, then adjust valve tappet clearance hot as outlined in paragraph 28.

VALVE CLEARANCE

All Models

28. Because the limited clearance between the head of the valve and top of piston valve clearance (Tappet Gap) should be checked and adjusted only with engine stopped. After any service to valve system, valve clearance should be adjusted cold before engine is started. Valve clearance should be rechecked with engine at operating temperature.

Diesel
Clearance Cold—
   Inlet & Exhaust............0.018 in.
Clearance Hot—
   Inlet & Exhaust............0.015 in.
Non-Diesel
Clearance Cold—
   Inlet ....0.022 in.
   Exhaust ....0.027 in.
Clearance Hot—
   Inlet ....0.020 in.
   Exhaust ....0.025 in.

Two position adjustment of all valves is possible as shown in Fig. 14 & 14A. To make the adjustment, turn the crankshaft until No. 1 piston is at TDC on compression stroke. Mark on crankshaft pulley (Fig. 15) indicates TDC and if clearances are nearly correct, both front rocker arms will be loose and both rear rocker arms will be tight with No. 1 piston on compression stroke. Adjust the six valves indicated in Fig. 14 to the correct clearances. Turn the crankshaft one complete rev-

olution until TDC marks are again aligned. This will position No. 6 piston at TDC on compression stroke. Adjust the remaining six valves shown in Fig. 14A.

TIMING GEAR COVER AND CRANKSHAFT FRONT OIL SEAL

All Models

29. **REMOVE AND REINSTALL.** To remove the timing gear cover, first remove front support casting as outlined in paragraph 9 or 10, then proceed as follows: Remove fan blade, fan belts and cap screw and washer from front end of crankshaft. Using a puller threaded into the tapped holes in crankshaft pulley hub as shown in Fig. 16, remove pulley from crankshaft.

CAUTION: Do not attempt to remove pulley by pulling on outer diameter as vibration dampener built into pulley will be damaged.

On non-diesel models, disconnect governor control cable and throttle link rod. On all models, remove screws from front end of oil pan that are threaded into timing gear cover.

NOTE: It is suggested to loosen all oil pan screws enough to lower front of oil pan or preferably remove oil pan.

Remove screws that attach timing gear cover to front of block and carefully pry cover forward to slide seal and cover forward off of crankshaft.

NOTE: Some of the attaching screws enter from rear and the timing gear cover is located on crankcase with dowel pins. Be careful not to lose the spring loaded pin from front end of diesel fuel injection pump driveshaft.

Crankshaft front oil seal can be renewed while timing gear cover is removed.

NOTE: Inner diameter of seal grips crankshaft and rotates with shaft.

When reinstalling timing gear cover, use new pan gasket front section and new timing gear cover gasket. Lubricate oil seals, be sure spring loaded pin is in place in fuel injection pump drive shaft and carefully install timing gear cover over crankshaft and the two dowel pins. Tighten the timing gear cover screws to 28-33 ft.-lbs., oil pan screws to 18-20 ft.-lbs. and crankshaft pulley retaining cap screw to 180 ft.-lbs.

TIMING GEARS

All Models

30. **VALVE TIMING.** Valves are properly timed when punch marked tooth of the crankshaft gear is in register with the punch mark between two

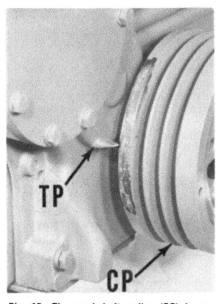

Fig. 15—The crankshaft pulley (CP) is provided with timing marks in degrees. When adjusting valve clearance (tappet gap), crankshaft should be set with TDC mark aligned with pointer (TP) on timing gear cover.

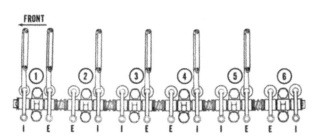

Fig. 14—With No. 1 piston at TDC on compression stroke, adjust tappet gap on valves shown with feeler gage. Refer to text for recommended clearance.

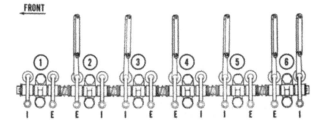

Fig. 14A—With No. 6 piston at TDC on compression stroke and No. 1 piston at TDC on exhaust stroke, adjust valves shown with feeler gage. Refer to text for recommended clearance.

Fig. 16—Do not pull crankshaft pulley by outer rim; thread pullers into hub of pulley as shown. Place correct size center plug over end of crankshaft after removing pulley retaining cap screw (CS) and washer.

Fig. 17—View of diesel engine with timing gear cover removed. Timing marks are the same for non-diesel engines. Thrust plunger (10) is in end of injection pump drive shaft.

teeth on the camshaft gear as shown in Fig. 17.

NOTE: Paragraphs 33 and 33A outline timing procedure for diesel fuel injection pump gears.

31. R&R CRANKSHAFT GEAR. The crankshaft gear is keyed and press fitted to the crankshaft. The gear can be removed by using a suitable puller after first removing the timing gear cover as outlined in paragraph 29.

New gear can be installed by heating it in oil for fifteen minutes prior to installation and drifting it on the crankshaft, or by pressing gear on crankshaft using crankshaft pulley retaining screw and suitable washers and

spacers. Be sure timing marks on crankshaft gear and camshaft gear are aligned as shown in Fig. 17. On diesel models, retime fuel injection pump as outlined in paragraph 33 or 33A.

32. R&R CAMSHAFT GEAR. The camshaft gear is keyed and press fitted to camshaft on all models except 200 and 7000 models. Models 200 and 7000 have a key and a retaining washer and cap screw. Gear (all models except 200 and 7000) can be removed and new gear installed in a press after camshaft is removed from engine. Camshaft gear on 200 and 7000 models can be pulled from camshaft after removing cap screw and retaining washer. Refer to paragraph 36 for removing camshaft, installing gear and installing camshaft thrust plate.

Timing marks on camshaft gear and crankshaft gear must be aligned as shown in Fig. 17.

INJECTION PUMP DRIVE GEAR

Diesel Models Except 7000

33. REMOVE AND REINSTALL. To remove the injection pump drive gear, drain cooling system, shut off fuel at tank and remove the radiator assembly. Remove injection pump timing window (TW—Fig. 19) and turn engine until pump timing marks are aligned as shown in Fig. 20 (Number one piston should be 26° BTDC on compression stroke for 180, 185 and 190XT models;

28° BTDC for 190 models without turbocharger, 24° BTDC for 200 models prior to serial number 2D-69516 and 18° BTDC for 200 models after serial number 2D-69515). Remove cover (C—Fig. 19) from timing gear cover and thrust plunger and spring from end of shaft. Remove nut from shaft and use a suitable puller to remove gear from tapered end of shaft.

CAUTION: If pump drive shaft is pulled out of pump, drive shaft seals may be damaged. To install the drive shaft and seals, the injection pump must be removed and seals installed after shaft is inserted through the adapter bushing.

To install injection pump drive gear, make certain that crankshaft is set at correct timing mark. (Number one piston should be 26° BTDC on compression stroke for 180, 185 and 190XT models; 28° BTDC for 190 models without turbocharger and 24° BTDC for 200 models prior to serial number 2D-69516 and 18° BTDC for 200 models after serial number 2D-69515). Make certain that pump timing marks are aligned as shown in Fig. 20 and install gear (8—Fig. 21) on pump drive shaft. The key in tapered end of pump drive shaft must engage slot in gear hub. With gear in place, timing marks (Fig. 20) must be aligned or closely aligned. Install lockwasher and gear retaining nut. Nut should be tightened to 35-40 ft.-lbs. Check pump timing as outlined in paragraph 70.

Model 7000

33A. REMOVE AND REINSTALL. To remove injection pump drive gear, remove injection pump timing window (TW—Fig. 19) and turn engine until pump timing marks are aligned as shown in Fig. 20. Number one piston should be 18° BTDC on compression. Detach right side engine panel and, if necessary for access to cover (C—Fig. 19), disconnect lower radiator hose. Remove cover (C), unscrew gear mounting screws and remove gear.

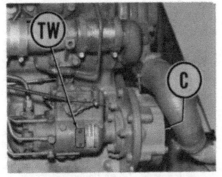

Fig. 19—View of diesel engine right side showing location of timing window (TW) and pump drive gear cover (C).

Fig. 20—View of diesel injection pump timing marks aligned.

1. Injection pump
2. "O" ring
3. Bushing
4. Pump mounting studs
5. Adapter cap screws
6. Adapter
7. "O" ring
8. Drive gear
9. Nut
10. Key

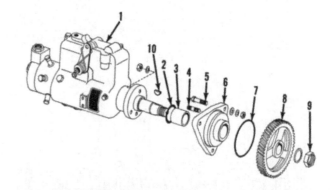

Fig. 21—Exploded view of injection pump adapter, drive shaft and gear used on 180, 185 190 and 190XT diesel models.

When installing gear, be certain the crankshaft is set with number 1 piston at 18° BTDC on compression and pump timing marks are aligned as shown in Fig. 20. After gear installation, check pump timing as outlined in paragraph 70.

INJECTION PUMP ADAPTER

Diesel Models Except 7000

34. **R&R AND OVERHAUL.** To remove the injection pump drive adapter, first remove drive gear as outlined in paragraph 33 and injection pump in paragraph 71A.

The bushing (3—Fig. 21) should be a press fit in adapter housing (6). Bushing journal of pump drive shaft should be 0.8735-0.8740 diameter and should have 0.001-0.0035 diametral clearance in bushing (3).

To reinstall, install adapter assembly on engine front plate, making certain that oil hole is toward top, then insert pump drive shaft through adapter bushing. Install new drive shaft seals and injection pump as outlined in paragraph 71A; then install drive gear as in paragraph 33. Refer to paragraph 70 to check and adjust pump timing.

INJECTION PUMP IDLER GEARS

Diesel Models

35. **R&R AND OVERHAUL.** To remove the fuel injection pump idler gears, first remove front support as outlined in paragraph 9 or 10, injection pump drive gear cover as in paragraph 33 or 33A and timing gear cover as in paragraph 29. Remove idler gear retaining cap screw (1—Fig. 22) and washer (2) and pull the idler gear assembly from idler shaft (7). To disassemble the unit, remove four cap screws (6) from rear face of gear (5).

NOTE: The bearing assembly (4) consists of two tapered roller bearing cones, two bearing cups and two spacers. If unit is to be reassembled reusing the bearing assembly, identify front and rear bearing cones as they are removed so that they can be reinstalled in their mating cups.

Idler shaft (7) is pressed into cylinder block and can be removed by using a slide hammer threaded into front end of shaft. Press or drive new idler shaft into cylinder block with flat on large diameter of shaft aligned with edge of engine front adapter plate. Oil hole in idler shaft must be up and aligned with slot in adapter plate.

If any part of the bearing unit (4) is damaged beyond further use or the bearings are worn so that idler gear end play is excessive, the complete bearing unit must be renewed.

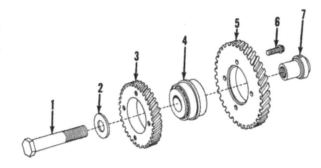

Fig. 22—Exploded view of diesel fuel injection pump idler gear assembly. Idler shaft (7) is pressed into front face of crankcase.

1. Cap screw
2. Retaining washer
3. Idler gear
4. Bearing assembly
5. Idler gear
6. Cap screws (4)
7. Idler shaft

To reassemble, place outer bearing spacer between the two idler gears and in the counterbore of each gear. Press bearing cups into idler gears with thick part of cup inward. Be sure that bearing cups are pressed tightly against the outer spacer. Place a bearing cone on idler shaft with taper forward and drive the cone back against shoulder on shaft. Align the single punch mark on camshaft and crankshaft gears, then install idler gears with large gear toward rear and the two marks on large idler gear aligned with the two marks on crankshaft gear. Install idler gear front bearing cone. Tighten cap screw to 95-105 ft.-lbs.

Reinstall fuel injection pump drive gear, aligning the single mark on small idler gear and single mark on fuel pump drive gear. Refer to paragraph 29 for installation of timing gear cover and paragraph 70 for injection pump timing.

HYDRAULIC PUMP DRIVE GEAR

All Models

Refer to paragraph 185 for information on the hydraulic pump drive gear.

CAMSHAFT AND BUSHINGS

All Models

36. **R&R CAMSHAFT.** To remove the camshaft, first remove oil pan, oil pump, timing gear cover, rocker arm assembly and push rods; then proceed as follows:

Use 5/8-inch diameter wooden dowels, 16 inches long inserted into cam followers to hold the followers up. On all models except Models 200 and 7000, remove screws attaching the thrust plate (4E, 4M or 4L—Fig. 23) to front of blocks by working through holes in camshaft gear, then withdraw camshaft. On Models 200 and 7000, remove retaining cap screw, washer and camshaft gear, then unbolt thrust plate and withdraw camshaft. It may be necessary to turn crankshaft slightly

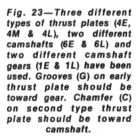

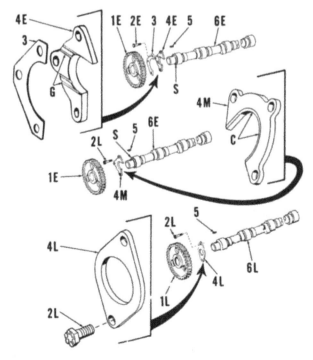

Fig. 23—Three different types of thrust plates (4E, 4M & 4L), two different camshafts (6E & 6L) and two different camshaft gears (1E & 1L) have been used. Grooves (G) on early thrust plate should be toward gear. Chamfer (C) on second type thrust plate should be toward camshaft.

C. Chamfer
G. Grooves
S. Shoulder
1E & 1L. Camshaft gears
2E & 2L. Screws
3. Lock plate for early screws
4E, 4L & 4M. Thrust plates
5. Woodruff key
6E & 6L. Camshaft

to prevent camshaft lobes from catching on connecting rods. Cam followers can be removed from bores in block and gear can be pressed from camshaft after the camshaft is withdrawn.

NOTE: Three different types of thrust plates (4E, 4M & 4L—Fig. 23) have been used.

Make certain that open types (4E & 4M) are not dropped as camshaft is removed. The earliest type thrust plate (4E) is identified by grooved surface (G) which should be toward gear (1E). The standard retaining screws (2E) and lock plate (3) can be discarded if later special self locking screws (2L) are installed. The second type thrust plate is similar to the first; however, the chamfered edge (C) should be toward camshaft front journal and three special lock screws (2L) are used. The first and second type thrust plates (4E & 4M) are located around shoulder (S) of cam shaft and gear (1E) should be pressed tightly against shoulder. The third type of thrust plate (4L) is retained by two special lock screws (2L) and completely circles the camshaft. The late thrust plate (4L) can be installed with either side toward gear. The late camshaft (6L) does not have a shoulder and gear (1L) should be pressed onto camshaft until front of gear hub is flush with end of camshaft.

NOTE: The (6L) camshaft is similar to the one used in 200 and 7000 models except the 200 or 7000 camshaft has a retaining washer and cap screw to hold gear on.

Camshaft journal diameter (all four journals) is 2.130-2.131. Desired camshaft to bushing clearance is 0.002-0.006. Bushings should be renewed as outlined in paragraph 37 if clearance exceeds 0.008 inch. Bushings are available in 0.010 undersize as well as standard. If camshaft journals are excessively worn or scored and camshaft is otherwise serviceable, journals may be ground to 2.120-2.121 and the 0.010 undersize bushings installed.

Camshaft end play is controlled by the thickness of thrust plate. Desired camshaft end play is 0.003-0.009. Renew thrust plate if end play exceeds 0.015 inch.

Backlash between crankshaft and camshaft gears should be 0.0015-0.0085 as measured on teeth of camshaft gear. New gear (or gears) should be installed if backlash exceeds 0.015 inch. On early camshafts (6E—Fig. 23) with shoulder (S), camshaft should have 0.0015-0.0030 interference fit in gear bore and gear should be pressed onto camshaft until hub is tight against shoulder. On later camshaft (6L) without shoulder, camshaft should have 0.0025-0.0040 inter-

ference fit in gear bore and gear should be pressed onto camshaft until front face of gear hub is flush with end of camshaft. On all models, heat gear in oil to 350-400°F. before pressing onto shaft.

The screws attaching the thrust plate to front of block should be tightened to 18-20 ft.-lbs. torque. On early models without self-locking screws, make certain that lock plate is installed and bend plate up against heads of screws after tightening.

On Models 200 and 7000, install cam gear washer and retaining screw and tighten screw to 72 ft.-lbs.

37. CAMSHAFT BUSHINGS. The camshaft is supported in four bushings. The camshaft should have 0.002-0.006 clearance in bushings. If clearance exceeds 0.008, camshaft and/or bushings should be renewed or camshaft journals ground to 2.120-2.121 and 0.010 undersize bushings installed.

To renew camshaft bushings, the engine must be removed from tractor and flywheel and rear adapter plate must be removed from engine. Press old bushings out and new bushings in with suitable bushing tools. Front bushing is 1-3/8-inches wide and the three rear bushings are one inch wide.

The front bushing should be flush to slightly recessed from front face of block. The rear bushing should be installed with front of bushing flush with bearing bore. Make certain that holes in all bushings are aligned with oil passages in block. Bushings are presized and should not require reaming if carefully installed; however, they should be checked after installation for localized high spots. Standard bushing diameter after installation should be 2.133-2.136

ROD AND PISTON UNITS

All Models

38. Piston and connecting rod units are removed from above after removing cylinder head, oil pan and rod bearing caps.

Connecting rod and rod cap mating surfaces are serrated and each rod and cap bear an assembly number in addition to the cylinder number. The cylinder number (1 through 6 from front to rear) is stamped on left (camshaft) side of piston, connecting rod and cap.

When reinstalling rod and piston units, be sure that cylinder numbers are in register and face towards camshaft; both bearing insert tangs must be towards same side of rod. Tighten the retaining screws to 40-45 ft.-lbs.

PISTONS, SLEEVES AND RINGS

All Models

39. The cam ground aluminum alloy pistons are fitted with three compression rings and one oil control ring. All rings are located above the piston pin. Pistons and rings are available in standard size only.

Install compression rings with side marked with dot or "T" toward top of piston. The top (chrome plated) ring can be installed with either side up if not marked. The second ring has a bevel cut in top of inner edge. The third compression ring has step cut in bottom of outer edge. If oil control ring is three piece type, expander should be installed in groove, then top and bottom rails should be installed over the expander with ends 180° apart from each other and 90° away from ends of expander. Some two piece oil control rings have slots toward bottom of piston while on others the oil return slot is between rails. Install expander in groove, then install the oil control ring with ends 180° from ends of expander.

With piston and connecting rod removed, use suitable puller to remove the wet type sleeve. Thoroughly clean all sealing and mating surfaces of block and cylinder sleeve. Insert sleeve in bore without sealing rings; sleeves should be free enough to be pushed into place and then be rotated by hand pressure. If sleeve cannot be inserted and turned by hand, more thorough cleaning is necessary. When sleeve is removed, check thrust sruface of the counterbore in top of cylinder block. If eroded or otherwise damaged, counterbore must be reseated.

Cylinder sleeve standout must be maintained at 0.002-0.005 above cylinder block surface. To check standout, measure thickness of sleeve flange with micrometer (do not include the 0.0445-0.0475 "fire wall" on top end of sleeve) and depth of counterbore in block with micrometer depth gage. Then, use special shims as necessary to provide the proper 0.002-0.005 standout. Shims are available in thicknesses of 0.005, 0.011, 0.015 and 0.020.

NOTE: Do not attempt to measure standout with straight edge and feeler gage after installing sleeve as rough edges on shims may give incorrect standout measurement.

When sleeve and bore in block are properly cleaned, install new sealing rings without lubricant of any kind on dry cylinder sleeve. On early sleeves, three sealing rings were used; the square (black) sealing ring in top seal groove, the round (black) sealing ring in the middle groove and the silicone (red) sealing ring in the lower groove. On later sleeves with only two grooves, two silicone (red) sealing rings are used on all models except Models 200 and 7000 which use one black ring in upper groove and one red (silicone) ring in lower groove. Prior to installing sleeve, brush the block bore and sealing rings with light motor oil; then, immediately install the sleeve. The sealing ring material expands rapidly upon contact with oil. Be extremely careful that the soft sealing rings are not cut by sharp edges of cylinder block.

Check the pistons, rings and sleeves against the following values.

DIESEL MODELS

Piston Ring Side Clearance:
Top (Chrome) Ring......0.005-0.0075
2nd Ring & 3rd Ring ...0.0035-0.0055
Oil Control Ring—
Rails (3 piece ring) 0.0015-0.007
Segment (2 piece ring) .0.0015-0.0035
Ring End Gap:
Top Ring (Models 200 &
7000) 0.019-0.029
Top Ring (all others) 0.017-0.032
2nd Ring—
Early 190* 0.017-0.032
Models 200 & 7000 .....0.015-0.025
All others 0.013-0.028
3rd Ring—
Early 190* 0.017-0.032
Models 200 & 7000 .....0.015-0.025
All others 0.012-0.028
Oil Control Ring—
Rails (3 piece ring) 0.013-0.058
Segment (2 piece ring):
Models 200 & 7000 .....0.010-0.020
All others 0.015-0.025
Piston Skirt to
Sleeve Clearance.......0.045-0.0070

*"One Ninety" tractors with 2800 engine before serial number 2D-05833.

NON DIESEL MODELS

Piston Ring Side Clearance:
Top Ring...............0.002-0.0035
2nd & 3rd Rings 0.002-0.004
Oil Control 0.002-0.007
Ring End Gap:
Top Ring—
G2500 engine 0.008-0.023
G2800 engine** 0.014-0.023
2nd & 3rd Rings—
G2500 engine 0.008-0.023
G2800 engine** 0.017-0.032
Oil ring 0.013-0.058
Piston Skirt to
Sleeve Clearance.......0.0015-0.004

**The G2800 engine is used on "190XT" models. Other models use G2500 engine.

PISTON PINS

All Models

40. Piston pins are retained in pistons by snap rings. Piston pin diameter is 1.2515-1.2517 for all models. Clearance in piston should be 0.0002 loose to 0.0006 tight for 2900 Diesel engine (190XT, 200 and 7000); 0.0001 loose to 0.0003 interference fit (at 70°F.) for all other models. Remove and install piston pin in a pin press or by heating piston in 180°F. hot water.

Desired clearance between piston pin and connecting rod bushing is 0.001-0.0017. Renew pin and/or bushing in connecting rod if clearance between pin and bushing exceeds 0.003. After installing new bushing, finish ream or hone the bushing inside diameter to 1.2527-1.2532.

CONNECTING RODS AND BEARINGS

All Models

41. Connecting rod bearings are precision, slip-in type. The bearing inserts can be renewed after removing oil pan and bearing caps. When installing new bearing inserts, be sure that linear projections (tangs) engage the milled slot in connecting rod and bearing cap and that cylinder numbers stamped on rod and cap are in register and face towards camshaft side of engine. Bearing inserts are available in standard size and undersizes of 0.002, 0.010, 0.020 and 0.040. Check the crankshaft crankpin journals and connecting rod bearings against the following values:

Crankpin diameter,
Std. 2.3720-2.3735
Rod bearing
clearance..............0.0009-0.0039

Rod side clearance 0.005-0.010
Rod bolt torque ....40-45 ft.-lbs.

CRANKSHAFT AND MAIN BEARINGS

All Models

42. The crankshaft is supported in seven precision, slip-in type main bearings. Main bearing inserts can be renewed after removing the oil pan, oil pump, oil tube and main bearing caps. Desired crankshaft end play is 0.004-0.009. Crankshaft end play is controlled by thrust flanges at each side of center main bearing.

To remove the crankshaft, first remove engine as outlined in paragraph 20. Then, remove clutch, flywheel, engine rear adapter plate, crankshaft pulley and timing gear cover. On diesel models, remove injection pump drive gear and shaft and injection pump idler gears. On all models, remove oil pan, oil pump, camshaft, engine front plate and rod and main bearing caps. Lift crankshaft from engine.

Check crankshaft and main bearing inserts against the following specifications:

Main journal Dia.,
(Std.) 2.7465-2.7480
Main bearing clearance,
desired 0.0016-0.0043
Main bearing bolt torque,
(ft.-lbs.) 130-140

Main bearing inserts are available in standard size and undersizes of 0.010, 0.020 and 0.040. All main bearing inserts are available as separate upper or lower half. Bearing lower halves do not have oil holes. Semi-finished main bearing caps available for service must be finished on cylinder block using a line boring bar. Diameter of bearing bore without inserts if 2.9368-2.9375.

CRANKSHAFT OIL SEALS

All Models

43. **FRONT SEAL.** The crankshaft front oil seal is located in the timing gear cover and can be renewed as outlined in paragraph 29.

44. **REAR SEAL.** The crankshaft rear oil seal (5—Fig. 24) is installed in the adapter plate (4) at rear of engine. To renew the seal, first remove engine from tractor as outlined in paragraph 20, then proceed as follows: Remove

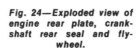
Fig. 24—Exploded view of engine rear plate, crankshaft rear seal and flywheel.

1. "O" ring
2. "O" rings
3. Pilot bearing
4. Rear plate
5. Crankshaft rear seal
6. Flywheel
7. Starter ring gear
8. Flywheel dowels
9. Sleeve

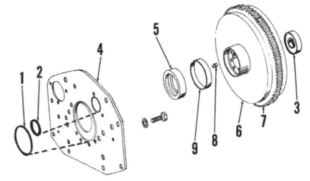

clutch, flywheel and engine rear adapter plate. Remove oil seal from the adapter plate and install new seal with lip of seal to front. Oil seal sleeve (9) on hub of flywheel should be renewed if nicked or worn. Beveled outside diameter of wear sleeve should be toward front (inside) of engine with flywheel installed. New sleeve should be pressed on flywheel flush with edge of flywheel hub.

NOTE: Do not heat or lubricate wear sleeve when installing. Renew the two "O" rings (1 & 2) located between adapter plate and engine, lubricate the oil seal and reinstall adapter plate.

Tighten adapter plate to cylinder block screws to 68-73 ft.-lbs and oil pan to adapter plate screws to 18-20 ft.-lbs Reinstall flywheel and tighten flywheel retaining screws to 68-73 ft.-lbs. Complete reassembly by reversing disassembly procedure.

FLYWHEEL

All Models

45. **REMOVE AND REINSTALL.** To remove the flywheel, first remove engine as outlined in paragraph 20 and unbolt clutch from flywheel. The flywheel is retained to crankshaft with four cap screws and two dowel pins. One cap screw hole is offset so that installation of flywheel is possible in one position only. To renew flywheel ring gear, flywheel must be removed from crankshaft. Inspect clutch friction surface of flywheel and crankshaft rear oil seal wear sleeve (9—Fig. 24) on flywheel hub. Heat new ring gear evenly until it will fit over flywheel and install with tooth bevel to front. Clutch pilot bearing in flywheel is a sealed unit.

When reinstalling flywheel, tighten retaining screws to 68-73 ft.-lbs. Complete the reassembly by reversing disassembly procedure.

OIL PAN (SUMP)

All Models

46. **REMOVE AND REINSTALL.** To remove the oil pan, it is first necessary to remove wide front axle assembly as follows: Support tractor under torque housing so that there is no weight on front axle pivot pins. On models with external steering cylinder, detach power steering cylinder from steering arm and front axle main member; it is not necessary to disconnect hoses from the cylinder. On models with dual internal steering cylinders, disconnect tie rods from center steering arm. On all wide front axle models, remove the front axle pivot pin and radius rod pivot bolt, raise front end of tractor to

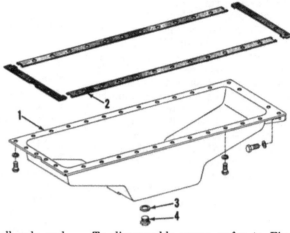

Fig. 25—The oil gasket (2) is made of interlocking pieces. Rear of pan is sealed by large "O" ring (1—Fig. 24).

1. Oil pan
2. Gasket
3. Seal
4. Drain plug

clear axle assembly and roll axle and front wheels forward out of the way. Place a safety support under front support casting, drain oil pan and unbolt pan from engine.

The oil pan is sealed at the rear by a large square-cut sealing ring (1—Fig. 24) that is located between adapter plate (4) and rear end of engine. Care should be taken not to damage this sealing ring when removing and reinstalling the oil pan as removal of clutch, flywheel and rear adapter plate is required to renew the sealing ring. Install all pan retaining screws loosely, tighten the pan to engine screws just enough to bring pan, gasket and cylinder block surfaces into contact. Then, tighten oil pan to rear adapter plate screws to 18-20 ft.-lbs. After tightening pan to rear adapter plate screws, tighten pan to engine screws to 18-20 ft.-lbs.

OIL PUMP AND RELIEF VALVES

All Models Except 200 and 7000

47. **R&R AND OVERHAUL.** Removal procedure will be evident after removing oil pan as outlined in paragraph 46.

To disassemble pump, refer to Fig. 26 and proceed as follows:

Remove screen (18), cover (16) and idler gear (12). Remove pin (9), and press shaft (13) out of gear (10). If necessary to renew idler shaft (11), press shaft out of pump body. Remove pin (4), retainer (1), spring (2) and 70-90 psi relief valve (3).

Check pump body, gears and shafts against the following specifications and renew any part excessively worn, scored or damaged.

Drive shaft diameter 0.6220-0.6225
Clearance, drive shaft
  to pump body 0.0015-0.003
Idler shaft diameter 0.6185-0.6190
Clearance, idler gear
  to idler shaft 0.001-0.0025
End clearance, both
  gears 0.002-0.004
Radial clearance
  gears to body 0.00075-0.00175

To reassemble pump, proceed as follows: Press idler shaft into pump body so that lower end of shaft (gage with idler gear) is 0.010 below flush with gasket surface of pump body. Press driver gear (15) onto drive shaft

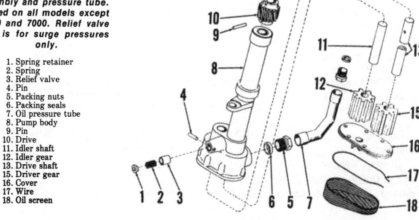

Fig. 26—Exploded view of early engine oil pump assembly and pressure tube. Used on all models except 200 and 7000. Relief valve (3) is for surge pressures only.

1. Spring retainer
2. Spring
3. Relief valve
4. Pin
5. Packing nuts
6. Packing seals
7. Oil pressure tube
8. Pump body
9. Pin
10. Drive
11. Idler shaft
12. Idler gear
13. Drive shaft
15. Driver gear
16. Cover
17. Wire
18. Oil screen

so that lower end of shaft is just below flush with lower end of driver gear. Idler shaft (11) should be 0.001-0.0025 interference fit in pump body. Driver gear (15) should be 0.001-0.0025 interference fit on drive shaft (13). Insert drive shaft into pump body and carefully install drive gear (10) on shaft. On early models, align pin holes in gear and shaft and install roll pin (9). On models with late type gear (10), gear is a press fit on shaft and pin (9) is not used. Gear should be heated slightly, then pressed onto shaft until distance from top surface of pump mounting tab to top of drive gear hub is 7-5/16 inches. Install idler gear and pump cover; then check pump for any binding condition and correct if necessary. Install screen (18), relief valve (3), spring (2), retainer (1) and pin (4).

When reinstalling the oil pump on diesel engines, be sure slot in oil pump drive gear engages drive pin on operation (hour) meter drive shaft on models so equipped. Tighten the pump retaining cap screw to 44-49 ft.-lbs. New sealing rings should be used on oil tube.

On non-diesel models, turn crankshaft until number one piston is at Top Dead Center on compression stroke (refer to Fig. 60). Turn distributor rotor until it points toward number one cylinder terminal on distributor cap and breaker points are just open. Install pump with slot in pump drive gear engaging drive tang on distributor drive shaft.

NOTE: The driving slot and tang are off-set from center line of shafts. Tighten pump retaining cap screw to 44-49 ft.-lbs. New sealing rings should

be used on oil tube. Check ignition timing as outlined in paragraph 83.

200 and 7000 Models

47A. R&R AND OVERHAUL PUMP. The oil pump can be removed after first removing the oil pan as outlined in paragraph 46. The pump is located on lower surface of block and is attached with two cap screws. Remove the suction tube assembly by removing two screws at oil pump and two screws at center main bearing cap. Refer to Fig. 26B for exploded view of pump.

To disassemble, first pull the pump drive gear using a screw type puller and two 3/8-16 bolts. Remove the pump cover and the gear assemblies. Remove roll pin (1—Fig. 26B) and relief valve washer, spring and piston.

Check pump body, gears, shafts and renew any parts excessively worn, scored or damaged. Pump gears and shafts (8 & 9) are available only with shafts.

When reassembling pump, and pump drive shaft bushing (11) is being renewed, press the bushing in flush with the front face of housing. Install gears and shafts assembly (8 & 9), new gasket (7) and cover (6) and tighten screws (5) to 18-20 ft.-lbs.

The pump gears must turn freely after pump is assembled and should have .004 to .009 inch end play. Preheat the drive gear (12) to 350-400 degrees F. and install on shaft. Drive gear (12) must have 0.040 to 0.060 inch clearance between gear and front of pump housing. Complete reassembly of pump by reversing disassembly procedure.

All Models

48. RELIEF VALVES. Two oil pressure relief valves are used. The relief

valve (3—Fig. 26 or 4—Fig. 26B) in oil pump body is non-adjustable and should by-pass oil at 70-90 psi. This is well above the normal oil pressure and opens only due to surge pressure when engine oil is cold.

The relief valve (Fig. 27) located on the right hand side of the engine just behind the diesel fuel injection pump or non-diesel governor, is adjustable to control normal engine oil pressure. To adjust the relief valve, loosen lock nut (2) and turn adjusting screw (3) in or out until oil pressure is 35 psi for all models except 200 and 7000 which should be 45 psi with engine running at 2200 rpm and with oil at normal operating temperature (180°-200°F.) Tractor is equipped with an oil pressure warning light instead of a pressure gage; therefore, a master pressure gage must be installed in place of oil pressure sending switch (1) when checking or adjusting system pressure. Adjust relief valve only when engine is running at normal operating temperature and speed. Do not attempt to adjust relief valve to compensate for worn oil pump or engine bearings.

GASOLINE FUEL SYSTEM

CARBURETOR

49. Gasoline tractors are equipped with either Holley or Zenith carburetors

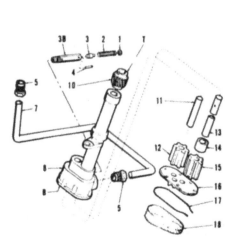

Fig. 26A—Exploded view of late type engine oil pump. Used on all models except 200 and 7000. Shafts (11 & 13) extend into cover (16) and surge pressure relief valve (1 thru 4) is different than early type.

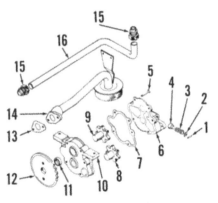

Fig. 26B—Exploded view of engine oil pump used on 200 and 7000 model tractors.

1. Roll pin
2. Washer
3. Spring (relief valve)
4. Piston (relief valve)
5. Place bolt
6. Cover
7. Gasket
8. Driven shaft assy.
9. Drive shaft assy.
10. Housing
11. Bushing
12. Gear (drive)
13. Gasket
14. Suction tube assy.
15. Tube nut
16. Pressure tube

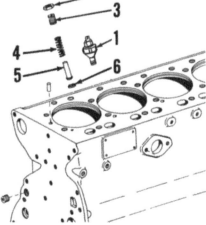

Fig. 27—The oil pressure sending switch and pressure relief valve are located on right side of engine.

1. Oil pressure sending switch
2. Lock nut
3. Adjusting screw
4. Spring
5. Oil pressure relief valve
6. Relief valve seat

shown in Fig. 29 or Fig. 30. On Holley carburetors, approximate setting is ½ turn open for idle mixture needle (51—Fig. 29); 5½ turns open for load adjustment needle (21). On Zenith carburetors, approximate setting is 1¼ turns open for idle mixture needle (19—Fig. 30).

On Holley carburetors, float setting should be 0.990 inch (Fig. 29A). Excessively rich mixture can be caused by power valve (36—Fig. 29) damage, retainer (39) detached from power

valve, broken spring (38) or ruptured diaphragm (43). Setting of needle (21) will not affect fuel leakage caused by preceding faults. Float spring (S—Fig. 29B) is installed on some models to provide smoother operation on rough ground. Refer to Fig. 29 and the following specifications for Holley carburetors used on G2500 and G2800 engines:

G2500 Engine

Venturi size .... 1-1/8 inches
Main jet (42) .... #73

G2800 Engine

Venturi size .... 1-5/16 inches
Main jet (42) .... #84

On Zenith carburetors, float setting should be 1-1/8 to 1-3/16 inches (Fig. 30A). Refer to Fig. 30 and the following specifications for Zenith carburetors used on G2500 and G2800 engines:

G2500 Engine

Venturi size (8) .... #28
Main jet nozzle (3) .... #90-3
Main jet (4) .... #29
Idle jet (11) .... #12
Well vent jet .... #21

G2800 Engine

Venturi size (8) .... #30
Main jet nozzle (3) .... #100-2
Main jet (4) .... #31
Idle jet (11) .... #12

FUEL PUMP

51. Gasoline models are equipped with an automotive type diaphragm fuel pump mounted on left side of engine and actuated by a cam on engine camshaft. Refer to Fig. 31 for exploded view of fuel pump. The pump repair kit available for service contains gaskets (2 & 9), screen (3), diaphragm (7), snap ring (11), and plug (14). Housing (5) should be renewed if check valves are damaged.

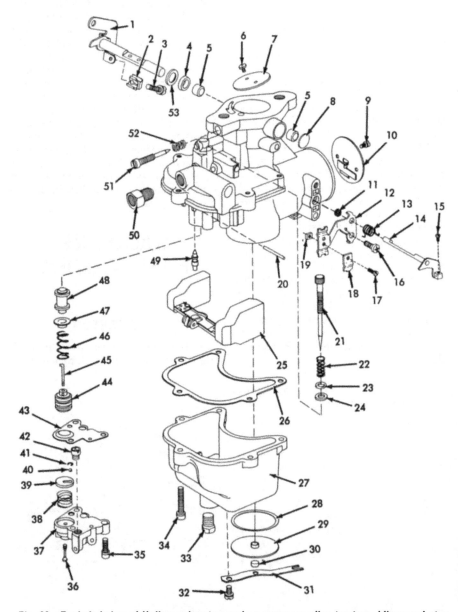

Fig. 29—Exploded view of Holley carburetor used on some gasoline tractors. Idle speed stop screw is shown at (3).

1. Throttle shaft	20. Float pivot pin	37. Metering valve body
3. Idle stop screw	21. Load adjustment needle	38. Power valve spring
4. Seal	23. Washer	39. Retainer
5. Bushing (2 used)	24. Seal	40. Pump inlet check ball
7. Throttle plate	25. Float	41. Retainer clip
8. Plug	27. Float bowl	42. Main jet
10. Choke plate	29. Plug	43. Diaphragm
11. Seal	30. Drip plug	44. Pump piston
12. Choke cable bracket	31. Spring clip	45. Link
13. Spring	33. Drain Plug	46. Pump operating spring
14. Choke shaft	36. Power valve	47. Spring retainer
		48. Vacuum piston
		49. Fuel inlet needle
		51. Idle mixture needle
		52. Spring
		53. Throttle shaft seal retainer

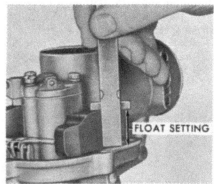

Fig. 29A—Float setting should be 0.990 inch when measured as shown on Holley carburetor.

Fig. 29B—View showing installation of float spring (S) installed on Holley carburetor.

LP-GAS SYSTEM

"One-Ninety XT" tractors are available with an LP-Gas system manufactured by the Ensign Products Section, American Bosch Arma Corporation. This system is designed to operate with fuel tank not more than 80% filled.

The American Bosch Model CBX carburetor and Model RDH regulator have three points of mixture adjustment, plus an idle speed stop screw.

ADJUSTMENTS

52. After overhauling or installing new carburetor or regulator, make following initial adjustments: Open idle mixture screw (K—Fig. 32) on regulator 1-1/16 turns. Open starting adjustment screw (14—Fig. 34) on carburetor 1¼ turns. Open load adjustment screw (17) on carburetor 4½ turns. Close choke and open throttle ½ way to start engine.

With initial adjustment made, start and run the engine until it reaches operating temperature, then place throttle control in low idle position and adjust idle stop screw on carburetor to provide a low idle speed of 400-450 rpm. Turn idle mixture needle (K—Fig. 32) on regulator either way as required to obtain the highest and smoothest engine operation. Readjust the carburetor idle stop screw, if necessary, to maintain low idle speed of 400-450 rpm.

With low idle speed readjusted, place throttle control in high idle position, turn main (load) adjustment screw (17—Fig. 34) in until engine begins to falter; then, back-out main adjustment screw until full power is restored and engine operates smoothly.

NOTE: In some cases, it may be necessary to vary the main adjustment slightly after load is placed on engine.

The initial starting adjustment should provide satisfactory starting performance; however, it may be varied if cold starting is not satisfactory.

CARBURETOR

53. The carburetor is serviced the same as a conventional gasoline type carburetor; that is the carburetor can be completely disassembled, cleaned and worn parts renewed. Refer to Fig. 34 for an exploded view. Pay particular attention to the economizer diaphragm assembly (20) and make certain that vacuum connections to the economizer chamber do not leak.

REGULATOR

54. **TROUBLE SHOOTING.** If engine will not idle properly, and if turning idle mixture adjusting screw on regulator will not correct the condition it will be necessary to disassemble regulator

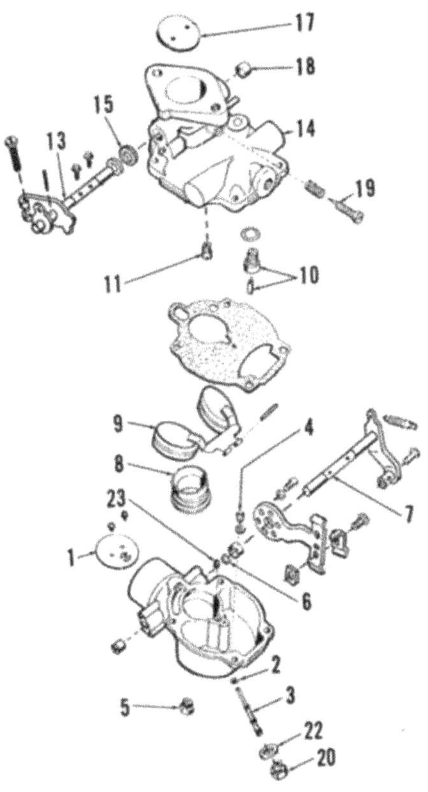

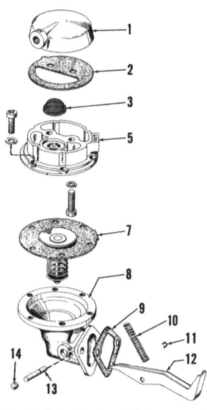

Fig. 31—Exploded view of fuel pump used on gasoline tractors.

1. Inlet housing
2. Airdome diaphragm
3. Inlet screen
5. Valve housing
7. Diaphragm
8. Body
9. Gasket
10. Spring
11. Snap ring
12. Cam lever
13. Pivot pin
14. Plug

Fig. 30—Exploded view of Zenith carburetor used on some gasoline tractors.

1. Choke plate
2. Gasket
3. Main jet nozzle
4. Main jet
5. Plug
6. Packing
7. Choke shaft
8. Venturi
9. Float
10. Inlet needle and seat
11. Idle jet
13. Throttle shaft
14. Throttle body
15. Packing
17. Throttle plate
18. Plug
19. Idle mixture needle
20. Plug
22. Gasket
23. Well vent jet

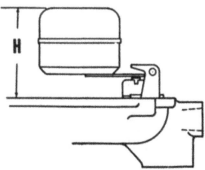

Fig. 30A—Float setting (H) should be 1-1/8—1-3/16 inches when measured as shown on Zenith carburetor.

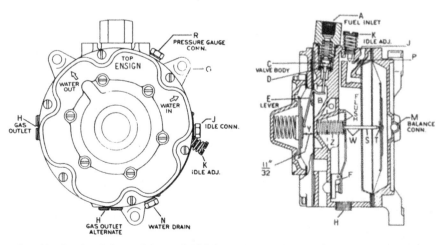

Fig. 32—Sectional views of the model RDH regulator used on LP-Gas equipped tractors.

unit and thoroughly clean the low pressure valve (F—Fig. 32).

To test condition of the high pressure valve (C), install a suitable pressure gage at pipe plug connection (R) in face of regulator cover. If gage pressure increases after a warm engine is stopped, the high pressure valve is leaking. Under normal operating conditions, the high pressure valve should maintain a pressure of 9-11 psi. If valve leaks, or does not maintain proper pressure, renew the high pressure valve.

NOTE: The high pressure valve and its gasket can be serviced without disconnecting or disassembling the regulator unit.

If, after standing for some time, the regulator unit is cold and shows moisture and frost, either the high pressure valve (C) or low pressure valve (F) is leaking or valve levers are not properly set.

55. **OVERHAUL.** Disassembly of the model RDH regulator is evident after an examination of unit and reference to Figs. 32 and 33. Wash all parts in suitable solvent and blow out all passageways with compressed air. When reassembling, renew all parts that show excessive wear and make certain that valve levers are set to the dimensions as follows. The high pressure lever (E—Fig. 32) should be set at approximately 11/32-inch from top of lever to face of plate (Y—Fig. 32). Bend lever if necessary to obtain this dimension.

When installing low pressure valve (27—Fig. 33) and spring (25), center the low pressure lever with push pin hole in center of partition plate (30). A rib is provided in recess of body (19) for the purpose of setting low pressure lever height. The top of lever, when valve is seated, should be flush with top of this rib. Bend lever if necessary to obtain correct setting.

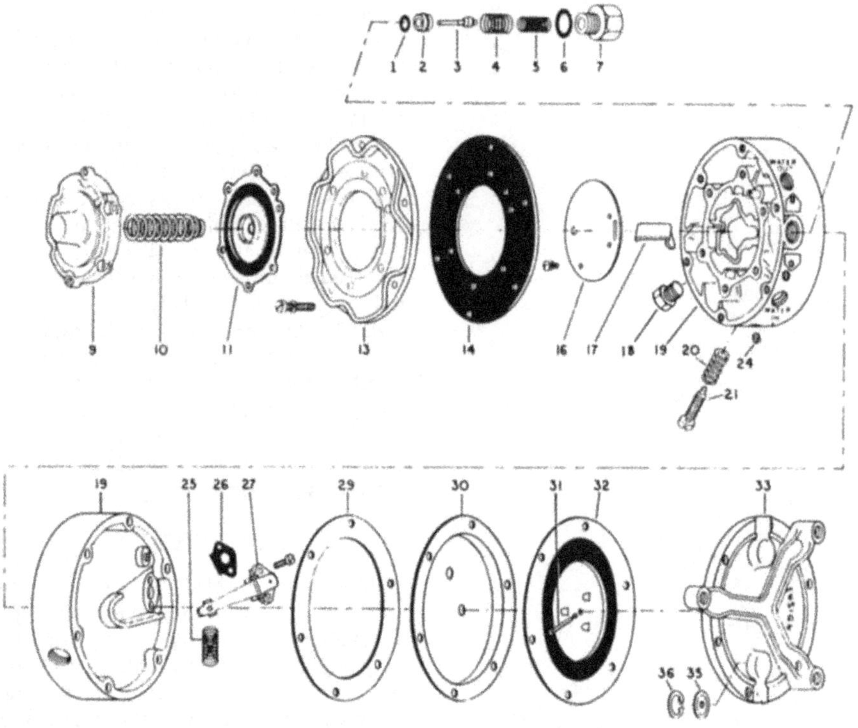

Fig. 33—Exploded view of the model RDH regulator used on LP-Gas equipped tractors.

1. "O" ring	9. Cover	19. Body	29. Gasket
2. Valve seat	10. Spring	20. Spring	30. Plate
3. High pressure valve	11. High pressure diaphragm	21. Idle screw	31. Push pin
4. Spring	13. Cover	24. Bleed screw	32. Low pressure diaphragm
5. Valve spring	14. Gasket	25. Valve spring	33. Support plate
6. Gasket	16. Plate	26. Gasket	35. Vent valve
7. Valve retainer	17. Valve lever	27. Low pressure valve	36. Snap ring

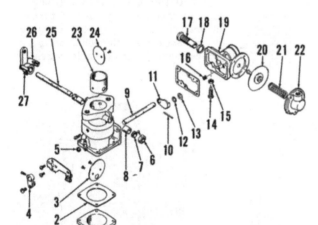

Fig. 34—Exploded view of American Bosch Series CBX carburetor used on LP-Gas engines.

1. Inlet elbow	9. Choke shaft	16. Economizer orifice	21. Spring
2. Gasket	10. Pin	17. Fuel adjusting screw	22. Diaphragm cover
3. Choke disc	11. Valve lever	18. Locknut	23. Venturi
4. Choke lever	12. Spring washer	19. Economizer body	24. Throttle disc
5. Seal	13. Washer	20. Economizer diaphragm	25. Throttle shaft
6. Collar	14. Starting adjustment		26. Throttle lever
7. Seal	15. Locknut		27. Idle stop screw
8. Bushing			

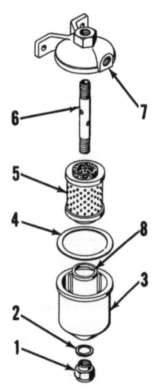

Fig. 35—Exploded view of LP-Gas filter.

1. Cap nut	5. Filter element
2. Seal ring	6. Stem
3. Filter bowl	7. Filter body
4. Sealing ring	8. Magnetic ring

FILTER

56. The filter (Fig. 35) used with the LP-Gas system is equipped with a felt filtering element and a magnet ring. When servicing the LP-system or on major engine overhauls, it is advisable to remove the lower part of filter and clean or renew filtering elements.

CAUTION: Shut off both liquid and vapor valves at fuel tank and run engine until fuel is exhausted before attempting to remove filter bowl.

DIESEL FUEL SYSTEM

When servicing any unit associated with the fuel system, the maintenance of absolute cleanliness is of upmost importance.

Probably the most important precaution that servicing personnel can impart to owners of diesel powered tractors is to urge them to use an approved fuel that is absolutely clean and free from foreign material. Extra precaution should be taken to make certain that no water enters fuel storage tanks. All diesel fuels contain some sulphur. When water is mixed with sulphur, an acid is formed which will quickly erode the closely fitting parts of the injection pump and fuel injection nozzles.

FILTERS AND BLEEDING

The fuel filtering system on all models except 200 and 7000, consists of fuel strainer and sediment bowl located under each side of the fuel tank and a single throw-away type filter on right side of engine. Both fuel sediment bowls under the fuel tank are provided with fuel shut-off valves.

The fuel filtering system on Models 200 and 7000, consists of a fuel shut off under the fuel tank, and a water separator with a replaceable filter and a single throw-away type fuel filter on right side of engine. Refer to Fig. 37.

60. **BLEEDING.** Each time the filter element is renewed, or fuel lines are disconnected, it will be necessary to bleed air from the system. Exact procedure may vary depending upon where air has entered the fuel system.

To bleed the filter, open the fuel tank shut-off valve and remove pipe plug from top of filter base. Operate primer pump (Fig. 36) until all air has escaped and fuel flows freely from bleed plug hole in filter base; then, reinstall plug.

To bleed the fuel line between fuel filter and injection pump, proceed as follows: Loosen inlet line at pump and operate primer pump until stream of fuel from loosened connection is free of air bubbles. Tighten connection while operating hand primer pump.

To bleed the high pressure nozzle lines, loosen connections at injectors and crank engine with starting motor. Tighten connections after fuel appears at all injectors.

61. **FILTERS.** All fuel sediment bowls (except Models 200 and 7000) should be checked daily or after each 10 hours of operation and the bowl and fuel strainer screen should be cleaned as necessary. Water separator on Models 200 and 7000 should be drained daily and element replaced once a year or more often if required.

The throw-away type filter element located on right side of engine should be renewed every 500 hours of operation. Poor fuel handling and storage facilities will decrease the effective life of filter elements. Filter should never remain in the fuel system until a decrease in power or engine speed is noticed because some dirt may enter pump and/or injector nozzles and result in severe damage.

INJECTION NOZZLES

The diesel engines are equipped with a direct type injection system and fuel leaves injection nozzles with enough force to penetrate the skin. When testing nozzles, keep clear of the nozzles spray.

62. **LOCATING A FAULTY NOZZLE.** If engine misfires or is rough running, check for defective nozzle as follows: With engine running at low idle speed, loosen the high pressure connection at each injector in turn. As in checking spark plugs, faulty injector is the one which least affects running, when its line is loosened.

If faulty nozzle is found and considerable time has elapsed since injectors have been serviced, it is recommended that all injectors be removed and serviced or that new or reconditioned units be installed.

63. **REMOVE AND INSTALL INJECTORS.** Before loosening any lines, wash the nozzle holder and connections with clean diesel fuel or kerosene. After disconnecting high pressure and leak-off lines, cover open ends of connections to prevent entrance of dirt or other foreign material. Remove nozzle holder stud nuts and carefully withdraw nozzles from cylinder head, being careful not to strike the tip end of nozzle against any hard surface.

Nozzle holder on 7000 tractor is located in a sleeve which is swaged in cylinder head. Sleeve must be split and driven out using tool ACTP 2057. To install sleeve, install two lubricated "O"

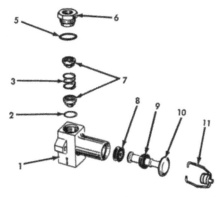

Fig. 36—Exploded view of primer pump used on diesel tractors. Check valves (7) must be installed so that fuel is pumped in direction of arrow on side of body (1).

1. Pump body
2. Gasket
3. Spring
5. Seal ring
6. Retainer nut
7. Valves
8. Plunger piston
9. Plunger guide
10. Plunger
11. Clamp assembly

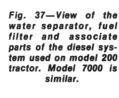

Fig. 37—View of the water separator, fuel filter and associate parts of the diesel system used on model 200 tractor. Model 7000 is similar.

rings (E—Fig. 38) on sleeve and insert sleeve in head. Do not use sealant on nozzle end of sleeve. Lubricate end of swage tool ACTP 2059 and check fit in nozzle end of sleeve. If swage does not slide easily into end of sleeve, install another sleeve. Assemble tools as shown in Fig. 38 and using a suitable press, force swage against sleeve until stop is contacted. Remove tools and dress nozzle end of sleeve but remove as little metal as possible.

Thoroughly clean nozzle recess in cylinder head before reinstalling nozzle and holder assembly. Use only wood or brass cleaning tools to avoid damage to seating surfaces and make sure the recess is free of all dirt and carbon. Even a small particle could cause the unit to be cocked and result in a compression loss and improper cooling of the fuel injector.

All models except 7000 models are equipped with a copper washer (11—Fig. 38A) to seal combustion pressure. Always renew copper washer and install with cup side down. Torque each of the two nozzle holder stud nuts in 2 ft.-lb. increments until each reaches the final torque of 11-13 ft.-lbs. for all models except 200 and 7000 which should be 9-12 ft.-lbs. This method of tightening will help prevent holder from being cocked in the bore.

64. **NOZZLE TESTING.** A complete job of testing and adjusting the nozzle requires the use of a special tester. Use only clean, approved testing oil in the tester tank. Operate the tester lever until oil flows; then, attach injector to tester and make the following tests:

65. **OPENING PRESSURE.** Close gage valve and operate tester lever several times to clear all air from the injector. Then open gage valve and while slowly operating tester lever, observe the pressure at which the injection spray occurs. This gage pressure should be 2750 psi for 190 tractors with 2800 engine, 3475 psi for 200 and 7000 tractors with 2900 engine, or 2900 psi for all other tractors with 2800 and 2900 engines. If the gage pressure is not as specified, remove cap nut (1—Fig. 38A), loosen lock nut (4) and turn adjusting screw (3) in or out as required to increase or decrease opening pressure. If opening pressure cannot be adjusted to the correct pressure, overhaul nozzle as outlined in paragraph 69.

66. **SPRAY PATTERN.** Operate the tester handle slowly and observe the spray pattern. All of the sprays must be similar and spaced equidistantly in a nearly horizontal plane. Each spray must be well atomized and should spread into a 1 inch cone at a 3 inch distance from injector tip. If spray pattern is not as described, overhaul nozzle as outlined in paragraph 69.

NOTE: Rapid operation of tester lever will frequently produce a spray pattern as described even if injector is faulty. Be sure to operate tester lever as slowly

as possible and still cause nozzle to open.

67. **SEAT LEAKAGE.** Wipe nozzle tip dry with clean blotting paper; then, operate tester handle to bring gage pressure to 500 psi on all models except 200 and 7000 which should be tested at 200 psi below opening pressure and hold this pressure for five seconds. If any fuel appears on nozzle tip, overhaul injector as outlined in paragraph 69.

68. **NOZZLE LEAK BACK.** Operate the tester handle to bring gage pressure to 2500 psi, then note time required for gage pressure to drop to 2000 psi. This time should be between 10 and 30 seconds.

If elapsed time is not as specified, nozzle should be cleaned or overhauled as outlined in paragraph 69.

NOTE: A leaking tester connection, check valve or pressure gage will show up in this test as excessively fast leak back. If, in testing a number of injector nozzles, all fail to pass this test, the tester rather than the injectors should be suspected.

69. **OVERHAUL.** Hard or sharp tools, emery cloth, crocus cloth, grinding compounds or abrasives of any kind should NEVER be used in cleaning of nozzles.

Wipe all dirt and loose carbon from injector assembly with a clean, lint free cloth. Carefully clamp injector fixture and remove the protecting cap (1—Fig. 38A). Loosen jam nut (4) and back off adjusting screw (3) enough to relieve load from spring (5). Remove nozzle cap nut (10) and nozzle assembly (9). Normally, the nozzle valve needle can easily be withdrawn from nozzle body. If it cannot, soak the assembly in fuel oil, acetone, or similar carbon solvent to facilitate removal. Be careful not to permit valve or body to come in contact with any hard surface.

If more than one injector is being serviced, keep the component parts of each injector separate from the others by placing them in a clean compartmented pan covered with fuel oil or solvent. Examine nozzle body and remove any carbon deposits from exterior surfaces using a brass wire brush. The nozzle body must be in good condition and not blued due to overheating.

All polished surfaces should be relatively bright without scratches or dull patches. Pressure surfaces (A, B and D—Fig. 39) must be absolutely clean and free from nicks, scratches or foreign material as these surfaces must register together to form a high pressure joint.

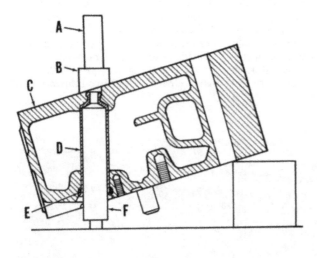

Fig. 38—View showing tool arrangement for injector holder sleeve installation on 7000 models. See text.

A. Swage-ACTP 2059
B. Swage stop-ACTP 2060
C. Cylinder head
D. Sleeve
E. "O" rings (2)
F. Swage adaptor-ACTP 2058

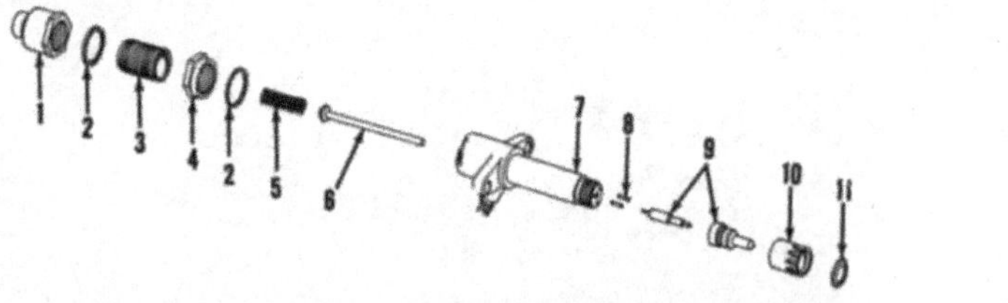

1. Cap
2. Copper washer
3. Adjusting screw
4. Lock nut
5. Spring
6. Spindle
7. Holder body
8. Dowel pins
9. Nozzle
10. Nut
11. Copper washer

Fig. 38A—Exploded view of typical Allis-Chalmers fuel injection nozzle and holder assembly. Washer (11) is not used on 7000.

Clean out the small fuel feed channels (G), using a small diameter wire as shown in Fig. 40. Insert the special groove scraper into nozzle body until nose of scraper locates in the fuel gallery. Press nose of scraper hard against side of cavity and rotate scraper to clean all carbon deposits from the gallery as shown in Fig. 41. Using seat scraper, clean all carbon from valve seat (J—Fig. 39) by rotating and pressing on the scraper as shown in Fig. 42.

Examine stem and seat end of nozzle valve and remove any carbon deposit using a clean, lint free cloth. Use extreme care, however, as any burr or small scratch may cause valve leakage or spray pattern distortion.

Before reassembling, thoroughly rinse all parts in clean diesel fuel and make certain that all carbon is removed from the nozzle holder nut. Install nozzle assembly and cap nut making certain that the valve stem is located in the hole of the holder body and the two dowel pins (8—Fig. 38A) enter holes in nozzle body. Tighten the holder nut to a torque of 20-25 ft.-lbs.

Install the spindle (6), spring (5), adjusting screw (3) and lock nut (4) using a new copper washer (2). Connect the injector to a nozzle tester and adjust opening pressure as outlined in paragraph 65. Use new copper gasket and install cap nut (1). Recheck nozzle opening pressure to be sure adjustment was not changed by tightening the lock nut and cap nut.

Retest the injector as outlined in paragraphs 66, 67 and 68. If injector fails to pass these tests, renew nozzle and needle assembly (9).

INJECTION PUMP

70. PUMP TIMING. The Roosa-Master DBGFC637-13AF injection pump used on "One-Ninety" tractors with 2800 diesel engines (without turbocharger) should be timed at 28 degrees BTDC. "One-Ninety XT" tractors (D2900 turbocharged diesel engine) with Roosa-Master DBGFC-637-31AF pump, 180 tractors with DBGFC637-75AF pump and 185 tractors with DBGFC637-76AF pump should all be timed at 26 degrees BTDC. Model 200 prior to serial number 2D-69516 tractors should all be timed at 24 degrees BTDC. Model 200 after serial number 2D-69515 and all 7000 models should be timed at 18 degrees BTDC. Timing marks are located on the crankshaft pulley as shown in Fig. 44.

Turn engine in normal direction of rotation until number one piston is coming up on compression stroke. Continue to turn engine slowly until the correct timing mark located on crankshaft pulley (28° BTDC for 190 tractors; 26° BTDC for 180, 185 and 190XT tractors, 24° BTDC for 200 tractors prior to serial number 2D-69516 and 18° for 200 tractors after serial number 2D-69515 and all 7000 tractors) is aligned with pointer on timing gear cover. Shut off fuel supply at tank and remove timing window cover (TW—Fig. 43). If timing marks on cam and governor drive plate are not aligned as shown in Fig. 45, loosen pump mounting bolts and turn pump so that marks are aligned. Retighten pump mounting stud nuts. Because of possible back-lash in the drive gears, pump timing should be rechecked after crankshaft is turned 2 complete revolutions.

NOTE: Timing procedure as outlined assumes normal timing check and adjustment on engine that is in operative condition. If engine will not start and improper installation of fuel injection pump is suspected or if pump cannot be rotated far enough in mounting slots to bring timing marks into alignment, refer to paragraph 33 or 33A for proper timing of fuel injection pump drive gear.

71. R&R FUEL INJECTION PUMP. MODEL 7000. Thoroughly clean pump,

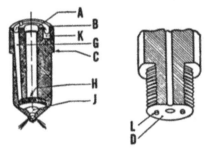

Fig. 39—Drawing showing text reference points on injector nozzle body and holder body.

A. Pressure surface
B. Pressure surface
C. Shoulder
D. Pressure surface
G. Fuel passage

H. Fuel gallery
J. Needle seat
K. Hole for dowel
L. Dowel

Fig. 40—Cleaning fuel passage in nozzle body.

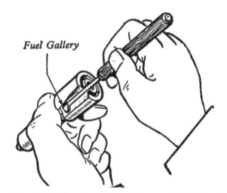

Fig. 41—Cleaning fuel gallery with hooked scraper.

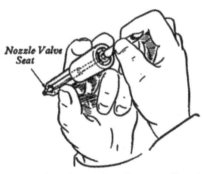

Fig. 42—Scraping carbon from needle valve seat.

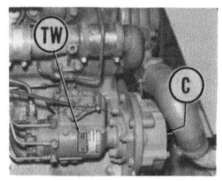

Fig. 43—Timing window (TW) cover location on fuel injection pump.

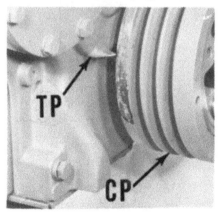

Fig. 44—View showing location of timing pointer (TP) on engine front (timing gear) cover. Align proper degree mark on crankshaft pulley (CP) with pointer.

fuel lines and fittings with diesel fuel, then remove fuel injection pump drive gear as outlined in paragraph 33A. Disconnect all fuel and control cables to pump, unscrew pump mounting screws and remove pump.

Install pump and refer to paragraph 33A for pump drive gear installation. Connect fuel lines and control cables but do not tighten fuel fittings until fuel system is bled as outlined in paragraph 60.

71A. ALL OTHER DIESEL MODELS. Remove lower side panel from right side of hood on all models except 200. It is not necessary to remove side panel on 200. Before loosening any connections, thoroughly clean the pump, fuel lines and fittings with diesel fuel or petroleum solvent; then, proceed as follows: Shut off fuel supply at tank, remove timing window cover (TW—Fig. 43) from fuel injection pump and turn engine so that pump timing marks are aligned in timing window as shown in Fig. 45. Pointer (TP—Fig. 44) on timing gear cover should be aligned with 28° mark on 190 tractors with 2800 engine; 26° mark on 180 , 185 and 190XT tractors, 24° mark on 200 tractors prior to serial number 2D-69516 and 18° BTDC for 200 tractors after serial number 2D-69515. On all models, disconnect the fuel supply line, throttle cable, shut-off cable and fuel return line from pump. Remove clamps from high pressure (nozzle) lines; then, disconnect lines from pump and loosen connections at nozzles. Remove pump mounting stud nuts and slide pump from drive shaft.

Before installing pump, make certain that the correct timing mark is aligned with pointer (TP—Fig. 44). "One-Ninety" models should be timed at 28° BTDC; 180, 185 and 190XT tractors should be timed at 26° BTDC, 200

tractors prior to serial number 2D-69516 should be timed at 24° BTDC and 200 tractors after serial number 2D-69515 should be timed at 18° BTDC.

Install new lip seals and "O" ring on injection pump drive shaft and apply lubricant to seals. Remove timing window cover (TW—Fig. 43) from outside of pump housing and turn pump so that timing marks (Fig. 45) are aligned. Carefully work the pump on shaft over the seals and attach pump to adapter with mounting studs and nuts so that timing marks are in alignment.

CAUTION: Be sure that lip of rear seal on pump shaft is not rolled or bent back as pump is worked onto shaft. If this occurs, remove pump and renew seal as the seal will have been damaged and early failure may result. Installation of pump on drive shaft can be simplified by using special seal compressor available from Roosa-Master.

Reconnect the high pressure lines leaving fittings loose both at nozzle and pump. Connect throttle cable, fuel supply line, fuel return line and shut-off cable. Reinstall clamps on high pressure (nozzle) lines and bleed complete fuel system as outlined in paragraph

60. Check throttle lever travel at pump and equalize lever overtravel by adjusting pivot pins on ends of throttle control cable. Throttle lever on pump is spring loaded.

72. SPEED ADJUSTMENTS. Actuate the throttle hand lever and observe overtravel of the spring loaded throttle arm after both the idle adjustment and high speed adjustment screws contact the stops. Overtravel in each direction should be approximately equal; if not, adjust pivot pin on end of throttle control cable and/or readjust position of throttle control cable in clamp to obtain equal over-travel. Then, check and adjust speed as follows:

Start engine and move hand throttle lever to low idle speed position; engine low idle speed should be 775 rpm for 200 and 7000 models and 675 rpm for all other models. Loosen locknut on idle speed adjustment screw and turn screw in to increase or out to decrease idle speed. Move hand throttle lever to high idle speed position and adjust the high idle speed screw to obtain 2200-2250 rpm for 180 models; 2400-2425 rpm for 185, 190 and 190XT models and 2375-2425 rpm for 200 and 7000 models.

DIESEL TURBOCHARGER

The diesel engine used in "One-Ninety XT," 200 and 7000 models is equipped with an exhaust driven turbocharger. Lubrication and cooling is provided by engine oil. After the engine is operated under load, the turbocharger should be allowed to cool by idling engine at 1000 rpm for 2-5 minutes. The engine should be immediately restarted if it is killed while operating at full load. After each 1000 hours of operation, the turbocharger should be inspected for cleanliness, freedom of rotation and bearing clearance. When servicing the turbocharger, extreme care must be taken to avoid damaging any moving parts.

TURBOCHARGER UNIT

Models So Equipped

73. REMOVE AND REINSTALL. Remove hood and disconnect air inlet hose from turbocharger. Remove turbocharger air outlet tube (6—Fig. 46 or 47) and exhaust outlet elbow (3). Disconnect oil inlet and return lines (1 & 7) from turbocharger, then unbolt and remove unit from exhaust manifold.

To reinstall, reverse removal procedure. Leave oil return line (7) disconnected and crank engine with starter until oil begins to flow from return port.

NOTE: Do not start engine until it is certain that turbocharger is receiving lubricating oil. Tighten stud nuts attaching turbocharger to exhaust manifold to 18-21 ft.-lbs. torque. Exhaust elbow mounting screws should be tightened to 28-33 ft.-lbs. torque.

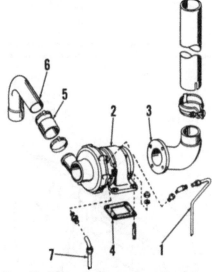

Fig. 46—View of Thompson turbocharger unit showing position of oil pressure line (1) and return line (7).

1. Lubrication line	5. Hose
2. Turbocharger	6. Tube to inlet
3. Exhaust elbow	manifold
4. Gasket	7. Oil return line

Fig. 45—Align timing marks on fuel injection pump governor weight retainer and cam as shown. Marks are visible after removing timing window cover (TW—Fig. 43) and rotating pump to bring mark on weight retainer in alignment with mark on cam.

Thompson Turbocharger

74. OVERHAUL. When servicing turbocharger as originally installed on 190XT tractors before engine serial number 2D-13744, it is advisable to use special tools available from Kent-Moore Organization, Inc. Some of the early turbochargers have been changed by installing later type parts and the special tools are not required. Attention will be called to difference in service procedure.

Remove turbocharger unit as outlined in paragraph 73. Scribe a mark across turbine housing (22—Fig. 48), clamp (20), bearing housing (13) and compressor housing (1) to aid reassembly and proceed as follows: Remove nut (19) and clamp (20); then, remove turbine housing (22). Remove cap screws attaching bearing housing (13) to compressor housing (1) and remove compressor housing. Hold blades of turbine wheel (18) with shop towel;

remove nut (3) and washer (4) from turbine shaft (L.H. threads). On early models, impeller (5E) is a tight fit on turbine shaft (18) and shaft must be pressed out of impeller as shown in Fig. 48A. On later models, impeller (5—Fig. 48) will slide easily off shaft.

CAUTION: On all models, do not allow turbine wheel and shaft to drop when impeller is removed.

Remove snap ring (6), then remove oil retainer plate (7 or 7E). On early type, retainer plate (7E) must be pressed out of bore using the Kent-Moore special tool. On later type (7), a ¾-inch driver can be used as shown in Fig. 48B to remove seal and retainer plate (7—Fig. 48), mating piece (10) and bearing (12).

It is recommended that late type seal (7) and mating ring (10) be renewed each time turbocharger is disassembled for service. The early type turbocharger can be changed to later type by installing new type parts (5, 7, 10 & 11).

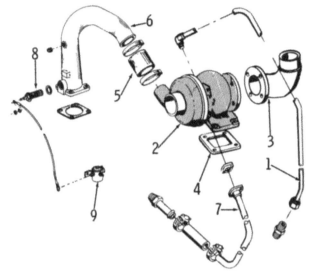

Fig. 47—View of Airesearch turbocharger unit showing position of pressure line (1) and return line (7).

1. Lubrication pressure line
2. Turbocharger
3. Exhaust elbow
4. Gasket
5. Hose
6. Tube to inlet manifold
7. Oil return line
8. Air heater element
9. Heater solenoid

Fig. 48A—View showing method of removing early type impeller using special Kent-Moore tool.

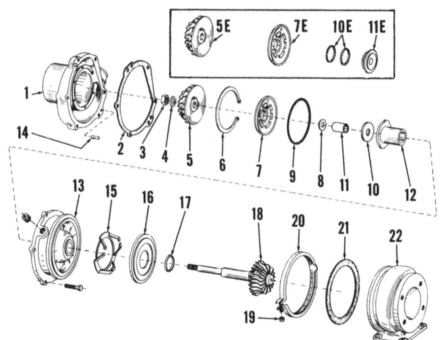

Fig. 48—Exploded view of Thompson turbocharger unit used on late 190XT diesel engines. Inset shows parts that are different on earlier unit.

1. Compressor housing	7E. Oil retaining plate	11E. Turbine shaft spacer	17. Turbine shaft seal
2. Gasket	8. Shims	12. Bearing	18. Turbine wheel & shaft
5 & 5E. Impeller	9. "O" ring	13. Bearing housing	20. Clamp
6. Snap ring	10. Mating ring	14. Groove pins	21. Gasket
7. Oil seal & retaining plate	10E. Turbine shaft seals	15. Spring ring	22. Turbine housing
	11. Shaft sleeve	16. Turbine wheel shield	

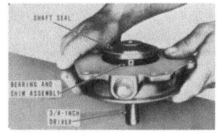

Fig. 48B—View showing method of removing oil seal and retainer from late type turbocharger. Parts should move out with only a small amount of pressure.

Clean all parts except late type oil seal (7) by washing in kerosene or diesel fuel. A nylon bristle brush may be used to clean carbon from parts.

CAUTION: Do not use wire brush, caustic cleaners, etc., on turbocharger parts. Inspect all parts for burring, eroding, nicks, breaks, scoring, excessive carbon build-up or other defects and renew all questionable parts.

On early type turbocharger, compressor impeller (5E) should be press fit on turbine shaft (18). Thrust surfaces of spacer (11E) must not be rough. Inspect grooves in spacer (11E) for excessive wear and make certain carbon is removed from bottom of grooves. Check O.D. of retainer plate (7E) for nicks that would damage "O" ring (9). Inspect bore in retainer plate for grooving caused by seal rings (10E) sticking and turning with shaft. Clean the bore chamfer to ease installation of seals (10E).

On late type turbocharger, inspect seal (7) carefully if the seal is to be reinstalled. The carbon face insert should move freely in and out and must not be scored or excessively worn. Mating ring (10) should not show evidence of wear or scuffing on either side. Remove any carbon from seal contact side. Examine sleeve (11) for burrs, scoring and wear. The sleeve is precision ground and any defect will distort the mating ring and cause seal leakage.

On all models, pay particular attention to blades of turbine wheel (18) and compressor impeller (5 or 5E) and to bushing (12). If bushing (12) and/or turbine shaft is worn excessively, the blades of turbine wheel and compressor impeller may have rubbed against housings. Shaft clearance in bushing should be measured at compressor impeller end of shaft using a dial indicator with unit assembled. If indicated clearance exceeds 0.022, bushing (12) and/or shaft (18) should be renewed. Inspect bore in bearing housing (13) for evidence of stuck seal ring (17).

If seal ring was sticking in shaft groove, the bore in housing will be grooved and should be renewed. Make certain that seal bore chamfer is clean and smooth to ease installation of shaft and seal ring.

To reassemble turbocharger, proceed as follows: Assemble bearing (12), mating ring (10) and sleeve (11) or bearing (12), and spacer (11E) on turbine shaft (18). Hold sleeve and mating ring (or spacer) tight against shoulder of turbine shaft and check end play of bearing with feeler gage as shown in Fig. 49. End play should be 0.004-0.006. If clearance exceeds 0.006, bearing (12—Fig. 48) should be renewed. **Record amount of end play for use in later step in reassembly.** Remove parts from turbine shaft.

Lubricate bearing (12) with motor oil and position in bearing housing bore with tabs on bearing between lugs in housing. Lubricate "O" ring (9) with silicone lubricant (or light grease) and position in bottom of retainer bore in bearing housing (13).

On early type, lubricate grooves in spacer (11E) and install new seal rings (10E) with end gaps opposite each other. Carefully slide spacer (11E) with seal rings (10E) into bore of retainer (7E). Install retainer and seal assembly in bearing retainer bore with one lug centered over each of the two tabs on bearing (12). Press retainer lightly into "O" ring (9) and install snap ring (6) with tapered side out.

On late type, lubricate mating ring (10) and position over hole in bearing (12). Install oil seal and retainer (7) in bearing retainer bore, press oil seal and retainer down lightly into "O" ring (9) and install snap ring (6) with tapered side out. When correctly assembled, lugs on inside of retainer will hold mating ring (10) in position.

CAUTION: Do not damage mating ring, "O" ring, seal and retainer by attempting to force installation with parts misaligned.

Check clearance between bearing and retainer as follows: Insert turbine shaft (18—Fig. 48) through bearing (12) and mating ring (10) or bearing and spacer (11E).

NOTE: Seal ring (17), spring ring (15) and shield (16) should not be installed for this check. On late type install sleeve (11) over end of shaft.

On all models, attach a dial indicator as shown in Fig. 49A. Measure and record the total amount of end play measured at sleeve (11—Fig. 48) or spacer (11E). If total end play is not within limits of 0.005-0.009, check for incorrect assembly. The end play of shaft in bearing (previously measured Fig. 49) should be 0.004-0.006 and end play of bearing between retainer and housing should be 0.001-0.003. If end play measurement is again within limits, **record the total shaft and bearing end play for use when selecting shims (8—Fig. 48).** Remove the turbine shaft (18) and proceed as follows:

Position impeller (5) over oil seal and retainer (7 or 7E). On late models, make certain that sleeve (11) is installed. On all models, install compressor housing (1) using a new gasket and three of the attaching screws torqued to 80 inch-pounds. Check end play of impeller in housing as shown in Fig. 49B. Subtract total shaft and bearing end play (found in preceding paragraph) from impeller end play. In final assembly, add shims (8—Fig. 48) of total thickness of 0.015-0.020 less than this determined value. As an example, if impeller to housing clearance is 0.049 and total shaft and bearing end play is 0.006; subtract as follows:

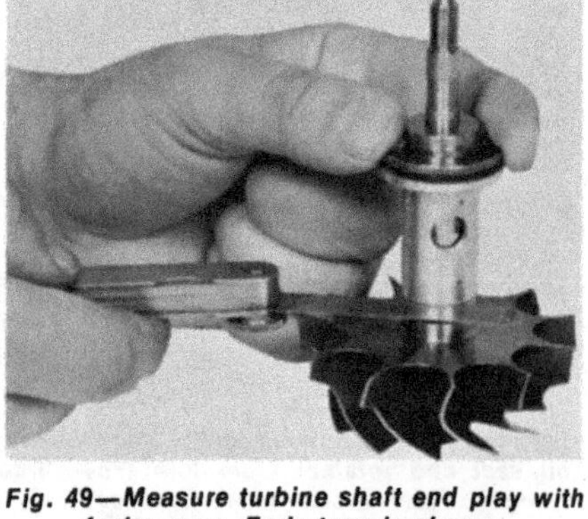

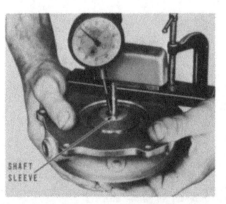

Fig. 49—Measure turbine shaft end play with feeler gage. Early type is shown.

Fig. 49A—Total end play of turbine shaft in bearing and bearing housing can be measured as shown. Late type turbocharger is shown but early type is similarly measured. It is usually necessary to push sleeve or spacer down, then push up on turbine shaft when measuring.

SHAFT SLEEVE

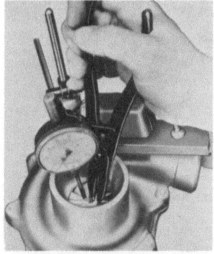

Fig. 49B—Measure impeller clearance in housing as shown. Turbine shaft is not installed when checking this clearance.

Impeller end clearance		0.049
Shaft and bearing end play		-0.006
		0.043
Determined value	0.043	0.043
Desired impeller clearance		-0.015 -0.020
Shim thickness required		0.028 0.023

Shims (8) are available in thicknesses of 0.010 and 0.015. The addition of one 0.010 and one 0.015 thick shim will be within the range of the shim thickness required in the preceding example.

Remove compressor housing (1) and impeller (5). Lubricate groove in turbine shaft (18) with motor oil and install new seal ring (17) in groove. Place plastic seal compressor (furnished with seal ring) over seal ring. Install spring ring (15) in bearing housing and place shield (16) over spring with projections on shield against the flat sections of the spring. Use two small "C" clamps to compress spring and hold shield in place. Insert tubine shaft (18) through hole in shield and bore of bearing. Plastic band used to compress seal ring (17) will be pushed off of seal ring as shaft is inserted in bearing and plastic will disintegrate from heat as soon as engine is started. Rotate turbine shaft to be sure it is a free fit in bearing. Place shim pack (8) (of thickness determined in previous step) on turbine shaft; then, install impeller onto turbine shaft. On early type, impeller must be pressed onto turbine shaft. On all models, install washer (4) and nut (3) on turbine shaft (L.H. threads) and tighten nut to a torque of 80-100 inch-pounds while holding turbine with shop towel. Check turbine shaft assembly for free rotation.

Reinstall compressor housing (1) and gasket (2) to bearing housing and tighten screws to 80-100 inch-pounds. Install turbine housing and gasket (21). Tighten clamp screw nut to 15-20 inch-pounds. Remove plug from bearing housing and fill reservoir with same type oil as used in engine. Protect all openings of turbocharger until unit is installed on tractor.

Rajay Turbocharger

74A. **OVERHAUL.** Remove turbocharger as outlined in paragraph 73. Scribe a mark across turbine housing (22—Fig. 50), bearing housing (13), housing flange (13A) and compressor housing (1) to aid reassembly and proceed as follows: Unscrew nut (19) and remove clamp (20) and turbine housing (22). Unscrew six countersunk screws retaining compressor housing (1) and separate housing from bearing housing flange (13A). Hold turbine wheel blades with a shop towel and

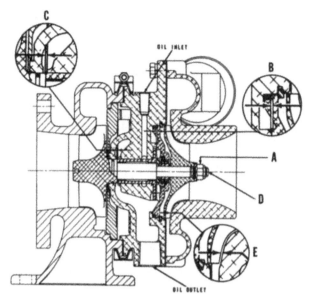

Fig. 49C—Cross section of Thompson turbocharger. Shaft side play in bearing, measured at (A) should not exceed 0.022. Shaft end play in bearing (C) is measured as shown in Fig. 49. Total shaft end play (D) is measured as shown in Fig. 49A. End clearance of bearing in housing (B) is difference of end play (D & C). Impeller to compressor housing clearance (E) is measured as shown in Fig. 49B.

remove nut (3) and washer (4)—nut (3) has left hand threads. Remove impeller (5) and shims (8) which may stick to impeller.

CAUTION: Do not allow turbine wheel and shaft to drop when impeller is removed.

Separate turbine wheel and shaft (18), turbine shield (16) and spring ring (15) from bearing housing (13). Be careful not to damage ring (17) ends when removing ring from turbine shaft. Remove bearing housing flange (13A) screws and separate flange from bearing housing (13). Remainder of disassembly is evident after inspection of unit and referral to Fig. 50.

Clean all parts except carbon insert (23) by washing in kerosene or diesel fuel. A nylon bristle brush may be used to clean carbon from parts.

CAUTION: Do not use wire brush, caustic cleaners, etc., on turbocharger parts. Inspect all parts for burring, eroding, nicks, breaks, scoring, excessive carbon build-up or other defects and renew all questionable parts.

Turbine end seal ring (17), "O" ring (9), mating ring (10) and carbon seal components (23, 24, 25 and 26) should be renewed whenever unit is disassembled. Carbon seal components (23, 24, 25 and 26) are available as a unit assembly only. Examine sleeve (11) for burrs, scoring and wear. Sleeve is precision ground and any defect will distort mating ring (10) and cause seal leakage. Note condition of turbine wheel blades, compressor impeller blades and bearing (12). If bearing (12) and/or turbine shaft is worn excessively, turbine wheel and compressor impeller

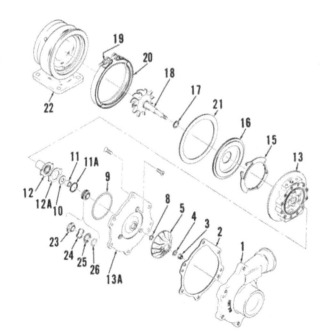

Fig. 50—Exploded view of Rajay turbocharger.

1. Compressor housing
2. Gasket
3. Nut
4. Washer
5. Impeller
8. Shims
9. "O" ring
10. Mating ring
11. Sleeve
11A. Spiral retaining ring
12. Bearing
12A. Wear ring
13. Bearing housing
13A. Bearing housing flange
15. Spring ring
16. Turbine wheel shield
17. Seal ring
18. Turbine wheel & shaft
19. Nut
20. Clamp
21. Gasket
22. Turbine housing
23. Carbon insert
24. Wave spring
25. Washer
26. "O" ring

blades may have rubbed against housings. Shaft clearance in bushing should be measured at compressor impeller end of shaft using a dial indicator with unit assembled. If indicated clearance exceeds 0.022, bushing (12) and/or shaft (18) should be renewed. Inspect bore in bearing housing (13) for evidence of stuck seal ring (17). If seal ring was sticking in shaft groove, the bore in housing will be grooved and should be renewed. Make certain that seal bore chamfer is clean and smooth to ease installation of shaft and seal ring.

To reassemble turbocharger, proceed as follows: Assemble bearing (12), mating ring (10) and sleeve (11) on turbine shaft (18). Hold sleeve and mating ring tight against shoulder of turbine shaft and check end play of bearing with feeler gage as shown in Fig. 49. End play should be 0.004-0.006. If clearance exceeds 0.006, bearing (12—Fig. 50) should be renewed. **Record amount of end play for use in later step in reassembly.** Remove parts from turbine shaft.

Assemble carbon seal components (23, 24, 25 and 26) and install assembly in bearing housing flange (13A) so "O" ring (26) is against flange. Install spiral retaining ring (11A) in groove of flange lugs. Press down evenly on carbon seal to be sure seal moves freely and is not binding.

Be sure wear ring (12A) is attached to bearing (12). Lubricate bearing (12) with motor oil and position in bearing housing bore with bearing tabs between bearing housing lugs. Lightly lubricate bearing (12) face, install mating ring (10) on bearing and lightly lubricate outside face of mating ring.

"O" ring (9) groove may be located in bearing housing (13) or flange (13A). Early models have "O" ring groove located in flange while later models have "O" ring groove located in bearing housing. Do not mix early and late style bearing housings (13) and flanges (13A). Lubricate "O" ring (9) with silicone lubricant (or light grease) and position in "O" ring groove. Refer to previously scribed marks and install flange (13A) on bearing housing (13). Tighten screws to 80-100 in.-lbs.

To check bearing clearance, proceed as follows: Install turbine wheel and shaft (18) without seal ring (17). Install sleeve (11) on turbine shaft. Attach a dial indicator as shown in Fig. 49A. Measure and record total amount of sleeve (11—Fig. 50) end play. If total end play is not within 0.005-0.009, check for incorrect assembly. Turbine shaft end play in bearing (previously measured in Fig. 49) should be 0.004-

0.006 and bearing (12—Fig. 50) end play between bearing housing (13) and flange (13A) should be 0.001-0.003. Record total shaft and bearing end play for use when selecting shims (8). Remove turbine shaft and proceed as follows:

Position impeller (5) over sleeve (11). Install compressor housing (1) using a new gasket and three of the attaching screws tightened to 80 inch-pounds. Check end play of impeller in housing as shown in Fig. 49B. Subtract total shaft and bearing end play (found in preceding paragraph) from impeller end play. In final assembly, add shims (8—Fig. 50) of total thickness of 0.015-0.020 less than this determined value. As an example, if impeller to housing clearance is 0.049 and total shaft and bearing end play is 0.006; subtract as follows:

Impeller end clearance		0.049
Shaft and bearing end play		-0.006
		0.0043
Determined value	0.043	0.043
Desired impeller clearance	-0.015	-0.020
Shim thickness required	0.028	0.023

Shims (8) are available in thicknesses of 0.010 and 0.015. The addition of one 0.010 and one 0.015 thick shim will be within the range of the shim thickness required in the preceding example.

Remove compressor housing (1) and impeller (5). Lubricate groove in turbine shaft (18) with motor oil and install new seal ring (17) in groove. Place plastic seal compressor (furnished with seal ring) over seal ring. Install spring ring (15) in bearing housing and place shield (16) over spring with projections on shield against the flat sections of the spring. Use two small "C" clamps to compress spring and hold shield in place. Insert turbine shaft (18) through hole in shield and bore of bearing. Plastic band used to compress seal ring (17) will be pushed off of seal ring as shaft is inserted in bearing and plastic will disintegrate from heat as soon as engine is started. Rotate turbine shaft to be sure it is a free fit in bearing. Place shim pack (8) (of thickness determined in previous step) on turbine shaft; then, install impeller onto turbine shaft. Install washer (4) and nut (3) on turbine shaft (L.H. threads) and tighten nut to 80-100 inch-pounds while holding turbine with

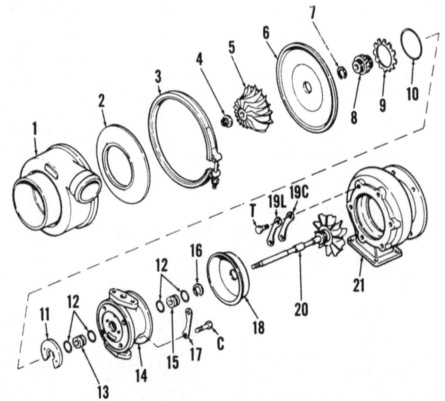

Fig. 51—Exploded view of Airesearch turbocharger. Use extreme care to prevent damage to parts.

C. Backplate screws	4. Locknut	10. Seal ring	17. Lock plates
T. Turbine housing screws	5. Compressor impeller	11. Thrust bearing	18. Turbine shroud
1. Compressor housing	6. Backplate	12. Bearing retainers	19C. Clamp plates
2. Diffuser	7. Seal ring	13. Bearing	19L. Lock plates
3. Clamp	8. Thrust collar	14. Center housing	20. Turbine wheel and shaft
	9. Spring	15. Bearing	21. Turbine housing
		16. Seal ring	

shop towel. Check turbine shaft assembly for free rotation.

Reinstall compressor housing (1) and gasket (2) to bearing housing and tighten cap screws to 80-100 inch-pounds. Install turbine housing and gasket (21). Tighten clamp screw nut to a torque of 15-20 inch-pounds. Remove plug from bearing housing and fill reservoir with same type oil as used in engine. Protect all openings of turbocharger until unit is installed on tractor.

Airesearch Turbocharger

74B. **OVERHAUL.** Remove turbocharger unit as outlined in paragraph 73. Mark across compressor housing (1—Fig. 51), center housing (14) and turbine housing (21) to aid alignment when assembling.

CAUTION: Do not rest weight of any parts of impeller on turbine blades. Weight of only the turbocharger unit is enough to damage the blades.

Remove clamp (3), compressor housing (1) and diffuser (2). Remove screws (T), lock plates (19L) and clamp plates (19C); then, remove turbine housing (21). Hold turbine shaft from turning by using the appropriate type of wrench at center of turbine wheel (20) and remove locknut (4).

NOTE: Use a "T" handle to remove locknut in order to prevent bending turbine shaft.

On some turbine shafts, an Allen wrench must be used at turbine end while others are equipped with a hex end and can be held with a standard socket. Lift compressor impeller (5) off, then remove center housing from turbine shaft while holding shroud (18) onto center housing. Remove back plate retaining screw (C), then remove back plate (6), thrust bearing (11), thrust collar (8) and spring (9). Carefully remove bearing retainers (12) from ends and withdraw bearings (13 & 15).

CAUTION: Be careful not to damage bearings or surface of center housing when removing retainers. The center two retainers do not have to be removed unless damaged or unseated. Always renew bearing retainers if removed from grooves in housing.

Clean all parts in a cleaning solution which is not harmful to aluminum. A stiff brush and plastic or wood scraper should be used after deposits have softened. When cleaning, use extreme caution to prevent parts from being nicked, scratched or bent.

Inspect bearing bores in center housing (14—Fig. 51) for scored surface, out of round or excessive wear. Bearing bore diameter must not exceed

0.6228 and maximum permissible out of round is 0.0003. Make certain bore in center housing is not grooved in area where seal (16) rides. Inside diameter of bearing (13 & 15) must not be more than 0.4019 and outside diameter must not be less than 0.6182. Thrust bearing (11) should be measured at three locations around collar bore. Thickness should be not more than 0.1720 or less than 0.1711. Inside diameter of bore in backplate (6) must not exceed 0.5015 and seal contact area must be clean and smooth. Compressor impeller (5) must not show signs of rubbing with either the compressor housing (1) or the back plate (6). Impeller should have 0.0002 tight to 0.0004 loose fit on turbine shaft. Make certain that impeller blades are not bent, chipped, cracked or eroded. Oil passage in thrust faces must not be warped or scored. Ring groove shoulders must not have step wear. Bearing area width must not exceed 0.1758 and width of groove for seal ring (7) must not exceed 0.0065. Clearance between thrust bearing (11) and groove in collar (8) must be 0.001-0.004, when checked at three locations. Inspect turbine shroud (18) for evidence of turbine wheel rubbing. Turbine wheel (20) should not show evidence of rubbing and vanes must not be bent, cracked, nicked, or eroded. Turbine wheel shaft must not show signs of scoring, scratching or overheating. Diameter of shaft journals must not be less than 0.3992 and out of round must not exceed 0.0003. Groove in shaft for seal ring (16) must not be stepped and diameter of hub near seal ring should be 0.682-0.683. Check shaft end play and radial clearance when assembling.

If bearing inner retainers (12) were removed, install new retainers using special Kent-Moore tools (JD-274).

CAUTION: Bore in housing may be damaged if special retainer installing tool is not used.

Oil bearings (13 & 15) and install outer retainers using the special tool. Position the shroud (18) on turbine shaft (20) and install seal ring (16) in groove. Apply a light, even coat of engine oil to shaft journals, compress seal ring (16) and install center housing (14). Install new seal ring (7) in groove of thrust collars (8), then install thrust bearing so that smooth side of bearing (11) is toward seal ring (7) end of collar. Install thrust bearing and collar assembly over shaft, making certain that pins in center housing engage holes in thrust bearing. Install new rubber seal ring (10), make certain that spring (9) is positioned in backplate (6), then install backplate making certain

that seal ring (7) is not damaged. Install lock plates (17) and screws (C), then tighten screws to 40-60 in.-lbs. torque. Install compressor impeller (5) and make certain that impeller is completely seated against thrust collars (8). Install lock nut (4) to 18-20 in-lbs, then use a "T" handle to turn locknut an additional 90 degrees.

CAUTION: If "T" handle is not used, shaft may be bent when tightening nut (4).

Install turbine housing (21) with clamp plates (19C) next to housing, tighten screws (T) to 100-130 ft.-lbs., then bend lock plates (19L) up around screw heads.

Check shaft end play and radial play at this point of assembly. If shaft end play (Fig. 52) exceeds 0.0042, thrust collars (8—Fig. 51) and/or thrust bearing (11) is worn excessively. End play of less than 0.001 indicates incomplete cleaning (carbon not all removed) or dirty assembly and unit should be disassembled and cleaned. Refer to Figs. 52B & 52C. If turbine shaft radial play exceeds 0.007, unit should be disassembled and bearings, shaft and/or center housing should be renewed. Maximum permissible limits of all of these parts may result in radial play which is not acceptable.

Make certain that legs on diffuser (2—Fig. 51) are aligned with spot faces on backplate (6) and install diffuser.

Fig. 52—View showing method of checking turbine shaft end play. Shaft end play should be checked after unit is cleaned to prevent false reading caused by carbon build-up.

Install compressor housing (1) and tighten nut of clamp (3) to 40-80 in-lbs. torque. Fill reservoir with engine oil and protect all openings of turbocharger until unit is installed on tractor.

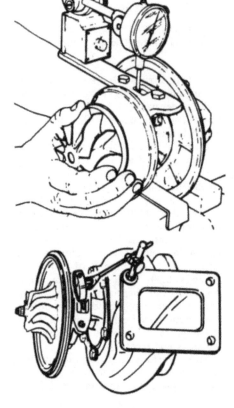

Fig. 52B—Turbine shaft radial play is checked with dial indicator through the oil outlet hole and touching shaft at (D—Fig. 52C). Two methods of attaching dial indicator to center housing are shown above.

OIL COOLER

"One-Ninety XT" and all 200 and 7000 Models

75. The engine oil on turbocharged models is used to cool and lubricate the turbocharger. These models are provided with an engine oil cooler (Fig. 55) located on right side of engine. Normal maintenance consists of renewing hoses (1 & 2) and cleaning or renewal of cooler (5). Make certain that heat shields (6) are in place. It may be necessary to use sealer on hoses.

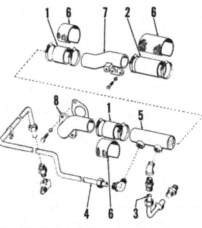

Fig. 55—Exploded view of engine oil cooler used on turbocharged "XT" diesel and all 200 and 7000 tractors.

1. Hose
2. Hose
3. Return tube
4. Inlet tube
5. Oil cooler
6. Hose heat shield
7. Water inlet
8. Water outlet

NON-DIESEL GOVERNOR

76. **SPEED ADJUSTMENT.** Before attempting to adjust the governed engine speeds, check the length of link rod (25—Fig. 57) between governor lever and carburetor. Disconnect link rod from carburetor and move carburetor throttle and governor lever to the wide open position. The link rod should be 1/32 to 1/16-inch too long. If rod length is incorrect, adjust at rod ends. Tighten locknuts at ends of link rod before reattaching.

Adjust low idle engine speed with idle stop screw on carburetor throttle shaft. Engine low idle speed should be 450-500 rpm for 180 tractors; 400-450 for all other non-diesel models. To adjust high idle speed, loosen locknut (22—Fig. 57) and turn plunger (21) in center of governor spring (20) on control cable end (23). High idle speed should be 2175-2225 rpm on 180 tractors; 2440-2460 rpm on all other non-diesel models.

77. **OVERHAUL.** The governor assembly shown in Fig. 57 is contained in the timing gear cover. Procedure for overhaul will be evident after removing the timing gear cover as outlined in paragraph 29. Bearing (11) is pressed into cylinder block.

COOLING SYSTEM

RADIATOR

78. **REMOVE AND REINSTALL.** Drain engine cooling system and remove grille. Remove hood top panel and tube from air cleaner to engine. Disconnect radiator hoses, then unbolt and remove radiator from tractor.

When reinstalling, make certain that sealing pads are installed around top of radiator. Cooling system capacity is 22.5 quarts on 7000 models and 18 quarts on all other models. Proper level is 2 inches below filler neck.

WATER PUMP

79. **REMOVE AND REINSTALL.** Drain cooling system and remove both hood side panels. Loosen alternator mounting bolts and remove fan belt. Remove fan retaining screws, lower fan in radiator shroud and remove pulley (1—Fig. 57A). "One-Ninety XT" diesel and 200 models have spacer (2) behind pulley. Unbolt inlet housing (15) from pump and remove by-pass tube between pump and thermostat housing. Remove screws attaching pump to

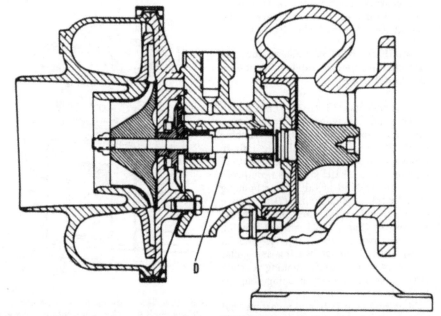

Fig. 52C—Cross-sectional drawing of Airesearch turbocharger. Shaft radial play is checked through oil outlet hole as shown at (D).

outlet housing (12) and withdraw pump from right side.

When reinstalling, always renew "O" ring (8). The fan and pulley retaining screws should be tightened to 30-35 ft.-lbs. Pump to outlet housing and inlet housing to pump screws should be tightened to 28-33 ft.-lbs.

80. **OVERHAUL.** Refer to Fig. 57A. Remove pulley hub (3) using a suitable puller. Remove retaining ring (4) on all models except 200 and 7000. Press shaft and bearing assembly (5) forward out of pump body (7) and impeller (11). Seal (9) can be pressed out toward rear of body.

Renew seal (9) and other parts as necessary. Apply gasket sealer to outer rim of seal and press seal into pump body. On all models except 200 and 7000, press slinger (S) onto rear of shaft until distance (D) is 1-15/16 inches. Models 200 and 7000 are not equipped with slinger.

Press shaft and bearing assembly (5) into front of pump body so that retainer (4) can be installed. Position plate (10) into recess in pump body and press impeller (11) onto rear end of shaft so that there is a 0.015 clearance between impeller vanes and body plate. Press pulley hub (3) onto front end of shaft with flange towards pump body. Front edge of hub flange should be 4 inches from rear surface of pump body (6).

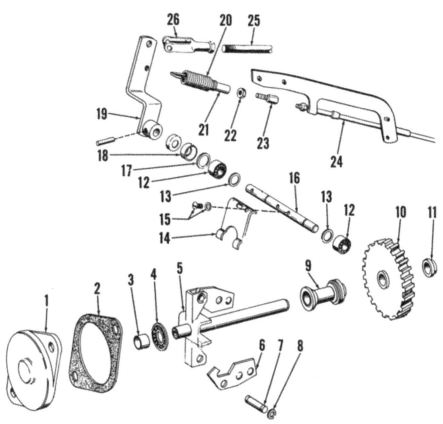

Fig. 57—Exploded view of non-diesel governor assembly. Unit is housed in timing gear cover.

1. Cover	8. Retainer ring	15. Screw (2 used)
2. Gasket	9. Thrust sleeve	16. Yoke shaft
3. Bushing	10. Drive gear	17. Retainer
4. Thrust bearing	11. Bushing	18. Oil seal
5. Shaft	12. Bearings	19. Governor lever
6. Weight (2 used)	13. Washers	20. Governor spring
7. Pin (2 used)	14. Governor yoke	21. Plunger
		22. Lock nut
		23. Cable end
		24. Throttle cable
		25. Carburetor link rod
		26. Yoke

IGNITION AND ELECTRICAL SYSTEM

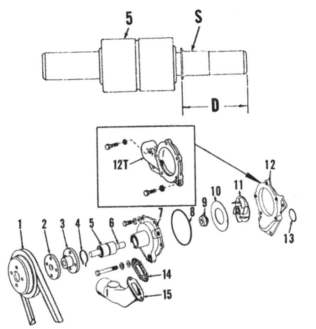

Fig. 57A—Exploded view of water pump. Spacer (2) and outlet housing (12T) are used on turbocharged "XT" and all 200 diesel models only. Refer to text for installation of slinger (S) and hub (3).

1. Pulley
2. Spacer
3. Hub
4. Retainer ring
5. Shaft and bearing
6. Water slinger
7. Body
8. "O" ring (4-7/8 inch)
9. Seal
10. Body plate
11. Impeller
12. Outlet housing
13. "O" ring (2¼ inch)
14. Gasket
15. Inlet housing

SPARK PLUGS

Non-Diesel

81. A-C 45XL, Autolite AG3-A, or Champion N-6 spark plugs should be used in gasoline engines. For LP-Gas, use A-C 42XL or Champion N-3 spark plugs. Set electrode gap to 0.025 for gasoline models, 0.020 for LP-Gas.

DISTRIBUTOR

Non-Diesel

82. Delco-Remy 1112653 or 1111311 ignition distributor is used on early gasoline models, Delco-Remy 1112667 distributor on early LP-Gas engines. Later LP-Gas and gasoline engines are equipped with Delco-Remy 1112668 distributor. The distributors used on early tractors are equipped with hour-meter drive; on later tractors, hour-meter is driven from rear of alternator. The ignition coil should be used **ONLY** with an external resistor. Refer to the following distributor specifications data.

1111311

Breaker contact gap 0.016
Cam Angle . . . . 31°-34°
Breaker arm spring pressure
(measured at center of
contact) . . . . 17-21 oz.
Advance data in distributor degrees
and rpm.
Start advance . . . . 0°-2° at 250 rpm
Intermediate
advance . . . . 6.5°-8.5° at 425 rpm
Intermediate
advance . . . . 12°-14° at 975 rpm
Maximum
advance . . . . 16°-18° at 1375 rpm

1112653

Breaker contact gap 0.016
Cam Angle . . . . 31°-34°
Breaker arm spring pressure
(measured at center of
contact) . . . . 17-21 oz.

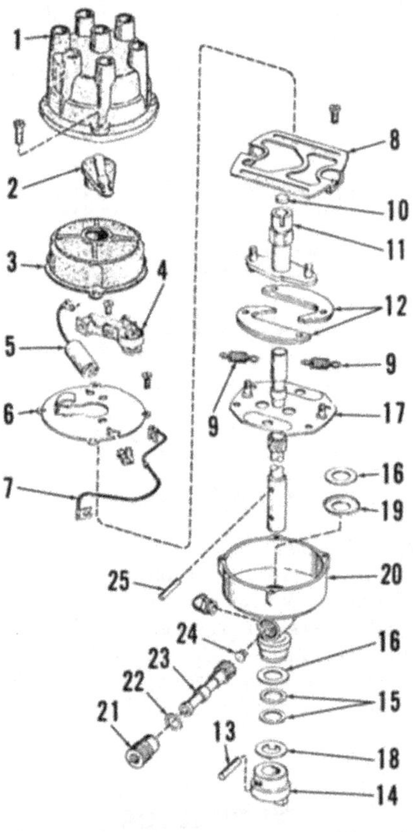

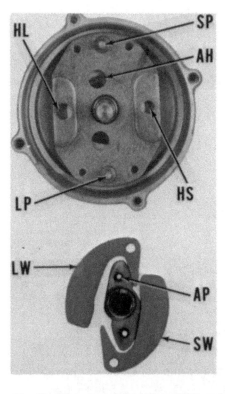

Fig. 58—Exploded view of the ignition distributor. Refer to text and Fig. 59 when assembling the advance weights to the main shaft. Thrust washer (18) and tachometer drive parts (21 thru 24) are not used on all models.

1. Cap
2. Rotor
3. Cover
4. Breaker points
5. Condenser
6. Breaker plate
7. Primary lead
8. Weight retainer
9. Advance springs
10. Oil wick
11. Breaker cam
12. Advance weights
13. Roll pin (1/8 x 7/8)
14. Drive coupling
15. Shims
16. Washer
17. Main shaft and weight plate
18. Thrust washer
19. Seal
20. Housing
21. Coupling
22. Washer
23. Tachometer drive shaft
24. Plug
25. Roll pin 1/8 x ¾)

Advance data in distributor degrees
and rpm.
Start advance . . . . . . 0°-2° at 250 rpm
Intermediate
advance . . . . 5°-7° at 400 rpm
Intermediate
advance . . . . 11°-13° at 850 rpm
Maximum
advance . . . . 16°-18° at 1200 rpm

1112667

Breaker contact gap 0.016
Cam Angle . . . . 31°-34°
Breaker arm spring pressure
(measured at center of
contact) . . . . 17-21 oz.
Advance data in distributor degrees
and rpm.
Start advance . . . . 0°-2° at 275 rpm
Intermediate
advance . . . . 4°-6° at 400 rpm
Maximum
advance . . . . 9°-11° at 550 rpm

1112668

Breaker contact gap 0.016
Cam Angle . . . . 31°-34°
Breaker arm spring pressure
(measured at center of
contact) . . . . 17-21 oz.
Advance data in distributor degrees
and rpm.

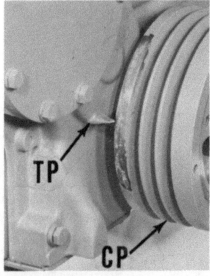

Fig. 59—When assembling the weights and breaker cam, parts must be assembled as shown. Refer to text.

LP & SP. Weight
    pivot pins
LW & SW. Centrifugal
    weights
AH. Advance stop
    hole used

AP. Long pin used
    as advance stop
HL. Large hole
HS. Small hole

Start advance . . . . . 1.5°-4.5° at 300 rpm
Intermediate
advance . . . . . 3.5°-6-5° at 370 rpm
Intermediate
advance . . . . . 5.1°-7.1° at 420 rpm
Maximum
advance . . . . . . 13°-15° at 1200 rpm

The main shaft (17—Fig. 58) should have 0.002-0.010 inch end play. Shims (15) are used to adjust end play and are available in thicknesses of 0.005 and 0.010 inch.

When assembling the distributor, make certain that weights and breaker cam are correctly installed. Turn the main shaft and weight plate until the larger hole (HL—Fig. 59) is on left as shown, then install two weights (LW) over the top pin (SP) so that the weight covers the larger hole (HL). Install the other weights (SW) on the lower pin (LP) with weight end covering the small hole (HS). This will correctly position the advance weights in relation to direction of rotation. The breaker cam has two spring attaching pins. The longer pin (AP) extends through cam plate and into hole (AH). Movement of the advance pin (AP) in the advance hole (AH) limits the amount of ignition advance. When assembling the breaker cam, make certain that the longer pin (AP) enters hole (AH) and not the similar hole at bottom.

Install the distributor as outlined in paragraph 83.

IGNITION TIMING

Non-Diesel

83. It is important that ignition timing be set using a power timing light and engine running at exactly

Fig. 60—View of timing marks on crankshaft pulley (CP) and timing pointer (TP).

2000 rpm on 180 tractor models; 2200 rpm on 190 and 190XT tractor. At this speed on gasoline models, timing flash should occur when 26 degree mark on crankshaft pulley is aligned with pointer on timing gear cover (Refer to Fig. 60). Ignition timing will advance beyond 26° if engine speed is higher than specified rpm. If timing is incorrect when engine is operating at specified rpm, loosen clamp screw at base of distributor, turn distributor body to correct the timing, then tighten clamp screw.

To install distributor and set initial timing, proceed as follows: Turn engine in normal direction of rotation until number one piston is coming up on compression stroke, then continue turning engine slowly until the TDC mark on crankshaft pulley is aligned with pointer on timing gear cover (Refer to Fig. 60). With point gap set at 0.016, install distributor with drive tang engaging notch in oil pump drive gear. (The driving notch and tang are off-set from center line of shafts.) Turn distributor body until rotor points toward number one cylinder terminal of distributor cap and breaker points are just open. This will provide initial timing; however, running timing MUST be checked as outlined in the previous paragraph.

ALTERNATOR, REGULATOR AND STARTING MOTOR

All Models

84. **ALTERNATOR AND VOLTAGE REGULATOR.** Delco-Remy alternator and regulator are used on all models. On early 190 and 190XT models, alternator number 1100720 was used and tachometer was driven by ignition distributor (gasoline models) or by a drive located in place of distributor on diesel models. On later 190 and 190XT models and all 180, 185 and 200 tractors, alternator number 1100735 is used and tachometer is driven by gear box located on rear of alternator. On 7000 models, alternator number 1102870 is used and regulator is solid state and contained in alternator housing. Refer to the following specification data:

Alternator
1100720 & 1100735
Field Current at 80°F.
Volts 12
Amperes 2.2-2.6
Cold Output at 14 Volts—
Amperes at 2000 rpm 21
Amperes at 5000 rpm 30
Hot Output (Rated)—
Amperes 32

1102870
Field Current at 80°F.
Volts 12
Amperes 4.0-4.5
Cold Output—
Amperes at 2000 rpm 22
Amperes at 5000 rpm 33
Hot Output (Rated)—
Amperes 37
*Maximum cold current output is obtained by connecting a carbon pile between battery terminals and adjusting carbon pile for maximum alternator current output.

Regulator

Regulator on 1102870 alternator is a non-adjustable solid state unit contained in alternator housing. If regulator malfunctions, regulator unit must be renewed. Refer to the following specifications for regulators used with alternators 1100720 and 1100735:

Relay Unit:
Air gap 0.015
Point opening 0.030
Closing voltage 3.8-7.2
Regulator:
Air gap
(lower points closed) 0.067*
Upper point opening
(lower points closed) 0.014
Voltage setting:

Temp., °F.**	Volts***
65	13.9-15.0
85	13.8-14.8
105	13.7-14.6
125	13.5-14.4
145	13.4-14.2
165	13.2-14.0
185	13.1-13.9

*Air gap setting of 0.067 is only a starting point; correct air gap is obtained by adjusting unit for proper voltage regulation.

**Ambient temperature measured ¼-inch away from voltage regulator cover; adjustment should be made only when at normal operating temperature.

***Regulated voltage when regulator is working on upper set of points; when regulator is working on lower set of points, voltage should be 0.1 to 0.4 volts less than given in table. Voltage setting may be increased up to 0.3 volts to correct chronic battery undercharge or decreased up to 0.3 volts to correct battery overcharging condition.

84A. **STARTING MOTOR.** Diesel tractors use Delco-Remy 1113657 starting motor and non-diesel tractors use 1107356 starting motor. Later 7000 models are equipped with an M50 Lucas starting motor. Refer to the following specification data:

Delco-Remy Starting Motor—
1113657
Brush Spring Tension .... 36 oz.
No Load Test—
Volts 9
Amperes 75-105
Rpm 5000-7000
†Includes solenoid
Lock Test—
Volts (approx.) 2.3
Amperes 600
Torque—Minimum .... 17 ft.-lbs.

1107356
Brush Spring Tension .... 40 oz.
No Load Test—
Volts 11.8
Amperes 72
Rpm 6025
Resistance Test—
Volts 3.5
Amperes 295-365

Lucas Starting Motor—
M50
Brush Spring Tension .... 42 oz.
No Load Test—
Volts 12
Amperes 100
Rpm 5500-7500
Lock Test—
Volts 12
Amperes 980
Torque 34

ENGINE CLUTCH

ADJUSTMENT

85. Clutch pedal linkage is properly adjusted when pedal has a free movement of 2½ inches on 180 and 185 models; 2 inches on 190, 190XT, 200 and 7000 models. Free movement is measured at the pedal pad. To adjust the linkage, remove the hood left side panel and disconnect yoke (Fig. 61 for 180 and 185 models; Fig. 61A for 190, 190XT and 200 models; Fig. 61B for 7000 models). Lengthen or shorten clutch pedal link rod as required until free travel is correct.

Fig. 61—View of clutch linkage used on 180 and 185 models. Refer to text for adjusting pedal free movement and safety starting switch.

On models so equipped the safety starting switch (Fig. 61 or Fig. 61B) must be adjusted to allow starting **ONLY** when the clutch pedal is depressed fully. To adjust, depress pedal and position the actuator so that button on switch is compressed 1/8-inch, then tighten the clamp screw in clip. The switch button must be com-pressed 1/16-inch to allow starting and may be damaged if compressed too far.

Model 7000 is equipped with a transmission brake which is actuated when clutch pedal is depressed. The transmission brake stops "Power Shift" clutch rotation to allow easier transmission shifting. Turn adjusting nut (N—Fig. 61C) so spring (S) length is two inches when clutch pedal is depressed fully. Transmission brake must be adjusted whenever clutch pedal is adjusted.

CLUTCH UNIT

86. REMOVE AND REINSTALL. The engine clutch can be removed from flywheel after first removing the engine as outlined in paragraph 20.

Reverse removal procedures to reinstall the clutch assembly. Install clutch disc with dampener springs toward engine. Use a suitable pilot tool to align clutch disc hub with the pilot bearing in flywheel. Tighten the clutch cover retaining screw to 25-30 ft.-lbs. on 190, 190XT, 200 and 7000 models; 45-50 ft.-lbs on 180 and 185 models.

87. OVERHAUL. Several variations of clutches are used and while some parts are the same or similar, it is important that correct parts are used.

On 180 and 185 tractor models, the clutch disc (1—Fig. 62) is 0.369-0.399 thick when new and the pressure plate (2) is identified by part number C4-11029-4X1. The 12 springs (3) are painted black, should have free length of 2.562 inches and should exert 150 lbs. pressure (minimum) when compressed to 1-13/16 inches.

Model 190 tractors before serial number 10208 are originally equipped with a round sintered metallic clutch disc (1), that is 0.281-0.301-inch thick when new. Before serial number 3066, nine brown clutch springs (3) were used. Nine lavender clutch springs (3) were used from serial number 3066 to 10208. Free length of the brown springs should be 2-13/16 inches and springs should exert 140 lbs. (minimum) when compressed to 1-13/16 inches. Free length of the lavender springs should be 2-9/16 inches and springs should exert 180 lbs. (minimum) when compressed to 1-13/16 inches. The later (lavender) springs should be used for service. All nine clutch springs must be the same type. The pressure plate (2) marked CL-11029-3-A is used only with the round, 0.281-0.301 thick clutch disc (1) and nine brown or lavender clutch springs (3). If type of pressure plate is questioned, measure thickness from friction surface to end of the lever attaching lug. On early type pressure plate, thickness should be approximately 2.840 inches. Thickness of later type plate is 2.750. Clutch can be changed to late type, by installing twelve black springs, later pressure plate and blade type clutch disc (0.372-0.388 thick).

Model 190 tractors after serial number 10207 and all 200 and 7000 models are equipped with a blade type clutch disc with four friction pads on each side. Thickness at friction pads is 0.372-0.388-inch when new. The pressure plate is approximiately 2.750 inches thick when measured from friction surface to end of lever attaching lug. The twelve springs (3) are painted black for identification, should have free length of 2-9/16 inches and should exert 150 lbs. (minimum) when compressed to 1-13/16 inches.

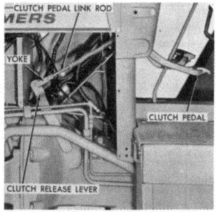
Fig. 61A—Typical view of clutch linkage used on 190, 190XT and 200 models. Refer to text for adjusting pedal free movement.

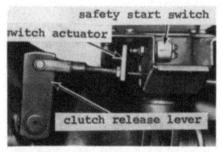
Fig. 61B—View of safety start switch on 7000 models. Refer to paragraph 85 for adjustment.

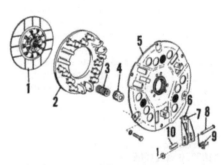
Fig. 61C—View of Model 7000 transmission brake actuating linkage. Refer to paragraph 85 for adjustment.

Fig. 62—Exploded view of clutch assembly. Round clutch disc (1) is shown. Springs are not installed in the 3 darkened holes when only 9 springs are used.

1. Driven disc
2. Pressure plate
3. Spring
4. Cup
5. Cover
6. Lever
7. Adjusting screw
8. Pin
9. Anti-rattle spring
10. Short pin

Reassemble clutch and disc on flywheel and adjust screws (7). Distance between cover plate (5) and release bearing contact surface of screws (7) should be the same for all three screws.

RELEASE BEARING

88. The release bearing (6—Fig. 63 or Fig. 63A) can be removed after first removing the engine as outlined in paragraph 20. Remove snap ring (4) from right end of shaft and disconnect operating rod from lever (1). Withdraw lever and shaft (1), removing keys (2K) as they slide out of fork (2). Bearing (6) can be pressed from hub (5).

When reassembling, make certain that bearing is pressed completely on hub (5). Straight side of release fork (2) should be toward rear.

CLUTCH SHAFT

Model 7000

89. To remove clutch shaft, remove "Power Shift" gears as outlined in paragraph 103 and PTO clutch as outlined in paragraph 168. Withdraw clutch shaft from rear of torque housing. Unscrew nut (22—Fig. 63) and remove snap ring (18) to remove gear (21) and bearing (20).

To install clutch shaft, reverse removal procedure. Tighten nut (22) to

40-50 ft.-lbs and stake nut. Be sure baffle (17) is installed with taper towards front of shaft.

All Other Models

89A. To remove the engine clutch shaft, it is necessary to remove the engine as outlined in paragraph 20. Remove snap ring (4—Fig. 63A), disconnect operating yoke and withdraw lever and shaft (1) from left side of torque housing. Remove release bearing and hub assembly (5 & 6) and fork (2). Unbolt and remove retainer (7), gasket (8) and seal (9). Remove the belt pulley, shifter assembly and coupling (10). Block the torque housing securely and split transmission from torque housing as outlined in paragraph 104. Remove the "Power-Director" outer (hollow) shaft as in paragraph 95. Remove snap ring (22) and thrust washer (21); then, press the clutch shaft (17) forward out of bearing cone (19).

Intermediate shaft and gears must be removed as in paragraph 96 if bearing (19) is to be renewed.

Reinstall in reverse of removal procedure. Snap ring (22) is available in several thicknesses from 0.093 to 0.137. Install snap ring of thickness required to provide clutch shaft (17) with 0.002-0.005 end play in bearings (15 & 19). Seal (9) should be installed with lip toward rear. Cap screws attaching retainer (7) to torque housing should be tightened to 35-40 ft.-lbs.

"POWER-DIRECTOR"

The "Power-Director" clutches are actuated by hydraulic pressure. With control lever in forward position, oil is directed to rear clutch piston, clutch is engaged and the transmission input shaft is driven at engine speed. With hand control lever positioned fully to the rear, front clutch is engaged and transmission input shaft is driven at reduced speed. With control lever in center (neutral) position, both clutches are disengaged providing live pto requirements.

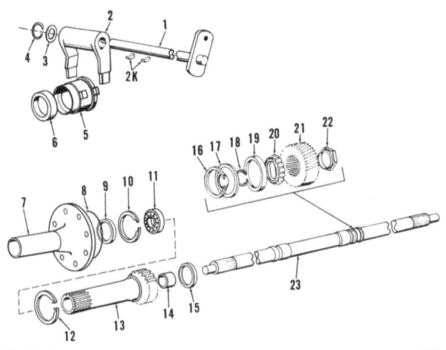

Fig. 63—Exploded view of engine clutch shaft and pto input shaft assemblies used on 7000 models.

1. Release lever and shaft	7. Front support plate	13. Pto clutch input shaft
2. Release fork	8. "O" ring	14. Bushing
3. Washer	9. Seal	15. Seal
4. Snap ring	10. Snap ring	16. Snap ring
5. Hub	11. Bearing	17. Oil baffle
6. Release bearing	12. Snap ring	

18. Snap ring	
19. Bearing cup	
20. Bearing cone	
21. Gear	
22. Nut	
23. Engine clutch shaft	

TESTS AND ADJUSTMENTS

All Models Except 7000

90. **CONTROL ROD.** The "Power-Director" clutch has three positions: High Range-Neutral-Low Range. The detents for these positions are located at the hand control lever and linkage must be correctly adjusted to make certain that control valve spools are correctly positioned when hand lever is in detent.

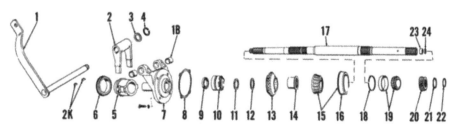

Fig. 63A—Exploded view of clutch shaft (17) and associated parts used on all models except 7000. Bevel gear (13) is used to drive belt pulley on models so equipped.

1. Release lever and shaft	5. Hub	12. Thrust washer	18. Snap ring
1B. Bushings (2 used)	6. Release bearing	13. Belt pulley drive gear	19. Bearing
2. Release fork	7. Retainer	14. Bushing	20. Drive gear
2K. Woodruff keys	8. Gasket	15. Bearing	21. Thrust washer
3. Washer	9. Seal	16. Snap ring	22. Snap ring
4. Snap ring	10. Coupling	17. Clutch shaft	23. Bearing
	11. Snap ring		24. Sleeve

190, 190XT AND 200 MODELS. Remove cotter pin (C—Fig. 64) or loosen locknuts (15). Move control lever (13) to the center (neutral) detent position. Remove the hood right side panel and connect a pressure gage as described in paragraph 91 for checking system pressure. Observe gage pressure with engine operating at 1600 rpm. If equipped with hydraulic pto clutch, make certain that pto control is in center (neutral) position. Adjust length of control rod (6) until lowest pressure (approximately 60 psi) is obtained. On late type linkage, do not change position of adjuster (16) when tightening nuts (15).

180 AND 185 MODELS. The control valve linkage can be adjusted after removing cover from below instrument panel. Move control lever to neutral detent position, then measure distance between punch mark on valve body and center line of connector pin (hole in valve spool). Distance should be 1½ inches. If incorrect, loosen lock nut and turn connector coupling at lower end of rod until distance is correct, then tighten locknut.

91. SYSTEM PRESSURE. Refer to the appropriate following paragraphs for checking relief pressure and testing for internal leakage.

190, 190XT AND 200 MODELS. Remove hood right rear side panel, disconnect by-pass tube from "Tee" fitting shown in Fig. 65 and install a plug in open end of by-pass tube. Connect a pressure gage to "Tee" fitting, start engine and set speed at 1600 rpm.

CAUTION: Do not turn steering wheel with by-pass tube capped.

With transmission "Power-Director" and pto hydraulic clutch in neutral positions, gage pressure should be approximately 60 psi. Engage Low Range of "Power-Director" (leave transmission in neutral) and observe gage pressure. Engage High Range and observe gage pressure. Pressure should be 255-270 psi on tractors **without**
hydraulic pto clutch; 295-310 psi for tractors **with** pto clutch. If pressure is erratic, relief valve may be sticking. If pressure is not the same in both Low and High Range, the control rod may not be correctly adjusted (paragraph 90) or pressure tubes and/or clutch seals are leaking. If pressure is incorrect but steady and same in both Low and High Ranges, remove plug (15—Fig. 65A) and add or deduct shims (16).

180 AND 185 MODELS. Remove cover from below instrument panel and disconnect "Tee" fitting (Fig. 66) from the power steering relief valve. Connect pressure gage to the "Tee" fitting, start engine and set speed at 2000 rpm.

CAUTION: Do not turn steering wheel with the power steering relief valve not connected to the "Tee" fitting.

The relief pressure can be checked with "Power-Director" and pto controls in neutral position. If tractor **is not**

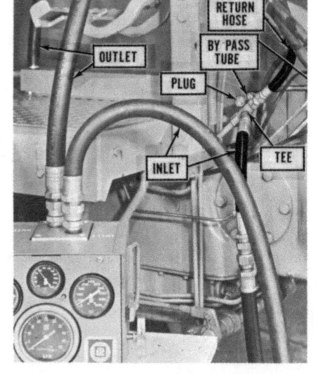

Fig. 65—View of OTC pressure and flow (volume) gage connected to "Tee" fitting for checking "Power-Director" system pressure on 190 and 200 tractors.

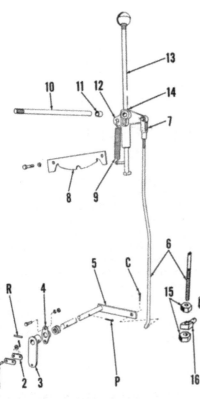

Fig. 64—Exploded view of "Power-Director" controls used on 190 and 200 models. Late type rod is equipped with adjuster (16) and lock nuts (15) at lower end as shown.

C. Cotter pin	8. Detent plate
P. Cotter pin	9. Detent spring
R. Roll pin	10. Shaft
1. Link pin	11. Shaft bushings
2. Link	(2 used)
3. Lever	12. Hub
4. Bearing block	13. Handle
5. Lever and	14. Handle bushings
cross shaft	(2 used)
6. Rod	15. Lock nuts
7. Clevis	16. Adjuster

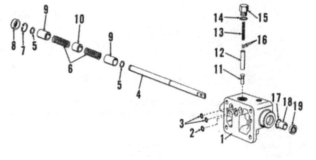

Fig. 65A—Exploded view of "Power-Director" control valve used on 190 and 200 tractors. Fig. 64 shows control linkage.

1. Valve body	
2. "O" ring	
3. "O" rings (2 used)	
4. Spool rod	
5. Snap rings (2 used)	
6. Springs	
7. Snap ring	
8. End plug	
9. Valve spools	
10. Center spool	
11. Relief plunger	
12. Spring guide	14. "O" ring	17. Snap ring
13. Spring	15. Relief valve plug	18. Bushing
	16. Shims	19. Seal

equipped with hydraulic pto clutch, pressure should be 255-270 psi. If tractor is equipped with hydraulic pto clutch, pressure should be 280-295 psi. To check for internal leakage, engage "Power-Director" High Range and Low Range and observe pressure. The hydraulic pto clutch and brake can also be checked for internal leakage by engaging and observing pressure. Normal pressure drop is approximately 10 psi. If pressure is erratic, relief valve may be sticking. If pressure is not the same in High and Low Range, the control rod may not be correctly adjusted (paragraph 90) or pressure tubes and/or clutch seals are leaking. If pressure is incorrect in Neutral, but steady, remove plug (1—Fig. 66A) and add or remove shims (2) as required. The addition of one shim will increase relief pressure approximately 15 psi. Relief valve parts can be removed from valve body for inspection and cleaning without removing unit from tractor.

CONTROL VALVE

All Models Except 7000

92. **R&R AND OVERHAUL.** The "Power-Director" control valve can be unbolted and removed after disconnecting tubes and control linkage.

To disassemble valve used on 190, 190XT and 200 models, remove end plug (8—Fig. 65A) and snap ring (7); then, slide spool assembly out end. Springs (6) should keep spools (9) compressed against snap rings (5). Before reinstalling make certain that new "O" rings (2 & 3) are correctly positioned in grooves in housing (1).

To disassemble valve used on 180 and 185 tractors, remove plugs (1—Fig. 66A) and withdraw components of relief valve and control valve. Seal (9) and tube seats (10 &11) can be removed from body if renewal is required. Relief valve spool (7) and body (8) are select fit and not available separately. Spool (7) should have 0.004-0.008 clearance in bore and should move and turn freely. Orifice (0.018-0.023 diameter) in spool (7) should be open and relief valve (6) should have 0.0003-0.001 clearance in spool and should not bind. When reassembling, renew "O" rings (3).

On all models, adjust control rod as outlined in paragraph 90 and relief pressure as outlined in paragraph 91 after valve is installed.

"POWER-DIRECTOR" CLUTCH

All Models Except 7000

93. **REMOVE AND REINSTALL.** To remove the "Power-Director" clutch assembly, it is necessary to split torque

tube from transmission as outlined in paragraph 104. Remove the two Allen screws (23—Fig. 67) and the ¼-inch diameter steel balls (located under

Allen screws), then slide clutch assembly from transmission input shaft.

NOTE: Remove one Allen screw, rotate clutch until hole is to bottom and

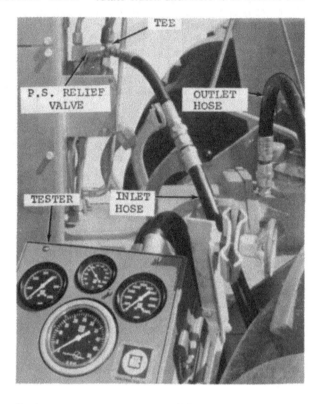

Fig. 66—View of OTC pressure and flow (volume) gage connected to "Tee" fitting for checking "Power-Director" system pressure on 180 tractor. Model 185 tractors are similar.

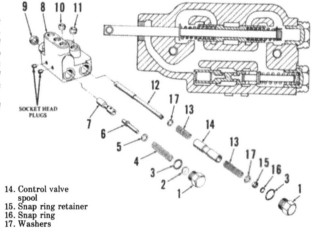

Fig. 66A—Exploded view of "Power-Director" control valve used on 180 and 185 tractors. The valve used for pto control is similar but is not fitted with relief valve and has hex head plugs on side, instead of socket head plugs (shown).

2. Shims
3. "O" rings
4. Relief valve spring
5. Washer
6. Relief valve
7. Relief valve spool
8. Valve body
9. Oil seal
10. Seat (½")
11. Seat 3/8")
12. Spool rod
13. Springs

14. Control valve spool
15. Snap ring retainer
16. Snap ring
17. Washers

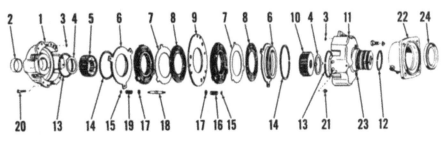

Fig. 67—Exploded view of "Power-Director" clutch.

1. Front housing
2. Bushing
3. "O" rings (2 used)
4. Thrust washer
5. Front hub
6. Pistons (2 used)
7. Drive plates (7 used)
8. Driving discs (9 used)
9. Center plate
10. Rear hub
11. Rear housing
12. Seal rings (3 used)
13. Inner seal rings (2 used)
14. Outer seal rings (2 used)
15. Snap rings (6 used)
16. Short return springs (3 used)
17. Washer (6 used)
18. Connecting pins (3 used)
19. Long return springs (3 used)
20. Cap screw (6 used)
21. Nuts (6 used)
22. Transmission input housing
23. Allen screws (2 used)
24. Seal

catch the steel ball before removing second Allen screw and ball to prevent loss.

Check bore of housing (22) for wear. If seal (24) and/or housing is renewed, make certain that seal does **NOT** cover drain hole in housing.

When reinstalling, be careful not to damage sealing rings (12). Make certain that one of the ¼-inch steel balls is located under each of the two Allen screws (23) and engage groove in transmission input shaft. Screws should be tightened to 5-7 ft.-lbs.

94. OVERHAUL. Refer to Fig. 67 and proceed as follows: Remove six cap screws and nuts (20 & 21) that attach housings together, compress return springs (16 & 19) and remove snap rings (15). Separate the assembly making certain that parts are not lost or damaged.

Check the internally splined driving plates (8) and driven plates (7) with external drive lugs for wear, cracks and evidence of overheating. Check bore in housing (22) for wear. If seal (24) and/or housing is renewed, make certain that seal does **NOT** cover drain hole in bottom of housing. If hole is covered, oil from "Power-Director" compartment may be pumped into transmission. When reassembling, all seal rings and "O' rings should be renewed. If bushing (2) is renewed, it should be pressed into housing 0.340-inch from machined surface (inside) of housing (1). Bushing (2) must be sized to 2.127-2.128-inch to provide 0.002-0.004-inch clearance on hollow shaft (20—Fig. 68).

Install inner seal ring (13—Fig. 67) on housing (1) making certain that ends overlap and lock together. Install outer seal ring (14) on piston (6) with ends overlapped and locked together; then, slide piston into housing (1). Install one plain washer (17) and one long return spring (19) over long end of each connecting pin (18). Attach the connect-ing pins to lugs on piston (6) with snap rings (15). Install thrust washer (4) with locating tabs on thrust washer engaging holes in housing (1). Position front (longer) hub (5) in front housing with flange side up. Install five driving discs (8) and four driven plates (7) alternately beginning with a driving disc (8). Position "O" ring (3) around oil passage in front housing (1) and install center plate (9) with oil passage hole over "O" ring. Install rear (short) hub (10) with shoulder engaging the flange on front hub. Install four driving discs (8) and three driven plates (7) alter-nately beginning with a driving disc (8). Install one plain washer (17) and one short return spring (16) over short end of each connecting pin (18). Install rear piston (6) over pins and install snap rings (15). It will be necessary to compress return springs (16 & 19) to install snap rings (15). Install outer seal ring (14) on rear piston (6) with ends overlapped and locked together. Install inner seal ring (13) on rear housing (11) with ends of seal overlapped and locked together. Use grease to stick thrust washer (4) and "O" ring (3) to rear housing (11) making certain that tabs on thrust washer engage holes in housing and "O" ring is over oil passage. Install rear housing being careful that thrust washer (4) and "O" ring (3) do not move and seal rings (13 & 14) are not damaged. Install cap screws (20) with heads toward front housing (1). Tighten nuts (21) to 15-20 ft.-lbs. When installing the three hub-seal rings (12), make certain that ends are overlapped and locked together.

"POWER-DIRECTOR" CLUTCH OUTER (HOLLOW) SHAFT AND INTERMDIATE GEARS

All Models Except 7000

95. OUTER (HOLLOW) SHAFT. To remove the outer (hollow) clutch shaft, it is necessary to first split the torque tube from the transmission as outlined in paragraph 104; then, proceed as follows: Remove nut (11—Fig. 68) and remove snap ring (13). Using a tool similar to that shown in Fig. 69 screwed onto the outer shaft (20—Fig. 68), pull the shaft (20), bearings (14 & 15) and spacer (16) out of the torque tube. After snap ring (21) is removed, thrust washer (22) and drive gear (23) can be removed.

NOTE: Two different (but similar) outer shafts (20) have been used and must be used with the correct accompanying parts to prevent damage.

Diameter, length and number of teeth are the same for both early and late outer shafts. The early outer shaft (20) can be identified by the absence of oil groove in outer diameter under spacer (16). The early spacer (16) is one piece, has oil grooves both inside and outside and is 0.687 thick. Bearing cups (15F & 15R) and bearing cones (14F & 14R) are interchangeable. The front snap ring (18) on early models is 0.093 thick. The late outershaft can be identified by the oil groove in outer diameter located under the spacer halves (16). The later spacer is two piece, with oil groove only on the outside diameter and is 0.580 thick. A wide bearing cone (14F) and cup (15F) is used at front. Thickness of cup (15F) is 0.765. The rear bearing cone (14R) and cup (15R) is the same as used on early models (0.625 thick). The front snap ring (18) on late models is 0.060 thick.

Bushings (17 & 19) are pressed into the outer shaft. Press ½-inch wide front bushing (19) into hollow shaft bore 3-25/32-inch from rear end of shaft. Press 1-inch wide rear bushing (17) into hollow shaft 21/32-inch from rear end of shaft. Ream or hone new bushings to give 0.001-0.003 clearance between bushings and engine clutch shaft bearing surface.

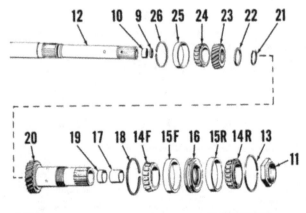

Fig. 68—Exploded view of "Power-Director" outer (hollow) shaft (20) and related parts. Refer to text identification of early and late parts (14F, 15F, 16, 18 & 20).

14. Bearing cones
15. Bearing cup
16. Spacer
17. Bushing
18. Snap ring
19. Bushing
20. Hollow (outer) shaft
21. Snap ring
22. Thrust washer
23. Drive gear
24. Bearing cone
25. Bearing cup
26. Snap ring

9. Sleeve
10. Bearing
11. Adjusting nut
12. Engine clutch shaft
13. Snap ring

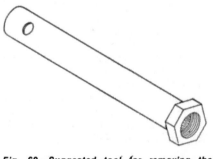

Fig. 69—Suggested tool for removing the "Power-Director" clutch outer (hollow) shaft can be made by welding a pipe to an extra nut (11—Fig. 68). This tool can be screwed on the shaft in place of the standard nut and pipe can be bumped rearward withdrawing the bearing cones, cups, spacer and outer shaft.

When reassembling, refer to the preceding note for identification of correct parts.

Install snap ring (18—Fig. 68) in groove in torque housing. Assemble front bearing cone (14F), cup (15F), spacer (16), rear bearing cup (15R), rear bearing cone (14R), and nut (11) on outer shaft. Do not tighten nut at this time. Install the assembly in bore of housing, drive bearing cups and spacer (16) against snap ring (18) and install as thick a snap ring (13) as possible in rear groove in torque housing. Snap ring (13) is available in

thicknesses of 0.094 to 0.109 in steps of 0.003. Adjust nut (11) to give 0.001-0.004 end play of outer shaft in bearings. Latest type nut (11) is self-locking.

96. INTERMEDIATE SHAFT AND GEARS. To remove the intermediate shaft and gears, it is first necessary to split the transmission from torque housing as outlined in paragraph 104. Hold shaft (6—Fig. 71) from turning and remove nut (1). Remove snap ring (8) and thread slide hammer adapter into threaded hole in rear of intermediate shaft. Bump shaft out to rear of torque housing as shown in Fig. 70. If necessary to renew front bearing cup (2F—Fig. 71), drive cup out toward rear. Rear bearing cup (2R) will be driven out as shaft is pulled from housing. To remove rear bearing cone from shaft first remove snap ring (7).

To reinstall, first drive front bearing cup (2F) in tight against shoulder in bore of housing. Drive rear bearing cone on rear end of intermediate shaft and install snap ring (7). Place front

bearing cone in cup and intermediate gears in housing. Small diameter hub of gears (4 & 5) should be toward ends of shaft and flat side of gear (5) should be against raised shoulder of gear (4) as shown in cross sectional view (Fig. 71). Insert shaft from rear into splines of gears and hold shaft against front bearing cone. Thread slide hammer adapter into rear end of shaft and bump shaft into front bearing cone. Install rear bearing cup in bore. Install nut (1) on front end of shaft and tighten to 50-60 ft.-lbs. Check to see if snap ring (7) is tight against rear bearing cone and that outer rims of gears are pulled tightly together; then, stake nut securely on shaft. Install snap ring (8) in groove behind rear bearing cup and seat cup against snap ring with slide hammer. Remove slide hammer and measure shaft end play with dial indicator. If end play is not within limits of 0.002-0.005, remove snap ring (8) and install another snap ring of proper thickness. Snap rings are available in thicknesses of 0.080 to 0.110 in steps of 0.002 inch.

Fig. 70—The "Power-Director" intermediate shaft can be removed using slide hammer as shown after removing snap ring (8) and nut (1) shown in Fig. 71.

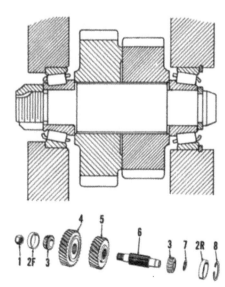

Fig. 71—Exploded view of "Power-Director" intermediate shaft and gears. On early models, gear (4) has 31 teeth; later models have 30 teeth.

1. Nut
2F & 2R. Bearing cups
3. Bearing cone
4. Driven gear
5. Driving gear
6. Intermediate shaft
7. Snap ring
8. Snap ring

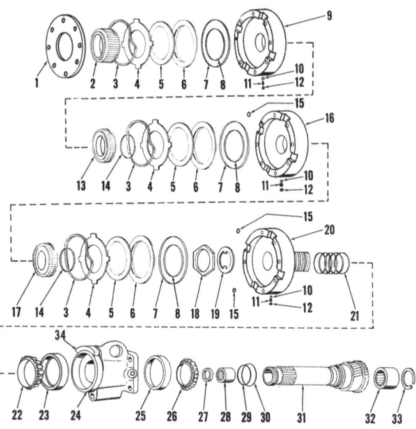

Fig. 72—Exploded view of "Power Shift" clutch assembly used on 7000 models.

1. End plate	11. Spring	19. Thrust washer	27. Seal
2. Low clutch hub	12. Plug	20. Intermediate clutch housing	28. Roller bearing
3. Wave washer	13. High clutch hub	21. Seal rings (5)	29. "O" ring
4. Friction plate	14. Snap ring	22. Bearing cone	30. "O" ring
5. Steel plate	15. "O" ring	23. Bearing cup	31. Transmission input shaft
6. Piston	16. High clutch housing	24. Sleeve	32. Roller bearing
7. Outer piston seal	17. Intermediate clutch hub	25. Bearing cone	33. Snap ring
8. Inner piston seal		26. Bearing cup	34. "Power Shift" brake
9. Low clutch housing	18. Nut		
10. Flyball			

"POWER SHIFT"

The "Power Shift" consists of three hydraulically actuated clutches and associated gears and shafts to provide high, intermediate and low speed operation in any selected transmission gear. "Power Shift" speeds may be selected by moving control lever without disengaging engine clutch. Power for intermediate speed operation is transferred directly from engine clutch shaft to intermediate clutch while power for high and low speeds is transferred from engine clutch shaft to a countershaft and two hollow shafts diregard the above.

Hydraulic oil for clutches is directed from power steering valve and passes through modulating, pressure regulating and control valves before providing control and lubricating oil for clutches.

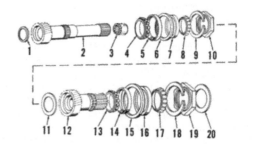

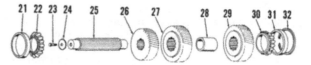

Fig. 73—Exploded view of "Power Shift" hollow shaft and countershaft assemblies. Screw (23) and washer (24) are used on both ends of countershaft (25).

1. Seal
2. High range hollow shaft
3. Roller bearing
4. Bearing cone
5. Bearing cup
6. Spacer
7. Bearing cup
8. Bearing cone
9. Snap ring
10. Nut
11. Seal
12. Low range hollow shaft
13. Bearing cone
14. Bearing cup
15. Spacer
16. Bearing cup
17. Bearing cone
18. Snap ring
19. Nut
20. Shim

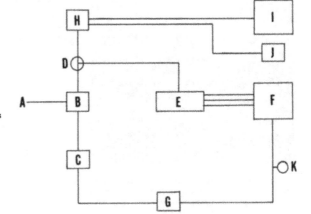

21. Bearing cup
22. Bearing cone
23. Cap screw (2)
24. Washer (2)
25. Countershaft
26. Driven gear

27. High range gear
28. Spacer
29. Low range gear
30. Bearing cone
31. Bearing cup
32. Snap ring

OPERATION

Model 7000

97. Power is transmitted by engine clutch shaft to drive gear (21—Fig. 63) and intermediate power shift clutch hub (17—Fig. 72). Drive gear (21—Fig. 63) rotates gear (26—Fig. 73), countershaft (25) and attached gears. Countershaft gears (27 & 29) drive low and high range hollow shaft (2 & 12) and attached low and high range clutch hubs (2 & 13—Fig. 72). Power shift clutch rotation speed is determined by hydraulically forcing either high, low or intermediate clutch piston (6) against clutch plates thereby driving clutch housings at same speed as clutch hub engaged by clutch plates. Clutch housings (9, 16 & 20) are bolted together and intermediate clutch housing (20) is splined with transmission input shaft (31).

Hydraulic oil flow for clutch actuation and lubrication is shown in Fig. 74. Oil is directed from the power steering valve and pump relief valve to modulator valve (B) and then to pressure regulator valve (C). Pressure regulator valve maintains system pressure at 210-230 psi. When system pressure is attained, modulator valve (B) spool directs oil to pto clutch and brake valve (H) and power shift control valves (E). Modulator valve (B) dampens oil pulsations during engagement and disengagement of power shift clutches. Power shift control valves (E) direct oil to one of the three power shift clutches (F). A control rack determines position of spools in control valves (E). After system oil pressure is attained, pressure regulator valve (C) opens to allow lubricating oil to pass through pressure relief valve (G) to power shift shafts

Fig. 74—Hydraulic flow diagram for "Power Shift" and pto.

A. From power steering valve
B. Modulator valve
C. Pressure regulator valve
D. Test port
E. "Power Shift" control valves
F. "Power Shift" clutches
G. Lube pressure relief valve
H. Pto clutch & brake valve
I. Pto clutch
J. Pto brake
K. Lube warning light sender

Fig. 75—View of right side of Model 700 torque housing.

and bearings (F). Pressure relief valve (G) maintains lubricating oil pressure at 125-140 psi.

TESTS AND ADJUSTMENTS

Model 7000

98. **CONTROL CABLE.** "Power Shift" control cable must be adjusted to synchronize engagement of shift lever and clutch valve detents. Detach clevis (C—Fig. 75) from "Power Shift" lever, push "Power Shift" lever all the way forward and move console power shift control lever to "III" position. Turn clevis until clevis and "Power Shift" lever holes are aligned and attach clevis to lever.

99. **SYSTEM PRESSURE.** To check "Power Shift" hydraulic pressure, unscrew test port plug (P—Fig. 75) and install a 400 psi gage. Start engine and note reading. Gage reading should be 210-230 psi with "Power Shift" in gear while dropping approximately 100 psi for less than a second when a different gear is selected. Shims are available to adjust pressure regulator valve setting.

To check lubricating oil pressure, unscrew oil pressure sending unit (S—Fig. 75) and install a 400 psi gage. Start engine and note gage reading. Lubricating oil pressure should be 4-140 psi. Disconnect lubricating oil line (small diameter line running under rear of torque housing) and install a flo-rater. With engine running, slowly close flo-rater. Pressure reading with flo-rater closed should be 125-140 psi. Shims are available to adjust relief valve pressure setting.

CONTROL VALVES

Model 7000

100. The "Power Shift" control valve assembly is located on right side of torque housing. Proceed as follows to remove valve assembly: Disconnect pto clutch and brake lines shown in Fig. 75. Detach "Power Shift" and pto cables and brackets. Detach brake return

spring and remove pto lever pivot bolt. Note spacer between lever and plate. Disconnect safety stop switch wires and remove pto lever. Unscrew nine cap screws securing valve housing and remove control valve assembly.

Disassemble control rack linkage shown in Fig. 76. Unscrew ten cap screws securing inner and outer valve bodies and separate valve bodies, gaskets and separator plate. Remove plate (29—Fig. 77) and control rack (28) from inner valve body (15). Remainder of disassembly is evident after inspection of components and referral to Fig. 77. Note location of pto plunger stop pin (7), pto spool stop (40) and modulator spool stop (33) as shown in Fig. 78.

To assemble pto control valve, install inner spring (42—Fig. 77), sleeve (43) and nut (44) on spool (41). Tighten nut (44) until bottomed on spool. Insert return spring (39) and spool (41) into outer valve body (12). Install outer

spring (45) and plunger (46). Hold plunger in while driving stop pin (See Fig. 78) in until it bottoms. Plunger should slide easily in bore. Install seal (47—Fig. 77) with lip in. Install spool stop (40) in annulus shown in Fig. 78 with bent edges toward plunger.

Install modulator valve as shown in Fig. 77. Spool stop (33) must be installed in annulus shown in Fig. 78. Install pressure regulator and lube pressure relief valves as shown in Fig. 77. Do not switch spools and springs. Pressure regulator spring (18) has 0.605 in. O.D. and 1.67 in. free length while lube pressure relief spring (21) has 0.595 in. O.D. and 1.54 in. free length. Pressure regulator valve spool (20) has a chamfer on rear land while lube pressure relief valve spool (23) has straight lands.

Blocking spools (9—Fig. 77) are used in valve bores A1, C1 and C2 called out in Fig. 77. Clutch valve spool (4) with longer bottom land must be installed

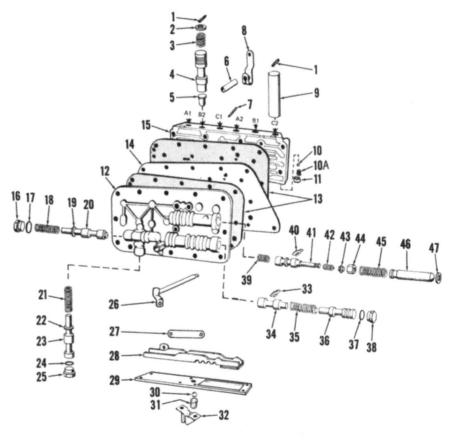

Fig. 77—Exploded view of "Power Shift" control valve assembly.

1. Pin	14. Separator plate	25. Plug	36. Modulator valve
2. Washer	15. Inner valve body	26. Control lever	plunger
3. Spring	16. Plug	& shaft	37. "O" ring
4. Clutch valve spool	17. "O" ring	27. Link	38. Plug
5. Actuator pin	18. Spring	28. Control rack	39. Spring
6. Spacer	19. Shim	29. Cover	40. Spool stop
7. Pto plunger stop pin	20. Pressure regulator	30. Detent ball	41. Pto valve spool
8. Clamping lever	valve spool	31. Spring	42. Spring
9. Blocking spool	21. Spring	32. Cap	43. Sleeve
10. Modulator check ball	22. Shim	33. Spool stop	44. Nut
10A. Spring	23. Lube pressure relief	34. Modulator valve	45. Spring
11. Plug	valve spool	spool	46. Pto valve plunger
12. Outer valve body	24. "O" ring	35. Spring	47. Seal
13. Gaskets			

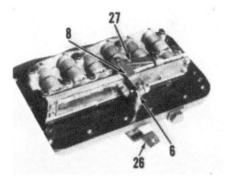

Fig. 76—Assemble control valve linkage as shown.

6. Spacer	26. Control lever & shaft
8. Clamping lever	27. Link

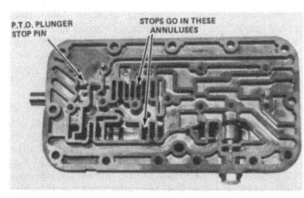

Fig. 78—View showing location of pto plunger stop pin and spool stops.

in valve bore B1. Clutch valve springs (3) should have 0.52 inch O.D. and 0.68 inch free length. Install modulator valve check ball (10) in inner valve body bottom adjacent to clutch valve bore C2. Lubricate control rack (28) sliding surfaces prior to assembly. Tighten plate (29) screws to 20 in-lbs. Tighten inner valve body (15) screws to 11 ft.-lbs. using sequence shown in Fig. 79. Install control valve assembly on torque housing and tighten retaining screws to 30 ft.-lbs. in sequence shown in Fig. 80. Refer to paragraphs 98 and 158 for adjustment of "Power Shift" and pto control cables.

"POWER SHIFT" CLUTCHES

Model 7000

101. **REMOVE AND REINSTALL.** Remove "Power Shift" control valve assembly as outlined in paragraph 100. Measure clearance between low range nut (19—Fig. 73) and low range clutch hub (2—Fig. 72). Clearance should be 0.013-0.027 inch. Clearance may change and require adjustment using shims (20—Fig. 73) if bearings, clutch hubs or clutch housings are renewed (renewal of low range hollow shaft, bearings or nut as noted in paragraph 103 will also affect clearance). Split torque housing from transmission as outlined in paragraph 105. Unbolt clutch sleeve (24—Fig. 72) and detach "Power Shift" clutch assembly from transmission housing. Note that clutch may require rotation so cutout portion of transmis-

sion input gear (31) is aligned with transmission countershaft gear.

102. **OVERHAUL.** Remove "Power Shift" clutch assembly as outlined in previous paragraph. Withdraw clutch hubs (2, 13 & 17—Fig. 72). Unscrew through bolts and disassemble clutch units. Unscrew nut (18), withdraw transmission input shaft (31) and separate intermediate clutch housing (20) and sleeve (24). Unscrew plugs (12) and remove springs (11) and flyballs (10).

Inspect components and renew if damaged or excessively worn. Brake pad (34) thickness should be 0.980-1.020 inches. Note the following points during assembly. Lubricate seal rings (21) prior to inserting intermediate clutch housing (20) in sleeve (24). Tighten nut (18) until input shaft (31) end play is 0.001-0.006 inch. Stake nut securely. Low range clutch housing (9) may be identified by single oil passage on backside while high range clutch housing (16) has two oil passages on backside. Clutch pistons (6), steel plates (5), friction plates (4) and wave washers (3) are identical. Install piston seals (7 & 8) with lip towards clutch housing. Be sure two "O" rings (15) are used in oil passages between intermediate and high range clutch housings and one "O" ring (15) is placed in oil passage between high and low range clutch housings. All clutch plates (4 & 5) must be dipped in oil prior to assembly. A steel plate (5) must be placed next to piston (6) followed by a

friction disc (4) and wave washer (3). This order must be followed in all clutch packs. Three steel plates, three friction plates and three wave washers are used in intermediate clutch assembly. Two each of the above are used in the high range clutch assembly while four each of the above are used in the low range clutch assembly. Use ACTP 3032 or other similar tool to compress clutch packs while installing end plate (1) and through bolts. Through bolts must be inserted with bolt head contacting end plate (1) and tightened to 38-45 ft.-lbs. Friction plates (4) should turn freely after bolts have been tightened. Be sure snap ring rings (14) are installed in high and intermediate range clutch hubs (13 and 17) and install clutch assembly as outlined in paragraph 101.

"POWER SHIFT" HOLLOW SHAFTS AND COUNTERSHAFT

Model 7000

103. **R&R AND OVERHAUL.** Remove pto brake as outlined in paragraph 170 and "Power Shift" clutch assembly as outlined in paragraph 101.

Remove screw (28—Fig. 73) and washer (24) in both ends of countershaft (25). Remove snap ring (32) and pull countershaft with bearing (30 & 31) out rear of torque housing. Remove gears (26, 27 & 29), spacer (28) and front bearing (21 & 22). Using tool ACTP 3027, unscrew shaft nut (19) and remove snap ring (18). Attach adapter tool ACTP 3030 to a suitable puller and pull low range hollow shaft (12) and bearings out of torque housing. Using tool ACTP 3025, unscrew shaft nut (10) and remove snap ring (9). Attach adapter tool ACTP 3028 to a suitable puller and pull high range hollow shaft (2) and bearings out of torque housing.

To reinstall high range hollow shaft assembly, install bearings, spacer (6) and nut (10) on shaft (2) but only hand tighten nut. Insert shaft into torque housing so spacer (6) is bottomed in housing. Install thickest snap ring (9) which will fit groove. Tighten nut (10)

Fig. 79—Use tightening sequence shown above during control valve assembly.

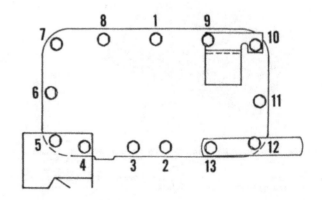

Fig. 80—Follow sequence shown at right when tightening valve body retaining screws.

until 4-10 in.-lbs. rolling torque is required to turn tool ACTP 3025 socket. Stake nut securely.

To reinstall low range hollow shaft, install bearings, spacer (15) and nut (19) on shaft (12) but only hand tighten nut. Insert shaft into torque housing so spacer (15) is bottomed in housing. Install thickest snap ring (18) which will fit groove. Tighten nut (19) until 4-10 in.-lbs. rolling torque is required to turn tool ACTP 3027 socket. Stake nut securely but do not allow displaced metal to contact shims (20).

High range countershaft gear (27) must be installed with smaller diameter hub next to spacer (28). If gears (26 & 29) were not renewed, they should be reinstalled to rotate in same direction indicated by wear pattern. Install countershaft (25) with larger bearing land towards rear of torque housing. Tighten screws (23) to 40-50 ft.-lbs. Countershaft end play should be 0.001-0.006 inch and is adjusted by varying thickness of snap ring (32). Snap ring (32) is available in thicknesses from 0.061 to 0.105 inch in increments of 0.004 inch.

If low range hollow shaft, bearings, spacer (15) or nut (19) was renewed, clearance between low range shaft nut (19) and low range clutch hub (2—Fig.

72) must be measured. Measure distance from low range shaft nut (19—Fig. 73) to face of torque housing as shown in Fig. 81 and add thickness of housing gasket. Measure distance from face of low range clutch hub to transmission housing face as shown in Fig. 82 and subtract measurement from sum of nut distance and gasket thickness. Result will be clearance between nut and clutch hub. Install shims (20—Fig. 73) to obtain 0.013-0.027 inch clearance. Shims are available in 0.010, 0.020 and 0.030 inch.

TRANSMISSION

SPLIT TRANSMISSION FROM TORQUE HOUSING

All Models Except 7000

104. Lower the lift arms and drain oil from "Power-Director" compartment. Drain **all** fuel from tank. Remove gear shift linkage covers and link rod (31—Fig. 87). Remove seat assembly, platform and tool box. Disconnect hoses from both lift arm cylinders, position control rod from right hand lift arm and traction booster linkage from "Snap Coupler" or torsion bar. Disconnect wires from clips on brake housings and remove both brake rods. Remove the hydraulic sump return line from filter. On 190 and 200 models, disconnect rods from the "Power-Director" cross shaft (5—Fig. 64). Remove cotter pin (P) from cross shaft and slide the rod out of torque housing. Remove the "Power-Director" control valve and plate under valve. On 180 and 185 models remove tubes leading to the manifold (M—Fig. 84), then unbolt and remove manifold and cover plate under

manifold. On all models, remove the two short tubes (T—Fig. 83) and push the "Power-Director" tubes and unions in to clear the tube bracket. Disconnect pto shaft linkage and on models with hydraulic clutch, detach valve bracket from right hand final drive housing. Disconnect oil lines from pump at the control valves.

On 190 and 200 models, loosen the two cap screws under front of fuel tank. The fuel tank prevents complete removal. Remove two cap screws at front of platform frame that attach frame to torque housing. Loosen the two cap screws attaching platform frame to lower rear hood support.

NOTE: It may be necessary to loosen hood side panels to reach nuts under hood rear support.

Fig. 84—View of "Power-Director" manifold (M) on 180 and 185 tractors.

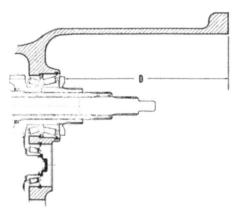

Fig. 81—Measure distance (D) as outlined in paragraph 103 and refer to Fig. 82.

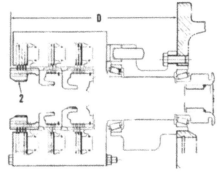

Fig. 82—Measure distance (D) from transmission housing face to low range clutch hub (2) as outlined in paragraph 103.

Fig. 83—The two short tubes (T) must be removed before torque housing is split from transmission.

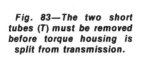

Attach a hoist to seat mounting nut plates on fuel tank and lift fuel tank, wheel guards and platform frame. A 4X4X14-inch wood block can be positioned as shown in Fig. 85 to hold platform assembly.

On 180 and 185 models, unbolt and remove fuel tank, fuel tank support and rear hood support. On all models so equipped, remove the "Snap-Coupler" assembly. On all models, block front of tractor to prevent tipping and support under rear of torque housing. Support front of transmission and remove screws attaching transmission to torque housing. Roll transmission assembly back, being careful not to bend the "Power-Director" tubes.

When reassembling, observe the following: Clean gasket surfaces of transmission and torque housing, then coat with No. 3 Permatex or equivalent and install new gasket. Install guide studs in torque housing and carefully move the transmission assembly forward. Rotate "Power-Director" clutch to align hub splines with input shafts. Be careful not to bend or damage "Power-Director" oil tubes.

NOTE: If housings are not correctly aligned, bushing (2—Fig. 67) in the front of "Power-Director" clutch may be damaged.

Tighten transmission to torque housing screws to 80-90 ft.-lbs. of torque.

On 190 and 200 models, lower the platform assembly making certain that rod and tubes are correctly positioned and not bent or caught. Install the platform frame attaching screws. On 180 and 185 models, install fuel tank, fuel tank support and rear hood support. On all models, install platform sections, seat assembly and tool box. Install the two short tubes (T—Fig. 83) on "Power-Director" oil lines. Coat both sides of cover plate gasket with No. 3

Permatex or equivalent and install cover plate as shown in Fig. 86. Make certain that oil lines (T) are correctly positioned. Install control valve using new "O" rings in tube bores and new gasket. Connect linkage and oil lines.

Model 7000

The tractor may be split between transmission and torque housing so cab remains with front or rear assemblies. Follow splitting procedure in paragraph 105 for access to torque housing or to paragraph 106 for access to transmission and rear end components.

105. **CAB REMAINS ON REAR SECTION.** Drain torque housing oil and remove fuel tank. Raise front of cab approximately two inches by removing two front cab isolator bolts; left bolt is in rectangular plate in front of clutch pedal while right bolt is underneath foam cube. Detach clutch and transmission brake rods from pedal assembly. Drive retaining roll pin forward and withdraw transmission brake shaft shown in Fig. 61C. Unhook clutch and brake return springs. Unbolt pedal assembly cross shaft. Drill two small holes through floor and use wire to hold cross shaft against floor. Support pedals with blocks. Disconnect pto and "Power Shift" control cables at torque housing end. Remove valve bracket screws (B—Fig. 75). Unbolt and prop up rear of hood center section. Remove air conditioner compressor and condenser, if so equipped, and move to cab. Drain engine coolant and disconnect heater hoses. Disconnect wiring, and tubing which will prevent tractor separation. Block front of tractor to prevent tipping and support rear of torque housing. Support front of transmission and remove screws attaching transmission to torque housing. Separate front and rear tractor assemblies.

Note the following when mating front and rear assemblies: "Power Shift" clutch hubs must fully engage clutch plates and shims (20—Fig. 73) must be in place. Brake pad and pin (34—Fig. 72) must be all the way forward. When mating transmission and torque housing do not back up as "Power Shift" clutch hubs may drop down and prevent further spline engagement. Two long transmission-to-torque housing cap screws are installed in two top holes. Tighten all transmission-to-torque housing cap screws to 90-110 ft.-lbs. Install transmission brake shaft shown in Fig. 61C with flat towards front.

106. **CAB REMAINS ON FRONT SECTION.** Drain torque housing oil and remove fuel tank. Remove cab floorboard, place transmission shift lever in neutral, unscrew transmission cover screws and remove transmission cover with shift assembly. Push park pawl against spring pressure so pawl will clear housing. Disconnect linkage from transmission brake shaft shown in Fig. 61C, drive retaining roll pin forward and withdraw transmission brake shaft. Detach brake rods and remove valve bracket screws (B—Fig. 75). Place remote control levers in raise position and lift arm control lever in "Traction Booster" position and disconnect control links from valves. Detach "Power Shift" and pto control cables from torque housing. Detach "Traction Booster" and position control cables at lower ends. Sufficiently disassemble valve stack operating linkage to allow withdrawal of "Power Shift" and pto control cable lower ends. Disconnect wiring and tubing which will prevent separation. Support cab and remove rear U-bolts (nuts are retained by Loctite). Block front of tractor to prevent tipping and support rear of torque housing. Support front of transmission and remove screws attaching

Fig. 85—A 4 x 4 x 14 inch wood block can be positioned as shown to hold platform up on 190 and 200 models.

HARDWOOD BLOCK (4" x 4" x 14")

Fig. 86—View of side cover installed and short tubes (T) correctly installed. If screw attaching control valve at point (X) is too long, leakage may occur behind side cover.

transmission to torque housing. Separate front and rear tractor assemblies.

Note the following when mating front and rear assemblies: "Power Shift" clutch hubs must fully engage clutch plates and shims (20—Fig. 73) must be in place. Brake pad and pin (34—Fig. 72) must be all the way forward. When mating transmission and torque housing do not back up as "Power Shift" clutch hubs may drop down and prevent further spline engagement. Two long transmission-to-torque housing cap screws are installed in two top holes. Tighten all transmission-to-torque housing cap screws to 90-110 ft.-lbs. Install transmission brake shaft shown in Fig. 61C with flat towards front. To install gear shift assembly, place shift lever and transmission gears in neutral position. Install gear shaft assembly with two special dowel (aligning) screws located in holes (H—Fig. 91). Tighten cover screws to 20-25 ft.-lbs.

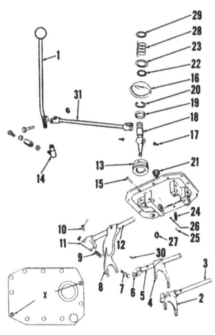

Fig. 87—Exploded view of transmission cover assembly. Note position of overshift washers (27) before disassembling.

1. Shift handle
2. Shift fork (3rd & 4th)
3. Shift rail
4. Shift fork (1st & 2nd)
5. Shift rail
6. Shift lug
7. Reverse latch plunger
8. Shift fork (reverse)
9. Spring
10. Lock screws (5 used)
11. Shift lug
12. Shift rail
13. Insert
14. Ball joint
15. Long plunger
16. Dust cover
17. Pivot pins
18. Shift lever
19. Pivot washer
20. Snap ring
21. Breather cap
22. Sealing ring
23. Washer
24. Detent spring (3 used)
25. Detent ball (3 used)
26. Short plunger
27. Overshift washers
28. Spring
29. Washer
30. Interlock pin
31. Link rod

SHIFTER ASSEMBLY

All Models Except 7000

107. R&R AND OVERHAUL. To remove the shifter assembly, shift the transmission into 2nd gear and unbolt cover from top of transmission. Two of the retaining cap screws are special dowel bolts.

Before disasembling, note the location of overshift limiting washers (27—Fig. 87) on shift rails.

To remove the gear shift lever (18), remove snap ring (20) and pivot washer (19), then pull lever out of cover. The two lever pivot pins (17) in cover are renewable after removing lever. To reinstall lever, reverse removal procedure.

Remove lock screw from reverse shifter fork (8), rotate shift rail ¼-turn, and catch detent ball (25) and spring (24) while sliding rail forward out of cover. Withdraw rail and fork. With reverse shift rail removed, long interlock plunger (15—Fig. 88) and interlock pin (30) may be removed from cover. Remove lock screw from first and second gear shifter fork (4—Fig. 87), rotate rail ¼-turn and catch detent ball and spring at rear of cover as rail is removed out to front of cover. Rotate third and fourth gear shift rail (3) and fork (2) ¼-turn and slide assembly forward out of cover while catching detent ball and spring at rear end of cover. Be sure not to lose short interlock plunger (26—Fig. 88). Renew bent, sprung or worn shifter forks or shift rails.

Wear on ends of interlock plungers (15 & 26—Fig. 88) and interlock pin (30) will allow more than one shift rail to be moved at a time. Plunger (15) should be 1.691 inches long; plunger (26) should be 0.566-inch long and pin (30) should be 0.551-inch long. Tighten nut on reverse latch plunger (7—Fig. 87) until cotter pin can be easily inserted.

To reassemble, reverse disassembly procedure. Insert detent spring and ball in cover and depress ball with punch while sliding shift rails in place. Washers (27) can be installed on shift rails to prevent overshift (past detents in shaft).

To install, place shift lever and transmission gears in 2nd gear position. Install cover gasket and cover assembly with the two special dowel (aligning) screws located in holes marked by (X—inset Fig. 87) at left front and right rear. Tighten all shift cover retaining screws to 20-25 ft.-lbs. torque.

Model 7000

108. R&R AND OVERHAUL. To remove shifter assembly, shift transmission to neutral and unbolt cover

from transmission. Two of the cap screws are special dowel bolts. Note: Shift lever may be removed without removing shift assembly.

To remove shift lever, remove snap ring (4—Fig. 89) and pivot washer (5), then pull lever out of cover. Two lever pivot pins (2) are renewable after removing lever. Pivot insert (6) is a press fit in cover and should be renewed if execssively worn. To reinstall lever, reverse removal procedure.

To disassemble shift assembly, remove lock screw from roller carrier (29—Fig. 89), rotate rail (18) ¼-turn and catch detent balls and springs in cover and reverse fork (19) while sliding rail forward out of cover. Remove interlock plunger (9B) and be careful not to lose interlock pin (10) in third and fourth gear shift rail (21). Withdraw third and fourth gear shift rail (21) while catching detent balls (24) and springs (23). Remove setscrew in first and second gear fork (27) and withdraw first and second gearshift rail (26) towards front while catching detent balls (24) and springs (23). Drive out roller carrier (29) roll pin and remove roller carrier (29) and shaft (28). Remove snap ring (11), shaft (13), park pawl (12), spring (32) and shims (33).

To adjust park pawl operation, install park pawl assembly excluding snap ring (11), roller carrier (29), shaft (28), reverse rail (18), detent spring (23) and ball (24). Install roller carrier (29) setscrew. Move reverse rail (18) to neutral position. Roller (31) should be within 0.010 of touching park pawl (12) at points (P—Fig 90). Adjust park pawl position using shims (33—Fig. 89).

To assemble shift assembly, install park pawl assembly and retain with snap ring (11). Install roller carrier (29) and shaft (28). Install two detent springs (23) and balls (24) in front and rear bosses and install first and second gear rail (26) and fork (27) next to park pawl shaft so fork (27) is between

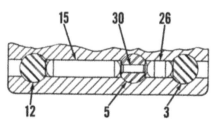

Fig. 88—Cut-away view of transmission cover showing location of interlock plungers (15 & 26) and interlock pin (30).

3. 3rd & 4th shift rail
5. 1st & 2nd shift rail
12. Reverse shift rail
15. Long interlock plunger
26. Short interlock plunger
30. Interlock pin

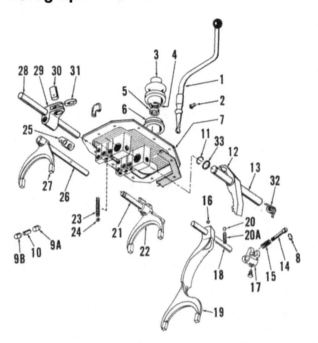

1. Shift lever
2. Pivot pins
3. Dust cover
4. Snap ring
5. Pivot washer
6. Insert
7. Cover
8. "E" ring
9A & 9B. Interlock plungers
10. Interlock pin
11. Snap ring
12. Park pawl
13. Shaft
14. Reverse plunger
15. Spring
16. Ball
17. Reverse lug
18. Reverse rail
19. Reverse shift fork
20. Ball
20A. Spring
21. Shift rail
22. Shift fork (3rd & 4th)
23. Spring
24. Ball
25. Lug
26. Shift rail
27. Shift fork (1st & 2nd)
28. Shaft
29. Roller carrier
30. Roller shaft
31. Roller
32. Spring
33. Shims

Fig. 89—Exploded view of shifter assembly used on Model 7000.

Early 190 and 190XT Models (Before Tractor Serial No. 9000)

109. INPUT SHAFT. To remove the input shaft, first split transmission from torque housing as outlined in paragraph 104 and remove "Power-Director" clutch as in paragraph 93. Disconnect oil tubes from bearing retainer (1—Fig. 92) and remove retainer. Input shaft (6) and bearing (5) can be pulled from transmission housing. If seal (3) and/or retainer (1) is renewed, make certain that new seal does not cover drain hole in retainer.

NOTE: If seal is pressed to bottom (front) of retainer bore, drain hole will be covered and oil from "Power-Director" compartment may be pumped into transmission compartment.

Before tractor serial number 8461, the pinion shaft pilot bearing (7) is retained by snap ring (8). Later models have thrust washer in this position.

When reinstalling, press input shaft and bearing assembly into transmission housing. Be careful not to damage pilot bearing (7). Lubricate input shaft and seal (3), install gasket (2) and carefully position retainer (1) over shaft. Tighten retainer cap screws to 40-50 ft.-lbs Reassemble tractor in reverse of disassembly procedure.

110. BEVEL PINION SHAFT. To remove the transmission bevel pinion shaft, first split transmission from torque housing as outlined in paragraph 104. Remove differential assembly as in paragraph 125, shifter cover as in paragraph 107 and input shaft as in paragraph 109. Remove bevel pinion bearing retainer (24—Fig. 92) and snap ring (8S). Bump or push bevel pinion shaft, together with bearing cups and cones (22), rearward and remove gears, collars and washers through top opening of transmission housing.

The mesh position of the bevel gears is controlled by thickness of snap ring

bosses with setscrew hole forward. Install fork setscrew and lug (25). Insert interlock plunger (9A)—plungers (9A and 9B) are identical. Install two detent springs (23) and balls (24) in front and rear bosses. Place interlock pin (10) in third and fourth gear rail (21) and install rail in cover. Install third and fourth gear fork (22) on end of rail (21) and install interlock plunger

(9B). Install one detent spring (23) and ball (24) in rear boss and detent ball (20) and spring (20A) in reverse fork (19). Place a detent ball (16) in groove of roller carrier shaft (28). Insert reverse rail (18) into cover and reverse fork (19). Detent ball (16) must align with hole in top of reverse fork (19) and roller carrier shaft (28) groove.

Note when installing shift cover assembly that two special cap screws are installed in holes (H—Fig. 91). Tighten shift cover retaining screws to 20-25 ft.-lbs.

TRANSMISSION OVERHAUL

Data on overhauling various components of the transmission are outlined in the following paragraph. The production changes noted may be on earlier models due to field changes.

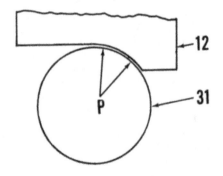

Fig. 90—Gap between park pawl (12) and roller (31) should be 0.010 at points (P). Refer to text.

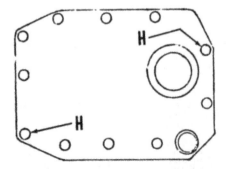

Fig. 91—Install special dowel capscrews in cover holes (H).

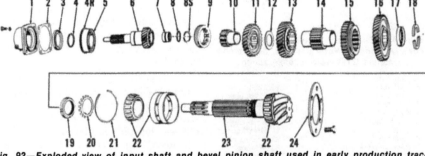

Fig. 92—Exploded view of input shaft and bevel pinion shaft used in early production tractors. Some early tractors may have been converted to the late type shown in Fig. 98.

1. Retainer
2. Gasket
3. Oil seal
4. Snap ring
4R. Snap ring
5. Bearing
6. Input shaft
7. Pilot bearing
8. Thrust washer
8S. Snap ring
9. 3rd & 4th coupling
10. Collar & bushing
11. 3rd gear
12. Thrust washer
13. 2nd gear
14. Collar & bushings
15. 1st & 2nd sliding gear
16. 1st gear
17. Retainer
18. Split collar
19. Bearing adjusting nut
20. Locking washer
21. Snap ring
22. Bearing
23. Bevel pinion
24. Retainer plate

(21) which is available in thicknesses of 0.177 to 0.191 in steps of 0.002. If necessary to renew snap ring only, be sure replacement is exactly the same thickness as original. However, if the bevel pinion shaft and differential ring gear, bearings or transmission housing is renewed, it will be necessary to follow the procedure outlined in paragraph 111 to select proper thickness snap ring (21).

If bevel pinion shaft, gears and/or bushings are renewed, it is advisable to first make a trial assembly on the bench of all parts located between and including the front snap ring (8S) and split sleeve (18). The front snap ring is available in several thicknesses. Select a snap ring (8S) which will eliminate all end play of the assembled parts when all are installed on the pinion shaft. If gear (11, 13 or 16) fails to rotate freely, it is an indication that the gear hub is wider than bushing (10 or 14) on which it rotates. To provide gear side clearance for free rotation, renew bushing or place a fine grade of emery cloth on a flat surface and lap hub of gear down to proper dimension to allow free rotation.

Assemble pinion shaft (23) and bearings (22) with wide roller bearing cone next to bevel pinion. Install pinion shaft with bearings in transmission housing without gears and adjust bearings by means of nut (19) so that 8-35 inch-pounds of torque is required to rotate shaft in bearings. After adjustment is complete, lock adjusting nut with one of the tabs of washer (20). Remove pinion shaft and bearing assembly after adjustment is complete. To facilitate reinstallation of pinion assembly in transmission housing, chill the bearing and shaft assembly in freezer or dry ice for at least 30 minutes prior to installation.

Install split sleeve (18) in groove just ahead of nut (19) and slide retainer ring (17) over the split sleeves. Enter pinion shaft and bearing assembly in transmission housing and proceed as follows: Install first gear (16) with

clutch jaws forward. Slide collar and bushing (14) on shaft with short bushing (1.252 in.) toward rear inside first gear hub. Install sliding gear (15) on collar with shift fork groove toward rear. Install second gear (13) over bushing (1.435 in.) with clutch jaws toward rear. Install thrust washer (12). Install third gear (11) with clutch jaws toward front and shifter coupling (9) with flat side toward rear (radius toward front). Bevel pinion shaft should now be fully in housing and tight against snap ring (21). Install splined collar (10) from front with bushing inside third gear hub (11). Install snap ring (8S) selected during trial assembly on bench or thickest snap ring that can be installed in groove. Install bearing retainer (24), tighten cap screws to 35-40 ft.-lbs. and safety wire the drilled heads.

Reinstall input shaft and retainer assembly. Reassemble tractor in reverse of disassembly procedure.

111. BEVEL PINION BEARING SNAP RING SELECTION. The mesh position of the bevel gears is controlled by the thickness of snap ring (21—Fig. 92) located in front of the bearing cup (22) in bearing bore of transmission housing. If bevel pinion (23) and ring gear set, bearing set (22) and /or transmission housing is renewed, the following procedure must be observed in selecting a new snap ring.

Assemble bearing assembly (22) on pinion shaft (23) with wide bearing and cone next to bevel pinion. Install pinion shaft and bearing assembly in transmission housing without gears. Adjust bearings by tightening nut (19) until 8-35 inch-pounds of torque is required to rotate shaft in bearings. After adjustment is complete, lock adjustment with one of the tabs of washer (20). Remove pinion shaft and bearing after adjustment is complete.

After adjusting bearing pre-load, proceed as follows: Measure distance

(D—Fig. 93) from rear face of **rear** bearing cone **inner race** to front edge of **front bearing cup**. Add to this measurement the figure etched on the rear face of the pinion shaft (Fig. 94). Subtract this sum from the figure stamped on the lower right hand corner of the rear face of transmission housing (Fig. 95). The result should be the thickness of snap ring (21—Fig. 92) used in reassembly of transmission. Snap rings are available in thicknesses of 0.177-0.191 in steps of 0.002.

As an example of above procedure, let measurement (D—Fig. 93) be 2.310. Add this to etched figure (Fig. 94) on rear of pinion gear, 5.345. This gives a sum of 7.655. Subtract this amount from the figure stamped on the transmission case (Fig. 95) which, in this example, is 7.840. Thus, 0.185 is the thickness of the snap ring that should be used for this particular set of bearings, bevel pinion and rear gear set, and transmission housing. Use closest thickness of snap ring to determined value.

112. COUNTERSHAFT AND GEARS. To remove countershaft and gears, the bevel pinion shaft must first be removed as outlined in paragraph 110. Remove plugs (5 &14—Fig. 96) from both ends of countershaft and snap ring (4) from front face of transmission housing. Press or bump shaft forward out of housing. Gears (6, 8, 9, 10 & 11), spacer (7) and rear bearing cone (12) can be removed out top opening. Front bearing cone (2) and cup (3) will be pulled with shaft. If necessary to renew rear bearing cup (13), drive cup out toward front.

End play of countershaft in bearings is controlled by varying the thickness of the snap ring (4) at front end of shaft. This snap ring is available in several thicknesses. If use of the thickest snap ring is not sufficient to control end play, the rear snap ring (15) may be removed and a thicker snap ring installed in that position;

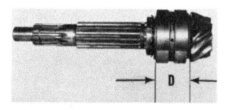

Fig. 93—View showing bearing assembled on transmission bevel pinion shaft. To determine proper thickness of snap ring (21—Fig. 92) required, measure distance from rear face of rear bearing cone inner face to front edge of front bearing cup (dimension "D") and refer to text.

Fig. 94—Cone measurement is etched on rear face of pinion gear and is used in determining thickness of snap ring (21—Fig. 92) for controlling pinion mesh position.

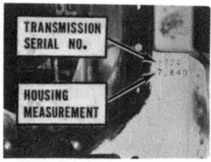

Fig. 95—Transmission housing measurement used in determining thickness of snap ring (21—Fig. 92) is stamped on rear face of housing below transmission serial number.

then, readjust end play with front snap ring. Since the bearing cones fit against the gear and spacer stack instead of against shoulders on countershaft, end play can be checked only when the shaft is installed with all gears and spacers in place and all end play removed from gears and spacers on the shaft.

To reinstall countershaft, proceed as follows: Drive front bearing cone onto end of countershaft. Install rear snap ring (normally 0.093 thick) and drive rear bearing cup in tightly against snap ring. Insert shaft through front bearing bore and install gears, spacers and rear bearing cone in following order: Install driven gear (6) with long hub to rear; spacer (7); third gear (8) with long hub to rear; second gear (9) with long hub to rear; reverse gear (10) with beveled edge of teeth to rear; place rear bearing cone (12) in cup and then install first gear (11) with spacer end toward front.

Press or drive the countershaft into the rear bearing cone. Install the front bearing cup and snap ring. Seat front bearing cup against snap ring (4) and remove all end play from countershaft gears and spacers by driving against inner race of front bearing cone with a hollow driver.

NOTE: Tightness of bearing cones on shaft will remove end play from gears and spacers.

Check end play of the complete shaft, gears and spacers assembly with a dial indicator while moving the assembly back and forth between the front and rear bearing cups. If end play is not within 0.002-0.005, remove snap ring (4) and install snap ring of proper thickness to bring end play within limits.

Apply sealer to rims of plugs (5 & 14) and drive plugs into bearing bores (flat side in) until rims of plugs are just flush with transmisssion housing. Reassemble tractor in reverse of disassembly procedure.

113. REVERSE IDLER GEAR AND SHAFT. The reverse idler gear and shaft can be removed after tractor is split between transmission and torque housing and both transmission and housing top covers are removed; however, in most instances it will be removed when bevel pinion shaft and countershaft assemblies are being overhauled as outlined in paragraph 110 and 112.

To remove reverse idler shaft, proceed as follows: Remove expansion plug located in line with idler shaft in front face of transmission housing. Remove cap screw (19—Fig. 96) and lockplate (16) from differential compartment. Thread slide hammer adapter into front end of shaft, pull shaft from housing and remove gear.

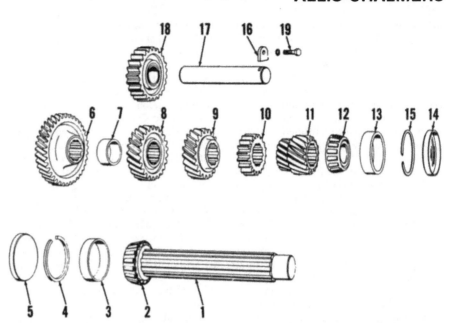

Fig. 96—Exploded view of countershaft and reverse idler gear used in early production tractors. Some early tractors may have been converted to late type shown in Fig. 103.

1. Countershaft		14. Rear plug
2. Bearing cone		15. Snap ring
3. Bearing cup	6. Driven gear	10. Reverse gear	16. Lock plate
4. Snap ring	7. Spacer	11. 1st gear	17. Reverse shaft
5. Front plug	8. 3rd gear	12. Bearing cone	18. Reverse idler
	9. 2nd gear	13. Bearing cup	

Reverse idler rotates on bronze bushing. When renewing bushing, press new bushing in flush with flat side of gear and hone bushing to 1.2375-1.2385. Service shaft (17) has lock notch at each end. Renew shaft if clearance between shaft and new bushing is excessive. Clearance should be 0.002-0.004.

NOTE: Threaded end of shaft is slightly larger diameter than other end.

Install shaft with end having threaded hole to front and with shifter groove of reverse gear to front. Install lockplate, cap screw and lockwasher at rear end of shaft. Apply gasket sealer to edge of expansion plug and drive plug into front face of transmission housing. When installing the differential and final drive (large) cover, use sealer on cap screw at center forward (open) hole.

All 180 and 185 Models, Late 190 and 190XT Models (Tractor Serial No. 9000 and up), All 200 Models

114. INPUT SHAFT. To remove the input shaft, first split transmission from torque housing as outlined in paragraph 104. Remove the "Power-Director" clutch as in paragraph 93 and the shifter cover as in paragraph 107. Unbolt and remove retainer (1—Fig. 98) being careful not to damage the "Power-Director" tubes. Rotate the input shaft until the cut-away portion

of the fourth speed shifter jaws (C—Fig. 97) is down and pull input shaft out toward front.

NOTE: The cut-away portion shown in Fig. 97 allows clearance past the countershaft driven gear.

If seal (3—Fig. 98) and/or retainer (1) is renewed, make certain that seal does not cover drain hole in bottom of retainer. Seal should be pressed in until flush with rear face of retainer.

NOTE: If seal is pressed to bottom (front) of retainer bore, drain hole will be covered and oil from "Power-Director" may be pumped into transmission compartment.

To reinstall, make certain that cut-away (C—Fig. 97) is down and press input shaft and bearing into transmission housing. Be careful not to damage pilot bearing (7—Fig. 98) or thrust washer (7T). Lubricate input shaft and seal (3), install gasket (2) and carefully position retainer (1) over shaft. Tighten retainer cap screws to 40-50 ft.-lbs.

Fig. 97—View of late production input shaft showing cut-away section of fourth speed shifter jaws. This cut-away must be down to clear countershaft drive gear before input shaft can be removed or installed.

Reassemble tractor in reverse of disassembly procedure.

115. BEVEL PINION SHAFT. To remove the bevel pinion shaft, first split transmission from torque housing as outlined in paragraph 104. Remove the differential assembly as in paragraph 125 and transmission input shaft as in paragraph 114. Remove bevel pinion bearing retainer (25—Fig. 98) and snap ring (8). Bump or push bevel pinion shaft (24), together with bearing assembly (23) out toward rear and remove gears, collars and bushings out top opening.

The mesh position of the bevel gears is controlled by thickness of snap ring (22). If necessary to renew snap ring **only**, be sure replacement is exactly same thickness as original. If the bevel pinion shaft and differential ring gear, bearing or transmission housing is renewed, it will be necessary to follow procedure outlined in paragraph 116 to select proper thickness snap ring.

If bevel pinion shaft, gears and/or bushings are renewed, it is advisable to first make a trial assembly on the bench of all parts located between and including the split sleeve (20) (used on 190 model) and front snap ring (8). The front snap ring is available in thicknesses of 0.083-0.107 in steps of 0.006. Select a snap ring which will eliminate all end play of the assembled parts

when all are installed on pinion shaft. If gear (11, 13, or 18) fails to rotate freely, it is an indication that gear hub is wider than bushing (9, 14 or 17) on which it rotates. To provide gear side clearance for free rotation, renew bushing.

NOTE: Bushing (14), bearing assembly (23), snap ring (21) and bevel pinion shaft (24) and ring gear are serviced in a kit only.

Assemble pinion shaft (24) on 190 only and bearings (23) with wide bearings next to bevel pinion. Adjust bearings by tightening nut (21) until 8-35 inch-pounds of torque is required to rotate shaft in bearings. After adjustment is complete, lock adjustment by staking nut to shaft. Adjustment can be accomplished with parts (21, 23 & 24) assembled before assembling in case. If rolling torque is checked with cord wrapped around bearing cups, make certain to correct for leverage advantage (2.0625-inch radius). To facilitate installation of pinion assembly, chill the pinion and bearings in freezer or dry ice for approximately 30 minutes prior to installation.

NOTE: Bearing (23) is a tight fit in transmission housing and gears, collars and spacers must be positioned on pinion shaft as the shaft and bearing assembly is pressed into transmission

housing. **If shaft is bumped into position, split sleeve (20) may fall out of groove.**

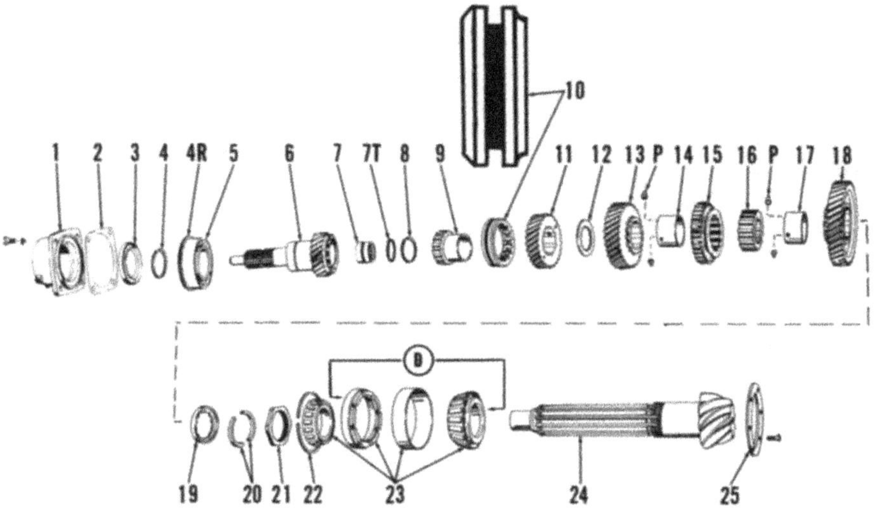

Because of the close tolerances, it is necessary to check all gears for freedom as the shaft is being pressed in. Make certain that all gears are free while pressing pinion shaft.

Install split sleeve (20) in groove just ahead of nut (21) and slide retainer (19) over split sleeve. Lubriplate may be used to hold retainer in position. Start the pinion shaft assembly in transmission compartment and install the short (1.391-1.394 inches) bushing (17) over shaft. The two retaining pins (P) must be installed from inside bushing and should be toward front. Lubriplate may be used to hold pins in position while installing. Install first speed gear (18) over bushing with shifter teeth toward front. Istall splined collar (16) and slide parts (18, 17 & 16) to rear of transmission compartment. Install sliding gear (15) on collar (16) with shifter groove toward rear. Install two pins (P) in wide (1.659-1.662 inches) bushing (14) and install bushing with pins toward front. Install second gear (13) over bushing with shifter teeth toward rear. Install 0.119-0.121 inch thick thrust washer (12) and make certain that all parts are toward rear of transmission compartment. Position third gear (11) over shaft with shifter teeth forward. Install splined collar (9) with bushing end inside hub of third gear. Install third-fourth speed shifter coupling (10) over splined collar with chamfered edge toward front and radius edge toward rear. Bevel pinion shaft (24) should be fully in housing and bearing (23) should be tight against snap ring (22) at this time. Install bearing retainer (25) and tighten screws to 35-40 ft.-lbs. Install snap ring (8) selected during trial assembly on bench or thickest snap ring that can be install in groove.

Reinstall input shaft and retainer as outlined in paragraph 114. Reassemble tractor in reverse of disassembly procedure.

116. BEVEL PINION SNAP RING SELECTION. Mesh position of the bevel gears (ring gear and pinion) is controlled by thickness of snap ring (22—Fig. 98). If bevel pinion (24) and ring gear set, bearing set (23) and/or transmission housing is renewed, observe the following for selecting new snap ring.

Assemble bearing (190 models only) assembly (23) on pinion shaft (24) with wide bearing cup and cone next to bevel pinion. Install pinion shaft and bearing assembly in transmission housing without gears. Adjust bearings by tightening nut (21) until 8-35 inch-

Fig. 98—Exploded view of late production input shaft and bevel pinion shaft. Some early tractors may have been converted to this type. Inset shows chamfered edge of 3rd and 4th speed shifter coupling toward front and radius edge toward rear. The four pins (P) are installed to keep bushings (14 & 17) from turning on shaft. Items 19 and 20 are not used on 200 model. Item 21 is a snap ring on 200 model. Location of measurement for 190 model shown in Fig. 99 is shown at (D).

1. Retainer	8. Snap ring	14. Bushing
2. Gasket	9. Splined collar	(1.659-1.662 inches)
3. Seal	with bushing	15. Sliding gear
4. Snap ring	10. Shift coupling	16. Spindle collar
4R. Snap ring	(3rd & 4th)	17. Bushing
5. Bearing	11. Third gear	(1.155-1.157 inches)
6. Input shaft	12. Thrust washer	18. First gear
7. Pilot bearing	(0.119-0.121 inch)	19. Retainer
7T. Thrust washer	13. Second gear	(0.420 inch)
(0.072-0.077		20. Split sleeve
inch)		(0.248-0.250 inch)
		21. Nut or snap ring
		22. Snap ring
		23. Bearing assembly
		24. Bevel pinion
		shaft
		25. Retainer

pounds of torque is required to rotate shaft in bearings. After adjustment is complete, lock adjustment by staking nut to shaft in two locations 180 degrees apart. Remove pinion shaft and bearing after adjustment is complete.

After adjusting bearing preload, proceed as follows: On all models measure distance (D—Fig. 99) from rear face of **rear** bearing cone **inner race** to front edge of **front bearing cup.** Add to this measurement (D) the number (measurement) etched on rear face of pinion shaft (Fig. 100). Subtract this sum from number (housing measurement—Fig. 101) stamped on lower right hand corner of transmission housing rear face. The result should be thickness of snap ring (22—Fig. 98) used in reassembly of transmission. Snap rings are available in thicknesses 0.177-0.191 in steps of 0.002.

As an **example** of above procedure, let measurement (D—Fig. 99) be 2.310. Add this to measurement (5.345 etched on end of pinion gear (Fig. 100). This gives a sum ot 7.655. Subtract this amount (7.655) from the housing measurement (Fig. 101) of 7.840. The result (0.185) is the thickness of snap ring (22—Fig. 98) that should be used. Use closest thickness of snap ring to determine value.

NOTE: Accurate measurement of bearing (D—Fig. 99) is extremely important.

117. COUNTERSHAFT AND GEARS. To remove countershaft and gears, the bevel pinion shaft must first be removed as outlined in paragraph 115. Remove plugs (2 & 12—Fig. 102) from transmission housing at both ends of countershaft. Plug (12) is not used at rear on 180 and 185 models. Remove snap ring (13) from behind rear bearing cup. Remove nut (3) from front end of countershaft. Press countershaft toward rear until rear bearing cup is out of transmission housing and front bearing cone is off shaft. Remove rear bearing cone (15) from the removed shaft. Remove snap ring (4) and front bearing cup (5) from housing.

Nut (3) at front of countershaft clamps gears and spacers together and adjustment of bearings is controlled by thickness of snap ring (4). Spacers (8) are both 0.8780-0.880 inch wide.

To reinstall, enter the shaft through rear bearing bore and slide one spacer (8) onto shaft. Install reverse gear (11) with beveled edge of teeth toward rear. Install second gear (10) with hub toward rear and flat side forward. Install third gear (9) with flat wide toward rear (against flat side of second gear). Slide spacer (8) over shaft and install countershaft driven gear (7) with long hub toward rear. Install rear bearing cup (14) and 0.093 inch thick

snap ring (13). Insert front bearing cone (6) through front bearing bore in housing and drive onto shaft using a hollow driver. Bearing cone (6) should be driven on shaft until gears and spacers (7, 8, 9, 10, 11 & 8) are tight on shaft. Drive bearing cup (5) into front bearing bore until all end play is removed from countershaft. Install nut (3) with larger flat surface toward bearing cone and tighten to 50-60 ft.-lbs. on all models except 200. On Models 200 tighten nut (3) to 80-100 ft.-lbs. Stake the nut on countershaft after tightening. Install the thickest snap ring (4) that can be installed in front of bearing cup (5). Seat both bearing cups (5 &14) against the retaining snap rings (4 & 13) by bumping shaft at both ends. Measure countershaft end play using a dial indicator and moving shaft back and forth between bearing cups. If end play is not within the limits of 0.002-0.006, remove snap ring (4) and install snap ring of proper thickness to bring end play within limits. Apply sealer to rim of front plug (2) and drive into front bearing bore (**flat side out**) until plug is flush with transmission housing. On 190 and 200 models apply sealer to rim of plug (12) and drive into rear bearing bore (flat side in) until rim is flush with transmission housing. Reassemble tractor in reverse of disassembly procedure.

118. REVERSE IDLER GEAR AND SHAFT. The reverse idler gear and shaft can be removed after the tractor is split between transmission and torque housing and both transmission housing top covers are removed; however, in most instances it will be removed when bevel pinion shaft and countershaft assemblies are being overhauled as outlined in paragraph 115 and 117.

To remove reverse idler shaft, proceed as follows: Remove expansion plug located in line with idler shaft in

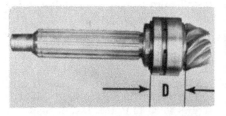

Fig. 99—View showing bearing assembled on transmission bevel pinion shaft on 190 models only. To determine proper thickness of snap ring (22—Fig. 98) required, measure distance from rear face of rear bearing cone inner race to front edge of front bearing cup (dimension "D") and refer to text.

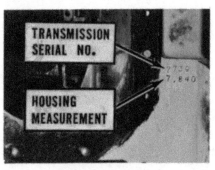

Fig. 101—Transmission housing measurement used in determining thickness of snap ring (22—Fig. 98) is stamped on rear face of housing below transmission serial number.

Fig. 100—Measurement etched on rear face of pinion is used in determining thickness of snap ring (22—Fig. 98) for controlling pinion mesh position.

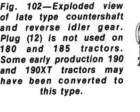

Fig. 102—Exploded view of late type countershaft and reverse idler gear. Plug (12) is not used on 180 and 185 tractors. Some early production 190 and 190XT tractors may have been converted to this type.

1. Countershaft and first gear
2. Front plug
3. Nut
4. Snap ring
5. Bearing cup
6. Front bearing cone
7. Driven gear
8. Spacers (2 used)
9. Third gear
10. Second gear
11. Reverse gear
12. Rear plug
13. Snap ring
14. Bearing cup
15. Rear bearing cone
16. Plug
17. Reverse idler gear
18. Idler shaft
19. Lock plate

front face of transmission housing. Remove screw (20—Fig. 102) and lockplate (19) from differential compartment. Thread slide hammer adapter into front end of shaft, pull shaft from housing and remove gear.

Service shaft (18) has lock notch at each end. Renew shaft if clearance between shaft and new gear is excessive. Clearance should be 0.002-0.004.

NOTE: Threaded end of shaft is slightly larger diameter than other end.

Install shaft with end having threaded hole to front and with shifter groove of reverse gear to front. Install lockplate, cap screw and lockwasher at rear end of shaft. Apply gasket sealer to edge of expansion plug and drive plug into front face of housing. When installing the differential and final drive (large) cover, use sealer on screw at center forward (open) hole.

Model 7000

119. **INPUT SHAFT.** Transmission input shaft is attached to "Power Shift" clutch assembly. Refer to paragraphs 101 and 102 for input shaft servicing.

120. **BEVEL PINION SHAFT.** To remove bevel pinion shaft, first split transmission from torque housing as outlined in paragraph 106. Remove differential assembly as outlined in paragraph 125 and transmission input shaft as directed in paragraph 101. Remove bevel pinion bearing retainer (22—Fig. 103) and snap ring (1). Slide components (2 through 5) forward until snap ring (6) is exposed. Lift snap ring (6) out of groove and move snap ring as far forward as possible. Slide components (8 through 10) forward on pinion shaft. Turn knurled, splined thrust washer (11) until splines match shaft splines and move washer forward. Push bevel pinion shaft and bearing assembly out rear of housing while removing components (2 through 12).

Mesh position of bevel gears is controlled by shims (17). If bevel pinion shaft and differential ring gear, bearing or transmission housing is renewed, it will be necessary to follow procedure outlined in paragraph 121 to select proper shims.

NOTE: Do not attempt to unscrew or tighten bearing nut (13) by placing tool against outer circumference of nut as seal (14) lip seals around nut.

Bearing preload is adjusted using nut (13). Tighten nut (13) to obtain rolling torque of 2 to 6 pounds measured around outer circumference of bearing cup (19). Stake nut (13) securely to shaft.

To install bevel pinion shaft assembly, install shims (17) in housing. Install

seal (14) with spring side forward and so seal is 0885-0.915 inches from shims (17). Insert pinion shaft and bearing assembly in housing and place first gear (12) on shaft with shift fork groove forward. Install knurled thrust washer (11) and collar (10) with inner cutout of collar towards washer. Install sliding gear (9) with shift fork groove rearward and second gear (8) with shift teeth rearward. Install thrust washer (7), snap ring (6) and thrust washer (5) approximately ½-inch onto shaft. Install third gear (4) with shift teeth forward, and collar (2). Install third-fourth speed shifter coupling (3) with chamfer forward. Press pinion shaft into housing while being careful not to bind gears. Work thrust washer (11) into pinion shaft groove and slide collar (10) over thrust washer. Move second gear (8) and position snap ring (6) in pinion shaft snap ring groove. Move remaining gear (4), collar (2) and shift coupling (3) rearward and install snap ring (1).

121. **BEVEL PINION SHIM SELECTION.** Mesh position of bevel gears (ring gear and pinion) is determined by thickness of shims (17—Fig. 103). If bevel pinion (21), ring gear set, pinion shaft bearings or transmission housing is renewed, observe the following for selecting shims.

Measure thickness of assembled bearing cup (19) and cone (20). Measure spacer thickness and add to bearing thickness. Add sum of measurements to number etched on rear face of pinion shown in Fig. 100. Subtract sum from housing measurement stamped on rear of housing shown in Fig. 101. Result will be shim (17) thickness.

As an **example** of above procedure, let combined bearing and spacer measurement be 1.763. Add this to measurement "6.019" etched on end of pinion gear (Fig. 100). This gives a sum of 7.782. Subtract this amount (7.782) from the housing measurement (Fig. 101) of "7.816." The result (0.034) is the

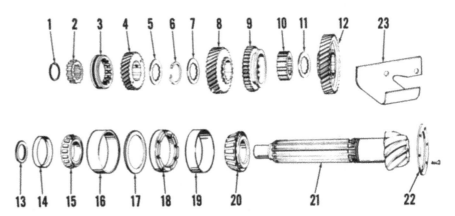

Fig. 103—Exploded view of Model 7000 bevel pinion shaft assembly. Transmission input shaft is shown in Fig. 72.

1. Snap ring	6. Snap ring	12. First gear	18. Spacer
2. Splined collar	7. Thrust washer	13. Nut	19. Bearing cup
3. Shift coupling (3rd & 4th)	8. Second gear	14. Seal	20. Bearing cone
4. Third gear	9. Sliding gear	15. Bearing cone	21. Bevel pinion shaft
5. Thrust washer	10. Splined collar	16. Bearing cup	22. Retainer
	11. Thrust washer	17. Shims	

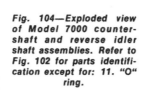

Fig. 104—Exploded view of Model 7000 countershaft and reverse idler shaft assemblies. Refer to Fig. 102 for parts identification except for: 11. "O" ring.

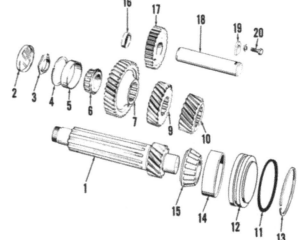

thickness of shim (17—Fig. 103) that should be used. Use closest thickness of shim to calculated value.

122. **COUNTERSHAFT AND GEARS.** To remove countershaft and gears, the bevel pinion shaft must first be removed as outlined in paragraph 120. Remove plugs (2 &12—Fig. 104) from transmission housing at both ends of countershaft. Remove nut (3) from front end of countershaft. Press countershaft toward rear until rear bearing cup is out of transmission housing and front bearing cone is off shaft. Remove rear bearing cone (15) from the removed shaft. Remove snap ring (4) and front bearing cup (5) from housing.

Nut (3) at front of countershaft clamps gears together and adjustment of bearing is controlled by thickness of snap ring (4).

To reinstall, press bearing cone (15) on shaft and install second gear (10) with hub toward rear and flat side forward. Install third gear (9) with flat side toward rear (against flat side of second gear). Install countershaft driven gear (7) with long hub toward rear. Install front bearing cone (6) and position assembly in housing. Install "O" ring (11) on plug (12). Drive bearing cup (14) and plug (12) into housing far enough to clear snap ring groove and install snap ring (13). Drive bearing cup (5) into front bearing bore until all end play is removed from countershaft. Install nut (3) with larger flat surface toward bearing cone and tighten to 80-100 ft.-lbs. Stake nut on countershaft after tightening. Install the thickest snap ring (4) that can be installed in front of bearing cup (5). Seat both bearing cups (5 &14) against the retaining snap rings (4 & 13) by bumping shaft at both ends. Measure countershaft end play using a dial indicator and moving shaft back and forth between bearing cups. If end play is not within the limits of 0.002-0.006, remove snap ring (4) and install snap ring of proper thickness to bring end play within limits. Apply sealer to rim of front plug (2) and drive into front bearing bore (**flat side out**) until plug stands out of housing 0.09-0.015 inches.

123. **REVERSE IDLER GEAR AND SHAFT.** The reverse idler gear and shaft can be removed after the tractor is split between transmission and torque housing and both transmission housing top covers are removed; however, in most instances it will be removed when bevel pinion shaft and countershaft assemblies are being overhauled as outlined in paragraphs 120 and 122.

To remove reverse idler shaft, proceed as follows: Remove expansion plug located in line with idler shaft in

front face of transmission housing. Remove cap screw (20—Fig. 104) and lockplate (19) from differential compartment. Drive shaft out rear of housing.

Renew shaft if clearance between shaft and new gear is excessive. Clearance should be 0.002-0.004.

NOTE: Forward end of shaft is slightly larger diameter than other end.

Drive shaft into housing being sure notch is properly aligned and with shifter groove of reverse gear to front. Install lockplate, cap screw and lockwasher at rear end of shaft. Tighten cap screw to 38-42 ft.-lbs. Apply gasket sealer to edge of expansion plug and drive plug into front face of housing.

DIFFERENTIAL, FINAL DRIVE AND REAR AXLE

DIFFERENTIAL

All Models

124. **BEARING ADJUSTMENT.** Adjustment of differential bearings is controlled by shims (3—Fig. 105 or Fig. 106) located between each bearing carrier (1) and transmission housing.

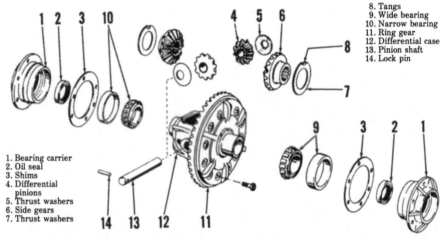

8. Tangs
9. Wide bearing
10. Narrow bearing
11. Ring gear
12. Differential case
13. Pinion shaft
14. Lock pin

1. Bearing carrier
2. Oil seal
3. Shims
4. Differential pinions
5. Thrust washers
6. Side gears
7. Thrust washers

Fig. 105—Exploded view of two pinion differential assembly. Refer to Fig. 106 for four pinion differential used on some models. Bull pinion inner bearing cup is pressed into outside of carriers (1). Right bearing (9) is wider than left bearing (10).

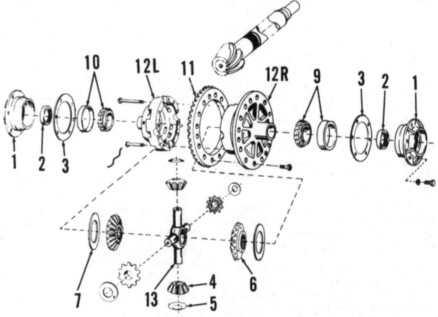

Fig. 106—Exploded view of four pinion differential used on some models.

1. Bearing carrier	5. Thrust washers	9. Wide bearing	12L & 12R. Differential
2. Oil seal	6. Side gears	10. Narrow bearing	housing valves
3. Shims	7. Thrust washers	11. Ring gear	13. Pinion shaft
4. Differential pinions			

Transferring shims from under one bearing carrier to under opposite bearing carrier will adjust backlash between the bevel ring gear and bevel pinion gear.

NOTE: Accurate measurement is impossible unless ring gear or pinion is removed.

To adjust differential bearings on 190, 190XT, 200 and 7000 tractors, remove bull gear as outlined in paragraph 137. On 180 and 185 models, remove both final drive units as outlined in paragraph 134. Then, on all models, proceed as follows: Vary number of shims (3—Fig. 105 or Fig. 106) between bearing carriers and transmission housing to remove all end play of differential unit, but permitting differential to turn without binding. With ring gear or bevel pinion shaft removed 8-12 inch-pounds torque should be required to rotate differential assembly in carrier bearings.

NOTE: A steady pull of 3 to 3½ pounds should be required to rotate the differential in bearings as measured at end of cord wrapped around differential case (12).

Transfer shims from under one bearing to opposite bearing carrier until bevel gear backlash is 0.007-0.012. Check backlash at several points to be sure that minimum and maximum backlash are within limits. Apply sealer to surface of bearing carrier that contacts transmission housing and to threads of retaining cap screws.

NOTE: Do not apply sealer to shim contact surfaces.

Adjustment of differential carrier bearings and bevel gear backlash will change bull pinion shaft end play. Refer to paragraph 132 (180 and 185 models) or 136 (190, 190XT, 200 and 7000 models) for adjustment of bull pinion shaft bearings.

125. **R&R AND OVERHAUL.** The two-pinion differential shown in Fig. 105 is used on some models. Late 190 XT, 200 and 7000 models are equipped with the four-pinion differential shown in Fig. 106. Removal procedure is the same for both types of differentials.

On 180 and 185 models, remove both final drive units as outlined in paragraph 134. On 190, 190XT, 200 and 7000 models, remove both rear axle shafts, bull pinion shafts and bull gears as outlined in paragraphs 136, 137 and 139. On all models unbolt and remove left differential bearing carrier. Unbolt right bearing carrier and support differential assembly while removing bearing carrier. Then remove differential assembly out rear opening of transmission housing. Refer to paragraph 126 or 127 for overhaul and installation.

126. To disassemble the two-pinion differential, drive out the lock pin (14—Fig. 105), then drive pinion shaft (13) from differential housing. Differential pinions (4), side gears (6) and thrust washers (5 & 7) can then be removed from differential housing. Renew all questionable parts.

If backlash between teeth of side gears and pinion gears is excessive, renew side gear thrust washers (7) and/or pinion gear thrust washers (5). If backlash is still excessive after renewing thrust washers, it may be necessary to renew side gears and differential pinions. Pinions and side gears are available in matched sets only. Hole in differential case (12) for lock pin (14) should be peened closed after pin is installed. For renewal of bevel ring gear, refer to paragraph 129.

Oil seals (2—Fig. 105) should be installed with lip facing differential assembly. The narrow bearing (10) should be on left side of differential. The wide bearing (9) should be right (ring gear) side. To reinstall differential, reverse removal procedure. Refer to paragraph 124 for carrier bearing and backlash adjustments. Sealer should be used on the threads of bearing carrier retaining screws. When installing differential and final drive (large) cover on 190 and 190XT models, use sealer on cap screw at center forward (open) hole. Observe the following torque values during reassembly.

180 & 185 Models
Bearing carrier to
 transmission........90-100 ft.-lbs.
Final drive sleeve to
 transmission.........200-210 ft.-lbs
Lift shaft housing to transmission—
 5/8-inch.............180-190 ft.-lbs
 ½-inch................80-90 ft.-lbs.

190 & 190XT Models
Bearing carrier to
 transmission...........45-50 ft.-lbs.
Brake housing to
 transmission.........100-110 ft.-lbs
Axle housing to
 transmission.........142-152 ft.-lbs.
Bull gear
 (axle inner end).......160-175 ft.-lbs
Brake cover plate retaining
 screws.................70-75 ft.-lbs.

Lift shaft housing to transmission—
 5/8-inch.............180-190 ft.-lbs
 ½-inch................80-90 ft.-lbs.

127. The four-pinion differential (Fig. 106) can be disassembled after removing the screws attaching the housing halves (12L & 12R) together. Before separating, the halves should be

marked to facilitate correct alignment when assembling.

Thickness of side gear thrust washers (7) should be 0.043-0.047 inch. Thickness of differential pinion gear thrust washers (5) should be 0.028-0.032 inch. Differential pinion gears (4) should have 0.004-0.006 inch clearance on pinion shafts (13).

Position one thrust washer (7), one side gear (6), differential pinions (4), thrust washers (5) and pinion (spider) shaft (13) in the right half of housing (12R), then install the second side gear, thrust washer and left half of housing (12L). Align the previously affixed marks on housing halves and install the retaining screws. Screws should be tightened to 60-70 ft.-lbs. then safe tied together with wire to prevent loosening. For renewal of bevel ring gear, refer to paragraph 129.

Oil seals (2) should be installed with lip facing differential assembly. The wide bearing (9) should be on right (ring gear) side and narrow bearing (10) should be on left side of differential. To reinstall, reverse removal procedure (paragraph 125). Sealer should be used on threads of bearing carrier retaining screws. Refer to paragraph 124 for carrier bearing and backlash adjustment. When installing differential and final drive (large) cover, use sealer on cap screw at center forward (open) hole. Observe the following torque values during reassembly.

190, 190XT, 200 & 7000 Models
Bearing carrier to
 transmission...........45-50 ft.-lbs.
Brake housing to
 transmission.........100-110 ft.-lbs.
Axle housing to
 transmission.........140-160 ft.-lbs.
Bull gear (inner
 axle end)............160-175 ft.-lbs.
Brake cover plate
 retaining screws........70-75 ft.-lbs.
Lift shaft housing to transmission—
 5/8-inch.............180-190 ft.-lbs.
 ½-inch................80-90 ft.-lbs.

MAIN DRIVE BEVEL GEARS

128. **MESH AND BACKLASH ADJUSTMENT.** The mesh position of the bevel pinion gear is controlled by the thickness of snap ring or shim (21—Fig. 92, 22—Fig. 98 or 17—Fig. 103) which is fitted to a particular transmission housing, bevel gear set and bevel pinion bearing. Refer to paragraph 111, 116 or 121 for selection of snap ring or shim.

To adjust the backlash between bevel pinion and bevel ring gear, refer to paragraph 124.

129. RENEW BEVEL GEARS. As the bevel pinion is an integral part of the transmission output shaft, refer to paragraph 110, 115 or 120 for service procedures.

NOTE: Main drive bevel pinion and ring gear are available only in a matched set.

The bevel ring gear is renewable when differential assembly is removed as outlined in paragraph 125. On factory assembled differential units prior to 190 tractor serial number 8461, the ring gear is riveted to differential housing. Special bolts and nuts are available for service installation. To remove rivets, drill 1/32-inch undersize from ring gear side. Install bolts with heads on ring gear side and tighten evenly to 70-75 ft.-lbs. Install cotter pin through castellated nuts and drilled cap screws.

On later models (after 190 tractor serial number 8460), the ring gear is attached to differential housing with special cap screws threaded into back of ring gear. Bevel ring gear retaining screws should be tightened evenly to 100-110 ft.-lbs. on models with two-pinion differential, 65-75 ft.-lbs torque on models with four-pinion differential.

FINAL DRIVE

180 and 185 Models

130. ADJUST WHEEL AXLE BEARINGS. Wheel axle shaft bearings are adjusted by adding or removing shims (27—Figs. 107 and 108) between end of wheel axle shaft and the pinned washer (28) to provide 0.002-0.005 preload. Screw (26) should be tightened to 170-180 ft.-lbs.

To adjust, add shims at (27) until some end play is obtained, tighten screw (26) to 170-180 ft.-lbs, then measure the end play. Remove shims equal to the measured end play plus 0.002-0.005.

EXAMPLE: If end play is 0.004, remove two 0.003 thick shims which will preload the bearings 0.002. All service shims are 0.003 thick; 0.010 and 0.003 thick shims are used during production.

131. RENEW WHEEL AXLE BEARINGS, SEAL AND/OR BULL GEAR. Support rear end of tractor and remove rear wheel and tire unit. Remove lower cover (23—Figs. 107 and 108) from final drive housing.

NOTE: No drain plug is provided; remove cover with oil it contains.

Disengage snap ring (21) holding bull gear (20) in position on axle shaft (16). Remove cap (29) from inner end of axle shaft and remove retaining cap screw (26), lockwasher, pinned washer (28) and shims (27). Then, while supporting

bull gear, bump axle shaft out of bull gear and final drive housing.

If not removed with axle shaft, remove the axle seal (17) and outer bearing cone (18) from housing. If necessary to renew bearing cups, drive cups from housing.

To reinstall removed parts, proceed as follows: Drive outer bearing cup (18) in tightly against shoulder in housing

and inner bearing cup (25) in tight against snap ring (24). Lubricate outer bearing cone and place cone in cup. Soak new seal (17) in oil, wipe off excess oil and apply gasket sealer to outer rim of seal. Install the seal with lip to inside of housing. Insert axle shaft through seal and outer bearing cone until spacer (19) can be placed on end of shaft, then position bull gear in

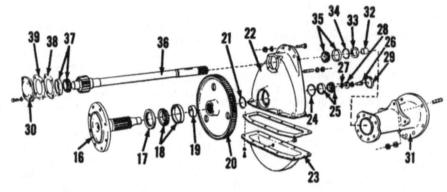

Fig. 107—Final drive and rear axle unit used on 180 and 185 models. Shims (38 & 39) control bull pinion shaft bearing adjustment. Shims (27) control wheel axle shaft bearing adjustment. Bushing (32) is pressed in the differential case.

16. Wheel axle shaft	23. Pan	29. Dust cap	34. Snap ring
17. Oil seal	24. Snap ring	30. Bearing retainer	35. Inner bearing
18. Outer bearing	25. Inner bearing	31. Bull pinion shaft	36. Bull pinion shaft
19. Spacer	26. Cap screw	housing	37. Outer bearing
20. Bull gear	27. Shims	32. Pinion shaft bushing	38. Shim (0.006 vellum)
21. Snap ring	28. Washer	33. Oil seal	39. Shim (0.015 steel)
22. Bull gear housing			

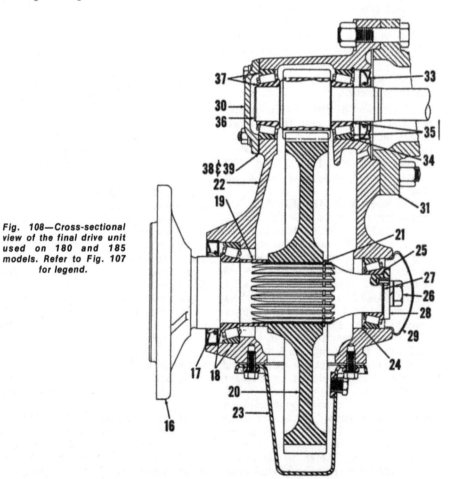

Fig. 108—Cross-sectional view of the final drive unit used on 180 and 185 models. Refer to Fig. 107 for legend.

housing with long hub to outside and push shaft on through spacer and bull gear until snap ring (21) can be placed over end of shaft. Press axle shaft inward until shoulder of shaft is against outer bearing cone and snap ring (21) can be installed in groove against inner side of bull gear hub. Install inner bearing cone (25), shims (27), pinned washer (28), lockwasher and retaining cap screw (26). Check adjustment of axle bearings as outlined in paragraph 130 and adjust bearings if necessary. Apply sealer to rim of cap and drive the cap into place. Install cover using new gasket. Tighten screws retaining cover (23) to a torque of 10-14 ft.-lbs. Fill housing with 1 quart of "Allis-Chalmers Special Gear Lube."

132. **ADJUST BULL PINION BEARINGS.** Bull pinion shaft should have 0.001-0.005 end play. To adjust bearings, remove the final drive and sleeve as outlined in paragraph 134. Vary the thickness of shims (38 & 39— Figs. 107 and 108) to provide pinion shaft with correct end play. Alternate vellum (paper) and steel shims and make certain that a paper shim is located on each side of shim stack for proper sealing. Tighten cap screws attaching the bearing retainer (30) to 29-32 ft.-lbs.

133. **RENEW BULL PINION, BEARINGS AND/OR SEAL.** Remove the final drive assembly as outlined in paragraph 134, then remove brake as outlined in paragraph 142.

After removing the bearing retainer (30—Figs. 107 and 108), the final drive pinion shaft (36) can be bumped out towards outside end of unit. Be careful not to catch inner bearing cone (35) on teeth of bull gear. If inner bearing cup or seal (33) is to be renewed, pinion shaft sleeve (31) must be unbolted and removed from final drive housing. Then remove seal, snap ring (34) and bearing cup. Bearing cones can now be renewed on pinion shaft.

To reassemble, proceed as follows: Drive inner bearing cup (35) in far enough to install snap ring (34); then, drive bearing cup back against snap ring. Drive bearing cones tightly against shoulders on pinion shaft and insert shaft into final drive housing.

NOTE: If seal (33) was not removed, tape inner end of shaft at Woodruff key and snap ring grooves to prevent damage to seal.

Install outer bearing cup, shims and bearing retainer. Pinion shaft end play should be 0.001-0.005. Add or remove shims (38 and 39) if end play is not within recommended limits. Alternate paper (vellum) shims (0.006 thick) and steel shims (0.015 thick) for proper sealing and use paper shim on each

side of shim stack. Soak new seal (33) in oil, wipe off excess oil and apply gasket sealer to outer rim of seal. Install seal over pinion shaft with lip towards pinion gear and drive seal into final drive housing flush with end of bore.

Apply shellac or equivalent setting sealer to contact surfaces of final drive housing and pinion sleeve. Tighten the retaining nuts to 210-220 ft.-lbs.

Install snap ring, Woodruff key and brake hub on pinion shaft and reinstall final drive unit to transmission, taking care not to damage seal in differential bearing carrier. Tighten the retaining nuts to 200-210 ft.-lbs. Reinstall wheel and tire unit and rear fender.

134. **R&R FINAL DRIVE UNIT.** Support rear end of tractor, remove rear wheel and tire unit and rear fender. Support final drive unit and unbolt pinion shaft sleeve from transmission housing. Carefully withdraw the final drive and pinion shaft from transmission and differential unit.

The brake assembly must be removed before separating final drive housing from the sleeve.

Reverse removal procedure to reinstall final drive unit taking care not to damage seal in the differential bearing carrier. Tighten the retaining nuts to 200-210 ft.-lbs.

135. **RENEW FINAL DRIVE PINION SHAFT SLEEVE.** Remove the final drive unit as outlined in paragraph 134. Remove the brake outer friction plate from inner end of pinion shaft sleeve. Be careful not to lose or damage shims between the friction plate and sleeve.

The pinion shaft sleeve (31—Fig. 107 and 108) may now be unbolted and removed from the final drive housing. Install the sleeve as follows: Apply shellac or equivalent setting sealer to contact surfaces of sleeve and final drive housing. Install sleeve to housing leaving out the two cap screws (X— Fig. 109). Ream the two cap screw

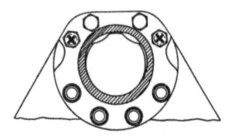

Fig. 109—When installing new bull pinion shaft housing (31—Figs. 107 and 108) on 180 and 185 models, holes indicated by "X" in new housing must be reamed to 0.623-0.625 after housings are bolted together. The two 0.624-0.626 dowel bolts should be installed in these two holes.

holes to 0.623-0.625 using holes in final drive housing as guides. Then, install the two cap screws (X). Tighten the retaining cap screws to 210-220 ft.-lbs. torque.

Install the brake outer friction plate on inner end of sleeve using same number of shims as removed during disassembly; then, check clearance between brake inner and outer friction plates as outlined in paragraph 142.

Reinstall final drive unit as outlined in paragraph 134.

190, 190XT, 200 and 7000 MODELS

136. **BULL PINION SHAFTS.** Bull pinion gear shafts are carried in tapered roller bearings located in differential bearing carriers and in brake housing. Number of shims used between differential bearing carrier and transmission housing to adjust differential bearing end play and bevel gear backlash will affect bull pinion shaft end play. Shims between brake housings and transmission housing are used to adjust bull pinion shaft end play.

To remove bull pinion shafts, proceed as follows: Split the transmission from the torque housing as outlined in paragraph 104 or 106. Remove both rear axles as outlined in paragraph in 139. Unbolt and remove brake housing covers. Pull brake disc and drum assemblies from bull pinion shafts. Unscrew brake adjusting cap screws and remove contracting brake bands from housings. Unbolt and remove brake housing from transmission housing. Pull the pinion shafts from differential.

Reassemble by reversing disassembly procedures. Always renew outer seal (4—Fig. 110). Spring loaded lip of seal should be toward inside (differential). Use proper number of shims between brake housing and transmission housing to allow 0.001-0.003 end play of bull pinion shafts. Shims are available in thicknesses of 0.003, 0.010 and 0.020.

137. **BULL GEARS.** To remove bull gears, remove bull pinion shafts as outlined in paragraph 136. Remove complete seat assembly and platform from transmission housing. Unbolt and remove top cover from rear of transmission housing. Lift bull gears from compartments in transmission housing. Reassemble as in paragraph 136 after first placing bull gears in transmission housing.

138. **AXLE SHAFT BEARING ADJUSTMENT.** Axle shaft bearing adjustment is controlled by shims (15—Fig. 110). Inner edge of seal retainer (13) contacts bearing cup (16). Axle shaft should be adjusted for reload as follows:

Support tractor and remove rear wheel and hub. Special bolts (Fig. 111) supplied with each new tractor are used to jack the hub out of wheel. Remove retainer (13—Fig. 110) and vary the thickness of shims (15) to provide axle shaft bearings with 0.006-0.008 preload on 200 and 7000 models or 0.002-0.005 preload on all other models. Tighten retainer cap screws to 70-75 ft.-lbs. and hub retaining screws to 180-190 ft.-lbs. for 190 and 190XT models and 160-175 ft.-lbs. for 200 and 7000 models.

139. R&R AXLE AND BEARINGS.
To remove either axle shaft (6—Fig. 110), first drain oil from final drive and transmission compartments. Disconnect and remove lift arm cylinders. Remove pto output shaft. Unbolt and remove lift shaft housing from rear face of transmission housing. Support rear of tractor solidly and remove rear wheel and hub. Special jack screws (Fig. 111) are available for pushing wheel from hub (Fig. 112).

Remove cap (19—Fig. 110), cap screw (18) and pinned washer (17 and 17A) from inner end of rear axle. Remove bearing retainer (13) from outer end of axle housing. Remove magnetic drain plug from final drive compartment to prevent damage to plug. Place a porta-power ram (or hydraulic jack) between inner end of axle and opposite wall of differential compartment. Press axle and outer bearing out of axle housing as shown in Fig. 113.

CAUTION: When pressing axle out, be sure that ram (or jack) is supported

by opposite wall and NOT by inner end of opposite rear axle or cap covering inner end of axle.

Axle shaft inner bearing (23 and 24—Fig. 110) can be renewed after removing bull pinion shaft as in paragraph 136 and bull gear as in paragraph 137.

Reverse removal procedures to reinstall axle and outer bearing. To pull axle through inner bearing cone (23), remove roll pin (17A) from washer (17) and use 2½-inch cap screw until cap screw bottoms in axle; then, use original cap screw (18) to complete operation. Reinstall roll pin (17A) in washer with pin inserted in off center hole in axle inner end. Tighten cap screw (18) to 160-175 ft.-lbs. Apply No. 3 Permatex or similar sealer to outer rim of plug (19) before installing. Apply sealer to both sides of lift shaft housing gasket. Tighten the ½-inch cap screws retaining lift shaft housing to 80-90 ft.-lbs. and 5/8-inch cap screws to 180-190 ft.-lbs. When installing differential and final drive top cover, observe the following: Clean all of the blind threaded holes in transmission housing. On all models except 7000, apply sealer to threads of 1¾-inch cap screw installed in open hole at center front and install short (1½ inch) cap screw at center rear. Tighten cover retaining screws on all models to 70-75 ft.-lbs. Adjust axle shaft bearings as outlined in paragraph 138.

140. R&R REAR AXLE HOUSING. To remove a rear axle housing, it is

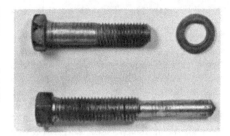

Fig. 111—Standard rear wheel hub cap screw and washer (top) compared with special cap screw (bottom) used for removing hub.

Fig. 112—Removing tapered, split hub from rear wheel. Special bolts (See Fig. 111) are used to press wheel off of tapered hub.

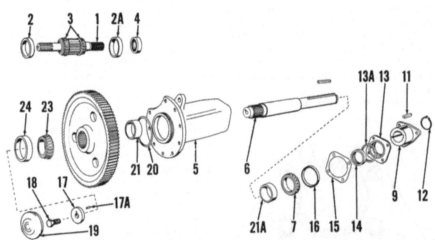

Fig. 110—Exploded view of final drive assembly used on 190, 190XT and 200 models. Bearing cup (2) is located in differential carrier; (2A) is located in brake housing; and (24) is located in inner wall of bull gear compartment in transmission housing. Refer to Fig. 114 for high clearances.

1. Bull pinion	6. Axle shaft	13A. "O" ring	19. Plug
2. Bearing cup	7. Bearing cone	14. Seal	20. "O" ring
2A. Bearing cup	9. Hub	15. Shims	21. Gear spacer
3. Bearing cones	11. Key	16. Bearing cup	21A. Bearing spacer
4. Seal	12. Snap ring	17. Washer	23. Bearing cone
5. Axle housing	13. Retainer	18. Cap screw	24. Bearing cup

Fig. 113—Hydraulic jack in position to press rear axle shaft out of bull gear and inner bearing cone. Refer to text for caution during this operation on 190, 190XT, 200 and 7000 models.

1. Final drive gear
2. Oil pan
3. Gasket
4. Snap ring
5. "O" ring
6. Bearing cup
7. Bearing cone
8. Gear spacer
9. Oil seal
10. Rear axle
11. Seal retainer
12. "O" ring
13. Idler shaft
14. Bearing cones
15. Bearing cups
16. Idler gear
17. Idler spacer
18. "O" ring
19. Intermediate pinion and shaft
20. Final drive housing
21. Bearing cone
22. Bearing cup
23. "O" ring
24. Studs
25. Drive shaft
26. Bearing cones
27. Brake housing seal
28. Bearing cups
29. Brake pivot pin
30. "O" ring
31. Brake housing
32. Shims (0.003, 0.010 & 0.020)
33. Drive gear
34. "O" ring
35. Washer
36. Cap
37. Driven gear
38. Bearing cone
39. Bearing cup
40. Gasket
41. Drain plug
42. Sleeve housing
43. Washer
44. Shims (0.003 & 0.010)
45. "O" ring
46. Bearing cone
47. Shims (0.003 & 0.010)
48. Washer and roll pin
49. Gasket
50. Rear axle cover
51. Gasket
52. Oil lever plug
53. Bearing cup
54. Snap ring
55. Snap ring
56. Oil seal
57. Bearing cup spacer

first necessary to remove the rear axle from housing as outlined in paragraph 139. Then, unbolt and remove rear axle housing from transmision housing.

Renew "O" ring (20—Fig. 110) and tighten retaining cap screws to 142-152 ft.-lbs. and nuts to 130-140 ft.-lbs. Reinstall rear axle as outlined in paragraph 139.

BRAKES

180 and 185 Models

141. **ADJUSTMENT.** To adjust the band/disc type brakes, detach brake rods from brake pedals and turn rods in or out to obtain 2½ inches free travel of pedal pads. Reattach rods to pedals.

142. **R&R BRAKE BANDS AND DISC ASSEMBLY.** Brake drum and disc assembly can be withdrawn from brake bands after removing final drive unit as outlined in paragraph 134.

Detach brake rods from brake pedals and unscrew rods from pivot pins (16—Fig. 115). Unhook the band return spring (20) from front and rear bands (7) and transmission housing. Thread slide hammer adapter into lower pin (18) and pull pin from housing. Pry upper pins (6) from transmission housing and remove brake bands.

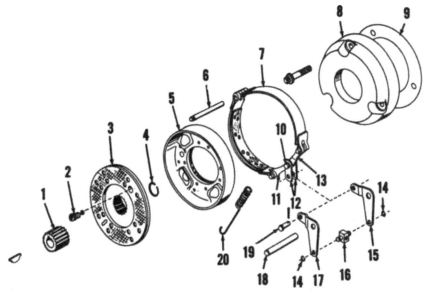

1. Splined hub
2. Retraction springs
3. Brake disc
4. Snap ring
5. Brake drum
6. Upper pin
7. Brake band
8. Outer friction plate
9. Shims
10. Clevis pin
11. Inner link
12. Link
13. Outer link
14. Snap ring
15. Outer lever
16 Pivot pin
17. Inner lever
18. Lower pin
19. Clevis pin
20. Band return springs (2)

Fig. 114—Cross-sectional view of final drive assembly used on 190 and 190XT high clearance tractors.

Fig. 115—Exploded view of band/disc type brakes used on 180 and 185 models. Outer friction plate (8) is attached to inner end of bull pinion shaft housing. Differential bearing carriers are machined for brake inner friction surface. See Fig. 116 for cross-sectional view.

A dimension of 2.034-2.044 inches (A—Fig. 116) should be maintained between the outer brake friction surface and the brake friction surface on the differential bearing carriers. To check this dimension, install final drive units **without** brake band or drum and disc units and measure distance between friction plate and bearing carrier brake friction surfaces with an inside micrometer or other accurate measuring instrument. If dimension is not within the limits of 2.034-2.044, vary number of shims (9) between the outer friction plate and the final drive pinion shaft sleeve to obtain the recommended dimension. Shims are available in thicknesses of 0.005-0.007.

Reassemble brake bands and the drum and disc unit, then install final drive in reverse of removal procedure. Disc part of drum and disc unit must be towards the differential bearing carrier. Ater reassembling tractor, adjust the brakes as outlined in paragraph 141.

143. OVERHAUL. To disassemble drum and disc unit (Fig. 119), insert slotted screwdriver tip through open end of springs and stretch the springs only far enough to unhook them. Remove disc and steel brake actuating balls from drum.

Linings are available separately from disc and drum. Renew band if linings are not reusable. Inspect friction surfaces of outer friction plate (8—Fig. 116) and differential bearing carrier (C) and renew friction plate or bearing carrier if friction surfaces are not suitable for further use.

Condition of the return springs (2 and 20—Fig. 115) is of utmost importance when servicing band/disc brakes. Renew any spring if coils of spring do not fit tightly together and be careful not to stretch springs any farther than necessary when reassembling brakes. Insufficient spring tension will allow brakes to drag.

190, 190XT and 200 Prior to Serial 200-4001 Models

144. ADJUSTMENT. To adjust either brake, turn the adjusting cap screw (Fig. 117) in or out as required to provide brake pedals with 3 inches of free travel. Adjusting screw is self locking. Be sure that brakes are adjusted so that pedals are aligned when both brakes are applied. Brake rods should be adjusted at front clevis ONLY to remove free play from linkage.

145. R&R AND OVERHAUL. To remove brake assemblies, unbolt and remove covers (1—Fig. 118). Drum and disc unit (Fig. 119) can be removed

from bull pinion shaft. To remove the contracting band, use slotted screw driver tip (or small hooked wire) to disengage the two springs (9—Fig. 118). Remove adjusting screw (12) and withdrawn band assembly from pivot

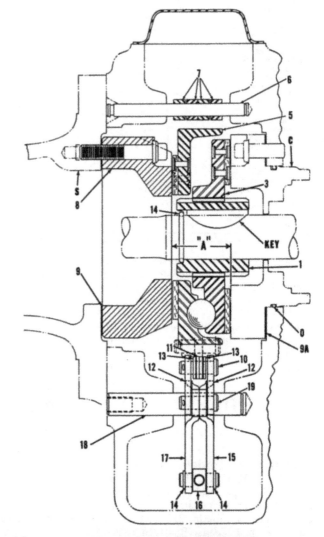

Fig. 116—Cross-sectional view of the band/disc brakes used on 180 and 185 models. Dimension "A" should be maintained at 2.034-2.044 inches. Refer to Fig. 115 for legend.

A. Carrier to friction plate dimension
C. Differential bearing carrier
O. "O" ring seal
S. Pinion shaft housing

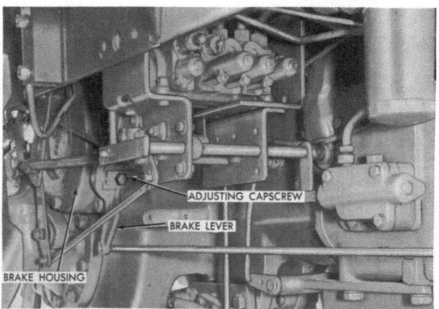

Fig. 117—View of 190 tractor right side showing brake adjusting cap screw. Travel at end of brake lever should be 5/8-inch. Free travel at pedal should be 3 inches.

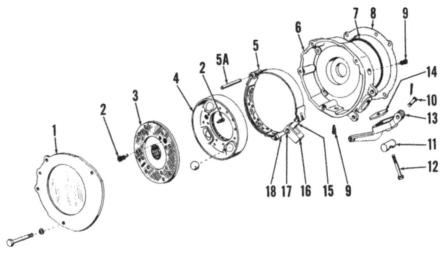

Fig. 118—Exploded view of brake assembly used on 190, 190XT and 200 prior to serial
number 200-4001 models. Pivot pin (5A) should be pressed into housing (6) until end is
almost flush with shim surface of housing.

1. Brake housing	5. Band assembly	9. Springs	14. Seal
cover	5A. Band pivot pin	10. Pin	15. Link (outer)
2. Springs	6. Brake housing	11. Bar	16. Yoke
3. Disc assembly	7. "O" ring	12. Adjusting screw	17. Pin
4. Drum assembly	8. Shims	13. Lever	18. Link (inner)

Insufficient spring tension will allow the brakes to drag as they will fail to return the bands and discs to proper position when brakes are released.

Reverse disassembly procedures to reassemble brakes. Splined disc (3) should be toward outside on all models. Cover (1) retaining screws should be tightened to 70-75 ft.-lbs. Adjust brakes as outlined in paragraph 144.

200 Model Serial Number 200-4001 and Up

146. **ADJUSTMENT.** To adjust either brake, turn the adjusting cap screw (Fig. 117) in or out as required to provide brake pedals with 3½ inches plus or minus ¼ inch of free travel. Be sure that brakes are adjusted so that pedals are aligned when both brakes are applied.

147. **R&R AND OVERHAUL.** To remove brake assemblies, remove housing cover, clevis pin from the adjusting yoke, outer disc, actuating disc assembly and inner disc. The need for further disassembly will be evident. Refer to Fig. 120.

Reverse disassembly procedure to reassemble brakes. Adjust brakes as outlined in paragraph 146.

Model 7000

148. **ADJUSTMENT.** To adjust either brake, disconnect brake rod and

pin (5A).

To disassemble drum and disc unit (Fig. 119), insert slotted screwdriver tip (or small hooked wire) through open end of return springs and stretch springs only far enough to unhook. Remove disc and steel balls from drum.

Linings are not available separately for brake band (5—Fig. 118). Inspect friction surfaces of brake housing (6) and cover (1) for scoring or scuffing and renew either if friction surface is not suitable for further use. If brake housing (6) is renewed, install new housing with proper number of shims (8) to limit bull pinion (differential) shaft end play to 0.001-0.003. Tighten

the housing retaining screws to 100-110 ft.-lbs.

Condition of return spring (2 & 9— Fig. 118) is of utmost importance when servicing band/disc brakes. Renew any spring if the coils of the springs do not fit tightly together and be careful not to stretch the springs any farther than necessary when reassembling brakes.

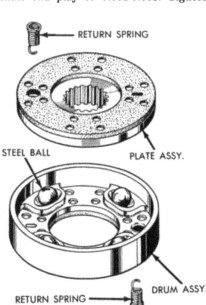

Fig. 119—View of brake drum and disc assembly used on 190, 190XT and early 200 models showing components parts.

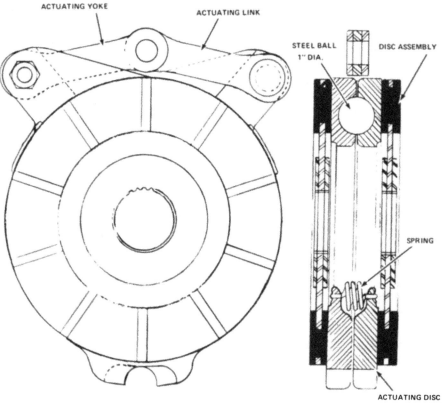

Fig. 120—Cross-sectional view of brake assembly used on model 200 tractor serial number 200-4001 and up and all 7000 models.

turn adjusting screw (Fig. 121) so a 15/16-inch gap is obtained at closest point between brake lever and brake housing. Attach brake rod and check pedal travel. Adjust brake rod to obtain 4 inches of pedal travel. Be sure brakes are adjusted so pedals are aligned for two pedal operation.

149. R&R AND OVERHAUL. Disconnect or move interfering linkage or hydraulic lines for access to brake housing. To remove brake assemblies, remove housing cover, clevis pin from adjusting yoke, outer disc, actuating disc assembly and inner disc. Further disassembly will be evident. Refer to Fig. 120.

Reverse disassembly procedure to reassemble brakes. Adjust brakes as outlined in paragraph 148.

BELT PULLEY

On some models, the belt pulley is used to drive the air conditioning compressor. An adapter and "V" belt pulley is attached to shaft (2—Fig. 122) instead of pulley (1).

PULLEY UNIT

150. GEAR ADJUSTMENT. To adjust backlash between belt pulley gears (13—Fig. 123 and 9—Fig. 122), first remove belt pulley as outlined in paragraph 151. Remove shims (8—Fig. 122) and reinstall unit. Tighten mounting screws finger tight and evenly so that gap between pulley housing (6) and torque housing is the same all the way around. Insert as many shims (8) between torque housing and pulley housing as possible. Remove and reinstall pulley using the selected amount of shims plus 0.030 thickness of shims to

provide proper backlash. Tighten the attaching screws to 90-100 ft.-lbs.

151. REMOVE AND REINSTALL. To remove the pulley assembly, disconnect oil line from pulley housing, remove six attaching cap screws; and withdraw unit. Make sure that shims (8—Fig. 122) are not lost or damaged.

If there is any reason to suspect improper gear backlash adjustment, refer to paragraph 150.

When reinstalling, tighten the attaching cap screws to 90-100 ft.-lbs.

152. OVERHAUL. To overhaul the removed belt pulley unit, refer to Fig. 122 and proceed as follows: Unstake and remove nut (11); then bump shaft from gear and inner bearing. The need and procedure for further disassembly will be evident after examination of the unit.

Inspect all parts for visible damage. Renew seal (3) and bearings (4 & 5) if questionable.

Reassemble in reverse of disassembly procedure. Tighten nut (11) to provide bearings with 4-10 inch-pounds preload and stake nut to shaft.

PULLEY DRIVE GEAR

153. The belt pulley drive gear (13—Fig. 123) located on the engine clutch shaft can be removed as follows: Remove the engine as outlined in paragraph 20, remove release bearing as outlined in paragraph 88 and belt pulley unit as outlined in paragraph 151. Remove belt pulley shifter fork as follows: Working through belt pulley

opening, remove the lock screw (10—Fig. 124) and set screw (7). Remove shifter shaft (2) out top of torque housing; then, remove shifter (8), spring and ball (6) and fork insert (9).

Remove front retainer (7—Fig. 123) and slide belt pulley shifter collar (10) from front end of engine clutch shaft. After removing snap ring (11) and

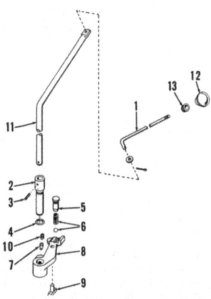

Fig. 124—Exploded view of belt pulley shift linkage. Insert (9) engages groove in coupling (10—Fig. 123).

1. Shifter rod	7. Set screw
2. Shifter shaft	8. Shifter
3. Roll pin	9. Insert
4. "O" ring	10. Lock screw
5. Spring retainer	11. Shift lever
6. Detent spring	12. Knob
and ball	13. Bushing

Fig. 122—Exploded view of the belt pulley assembly. The pulley drive gear and shifter collar are shown in Fig. 123.

1. Pulley	4. Bearing cone	7. "O" ring	
2. Pulley shaft	5. Bearing cup	8. Shims	10. Washer
3. Oil seal	6. Pulley housing	9. Pulley gear	11. Nut

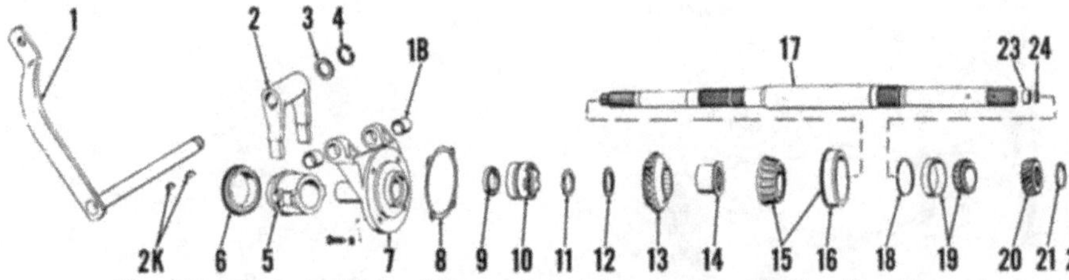

Fig. 123—Exploded view of the engine clutch shaft and belt pulley drive.

1. Release lever	5. Hub	12. Thrust washer	18. Snap ring
and shaft	6. Release bearing	13. Belt pulley	19. Bearing
1B. Bushings	7. Retainer	drive gear	20. Drive gear
(2 used)	8. Gasket	14. Bushing	21. Thrust washer
2. Release fork	9. Seal	15. Bearing	22. Snap ring
2K. Woodruff keys	10. Coupling	16. Snap ring	23. Bearing
3. Washer	11. Snap ring	17. Clutch shaft	24. Sleeve
4. Snap ring			

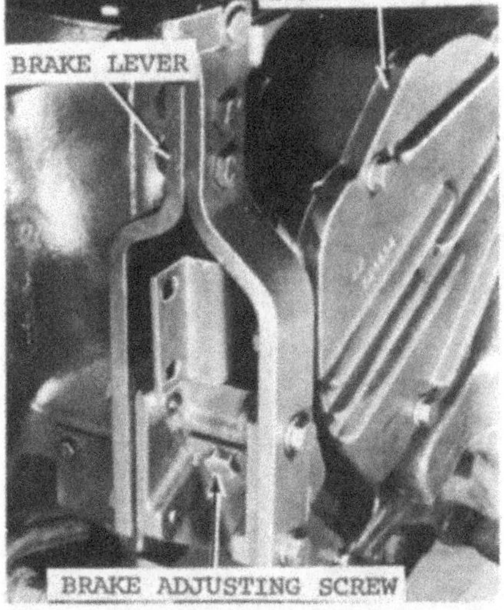

Fig. 121—View showing location of brake adjusting screw on 7000 models.

spacer (12) remove belt pulley drive gear (13) and bushing (14) from clutch shaft.

To reinstall, reverse removal procedures. Drive retainer (5—Fig. 124) up far enough to allow installation of ball (6) after shifter fork is installed. Then drive retainer down flush with top of torque housing.

NOTE: On some models, spring retainer (5) is used as a hydraulic return fitting and tube must be detached before moving retainer.

POWER TAKE-OFF

One-Eighty, One-Ninety, One-Ninety XT and Two-Hundred models are available with single speed pto, either with hydraulic clutch and brake or without clutch. A two-speed pto is standard on 7000 models. Different driver gear (3—Fig. 137) and driven gear (11—Fig. 132 or 18—Fig. 133) are installed to change pto speed from 540 to 1000 rpm on 180, 190 and 190XT models. The rear pto shaft (4—Fig. 130) and clutch (16—Fig. 133) are also different depending upon pto speed.

The single speed (540 rpm) pto with hydraulic clutch and brake available on 185 tractors is the same as used on 180 models. A convertible, two-speed (540 and 1000 rpm) pto, is available on 185 and 200 tractors and standard on 7000 tractors. The front drive and hydraulic clutch is similar to the single speed however, a gear reduction assembly (Refer to Fig. 139) is located in the lift arm housing.

Refer to the appropriate following paragraphs for testing and servicing.

TESTS AND ADJUSTMENTS

190, 190XT and 200 Models With Hydraulic Clutch and Brake

154. **CONTROL ROD.** The pto clutch is provided with three positions: Clutch Engaged—Neutral—Brake Engaged. The detents for these positions are located at the control lever. Length of control rod (13—Fig. 125) must be adjusted to provide correct position of valve spool.

Remove the side panel from control console and loosen both jam nuts (8). Move control lever (6) to the Neutral detent (center) position. Remove the hood right side panel and connect a pressure gage as described in paragraph 155 for checking system pressure. Observe gage pressure with engine at 1600 rpm and "Power-Director" in NEUTRAL (center) position. Adjust length of pto control rod

(13) by turning nuts (8) until the lowest gage pressure is obtained. Do not change position of adjuster (10) when tightening nuts (8).

155. **SYSTEM PRESSURE.** To check the pto clutch hydraulic system pressure, proceed as follows: On 190 and 190XT models remove the hood right rear side panel, disconnect by-pass tube from "Tee" fitting shown in Fig. 126 and install a plug in open end of by-pass tube. On 200 model pressure may

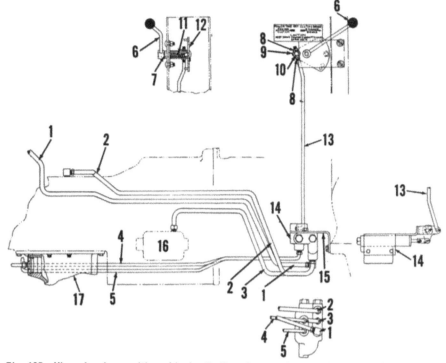

Fig. 125—View showing position of hydraulic lines for pto clutch and control rod adjustment points on 190, 190XT and 200 models.

1. Input tube	6. Control handle	11. Detent spring & ball	15. Bracket
2. Tube to sump	7. Snap ring	12. Detent plate	16. "Power-Director"
3. Tube to "Power-Director" valve	8. Adjusting nuts	13. Rod	control valve
4. Tube to pto clutch	9. Adjuster	14. Pto control valve	17. Pto front shaft
5. Tube to pto brake	10. Snap ring		and clutch housing

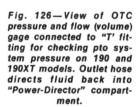

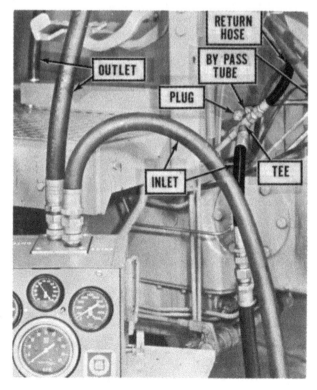

Fig. 126—View of OTC pressure and flow (volume) gage connected to "T" fitting for checking pto system pressure on 190 and 190XT models. Outlet hose directs fluid back into "Power-Director" compartment.

be checked by removing the 1/8-inch pipe plug nearest the rod end of the pto valve. It will be necessary to use a 90 degree elbow #915399 and a 1/8-inch pipe nipple 3 or 4 inches long threaded into the elbow. Connect a pressure gage to "Tee" or elbow fitting, start engine and set speed at 1600 rpm.

CAUTION: Do not turn steering wheel with by-pass tube capped. With transmission "Power-Director" and pto hydraulic clutch in neutral positions, gage pressure should be approximately 60 psi.

Engage pto clutch and power director and observe gage pressure. Pressure should be 295-320 psi. If pressure is erratic, relief valve may be sticking. If pressure is not the same in both Clutch and Brake positions, the control rod may not be correctly adjusted (paragraph 154) or seals may be leaking. If pressure is incorrect but steady and same in both Clutch and Brake position, remove plug and add or deduct shims (1—Fig. 128). "Power Director" system can be checked with same tester connections as outlined in paragraph 91.

180 and 185 Models With Hydraulic Clutch and Brake

156. **CONTROL ROD.** The control valve linkage can be adjusted after removing cover from below instrument panel. Move control lever to neutral detent position, then measure distance between punch mark on valve body and center line of connector pin (hole in valve spool). Distance should be 1½ inches. If incorrect, loosen locknut and turn the connector coupling at lower end of rod until distance is correct, then tighten locknut.

157. **SYSTEM PRESSURE.** The system can be tested as outlined in paragraph 91 to check for internal leakage and "Power-Director" relief setting. Pressure is regulated by "Power-Director" relief valve and pto system is not equipped with a separate relief valve.

Model 7000

158. **CONTROL CABLE.** To coordinate movement of control lever and pto valve lever, move control lever to disengaged (rearmost) position. Gap between pto valve plunger end (S—Fig. 127) and pto valve lever (L) should be 0.001-0.020 inch. Gap is adjusted by detaching and turning cable yoke (Y).

159. **SYSTEM PRESSURE.** To check pto clutch hydraulic system pressure, unscrew test port plug (P—Fig. 75) and install a 400 psi gage. Place pto control lever in "DISENGAGED" position and Power Shift lever in "PARK". Start

engine and note gage reading. Move pto control lever to "ENGAGED" position and note gage reading. Pressure should be 210-230 psi with pto control lever in either position. If pressure is erratic, pressure regulator valve may be sticking. Low pressure in both positions may be due to improper pressure regulator valve setting; refer to paragraph 99. If pressure is not the same in both "ENGAGED" or "DISENGAGED" positions, the control cable may be improperly adjusted (see paragraph 158) or seals may be leaking. "Power Shift" system may be checked

with same connections as outlined in paragraph 99.

CONTROL VALVE

All Models With Hydraulic Clutch and Brake Except Model 7000

160. **R&R AND OVERHAUL.** On 180 and 185 models, control valve is located under the instrument panel. On 190, 190XT and 200 models control valve is mounted on the right side of transmission housing. On all models, valve can be removed after disconnecting the hy-

Fig. 127—Turn joke (Y) to adjust gap between pto plunger (S) and lever (L).

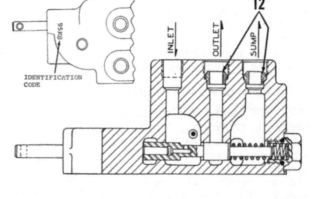

Fig. 128—Cross-sectional view of pto control valve used on 190, 190XT and 200 tractors. Top view shows relief valve and two of the four tube seats. Later valves use snap ring (9) and retainer (9A) as shown in Fig. 129.

1. Shims
2. Relief valve spring
3. Valve spool
4. Valve plunger
5. Seal
6. Washers (2 used)
7. Springs (2 used)
8. Control valve spool
9. Snap ring
10. "O" rings
11. Plugs (2 used)
12. Tube seats

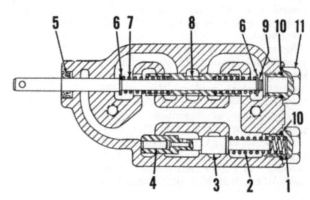

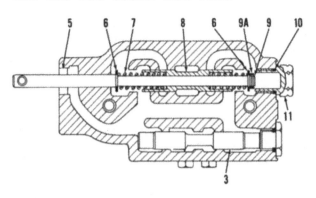

Fig. 129—Cross-sectional view of pto control valve used on 180 and 185 models. Valve spool (3) and orifices in body are different than valve used on 190 and 190XT tractors. Snap ring (9) and retainer (9A) are used on later valves.

draulic tubes and unbolting from bracket.

NOTE: Valves and spools used for controlling hydraulic pto and "Power-Director" on 180 and 185 models and the pto on 190, 190XT and 200 models are similar but, NOT interchangeable. Valves are stamped with an identification code and it is important that correct valve is installed. Valves with identification code beginning with "7" (such as 75D8) should only be used for "Power-Director" control valves on 180 and 185 model tractors. Valves with identification code beginning with "8" (such as 84D8) should only be used as control valve for pto on 180 and 185 model tractors. Valves with identification code beginning with "9" (such as 95K8) should only be used as pto control valve on 190, 190XT and 200 model tractors.

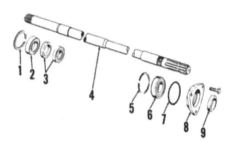

Fig. 130—View of pto rear (output) shaft, bearings and seals used on models with single speed pto. Bearing (2) and seals (3) are in front face of transmission housing. On some models, a bushing is used instead of bearing (2). A single seal (3) is used on 7000 models.

1. Snap ring	
2. Bearing	6. Bearing
3. Lip-seals	7. "O" ring
4. Rear-shaft	8. Retainer
5. Snap ring	9. Seal

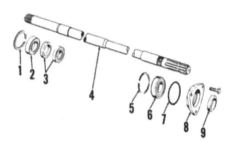

Fig. 131—View of intermediate shaft used on 185, 200 and 7000 models with two speed pto. Lips of seals (3) should face away from each other. Model 7000 uses only one seal (3). A bearing is used in place of bushing (2) on 200 models.

2. Bushing	
3. Oil seals	10. Coupling
4. Intermediate shaft	11. Snap ring

Spools can be removed, after removing both plugs (11—Fig. 128 or 129) 129). Make certain that all orifices in valve body and valve spools are clean and open. Always renew "O" rings (10) under plugs when reassembling.

When reinstalling, adjust control rod as outlined in paragraph 154 or 156. Paragraph 155 outlines procedures for adjusting system pressure on 190, 190XT and 200 models, by adding or subtracting shims (1).

Model 7000

161. **R&R AND OVERHAUL.** The pto control valve is contained in control valve housing located on right side of torque housing. Refer to paragraph 100 for servicing pto control valve.

REAR SHAFT

All Models With Single Speed Pto

162. The pto rear (output) shaft can be removed as follows: Drain oil from "Power-Director" or "Power Shift" (hydraulic sump), transmission and differential compartments. Remove cap screws attaching retainer (8—Fig. 130)

to rear face of lift shaft housing and pull shaft out.

Bearing (6) and rear seal (9) can be renewed at this time. To renew bearing (2) and seals (3), transmission must be split from torque housing as outlined in paragraph 104.

The 1.5615-1.5625 diameter shaft journal except on Models 200 and 7000 should have 0.0015-0.0045 clearance in bushing. On all models except 7000, lip of front seal (3) should be toward front; lip of rear seal (3) should be toward rear. Install seal (3) on 7000 models so spring is on inner side.

When reinstalling pto shaft, take care not to damage seals (3) and/or bearing (2).

INTERMEDIATE SHAFT

185, 200 and 7000 Models With Two-Speed Pto

163. The intermediate pto shaft connects the front shaft (24—Fig. 133) to the 1000 rpm shaft (2—Fig. 139) and is located much the same as the rear pto shaft on single speed models. The shaft can be withdrawn after removing the lift arm housing. Transmission must be split from torque housing as outlined in paragraph 104 to renew seals (3—Fig. 131) and bushing or bearing (2) at front of intermediate shaft. Lips of seals (3) should face away from each other on 185 and 200 models. Install seal on 7000 models with spring on inner side.

FRONT SHAFT AND HOUSING

Pto clutch on 7000 models is driven by an input shaft attached to engine clutch with pto brake acting against front shaft gear. Refer to paragraphs

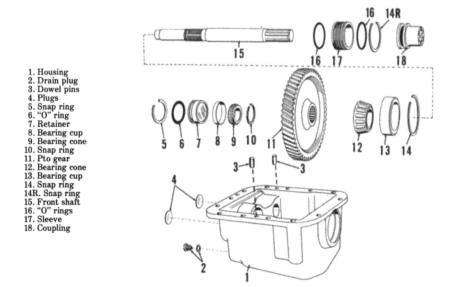

1. Housing
2. Drain plug
3. Dowel pins
4. Plugs
5. Snap ring
6. "O" ring
7. Retainer
8. Bearing cup
9. Bearing cone
10. Snap ring
11. Pto gear
12. Bearing cone
13. Bearing cup
14. Snap ring
14R. Snap ring
15. Front shaft
16. "O" rings
17. Sleeve
18. Coupling

Fig. 132—Exploded view of pto front shaft and housing for models without clutch. Coupling (18) remains in "Power-Director" compartment.

168 and 170 for pto clutch and front shaft service. Pto clutch and front shaft on all other models are driven by idler gears. Refer to following paragraphs for front shaft and housing service on all models except 7000.

Models Without Clutch

164. REMOVE AND REINSTALL. To remove the front shaft and housing, first drain oil from the "Power-Director" (hydraulic sump) compartment and engage the pto.

NOTE: If pto coupling (18—Fig. 132) is not engaged with rear pto shaft, coupling will fall to bottom of "Power-Director" compartment.

Remove the two dowel pins (3) and the retaining cap screws. Slide housing forward until free of sleeve (17) and withdraw from torque housing.

When reinstalling, always renew "O" rings (16). Make certain that mating surfaces of housing (1) and torque housing are clean because no gasket is used. Apply an even coat of Permatex Form-a-Gasket or equivalent and reinstall assembly. Retaining cap screws should not be tightened until dowel pins (3) are installed. Cap screws should be tightened to 35-40 ft.-lbs.

165. OVERHAUL. Remove snap ring (5—Fig. 132) and press shaft (15) forward until retainer (7) and bearing cup (8) are out of housing. Pull bearing

cone (9) from shaft and remove snap ring (10). Press shaft (15) toward rear out of gear (11) and bearing cone (12).

When reassembling, position gear (11) in housing with long hub toward front and position bearing cone (12) in cup (13). Press shaft through bearing (12) and gear (11) until snap ring (10) can be installed in front of gear. Press shaft to rear until bearing cone (12) is tight against gear (11) and gear is tight against snap ring (10). Press bearing cone (9) on front of shaft until tight against shoulder. Install bearing cup (8) and retainer (7) with new "O" ring (6) in groove. Install a snap ring (5) of correct thickness to provide shaft with 0.001-0.003 end play. Snap ring (5) is available in thicknesses of 0.069-0.109 in steps of 0.002.

Models With Hydraulic Clutch and Brake Except Model 7000

166. REMOVE AND REINSTALL. To remove the front shaft housing, first drain oil from the "Power -Director" (hydraulic sump) compartment and engage the pto sliding coupling (25—Fig. 133).

NOTE: If pto sliding coupling is not engaged with rear pto shaft, coupling will fall to bottom of "Power-Director" compartment.

Disconnect hydraulic tubes from plugs (6 & 11) and remove the two dowel pins (15). Remove the retaining cap screws, slide housing forward until clear of sleeve (28) and withdraw from torque housing.

When reinstalling, always renew "O" rings (27). Make certain that mating surfaces of housing (14) and torque housing are clean because no gasket is used. Apply an even coat of Permatex Form-a-Gasket or equivalent and reinstall assembly. Retaining cap screws should not be tightened until dowel pins (15) are installed. Cap screws should be tightened to 35-40 ft.-lbs.

167. OVERHAUL. Remove snap ring (9—Fig. 133) and pull plug (11) from housing. Remove snap ring (26) from rear. Press shaft (24) toward rear until shaft is out of front bearing cone (13) and rear bearing cup (29) is out of housing. After shaft is withdrawn, clutch and gear assembly (16, 17 & 18) and front bearing cone (13) can be lifted from housing. Brake pistons (4) can be withdrawn after removing snap rings (8) and plugs (6).

The clutch assembly can be disassembled after removing snap ring (8—Fig. 134). The clutch used on all models except 190 and 190XT tractors with 540 rpm pto is equipped with six driving discs (6) and five driven plates (5). The clutch used on 190 and 190XT

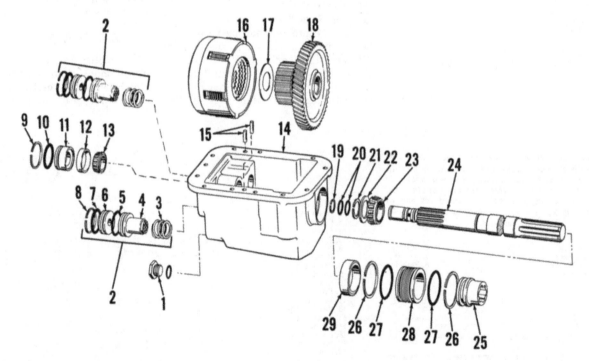

Fig. 133—Exploded view of front shaft, brake and clutch housing. Refer to Fig. 134 for exploded view of clutch assembly (item 16). Coupling (25) remains in "Power-Director" compartment.

1. Drain plug	6. Plug	11. Plug	16. Clutch assembly	20. "O" rings	25. Coupling
2. Brake assemblies	7. Plug "O" ring	12. Bearing cup	17. Thrust washer	21. Snap ring	26. Snap rings
(2 used)	8. Snap ring	13. Bearing cone	18. Gear and	22. Thrust washer	27. "O" rings
3. Release spring	9. Snap ring	14. Housing	clutch hub	23. Bearing cone	28. Sleeve
4. Brake piston	10. Plug "O" ring	15. Dowel pins	19. Seal ring	24. Front shaft	29. Bearing cup
5. Piston "O" ring					

models with 540 rpm pto is equipped with eight driving discs (6) and seven driven plates (5). On clutches with six discs (6) and five plates (5), pins (10) and springs (9) are shorter and snap ring (8) is installed in the inner groove of drum (1). Wave washers (11) are used in place of springs (9) and pins (10) on Model 200 after serial number 8608. Six wave washers are installed in two-speed clutch while eight wave washers are used in single-speed clutch.

On some models, seals (19—Fig. 133) and (2 & 3—Fig. 134) are glass filled Teflon. Steel expanders are used behind Teflon seals (2 & 3—fig. 134). Teflon sealing rings are easily broken when cold and must be heated before installing. To assemble, install the steel expanders in grooves, heat Teflon sealing rings in water or oil to 150-200°F., then work the sealing rings into grooves using a blunt tool with no sharp edges. Make certain that all seal contacting areas of clutch drum (1) and piston (4) are smooth and well lubricated, then slide piston into clutch drum.

To reassemble, position the clutch and gear assembly (16, 17 & 18—Fig. 133) in housing. Install rear bearing cone (23), thrust washer (22) and snap ring (21) on shaft (24). Install new "O" rings (20) in groove at each side of oil passage on end of shaft. Slide shaft through housing bore, gear and hub (18), thrust washer (17) and clutch drum (1—Fig. 134). Make certain that shaft does not catch on thrust washer (17—Fig. 133). Install rear bearing cup (29) and snap ring (26). Press shaft toward rear to seat bearing cup (29) against snap ring (26). Install front bearing cone (13) and cup (12). Install seal ring (19) in groove at front end of shaft.

NOTE: If Teflon seal is used at (19), install plug (11) without "O" ring (10) to compress the seal ring into groove while cooling.

Renew "O" ring (10) and carefully press plug (11) into housing. Install snap ring (9) of correct thickness to provide shaft (24) with 0.002-0.005 end play. Snap ring (9) is available in thicknesses of 0.069-0.109 in steps of 0.002. Press shaft forward until plug (11) is tight against snap ring (9) before measuring end play.

PTO CLUTCH AND INPUT SHAFT

Model 7000

168. REMOVE AND REINSTALL. To remove pto clutch, remove engine clutch release bearing and shaft as

outlined in paragraph 88. Unbolt and remove front support plate (1—Fig. 135) and input shaft (7) assembly. Remove snap rings (19) and withdraw pto clutch assembly from torque housing.

To install pto clutch and input shaft, reverse removal procedure. Install two snap rings (19) so clutch shaft (27) end

play is 0.001-0.006 inch. Snap rings (19) are available in thicknesses from 0.030 to 0.078 inches in increments of 0.004 inch. Tighten front support plate retaining screws to 45-50 ft.-lbs.

169. OVERHAUL. Remove input shaft and pto clutch as outlined in previous paragraph. Remove snap ring (6—Fig. 135) to separate input shaft

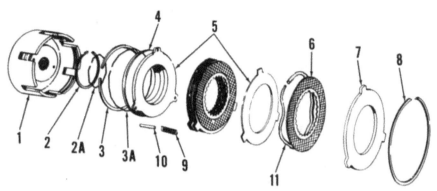

Fig. 134—Exploded view of hydraulic pto clutch. Wave washers (11) are used in place of springs (9) and pins (10) on Model 200 after serial number 8608.

1. Clutch drum
2. Piston inner ring
2A & 3A. Expanders
3. Piston outer ring
4. Clutch piston
5. Drive plate
6. Driving disc
7. End plate
8. Snap ring
9. Release spring (3 used)
10. Spring guide pin (3 used)
11. Wave washers

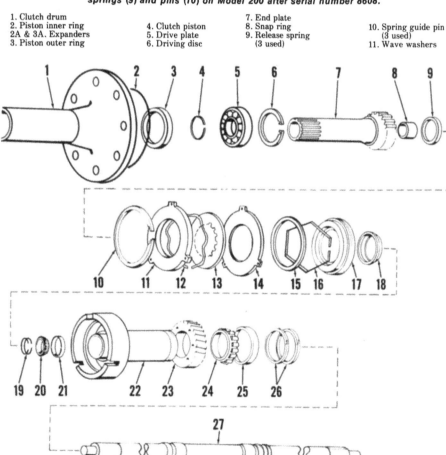

Fig. 135—Exploded view of hydraulic pto clutch used on Model 7000.

1. Front support plate
2. "O" ring
3. Seal
4. Snap ring
5. Bearing
6. Snap ring
7. Input shaft
8. Bushing
9. Seal
10. Snap ring
11. End plate
12. Wave springs
13. Fiber plates
14. Steel plate
15. Outer piston seal
16. Expander
17. Piston
18. Inner piston seal
19. Snap rings
20. Bearing cone
21. Bearing cup
22. Clutch drum & shaft
23. Gear
24. Bearing cone
25. Bearing cup
26. Seal rings
27. Engine clutch shaft

from front support plate. Remove snap ring (10) and disassemble pto clutch assembly. Note that Allen screws retaining gear (23) are behind piston (17).

To assemble pto clutch, reverse disassembly procedure. Tighten Allen screws retaining gear (23) to 38-45 ft.-lbs. Three steel plates (14), three fiber plates (13) and three wave springs (12) are used. A steel plate (14) is placed next to piston (17) followed by a friction disc (13) and wave spring (12). Follow this sequence for remaining plates, discs and wave springs. Install seals (3 and 9) with lip towards rear.

FRONT SHAFT AND PTO BRAKE

Model 7000

170. **R&R AND OVERHAUL.** To remove front pto shaft and pto brake, remove pto clutch as outlined in paragraph 168 and split tractor between transmission and torque housing as outlined in paragraph 105. Remove

torque housing oil pan and pull coupling (14—fig. 136) off front shaft (12). Remove plug (1), hold shaft with ACTP 3031 spline socket and unscrew cap screw (2). Remove washer (3) and reinstall cap screw (2). Apply force to cap screw and drive front shaft (12) rearward out of gear (8). Remove gear (8) and oil deflector (7 and 9). Oil deflector components (7 and 9) are clipped together. Remove pto brake components (15 through 18) from torque housing.

To assemble and reinstall pto brake and front shaft assemblies, reverse disassembly procedure. Install pto brake plug (17) and piston (15) with small end forward. Assemble oil deflectors (7 and 9) around gear (8) so short gear hub is forward and deep oil deflector section (9) is on pto brake side of gear (8). Front shaft (12) should be pulled and not driven into position by tightening cap screw (2). Tighten cap screw (2) to 45-50 ft.-lbs. Front shaft end play should be 0.001-0.006 inch and is adjusted by varying thickness of snap ring (4). Snap ring (4) is available in thicknesses from 0.073 to 0.125 inch in increments of 0.004 inch.

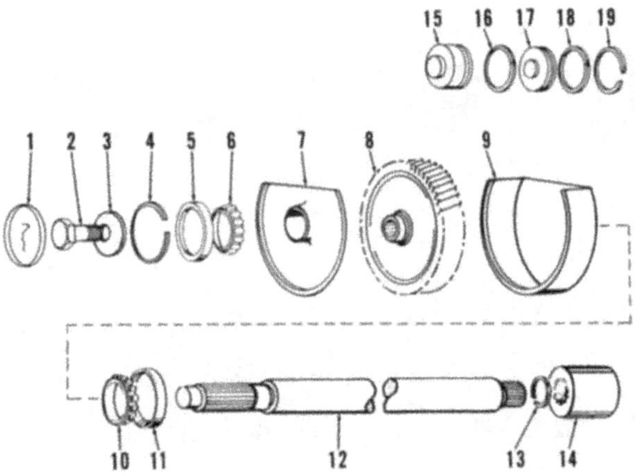

FRONT IDLER GEARS AND SHAFT

All Models Except 7000

171. **REMOVE AND REINSTALL.** Remove the pto front shaft and housing as outlined in paragraph 164 or 166. Remove nut (1—Fig. 137), washer (2), gear (3) and spacer (4) through bottom opening. Split transmission from torque housing as outlined in paragraph 104. Remove nut (14), and bump shaft (8) forward out of bearing cones (11).

When reinstalling, tighten nut (14) until shaft end play is 0.001-0.003 and stake nut (14) to shaft (8). Nut (1) should be staked to shaft (8) after tightening to 140-150 ft.-lbs.

COUPLING AND SHIFTER

All Models Except 7000

172. **REMOVE AND REINSTALL.** To remove the pto shift coupling (18—Fig. 132 or 25—Fig. 133) and shifter lever (2—Fig. 138), it is necessary to split transmission from torque housing as outlined in paragraph 104. Further procedure will be evident after an examination of the unit and reference to Fig. 132, 133 and 138.

Be sure to wire the pin (6—fig. 138) in place before reattaching transmission to torque housing.

REAR DRIVE ASSEMBLY

185, 200 and 7000 Model With Two Speed Pto

173. **R&R AND OVERHAUL.** The rear two speed drive assembly is contained in the lift arm housing. The unit can be disassembled as follows after the lift arm housing is removed from tractor: Remove output shaft (1—fig. 139) and snap ring (7), then press 1000 rpm shaft (2) out of bearing (6). Gear (3), spacer (4) and bearing (6) will remain in housing until countershaft and gears are removed. Remove

Fig. 136—Exploded view of Model 7000 pto driven gear, front shaft and brake assemblies. Pto brake assembly is located just forward of transmission countershaft.

1. Plug
2. Cap screw
3. Washer
4. Snap ring
5. Bearing cup
6. Bearing cone
7. Front deflector
8. Gear
9. Rear deflector
10. Bearing cone
11. Bearing cup
12. Front pto shaft
13. Snap ring
14. Coupler
15. Pto brake
16. Seal ring
17. Seal plug
18. Seal ring
19. Snap ring

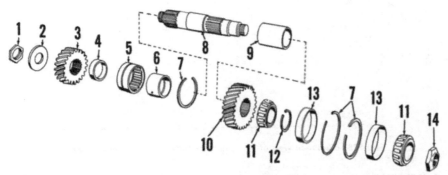

Fig. 137—Exploded view of idler gears and shaft. Driver gear (3) meshes with gear (11—Fig. 132 or 18—Fig. 133).

1. Nut	5. Roller bearing	8. Idler shaft	11. Bearing cones
2. Washer	6. Bearing inner race	9. Spacer	12. Snap ring
3. Drive gear	7. Snap rings	10. Driven gear	13. Bearing cups
4. Spacer			14. Nut

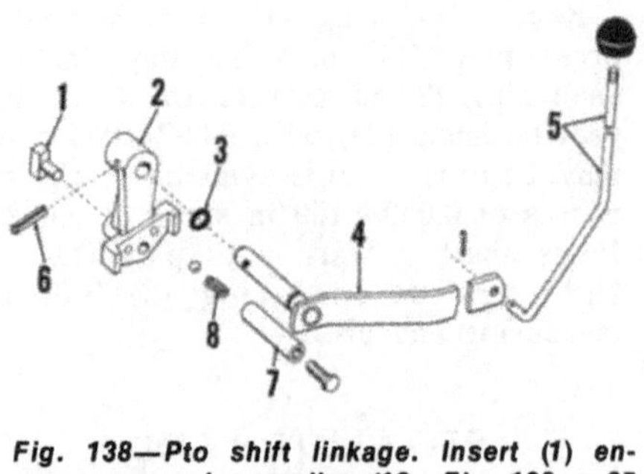

Fig. 138—Pto shift linkage. Insert (1) engages groove in coupling (18—Fig. 132 or 25—Fig. 133).

1. Insert
2. Shift arm
3. "O" ring
4. Shift lever
5. Shift rod
6. Roll pin
7. Spring retainer
8. Detent spring and ball

screws attaching shield brackets (25) and retainer (24) to housing, then parts (9, 10, 11, 12, 13, 14, 15, 24, 25, 27, 28, 29 & 30) can be withdrawn as a unit. To disassemble the hollow shaft unit, press gear (11) toward rear (flange of hollow shaft), remove snap ring (10), then parts can be pressed from shaft. To remove countershaft and gears, remove plug (23), drive rear bearing cup (16R) forward slightly, then remove snap ring (22). Rear of shaft is tapped with 3/8-16 N.C. threads for using a slide puller when removing shaft (18). Remainder of disassembly is evident.

Before assembling other parts, it is important that parts (3, 4, 8 and 16) are located in housing. Make certain that bearing cups (8 & 16) are correctly seated. Spacer (4) and gear (3) must be positioned in bottom of housing before installing countershaft. Press rear bearing cone (17R) onto end of shaft (18) which has threaded puller hole. Start shaft through rear hole, install gear (21) and spacer (20), then locate front bearing cone (17) in cup and install gear (19) between spacer (20) and front bearing cone. Slide the shaft through gear (19) and use slide hammer to press shaft into front bearing cone. Press rear bearing cup (16R) into bore and install correct thickness of snap ring (22) that will provide shaft with 0.002-0.006 end play with cups (16 & 16R) seated. Install plugs (23) with sealer on outer diameter and cupped side toward inside.

Before installing, the hollow shaft must be assembled as shown in Fig. 140. Press the roller bearing (29) into bore with lettered side out (toward rear) until distance (A) is 1-3/8 inches. Press oil seal (30) into bore with lip toward inside (roller bearing) until distance (B) is 5/32-inch. Press bearing cup (13) into retainer (24) until seated in bore. Press oil seal (27) into retainer with lip toward inside (bearing cup) until distance (C) is 1/16-inch. Lubricate seal (27) and surface of hollow shaft, then position retainer over hollow shaft. Press bearing cone (12) onto shaft, then install gear (11) and snap ring (10).

CAUTION: Be careful not to damage seal (27) when installing bearing cone (12).

After snap ring (10) is located in groove, press retainer (24), bearing (12 & 13) and gear (11) toward front until gear is seated against snap ring. Press bearing cone (9) onto front of hollow shaft until seated against shoulder.

To install hollow (540 rpm) shaft unit, position original shims (15—Fig. 139) over retainer (24), then locate "O" ring (14) in groove of retainer. Push

shaft unit into bore of lift shaft housing, install shield bracket (25) and tighten the four retaining screws to 70-75 ft.-lbs. Bearing adjustment is accomplished by varying the thickness of shims (15). Rolling torque of the hollow (540 rpm) shaft should be 20-35 inch-pounds with bearing correctly adjusted. Rolling torque can be measured by hollow shaft (28) which has a radius of 2.875 inches or by wrapping cord

around output shaft (1) which has radius of 0.6875 inches.

If rolling torque is measured with cord wrapped around hub of hollow shaft (28), spring scale should indicate 7-10 lbs. If rolling torque is measured with cord wrapped around output shaft (1), spring scale should indicate 30-43 lbs. Bearing adjusting shims (15) are available in thicknesses of 0.010, 0.012, 0.015 and 0.018.

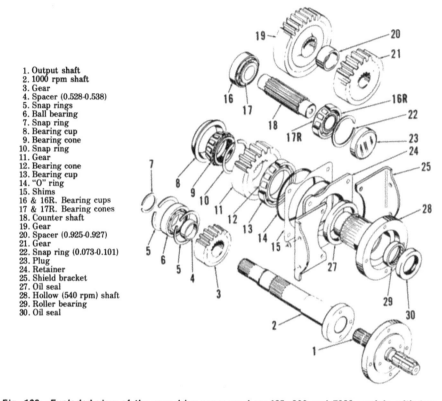

1. Output shaft
2. 1000 rpm shaft
3. Gear
4. Spacer (0.528-0.538)
5. Snap rings
6. Ball bearing
7. Snap ring
8. Bearing cup
9. Bearing cone
10. Snap ring
11. Gear
12. Bearing cone
13. Bearing cup
14. "O" ring
15. Shims
16 & 16R. Bearing cups
17 & 17R. Bearing cones
18. Counter shaft
19. Gear
20. Spacer (0.925-0.927)
21. Gear
22. Snap ring (0.073-0.101)
23. Plug
24. Retainer
25. Shield bracket
27. Oil seal
28. Hollow (540 rpm) shaft
29. Roller bearing
30. Oil seal

Fig. 139—Exploded view of the rear drive gears used on 185, 200 and 7000 models with two speed pto. The gears are contained in the lift arm housing.

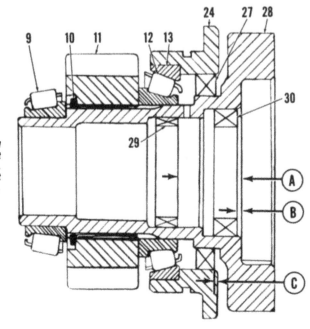

Fig. 140—Cross-sectional view of the hollow shaft assembly. Refer to Fig. 139 for legend and to text for measurements (A, B & C).

Slide the 1000 rpm shaft (2) through the hollow shaft (28), into gear (3) and through spacer (4). Install ball bearing (6) and snap rings (5) either before or after shaft (2). Identification marks on ball bearing (6) should be toward front and snap ring (7) must be installed in groove of shaft to hold shaft in bearing.

One of the holes for attaching output shaft (1) is unequally spaced to prevent incorrect installation.

HYDRAULIC POWER LIFT SYSTEM

TEST AND ADJUSTMENTS

All Models

174. "SNAP COUPLER" PRELOAD ADJUSTMENT. Remove any weight or implement attached to drawbar. Remove drawbar stop (10—Fig. 141) and lock (9). Loosen adjusting nut (N) until tension is removed from spring (13); then, retighten nut to preload spring 5/32 to 7/32-inch. Lock adjustment with cotter pin. Install stop (10) and tighten lock (9) against stop; then, back off lock (9) one flat (1/6 turn) and install cotter pin. It may be necessary to adjust "Traction Booster" linkage as outlined in paragraph 176.

175. TORSION BAR ADJUSTMENT. Remove any weight or implement attached to the three point hitch. Loosen locknut and back the preload adjusting screw (Fig. 142) out until torsion bar tube (3) is free to turn in the support brackets. Then, turn adjusting screw in just far enough to eliminate all free movement of the torsion bar tube and tighten the lock nut while holding the screw in this position.

176. "TRACTION BOOSTER" (DRAFT) ADJUSTMENT. Remove any weight or implement attached to 3 point hitch and/or drawbar. Adjust "Snap Coupler" preload as in paragraph 174 or torsion bar as in paragraph 175; then, proceed as follows:

180 AND 185 MODELS. Move the lift arm lever (1—fig. 143) to the "Traction Booster" position. Move the position control lever (2) all the way forward and the "Traction Booster" lever (3) all the way to the rear. Set engine speed at 1000 rpm and observe position of lift arms. If lift arms are not horizontal (straight back), turn the two adjusting screws (B—Fig. 145) until lift arms are as near horizontal as possible. **NOTE: Turn both nuts equally to prevent damaging the linkage.**

Four adjusting holes (1, 2, 3 & 4) are provided to adjust the sensitivity in relation to movement of the torsion bar. For heavy draft loads with fully mounted implements, it may be desirable to decrease movement by locating rods in holes (2 & 3). Holes (1 & 4) can be used for lighter draft loads with semi-mounted implements and linkage will be more sensitive. A combination of holes (1 & 3) or holes (2 & 4) will provide medium (intermediate) sensitivity.

The "Traction Booster" system is equipped with adjustable feed back linkage. Normally, the link is attached in the bottom hole (1—Fig. 146); however, two other positions are provided. When operating over extremely uneven ground, it may be desirable to attach the feed-back line in the upper hole (3) as shown or in the intermediate hole (2).

190, 190XT AND 200 MODELS. Move lift arm control lever (4—Fig. 144) to "Traction Booster" detent position, position control lever (5) all the way down and "Traction Booster" control lever (6) all the way up. Adjust

"Traction Booster" link rod (Fig. 147 or 149) until it is exactly the correct length without end play. Shorten the

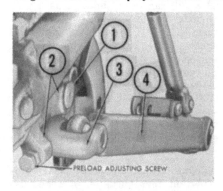

Fig. 142—View of the torsion bar preload adjusting screw and locknut. Refer to paragraph 175 for adjustment precedure.

1. Torsion bar
2. Left hand torsion bar support
3. Torsion bar tube
4. Left hand draft arm

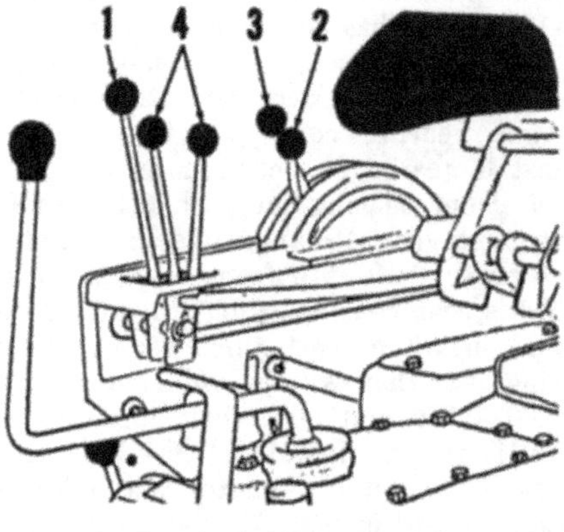

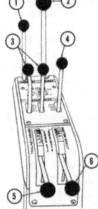

Fig. 143—The hydraulic control levers for 180 and 185 tractors with "Traction Booster" are as shown. Forward position of lift arm lever (1) is lower, next is "Traction Booster" position, then hold position all the way to rear is lift position. The forward position of both the position control lever (2) and "Traction Booster" lever (3) will provide maximum depth. Remote ram control levers (4) have four positions; float, lower, hold and lift.

Fig. 144—View of control console showing position of the control levers on 190, 190XT and 200 tractors.

1. Throttle lever
2. "Power-Director" lever
3. Remote ram control levers
4. Lift arm control lever
5. Position control lever
6. "Traction Booster" control lever

Fig. 141 —Exploded view of the "Snap-Coupler" assembly available on some models.

N. Adjusting nut
1. Coupler bell
2. Pivot shaft
3. Pivot shaft
4. Brackets
5. Pivot screws
6. Spring rod
7. Support
8. Sensing linkage bushing (2 used)
9. Lock
10. Drawbar stop
11. Bearing
12. Spring retainer
13. Drawbar spring
14. Spring pilot
15. "Snap-Coupler" release handle
16. Coupler hook
17. Spring
18. Release spring
19. Snap rings (2 used)
20. Release link rod
21. Release lever
22. Bracket

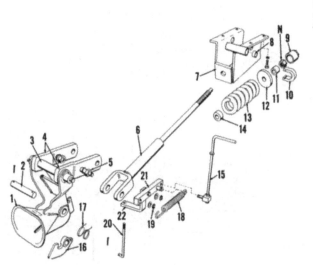

link rod 1½ turns by turning yoke onto rod (or rod into pivot) and reconnect. The lift arms should not raise with engine running at idle speed.

NOTE: Length of rod (11—Fig. 148) should be adjusted to provide spring length of 11 inches.

Four adjusting holes (4—Fig. 148) are provided to adjust the amount of feed back in relation to movement of tractor lift arms. The upper hole provides the minimum amount of feed back and the lower hole provides the maximum amount of feed back. The lower hole should be used for implements such as plows where minimum movement of lift arms is desired.

7000 MODELS. To adjust "Traction Booster" linkage, remove all loads from draft arms and adjust torsion bar as outlined in paragraph 175. Refer to Fig. 150 and turn nut (N) so length of overtravel spring (S) is 3-1/16 inches. Move lift arm control lever to "Traction Booster" detent position and move "Traction Booster" control lever to rear position. Adjust "Traction Booster" control cable yoke end so there is 5½ inches from yoke pin (P—Fig. 151) to attachment point (A). With engine running at 1000 rpm, lift arms should be in horizontal position. If not, raise lift arms by turning yoke to shorten cable or lower lift arms by turning yoke to lengthen cable.

177. **POSITION CONTROL ADJUSTMENT.** Refer to the appropriate following paragraphs for checking and adjusting.

180 AND 185 MODELS. With engine running at 1000 rpm, move the lift arm control lever (1—Fig. 143) to the "Traction Booster" position. Move "Traction Booster" control lever (3) all the way forward and the position control lever (2) all the way to the rear. Ends of lift arms should raise to within 1-inch of full raised position. If

incorrect, turn nut (N—Fig. 146) as required.

190, 190XT AND 200 MODELS. First check length of spring (11—Fig. 148). Length of rod should be adjusted to provide spring length of 11 inches. With engine running at low idle speed, move the lift arm control lever (4—Fig.

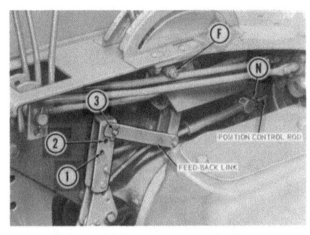

Fig. 146—View of adjustment points located under control panel of 180 and 185 tractors. Refer to text for explanation and adjustment procedure.

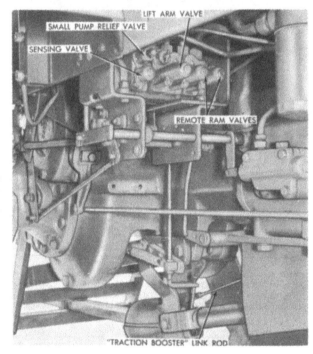

Fig. 147—View of "Traction Booster" link rod for 190 and 190XT models with "Snap-Coupler". Refer to text for adjustment procedure.

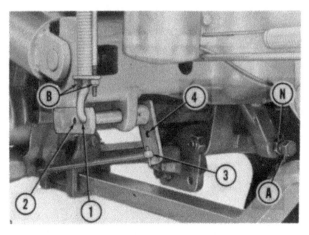

Fig. 145—View of adjustment point on 180 and 185 tractors. Refer to text for procedure.

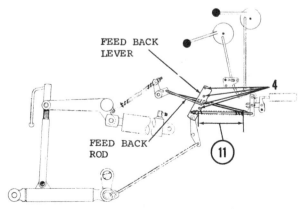

Fig. 148—Drawing of "Traction Booster" and position control linkage used on 190, 190XT and 200 tractors. The feed back lever is located near the right brake assembly.

144) to the "Traction Booster" detent position, move the "Traction Booster" control lever (6) all the way downward, move the position control lever (5) all the way up. Turn the position control adjustment nut (N—Fig. 152) out until lift arms raise, then with lift arms at top of travel, turn the adjusting nut (N—Fig. 152) onto rod until pressure is below ½ of scale on the "Traction Booster" gage.

7000 MODELS. To adjust position control on 7000 models, turn feedback rod (R—Fig. 151) so spring (S) length is 5 inches. Move lift arm control lever to "Traction Booster" detent and move

"Traction Booster" control lever to forward position. Place position control lever in rearmost location. Adjust position control cable yoke end so there is 5½ inches from yoke pin (Y) to attachment point (T). With engine running at 1000 rpm, turn position control rod nuts (R—Fig. 150) so lift arms raise to within 1 inch of full raised position.

178. LEVER FRICTION ADJUST-MENT. With the engine stopped,

completely lower the lift arms. On 180 and 185 models, move the position control lever (2—Fig. 143) to the full rearward position. On 190, 190XT and 200 models, move the position control lever (5—Fig. 144) to full forward (up) position. On 7000 models, move position control and "Traction Booster" levers to rear positions. On all models, if the position control lever will not stay in this position, tighten the friction adjusting nut. Friction adjusting nuts

Fig. 149—View of "Traction Booster" link rod for 190, 190XT and 200 models with 3-point hitch. Refer to text for adjustment procedure.

Fig. 151—View of Model 7000 valve stack and associated linkage.

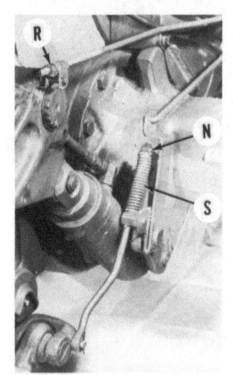

Fig. 150—Overtravel spring (S) length should be 3-1/16 inches. Turn nut (N) to adjust spring length. Turn position control rod nuts (R) to adjust lift arm raised position. See text.

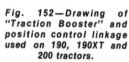

Fig. 152—Drawing of "Traction Booster" and position control linkage used on 190, 190XT and 200 tractors.

N. Position control adjusting nut
5. Position control lever
6. "Traction Booster" lever
7. Position control rod
8. Position control rod
9. "Traction Booster" rod (for 3-point hitch)
11. Friction adjusting bolt

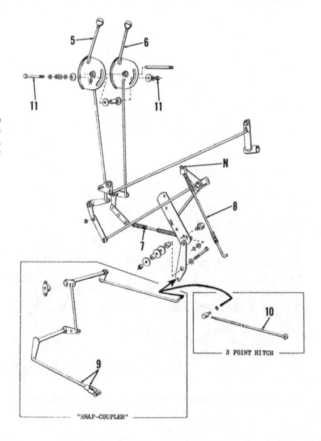

on 7000 models are located on side of control console. Friction adjusting nuts (F—Fig. 146 or 11—Fig. 152) are located behind control console side panel on other models.

179. LOWERING RATE ADJUSTMENT. The rate of lowering can be adjusted by turning the adjusting screw (56—fig. 151 or 153) in to slow the lowering rate or out to increase the speed of lowering. Normal setting is accomplished by turning needle in until it seats, then backing screw out ¾-turn. On 180 and 185 models, the adjusting needle is located at bottom of lift arm valve body, just ahead of the lift arm ram outlet connection. On 190, 190XT and 200 models the adjusting needle is located at top of lift arm valve body, just behind the lift arm ram outlet connection. On 7000 models the adjusting needle is located on side of valve body as shown in Fig. 151.

NOTE: On all models, the high volume bleed-off adjustment screw (34—Fig. 151 or 153) should NOT be mistaken for the rate of lowering adjustment screw (56). Normal setting for the high volume bleed-off screw (34) is 4 turns open.

180. SYSTEM RELIEF PRESSURE. The hydraulic system relief pressure can be checked at a remote cylinder connection as follows: Install a 3000 psi gage in a remote cylinder (ram) connection and pressurize that port.

NOTE: Control valve must be held in position when checking pressure.

Gage pressure should be 2250-2350 psi with engine running at 2000 rpm. If pressure is incorrect, remove cap nut (93—Fig. 153), loosen locknut (92) and turn the adjusting screw (90) as required to obtain 2300 psi. Refer to paragraph 183 for complete system check.

181. "TRACTION BOOSTER" RELIEF VALVE. Pressure in the "Traction Booster" system is controlled by the relief valve (15 through 32—Fig. 153). To check the system pressure, disconnect a hose from one of the tractor lift arm rams (cylinder) and connect a gage to the hose. With engine running at 2000 rpm, actuate the "Traction Booster" sensing valve. Gage pressure should be 2100 psi. If pressure is incorrect, remove cap nut (15), loosen locknut (17) and turn adjusting screw (18) as required to obtain 2100 psi. Refer to paragraph 183 for complete "Traction Booster" and lift system check.

182. CONTROL LEVER RELEASE PRESSURE. When engine is running at normal operating speeds, the remote cylinder control levers should automatically return to neutral position when remote cylinder reaches end of stroke

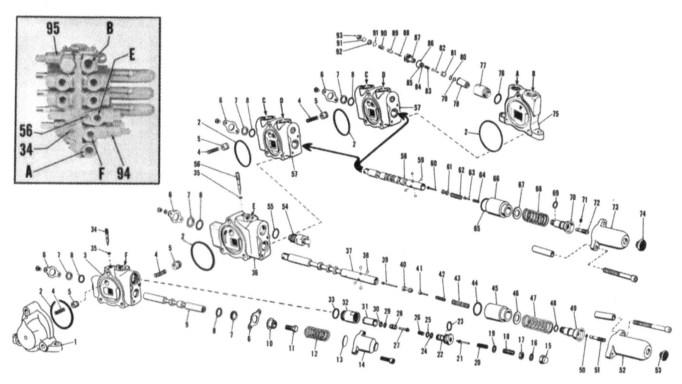

Fig. 153—Exploded view of "Traction Booster" and hydraulic power lift control valves. Late production units may have outlet port (A) in the outlet housing (1) as shown at inset. Side of valves shown in the inset is on bottom for 180 and 185 models, on top for 190, 190XT and 200 models, and on side for 7000 models. Refer to Fig. 151 for view of valves on 7000 models.

A. Outlet port to sump
B. Inlet from lift pump
C & D. Double acting remote cylinder ports
E. Port to lift arm rams
F. Inlet from "Traction Booster" pump
1. Outlet housing
2. "O" rings
3. "Traction Booster" valve housing
4. Check valve springs
5. Check valves
6. Seal plates
7. Seal wiper rings
8. "O" rings

9. "Traction Booster" sensing valve
10. Spring seat
11. Socket head screw
12. Valve spring
13. Spacer
14. Cover
15. Cap nut
16. Copper washer
17. Lock nut
18. Adjusting screw ("Traction Booster" relief valve)
19. Copper washer
20. Spring
21. Plunger
22. Plug
23. "O" ring
24. "O" ring
25. Back-up ring
26. Spring

27. Relief valve ("Traction Booster")
28. Piston
29. "O" ring
30. Back-up ring
31. Valve sleeve
32. "Traction Booster" relief valve cap
33. "O" ring
34. High volume bleed-off adjusting screw
35. "O" rings
36. "Traction Booster" —lift arm valve housing
37. Valve spool
38. Steel balls
39. Poppet
40. Cam
41. Spring guide
42. Spring

43. Detent spring
44. "O" ring
45. Sleeve
46. Washer
47. Plunger spring
48. "O" ring
49. Spring seat
50. "O" ring
51. Adjusting screw (for self-cancelling)
52. Cover
53. Rubber plug
54. Shut-off valve
55. "O" ring
56. Lift arm rate of lowering adjusting screw
57. Remote cylinder control housing
58. Valve spool
59. Steel balls
60. Poppet

61. Cam
62. Detent spring
63. Spring guide
64. Spring
65. "O" ring
66. Sleeve
67. Washer
68. Plunger spring
69. "O" ring
70. Spring seat
71. "O" ring
72. Adjusting screw (for self-cancelling)
73. Cover
74. Rubber plug
75. Inlet housing
76. "O" ring
77. Lift system relief valve cap
78. Valve sleeve
79. Back-up ring
80. "O" ring

81. Piston
82. Relief valve (hydraulic lift system)
83. Spring
84. Back-up washer
85. "O" ring
86. "O" ring
87. Plug
88. Plunger
89. Spring
90. Adjusting screw (hydraulic lift system relief valve)
91. Copper washers
92. Lock nut
93. Cap nut
94. "Traction Booster" relief valve location
95. Lift system relief valve location

and the lift arm control lever should return to hold position from raising postion when lift arms reach fully raised position. If the controls do not return to neutral or hold position, remove the rubber cap (53 or 74—Fig. 153) and turn adjusting screw (51 or 72) out just enough to allow valve to release. If controls release too soon, turn adjusting screw in.

183. **COMPLETE SYSTEM CHECK.** An OTC Y81-21 or equivalent hydraulic tester can be used to check the complete "Traction Booster" and power lift hydraulic system. Refer to the appropriate following paragraphs for attaching tester and checking system.

180 AND 185 MODELS. To connect the hydraulic tester, disconnect one pressure line for lift arm rams from the "T" fitting and connect the inlet hose to tester as shown in Fig. 154. Attach outlet hose from tester to the filler opening.

NOTE: When testing, lift arm control lever must be held in lift position or remove plug (53—Fig. 153) and turn screw (51) in until the spool will not automatically return to hold position.

Open the hydraulic tester valve fully, move the lift arm control lever to rear (lift position). Operate the engine at 2000 rpm. When hydraulic fluid reaches 102°-135° F., close tester valve completely, If relief pressure is not 2250-2350 psi, remove cap nut (93), loosen locknut (92) and turn the adjusting screw (90) as required to obtain 2300 psi. Open tester valve until pressure is 2000 psi and observe volume of flow. Lift pump volume at 2000 psi should be 10.5 gpm with engine speed of 2000 rpm and approximately 3.4 gpm with engine speed of 800 rpm. Reset release pressure as outlined in paragraph 182.

To check the "Traction Booster" system it is important that the high volume bleed screw (34) is **at least 4** turns out from seated position. Move the lift arm control lever to "Traction Booster" detent position. Shorten the "Traction Booster" linkage (B—Fig. 145) until the sensing valve (9—Fig. 153) is pushed into valve housing as "Traction Booster" control lever is moved. Open tester valve and operate engine at 2000 rpm. Close the valve on tester and check "Traction Booster" relief pressure. If relief pressure is not 2050-2150 psi, remove cap nut (15—Fig. 153), loosen locknut (17) and turn adjusting screw (18) as required to obtain 2100 psi. Open tester valve until pressure is 1800 psi and check pump volume. Pump volume at 1800 psi should be 3.4 gpm with engine speed at 2000 rpm and approximately 0.7 gpm at engine speed of 800 rpm. After checks

and adjustments are completed, turn high volume bleed-off screw (34) in until it seats, then back screw out 4 turns. Adjust the "Traction Booster" linkage as outlined in paragraph 176.

190 AND 190XT. To connect the hydraulic tester, remove the right front section of platform. Refer to Fig. 155.

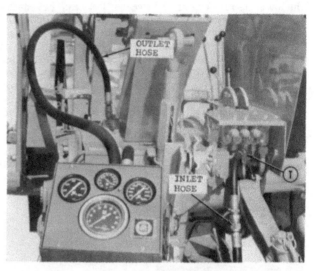

Fig. 154—View of hydraulic tester connected for testing "Traction Booster" and lift systems on 180 and 185 models.

Disconnect "Traction Booster" gage line from "Tee" fitting (T—Fig. 155) on lift ram pressure line and connect the inlet hose to tester to the "Tee." Remove "Power-Director" compartment filler plug and position outlet hose from the tester in filler opening to sump.

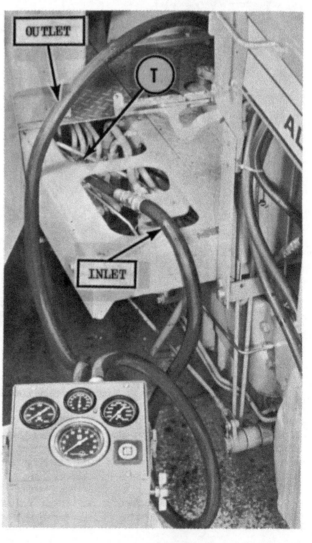

Fig. 155—View of hydraulic tester connected for testing "Traction Booster" and lift systems on 190 and 190XT tractors. Inlet hose to tester is connected to "Traction Booster" gage tee fitting (T). Outlet hose is connected to sump filler opening.

To check power lift system, first remove the rubber plug (53—Fig. 153) and turn adjusting screw (51) in until spool will not automatically return to hold position. Open hydraulic tester valve fully, move lift arm control lever (4—Fig. 144) to "Raise" position (rear) and operate engine until hydraulic fluid temperature reaches 100°F. Set engine speed at 2000 rpm and close tester valve completely. If relief pressure is not 2250-2350 psi, remove cap nut (93—Fig. 153), loosen locknut (92) and turn adjusting screw (90) as required to obtain 2300 psi.

After setting relief pressure, open tester valve and set engine speed at 2200 rpm. Turn tester valve in (partially closed) until pressure is 2000 psi. Volume of flow should be 12 gpm for new pump. Reset the control lever release pressure as outlined in paragraph 182.

To check the "Traction Booster" system, it is necessary to back-out the high volume bleed screw (34—Fig. 153) five turns from seated (closed) position. Shorten the "Traction Booster" linkage (Fig. 147 or 149) until sensing valve (9—Fig. 153) is pushed into the valve housing. Position lift arm control lever (4—Fig. 144) in "Traction Booster" detent, and move "Traction Booster" control lever (6) to full raised position. To check "Traction Booster" relief pressure, completely close valve on hydraulic tester. If relief pressure is not 2050-2150 psi, remove cap nut (15—Fig. 153), loosen locknut (17) and turn adjusting screw (18) as required to obtain 2100 psi. Open the tester valve and operate engine at 2000 rpm. Tester will show false reading (increased volume) due to partial flow of lift pump until pressure is increased. Close the valve on tester until pressure is 1800 psi and observe "Traction Booster" pump volume which should be 2.5 gpm. If volume is less than 1.0 gpm, "Traction Booster" pump should be overhauled. After checks are completed, turn the high volume bleed-off adjusting screw (34—Fig. 153) in until it seats, then back screw out 4 turns. To provide faster rate of lift, turn adjusting screw in. Adjust the "Traction Booster" linkage as outlined in paragraph 176.

200 AND 7000 MODELS. Lower lift arms and disconnect lift arm pressure line from "Tee" located between lift arm cylinders. Connect flo-rater inlet to lift arm pressure line and route flo-rater outlet into one of the tractor remote outlets.

NOTE: Be sure control lever for remote outlet is in float position.

To check "Traction Booster" circuit move lift arm control lever to "Traction Booster" detent and move position control lever all the way to rear so sensing valve plunger is pushed in all the way. Back out high volume bleed-off screw (34—Fig. 151) to six turns (normally set at four turns out). "Traction Booster" flow should be 3 gpm at 1800 psi at 2200 engine rpm. "Traction Booster" relief valve can be checked by closing flo-rater. Refer to paragraph 181 for relief valve adjustment.

To check lift system move lift arm control lever to lift position. Slowly increase engine rpm. Lift system flow should be 13.15 gpm at 2000 psi at 2000 engine rpm. Relief valve can be checked by closing flo-rater. Refer to paragraph 180 for relief valve adjustment. Be sure to reset high volume bleed-off screw (34—Fig. 151) after tests are completed.

Pump output for "Traction Booster" circuit can be checked by connecting flo-rater to fitting in top middle of pump while pump output for lift circuit can be checked by connecting flo-rater to fitting at top end of pump. Route flo-rater outlet into one of the tractor remote outlets.

NOTE: Be sure control lever for remote outlet is in float position.

NOTE: These pump output tests do not have a relief valve in the circuit and caution should be used.

Pump "Traction Booster" output should be 3 gpm at 1800 psi at 2200 engine rpm. Pump lift circuit output should be 13.15 gpm at 2000 psi at 2200 engine rpm.

PUMP

Several different (but similar) pumps are used. The gear type pumps are either one section (steering), two section (steering, lift and "Traction Booster" systems). Oil flow from the power steering section of the pump is also used to engage "Power-Director" clutches, pto clutch and pto brake.

184. R&R HYDRAULIC PUMP. Before removing the hydraulic pump be sure to clean the pump and connections thoroughly. If inlet tube is not to be removed, it will not be necessary to drain

Fig. 156—Exploded view of hydraulic pump drive adapter.

1. Gear and shaft	4. Adapter
2. Bearing assembly	5. Nut
3. "O" ring	6. Gasket

the transmission. Disconnect the hydraulic lines from the pump (and inlet manifold on 190 and 190XT models), then unbolt and remove the pump from the pump drive adapter. Be sure to cap or cover all hydraulic connections and openings after removing pump from tractor.

NOTE: On 190 and 190XT models, the manifold unit can be removed for service without removing the pump.

When reinstalling pump, use a new gasket between pump and pump drive adapter and tighten the two pump retaining cap screws to 28-33 ft.-lbs.

185. R&R AND OVERHAUL PUMP DRIVE ADAPTER. Several different adapter housings (4—Fig. 156) have been used and must be correctly matched to the type of engine front adapter plate. The bearing (2) must also be correctly matched to the adapter housing (4). The gear and shaft (1) must be removed from front, before removing adapter housing (4) on early models. On later models, the hole in engine front adapter plate is large enough to withdraw the pump drive gear from the rear. (Before removing either early or late type, remove the hydraulic pump as outlined in paragraph 184). The following procedure outlines disassembly and removal with early type engine front adapter plate but can also be used for later type except that pump drive adapter can be removed as a unit before disassembling.

Remove the pump drive shaft cover from the front side of the engine timing gear cover. Unstake the nut (5) from pump drive shaft and gear (1) and remove nut from shaft. Push the shaft and gear forward out through opening in timing gear cover; the front bearing cone and roller assembly and inner spacer from bearing unit (2) will be removed with the shaft and gear. Remove the rear bearing cone and roller assembly from rear of adapter housing. Unbolt and remove the adapter housing from the engine front plate.

NOTE: Two adapter housing retaining cap screws are located inside the timing gear cover. To remove these cap screws with timing gear cover in place, re-shape the handle and grind the head of a 9/16-inch box end wrench and remove the cap screws by working through opening in front of timing gear cover.

Remove the front bearing cone and roller assembly and the inner spacer from the pump drive shaft and gear and remove the two bearing cups and outer spacer from the adapter housing if necessary. If any of the bearing cups or cone and roller assemblies are worn or damaged beyond further use, it will be necessary to renew both bearing cups, both cone and roller assemblies and the two spacers as a complete unit. When the pump drive adapter is properly assembled with a new bearing unit, the pump drive shaft end play should be 0.002-0.006.

To reassemble and reinstall the pump drive adapter, proceed as follows: Insert outer bearing spacer in the groove in bore of adapter housing. Be sure that there are no burrs on the outer spacer or the adapter housing bore; then, press the two bearing cups into the housing tightly against the spacer with the thick side of both cups toward the spacer. Place a new "O" ring (3) in the groove in adapter housing and install housing on engine front plate. Tighten the retaining cap screws to a torque of 28-33 ft.-lbs. Special seal washers should be used on screws located inside the timing gear cover. Install a bearing cone and roller assembly on pump drive shaft tightly against the gear and place the bearing inner spacer against bearing cone. Insert the shaft through opening in front of timing gear cover and install the rear bearing cone and roller assembly on the shaft. Install the retaining nut (5) and tighten the nut to a torque of 240-260 ft.-lbs. Stake the nut to the shaft. The shaft should than have 0.002-0.006 end play; bump the shaft forward, then to rear before checking end play.

NOTE: If the bearing unit is being reused, the bearing cone and roller assemblies should be reinstalled with their mating cups. Bearing cups fit in early type adapter housing with a zero to 0.002 loose fit. On later adapter housing, bearing cups should have 0.002 interference fit. On all models, bearing cones fit on pump drive shaft with a 0.0005 interference to 0.001 clearance fit.

Reinstall cover on front side on timing gear cover and tighten the retaining cap screws to 28-33 ft.-lbs.

186. OVERHAUL HYDRAULIC PUMP. The pump will be a one, two or three section gear type pump depending upon the type of hydraulic system used on tractor. Service instructions which follow are for the three section pump used on 190 and 190XT tractors equipped with "Traction Booster." However, service procedures are similar for single and two section pump. Difference will be noted where applicable.

187. DRIVE SHAFT SEAL. To service the pump drive shaft seal, first remove pump assembly as outlined in paragraph 184; then, proceed as follows:

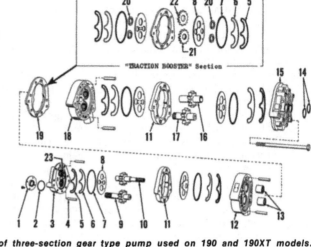

1. Seal body
2. "O" ring
3. Body
4. Dowel pins (6 used)
5. Sealing rings (6 used)
6. Back-up rings (6 used)
7. Seal ring (6 used)
8. Wear plates (6 used)
9. Lift pump gear
10. Lift pump gear
11. Gear plates (2 used)
12. Support plate
13. Bearings (8 used)
14. "O" rings (2 used)
15. Cover plate
16. Power steering pump gear
17. Power steering pump gear
18. Support plate
19. Gear plate
20. Back-up washers (4 used)
21. "Traction Booster" pump gear
22. "Traction Booster" pump gear
23. "O" ring (4 used)

Fig. 157—Exploded view of three-section gear type pump used on 190 and 190XT models. Small gears (21 & 22) are for "Traction Booster" system.

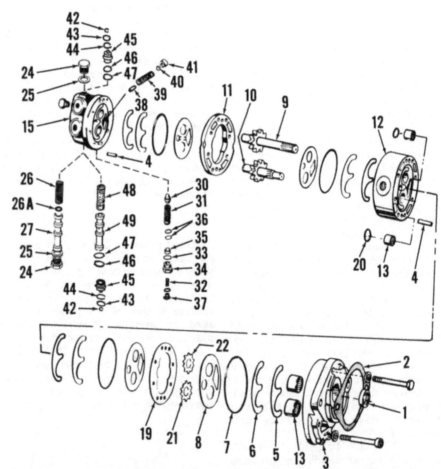

Fig. 158—Exploded view of two section pump used on 200 and 7000 models.

1. Seal	13. Bearings	27. Flow divider valve spool	39. Spring
2. Gasket	15. Cover plate	30. Power steering relief valve	40. Check valve ball
3. Body	19. Gear plate	31. Spring	41. Check valve seat
4. Dowel pins	20. Back-up washers	32. Adjusting screw	42. Snap ring
5. Sealing rings	21. Traction Booster pump idler gear	33. "O" ring	43. Washer
6. Back-up rings	22. Traction Booster pump drive gear	34. Plug	44. Snap ring
7. Seal rings	24. Plugs	35. Guide	45. Plug
8. Wear plates	25. "O" rings	36. Shims	46. Washer
9. Drive shaft and pump gear	26. Spring	37. Lock nut	47. "O" ring
10. Idler shaft and pump gear	26A. Orifice disc	38. Guide pin	48. Spring
11. Gear plate			49. Flow divider valve spool
12. Support plate			

On 190 and 190XT models, remove the four screws retaining seal plate (1—Fig. 157) to pump front cover and remove the seal plate and "O" ring (2). Shaft seal is serviced only as an assembly with the seal plate. Install new "O" ring in pump body, lubricate the seal and shaft and carefully install seal plate with flat side out.

On 200 and 7000 models, seal (1—Fig. 158) can be removed and installed after removing the pump. Spring loaded lip of seal should be toward inside and seal should be pressed into bore until seal bottoms in bore.

On 180 and 185 models, seal (1—Fig. 159) can be removed and installed after removing the pump. Spring loaded lip of seal should be toward inside and seal should be pressed into bore until seal bottoms in bore.

188. POWER STEERING FLOW CONTROL AND RELIEF VALVES. On 190 and 190XT models, the inlet manifold (Fig. 160) containing the power steering flow control and relief valves, can be removed from the pump without removing pump from tractor. The flow control valve (4) should limit the volume of oil for power steering system to 4.75-5.75 gpm at 1200 psi and at engine speeds between 1100-2200 rpm. Refer to paragraph 13 for checking pressure and volume. Disassembly procedure of inlet manifold is evident.

Flow divider valve on Model 200 may be retained with threaded plugs (24—Fig. 158) or by plugs (45) and snap rings (44). Spring (48) length is 2.82 inches for units using plugs (45) retained by snap rings (44). Threaded plugs (24) which are deep internally require a spring (26) 3.80 inches long while internally shallow plugs (24) require a 2.82-inch spring.

Two different flow divider valve spools (27) and springs (26) have been used on 7000 models. Flow divider valve spool (27) on models prior to serial number 7000-1035 has four 0.113 diameter holes. Install spool so 0.113-inch holes are nearer spring (26) which is 2.82 inches long. Spool (27) on

models after serial number 7000-1034 does not have aforementioned holes in spool but orifice disc (26A) must be installed in spool under spring (26) which is 2.75 inches long. Pumps using spool with disc (26A) may be identified by a groove milled in pump discharge area of end cover (15).

On all models, be sure that flow divider valve spool (27—Fig. 158 or 4—Fig. 160) slides freely in its bore. Small burrs can be removed with crocus cloth. Renew flow divider valve and/or manifold if excessively worn or deeply scored. Use new sealing "O" ring (2 or 25) when installing flow control valve plug.

Inspect relief valve (7 or 30) and the valve seat in inlet manifold for nicks, excessive wear or grooving. Renew relief valve and/or manifold (pump rear cover on 200 and 7000 models) if serious defects are noted. Check spring free length against that of new spring; renew the spring if short, worn, distorted or cracked. Recheck system pressure as outlined in paragraph 13 or 13B before adding or deleting shims (9 or 36) on shim adjusted relief valves. Use new plug seal (10 or 36) when reinstalling relief valve plug.

Tighten pump through bolts to 35 ft.-lbs. Tighten four pump inlet manifold retaining screws on 190 and 190XT models to 18-22 ft.-lbs.

189. PUMP BODY AND GEARS. After removing pump as outlined in paragraph 184, thoroughly clean outside of pump and mark each section of pump for correct assembly. Remove the screws attaching pump sections together, then bump the drive shaft against work bench or wood block to separate the pump sections. Do not pry sections apart as this may damage the sealing surfaces. Refer to Fig. 157 for pump used on 190 and 190XT models; or Fig. 159 for pump used on 180 and 185 models.

Renew needle bearings if needles are loose or scored. Renew the pump as an assembly if excessive defects render renewing individual parts impractical. Inspect wear plate (8), gear plate (11 & 19) and gears (9, 10, 16, 17, 21 & 22)

for wear or damage. Refer to the following specification data:

190 and 190XT Models
Refer to Fig. 157.
Power Steering Pump—
  Thickness of gears
    (16 & 17) 0.5005-0.5010
  Thickness of
    plate (11) 0.5011-0.5016
  Gear to plate
    clearance............0.0001-0.0011
Lift System Pump—
  Thickness of gears
    (9 & 10) 0.5005-0.5008
  Thickness of
    plate (11) 0.5011-0.5016
  Gear to plate
    clearance............0.0003-0.0011
"Traction Booster" Pump—
  Thickness of gears
    (21 & 22) 0.2005-0.2010
  Thickness of
    plate (19) 0.2011-0.2016
  Gear to plate
    clearance............0.0001-0.0011

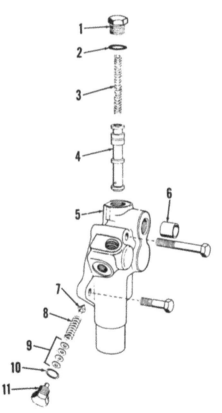

Fig. 160—On 190 and 190XT models, the hydraulic pump inlet manifold contains power steering flow control and pressure relief valves.

1. Plug	6. Bushing
2. "O" ring	7. Relief valve
3. Spring	8. Spring
4. Flow control	9. Shims
spool	10. Seal
5. Manifold	11. Plug

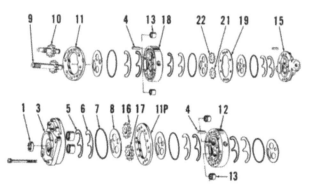

Fig. 159—Exploded view of three-section gear type pump used on 180 and 185 models. Gears (9 & 10) are for lift system, gears (16 & 17) are for power steering and gears (21 & 22) are for "Traction Booster" system. Refer to Fig. 157 for legend.

200 and 7000 Models

Refer to Fig. 158.

Power Steering & Lift Pump—
Thickness of gears
(9 & 10)............0.9005-0.9008
Thickness of
plate (11) 0.9011-0.9016
Gear to plate
clearance...........0.0003-0.0011
"Traction Booster" Pump—
Thickness of gears
(21 & 22)...........0.1501-0.1504
Thickness of
plate (19) 0.1511-0.1516
Gear to plate
clearance...........0.0007-0.0015

180 and 185 Models

Refer to Fig. 159.

Power Steering Pump—
Thickness of gears
(16 & 17)...........0.3005-0.3008
Thickness of
plate (11P) 0.3011-0.3016
Gear to plate
clearance...........0.0003-0.0011
Lift System Pump—
Thickness of
gears (9 & 10) 0.5005-0.5008
Thickness of
plate (11) 0.5011-0.5016
Gear to plate
clearance...........0.0003-0.0011
"Traction Booster" Pump—
Thickness of
gears (21 & 22) 0.1505-0.1508
Thickness of
plate (19) 0.1511-0.1516
Gear to plate
clearance...........0.0003-0.0011

If renewing needle bearings, drive or press on lettered end of bearing cage only. Be very careful to keep the needle bearing assemblies clean; it is very difficult to remove dirt or foreign material from within the bearing.

190. When reassembling pump, refer to Fig. 157, 158 or 159 and proceed as follows: Install the "E" shaped neoprene rubber sealing ring (5) in groove with flat side down. Install nylon or steel back-up ring (6) in groove at outer side of rubber ring with flat side of back-up ring up. Install wear plate sealing ring (7) in large oval groove with flat side down. On 190 and 190XT models, place small "O" ring (23) in recess of plate. Diamond shaped openings in wear plates (8) should be on inlet side of pump. If only one wear plate has diamond shaped opening, install this wear plate next to cover (15—Fig. 157). If wear plates have two small (1/8-inch) holes, this side should be toward "E" rings (5 & 6). If so equipped, the small 1/8-inch holes (H—Fig. 161) in gears plates should all be on suction side of pump and aligned with similar holes in support plates. On all models, brass side of all wear plates (8—Fig. 157, 158 or 159) should be toward gears.

On all models, screws attaching pump together should be tightened evenly to 35 ft.-lbs. while turning the shaft to check for binding. Refer to paragraph 187 for installation of drive shaft seal. On 190 and 190XT models, refer to paragraph 188 for servicing oil inlet manifold. The oil manifold attaching screws should be tightened to 18-22 ft.-lbs.

CONTROL VALVES

191. **R&R AND OVERHAUL.** The control valve assembly can be removed from bracket after disconnecting hydraulic lines and control rods.

Individual sections of the control valve assembly (Fig. 153) can be overhauled. Valve spools and housings (36 & 37; 57 & 58) are not available separately and if damaged, the complete section of valve must be renewed. Refer to paragraph 174 and following for system checks and adjustments.

LIFT ARMS (ROCKSHAFT)

192. **SYSTEM ADJUSTMENTS.** For satisfactory operation of the lift system, the control linkage and valves

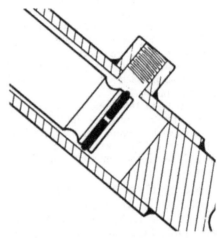

Fig. 163—When disassembling lift arm cylinder, align retaining ring with hose port as shown.

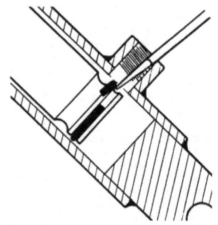

Fig. 164—Working through port opening, move retaining ring into the deep second groove.

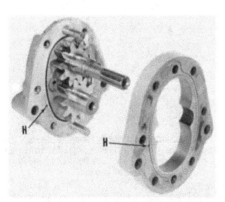

Fig. 161—Partly assembled view of hydraulic pump. Install gear plate with holes (H) in gear plate and pump cover aligned.

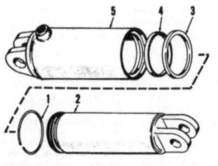

Fig. 162—Exploded view of three-point hitch lift arm cylinder used on 200 and 7000 models. Refer to Figs. 163 through 166 for removal and installation of piston rod.

1. Retaining ring
2. Piston rod
3. Wiper
4. Seal
5. Cylinder

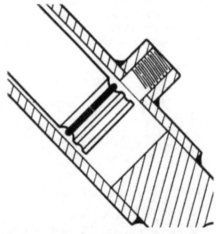

Fig. 165—Piston rod can be removed and installed with retaining ring in the deep second groove.

should be checked and adjusted as outlined in paragraph 174 through 183.

193. R&R LIFT CYLINDERS (RAMS). To remove the lift cylinders, first move the hitch to fully lowered position and block up under rear ends lift arms to take weight off the lift cylinders. Disconnect hydraulic lines from cylinders and remove the cylinder attaching pins. Pins are retained by snap rings.

194. OVERHAUL LIFT CYLINDERS. On all except 200 and 7000 models, unscrewing piston rod bearing retaining nut (8—Fig. 167 or 168) with pin type spanner wrench will allow the piston rod, nut, bearing and piston assembly to be removed from cylinder tube.

Using two pin type spanner wrenches, hold rear side (5) of piston and unscrew head end (2) from piston rod (10). Remove "O" ring (1) from piston rod and unscrew remaining part of piston from rod. Withdraw rod from bearing (7) and retaining nut.

Inspect cylinder tube (6) for wear or scoring and hone or renew cylinder tube if necessary. Clean the breather screen (11) in vent hole near open end or cylinder tube.

Install new seal (9) in piston rod bearing retaining nut. Lip of seal is towards outer side of nut (8). Install retaining nut on piston rod, outer side first.

On 190 and 190XT models, install back-up washer (12—fig. 167) in bearing (7), install "O" ring (14) toward inside of bearing and install "O" ring (13) on groove of bearing.

On all models, slide bearing (7—Fig. 167 or 168) on rod with chamfer toward piston end. Screw the piston retainer nut (5) on rod and install new "O" ring (1). Install new seals (3) and wearstrips (4) on piston. Install and securely

tighten head end of piston and stake end of piston rod with center punch.

Lubricate cylinder tube and piston, then carefully install piston and rod assembly. Securely tighten bearing retaining nut with spanner wrench.

193. On 200 and 7000 models, disassemble the lift arm cylinders as follows: Align retaining ring (1—Fig. 162) with cylinder port with end of ring slightly to one side of port opening as shown in Fig. 163. Insert screwdriver or similar tool through port opening and work retaining ring into the deep second groove in piston rod (Fig. 164 and Fig. 165). Piston can now be withdrawn.

Install seal (4—Fig. 162) with thick portion toward inside of cylinder (5). Install wiper (3) and position the retaining ring (1) in the deep second groove as shown in Fig. 165. Lubricate

seal, wiper and piston, then insert piston into cylinder. Pull the retaining ring out of deep groove and into groove at end as shown in Fig. 166. Pull piston against stop after retaining ring is repositioned to make sure that ring is fully seated.

REMOTE CYLINDER

194. 2½-INCH REMOTE RAM (CYLINDER). Refer to Fig. 170 for cross-sectional view of this unit. The 2½-inch ram may be used for single acting application when equipped with vent (6) as shown, or may be used for double acting applications by removing the vent and installing a hose in that port.

To disassemble ram, remove snap ring (9), spacer (8), snap ring (7) and withdraw piston rod, piston, and piston

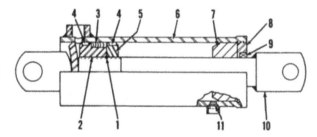

Fig. 167—Cross-sectional view of lift arm cylinder used on 190 and 190XT models.

1. "O" ring
2. Piston head
3. "V" ring packing
4. Wearstrips
5. Piston retainer nut
6. Cylinder
7. Bearing
8. Retainer nut

9. Seal
10. Rod
11. Breather

12. Back-up washer
13. "O" ring
14. "O" ring

Fig. 168—Cross-sectional view of lift arm cylinder used on 180 and 185 models. Refer to Fig. 167 for legend.

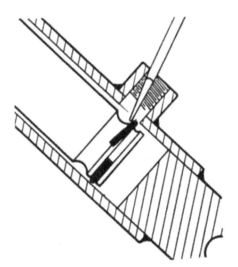

Fig. 166—When assembling, work retaining ring from deep groove back into end groove.

Fig. 169—Exploded view of three point hitch draft links and torsion bar unit. Wear plates (9) on draft links (1 & 6) are renewable. Side sway of draft links is controlled by renewable guide plates (18) attached to torsion bar supports (13 & 17). The Category II hitch can be converted to use Category I implements.

1. Draft link (R.H.)
2. Swivel end
3. Pin
4. Spring
5. Latch
6. Draft link (L.H.)
7. Swivel end
8. Adj. lift link (winging screw)
9. Wear plates
10. Special bolt
11. Torsion bar
12. Snap ring
13. Support (L.H.)

16. Bushings
17. Support (R.H.)
18. Guide plates
20. Bushings

21. Linch pins
22. Pin
23. Torsion bar tube
24. Adapter bushings

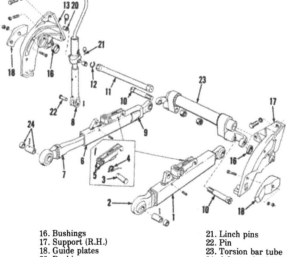

rod support assembly from cylinder. To renew piston seals, install one back-up ring (3) at each side of "O" ring (4). Further disassembly and overhaul procedure is evident from inspection of unit and reference to Fig. 170.

195. 3-INCH REMOTE RAM (CYLINDER). Refer to Fig. 171 for cross-sectional view. The 3-inch ram may be used for single acting applications with breather (4) installed as shown, or for double acting application by removing breather and connecting hose to that port.

To disassemble, remove nut (7) and withdraw piston, rod and bearing from cylinder. One back-up washer (8) is installed on each side of "O" ring (9).

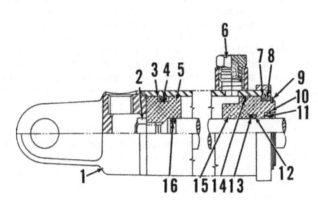

Fig. 170—Cut-away view of the 2½ inch remote ram which is available.

1. Cylinder
2. Nut
3. Back-up rings (2)
4. "O" ring
5. Piston
6. Breather vent
7. Snap ring
8. Spacer
9. Snap ring
10. Snap ring
11. Wiper
12. Back-up ring
13. "O" ring
14. "O" ring
15. Support
16. "O" ring

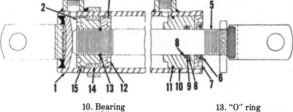

Fig. 171—Cross section of 3-inch remote ram cylinder that is available.

1. Cylinder
2. Piston
3. Inlet port
4. Breather
5. Rod
6. Wiper seal
7. Retainer nut
8. Back-up washers (2 used)
9. "O" ring
10. Bearing
11. "O" ring
12. Piston retaining nut
13. "O" ring
14. "V" ring packing
15. Wear strips

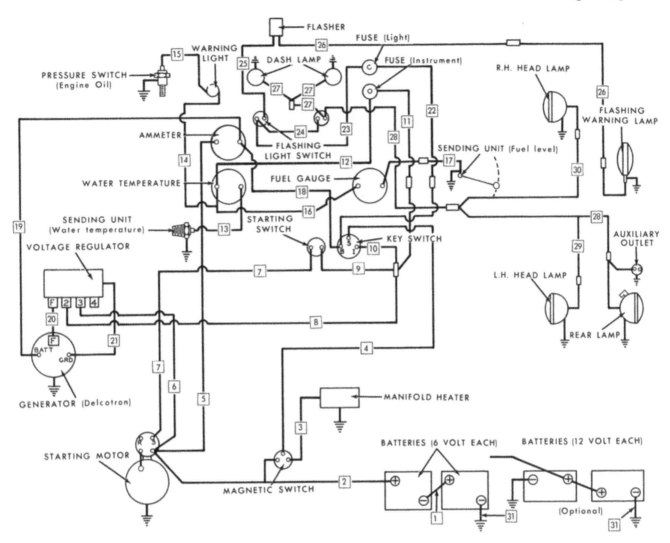

Wiring diagram for 190 and 190XT diesel tractors after serial number 190-13014. Wiring is similar for other diesel models except 7000.

1. Black	7. White		16. Purple	21. Black
2. Black	8. Brown/white stripe	13. Yellow w/black	17. White w/black	22. Orange
3. Black	9. White	stripe	stripe	23. Black
4. Red w/white stripe	10. White	14. Purple	18. Red	24. White
5. Red	11. Purple to red	15. Dark green	19. Orange	25. Tan w/white stripe
6. Brown	12. Black	w/white stripe	20. Light green	26. Tan w/white stripe

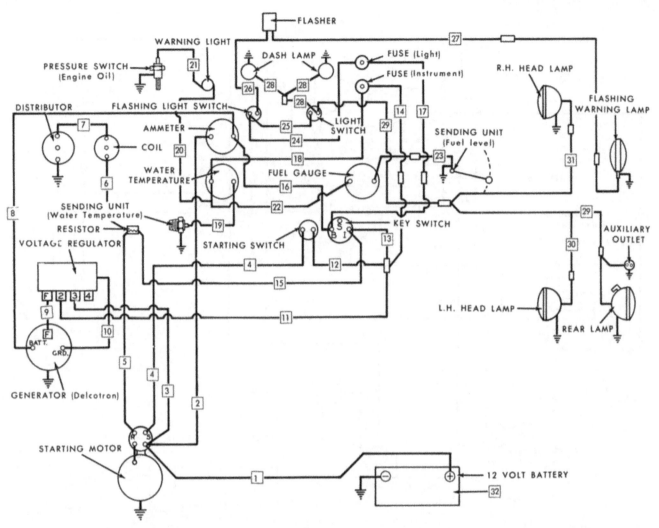

Wiring diagram for 190 and 190XT non-diesel tractors after serial number 190-13014. Wiring is similar for other non-diesel models.

1. Black
2. Red
3. Brown
4. White
5. Red w/white stripe
6. Red w/white stripe
7. Breaker point to coil wire
8. Orange
9. Light green
10. Black
11. Brown w/white stripe
12. White
13. White
14. Purple to red
15. Red w/white stripe
16. Red
17. Orange to red
18. Black
19. Yellow w/black stripe
20. Purple
21. Dark green w/white stripe
22. Purple
23. White w/black stripe
24. Black
25. White
26. Tan w/white stripe
27. Tan w/white stripe

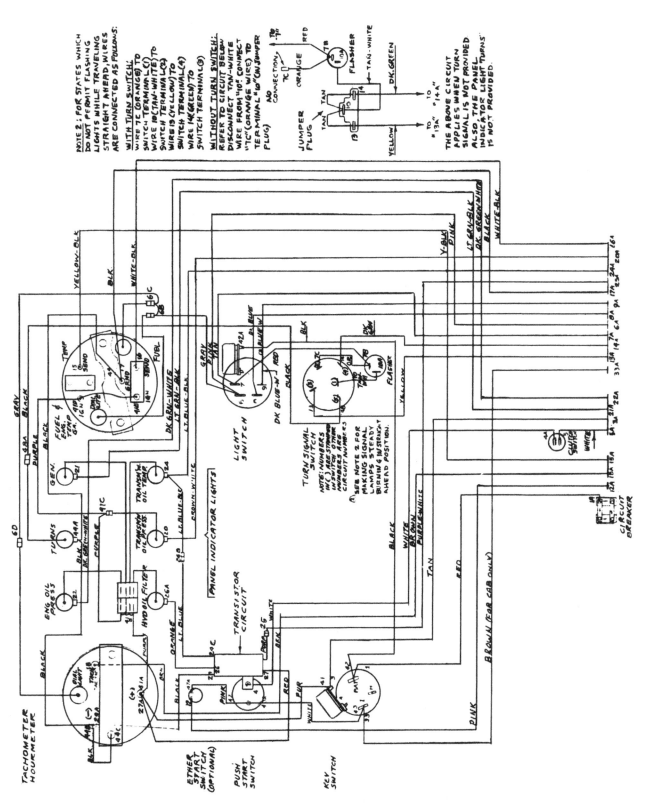

Wiring diagram for Model 7000 instrument panel.

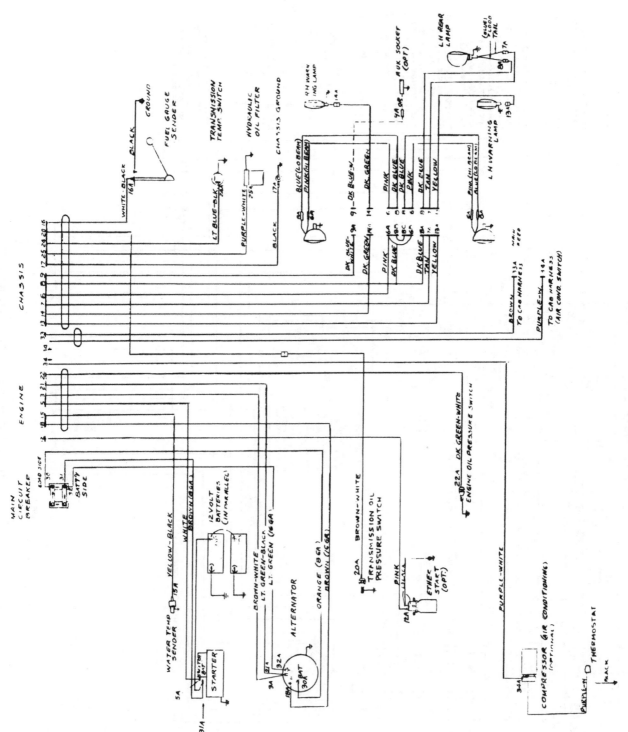

Wiring diagram for Model 7000 engine and chassis.

ALLIS-CHALMERS

Models ■ D-21 ■ D-21 Series II ■ Two-Ten ■ Two-Twenty

Previously contained in I&T Shop Service Manual No. AC-21

SHOP MANUAL
ALLIS-CHALMERS

MODELS D-21, D-21 SERIES II, TWO-TEN AND TWO-TWENTY

Tractor serial number is located on the front flange of the torque (clutch) housing at left hand side of tractor. Engine serial number is located on nameplate which is attached to the upper rear corner of cylinder block on the left hand side of tractor.

Model D-21 tractors were available with a naturally aspirated diesel engine. D-21 Series II, Two-Ten and Two-Twenty tractors are equipped with a turbo-charged diesel engine. All models are available with an adjustable, wide front axle. Two-Twenty models are also availalbe with a driving front axle (Front wheel Assist).

INDEX (By Starting Paragraph)

CONDENSED SERVICE DATA
GENERAL

Liquid Capacities	D-21	D-21 Series II	210 220
Cooling System	21 Qts.	31 Qts.	21 Qts.
Crankcase (with filter)	12 Qts.	12 Qts.	†21 Qts.
Fuel Tank	52 Gals.	52 Gals.	51 Gals.
Transmission (approx.)	21 Gals.	21 Gals.	27 Gals.

†Capacity is 15 qts. for 220 model.

Front Wheels
Toe-In .... ——— $1/16$ - $1/8$ Inch———
Speeds
Number Forward......................... ———8———
Number Reverse .... ———2———
Electrical System
Voltage................................. ———12———
Ground Polarity .... ———Negative———
No. of Batteries .... ———2 or 4———
Battery Voltage, each................... ———12———

CONDENSED SERVICE DATA (Cont.)

DIESEL ENGINE

General	D-21	D-21 Series II	210 220
Make.................		Own	
Model	3400	3500	3500
No. of Cylinder	6	6	6
Bore—Inches	4 1/4	4 1/4	4 1/4
Stroke—Inches.........	5	5	5
Displacement–Cubic Inches	426	426	426
Compression Ratio.......	16:1	16:1	16:1
Firing Order		1-5-3-6-2-4	

Compression Pressure @			
150 RPM*.............	400 psi	400 psi	400 psi
600 RPM*.............	500 psi	500 psi	500 psi

*Compression pressures given are approximate and at sea level.

Tappets & Valves			
Tappet Gap, Hot.........	0.015	0.015	0.015
Cold..................	0.018	0.018	0.018

Valve Face & Seat Angle—	D-21	D-21 Series II	210 220
Intake.................	30°	30°	30°
Exhaust	45°	45°	45°
Diesel System			
Pump Make		Roosa-Master	
Injection Timing, Static . . .	34°BTDC	34°BTDC	34°BTDC
Timing Mark on		Crankshaft Pulley	
Nozzles Make		Own	
Opening Pressure.......	2750 psi	2900 psi	3125 psi
Governed Speed			
Low Idle RPM..........	675	675	675
High Idle RPM	2400	2400	2400
Rate Load RPM	2200	2200	2200
Tightening Torques (Ft.-Lbs.)			
General Recommendations		See End of Shop Manual	
Cylinder Head		See Paragraph 21	
Crankshaft Pulley	200-220	200-220	200-220
Flywheel	95-105	95-105	95-105
Main Bearing Cap Screws .	170-190	170-190	170-190
Rod Cap Screws		See Paragraph 38	

FRONT AXLE SYSTEM
(Models without Front Wheel Assist)

SPINDLES

1. **R&R SPINDLES.** To remove front spindle (10—Fig. AC1), support front of tractor, remove front wheel and proceed as follows: Remove snap ring (6) and pull steering arm (8) from spindle. Remove key (K) from spindle and withdraw spindle from bottom of axle extension (5). Remove thrust washers from spindle.

2. **SPINDLE BUSHINGS.** Spindle bushings can be renewed after removing spindle as outlined in paragraph 1. Remove bushings (7) using suitable drift punch. New bushings are pre-sized and should not require reaming if carefully installed. Renew thrust washers if necessary before reinstalling spindle.

TIE-RODS AND TOE-IN

3. Refer to Figs. AC1 and AC2. Toe-in of front wheels should be 1/16 to ⅛-inch, and can be adjusted by lengthening or shortening tie-rods (12) equally as follows: Remove bolt (B) from outer ends of tie-rods and loosen clamp bolts (C) at inner ends. Turn the tie-rod tubes (12) in or out on inner tie-rod ends (13) to obtain correct adjustment. Reinstall and tighten bolt (B), then tighten clamp bolt (C) and recheck toe-in. Readjust if necessary.

Tie-rod end sockets are non-adjustable automotive type and except for dust cover, must be renewed as a complete unit.

AXLE EXTENSIONS

4. To renew either axle extension (5—Fig. AC1 or Fig. AC2), remove spindle as outlined in paragraph 1; then, remove tread width adjusting bolts and withdraw axle extension from main member.

STEERING ARM

5. To renew the steering arm (15—Fig. AC1) or steering arm bushings (16), proceed as follows: Disconnect steering cylinder piston rod and tie-rods from the steering arm, then remove steering arm from pivot pin (P) on the crossbar in front axle main member.

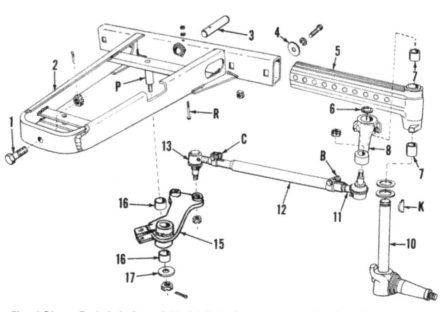

Fig. AC1 — Exploded view of Model D-21 front axle assembly. Bolt (R) retains front pivot pin (3). To adjust front wheel toe-in, remove bolt (B), loosen clamp (C) and turn tie rod tube (12) in or out as required.

B. Bolt	1. Rear pivot bolt	6. Snap ring	12. Tie rod tube
C. Clamp	2. Axle main member	7. Spindle bushings	13. Inner tie rod end
K. Key	3. Front pivot pin	8. Spindle arm	15. Steering arm
P. Pivot for (15)	4. Flat washer	10. Spindle	16. Bushings
R. Retainer bolt	5. Axle extension	11. Outer tie rod end	17. Retainer washer

Fig. AC2 — View showing front axle extension. Toe-in should be correct for each tread width position when bolt (B) is located in proper notch (N) in tie rod. Refer to Fig. AC1 for legend.

Drive the bushings (16) out of steering arm using a suitable bushing driver. New bushings are pre-sized and should not require reaming if carefully installed.

The steering arm pivot pin (P) is an integral part of the axle center (main) member and radius rod assembly.

AXLE CENTER (MAIN) MEMBER AND PIVOT PINS

6. The axle center (main) member and the radius rods are welded one-piece assembly (2—Fig. AC1 or Fig. AC2). The center member pivots on a bolt (1) at rear end of assembly and a pin (3) at front end. The bolt and pin pivot in renewable bushings in the clutch (torque) housing and in the front support casting.

To renew the pivot bolt or pivot pin, proceed as follows: Support tractor under the torque housing so that no weight is carried on the front axle. Then, remove the retaining bolt (R—Fig. AC1) and drive the front pivot pin out towards front of tractor. The plug (3—Fig. AC3) with lubricating fitting will be driven out with pin. Then, raise front of tractor until front support is clear of axle and drive bushing (2) out to rear with suitable driver. New bushing is pre-sized and should not require reaming if carefully installed. Install pivot pin and retaining bolt, then drive the plug (3) into front support with a hollow driver so that plug is flush with casting.

To remove rear pivot bolt, support rear (radius rod) end of axle main member and remove cotter pin and

slotted nut from pivot bolt. Then, remove pivot bolt from axle and torque housing. Drive bushing out of torque housing. New bushing is pre-sized and should not require reaming if carefully installed. Install pivot bolt through torque housing, install slotted nut and tighten nut to a torque of 150 Ft.-Lbs. If slot in nut does not align with hole in bolt, continue to tighten nut until cotter pin can be inserted.

To renew the axle center (main) member, remove both axle extensions as outlined in paragraph 4, disconnect rear end of power steering cylinder, remove steering arm from pivot pin and remove the front axle pivot pin and rear pivot bolt.

FRONT SUPPORT

7. The front support (1—Fig. AC3) is a heavy one-piece casting. The front axle pivot bushing (2) and plug (3) can be renewed as outlined in paragraph 6 without removing the front support. To remove front support for other jobs, proceed as follows: Drain the cooling system. Remove hood and air cleaner to intake manifold tube. On Two-Twenty models, disconnect hydraulic lines to the oil cooler. On all models, disconnect both radiator hoses, then unbolt hood front support panels from side rails, and remove the panels, radiator and air cleaner as a unit. Disconnect front of steering cylinder from steering arm (15—Fig. AC1). Support the tractor under torque housing so there is no weight on front axle pivot pin. Attach a hoist to front support at lifting point provided, remove cap screws retaining side rails to front support and remove the front support from side rails and front axle pivot pin. NOTE: If front support casting binds between the side rails, loosen the bolts retaining side rails to engine front support plate. Any front end weights attached to front support casting should be removed before unbolting and removing the front support.

POWER STEERING SYSTEM

All models are equipped with hydrostatic power steering system that has no mechanical linkage between the steering wheel and tractor front wheels. Refer to Fig. AC4 for drawing showing the steering system.

Power for steering is supplied by a gear type pump that is driven from the engine timing (camshaft) gear. On models with a hydraulic lift system, the hydraulic and power steering pumps are an integral unit. Transmission oil is utilized as fluid for the system.

The control valve unit (1—Fig. AC4) contains a rotary metering motor, a commutator feed valve sleeve and a selector valve spool. In the event of engine or hydraulic power failure, the metering motor becomes a rotary hand pump to actuate the power steering cylinder when the steering wheel is turned. A check valve within the control

valve housing allows recirculation of fluid within the control valve and steering cylinder during manual operation. NOTE: The maintenance of absolute cleanliness of all parts is of the utmost importance in the operation and servicing of the hydraulic power steering system. Of equal importance

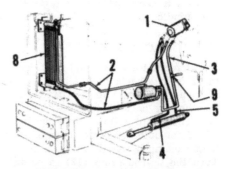

Fig. AC4—Drawing showing components of hydraulic power steering system. No mechanical linkage is used between the steering wheel and tractor front wheels. Note that the components are not located in position in which they are installed on tractor. Cooler (8) is not used on D-21 models.

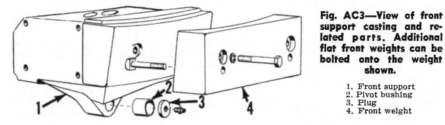

Fig. AC3—View of front support casting and related parts. Additional flat front weights can be bolted onto the weight shown.

1. Front support
2. Pivot bushing
3. Plug
4. Front weight

1. Control valve unit
2. Pressure tube
3. Return tube
4. "Right turn" tube
5. "Left turn" tube
8. Oil cooler
9. "T" fitting to PTO valve

is the avoidance of nicks or burrs on any of the working parts. Do not use cloth shop towels in cleaning the system internal parts; use only lint free power shop towels.

TROUBLE SHOOTING

8. Before attempting to adjust or repair the power steering system, the cause of any malfunction should be located. Refer to the following paragraphs for possible causes of power steering system malfunction:

Irregular or "Sticky" Steering. If irregular or "sticky" feeling is noted when turning the steering wheel with forward motion of tractor stopped and with engine running at rated speed, or if steering wheel continues to rotate after being turned and released, foreign material in the power steering fluid (transmission oil) is the probable cause of trouble. Clean or renew all hydraulic system filters. It may be necessary to also drain the transmission oil and refill with clean oil. If trouble is not corrected, the power steering valve assembly should be removed and serviced; refer to paragraph 12.

Steering Cylinder "Hesitates". If steering cylinder appears to pause in travel when steering wheel is being turned steadily probable cause of trouble is air trapped in the power steering cylinder. Bleed the cylinder as outlined in paragraph 9.

Slow Steering. Slow steering may be caused by low oil flow from pump. Check time required for full stroke travel of power steering cylinder; first with tractor weight on the front wheels, then with front end of tractor supported by a jack. If time between the two checks varies considerably, overhaul the power steering pump as outlined in paragraph 139.

Loss of Power. Loss of steering power may be caused by system relief setting being too low. Check and adjust relief valve setting as outlined in paragraph 10A or 10B.

System Overheats. Overheating of system may be caused by neutral (by-pass) pressure being too high. Check neutral pressure as outlined in paragraph 10A. If neutral pressure is excessive, check for obstructions in system such as kinked tube, etc. If neutral pressure is erratic, check for binding of control valve shaft, spool and/or sleeve.

LUBRICATION AND BLEEDING

9. Transmission lubricating oil is utilized as the power steering fluid; refer to note preceding paragraph 80. The power steering system is usually self-bleeding. With engine running at high idle speed, cycle system through several full strokes of the cylinder. In some cases, it may be necessary to loosen connections at cylinder to bleed trapped air.

SYSTEM OPERATING PRESSURE AND RELIEF VALVE

10. Tests of the power steering system will disclose whether the pump, relief valve or some other unit in the system is malfunctioning. To make such a test, proceed as outlined in the appropriate following paragraph:

D-21 and D-21 Series II

10A. Power steering relief pressure can be checked by connecting a high pressure (3000 psi) hydraulic gage into the pressure line between the pump and power steering control valve. Turn the steering wheel to one end of its travel and hold in that position only long enough to read the pressure gage. CAUTION. Never hold steering wheel in this position for longer than a few seconds at a time; overheating of oil and possible damage to pump will occur if wheel is held in this position for an extended period of time. System relief pressure should be 1500 PSI at 2200 engine RPM. Pressure relief valve is located in manifold of hydraulic pump and can be adjusted as outlined in paragraph 141.

Check the neutral (by-pass) system pressure as follows: With pressure gage connected as outlined in preceeding paragraph, slowly turn steering wheel in each direction and stop wheel in position that shows lowest gage reading; this reading is the neutral (by-pass) pressure. Turn wheel to limit stop in either direction and hold wheel against stop for one to two seconds, then release the wheel; gage pressure should drop to nearly the same as neutral pressure reading. If gage remains considerably higher than neutral pressure reading, a binding control shaft or foreign material between control valve sleeve and valve spool is indicated.

A hydraulic tester (such as OTC Y81-21) can be connected as shown in Fig. AC 4A to check pump volume and relief pressure. The outlet hose should be routed to sump as shown. With engine operating at approximately 1600 RPM, close the tester valve until pressure is approximately 1200 psi. When oil temperature reaches 100° F., the volume of flow should be 5.25 GPM with engine speed between 1050 and 2200 RPM. With tester valve completely closed, pressure should be 1500 psi. Add or remove shims under pressure relief valve plug to change pressure. The pump and/or flow control valve should be serviced if volume of flow is incorrect.

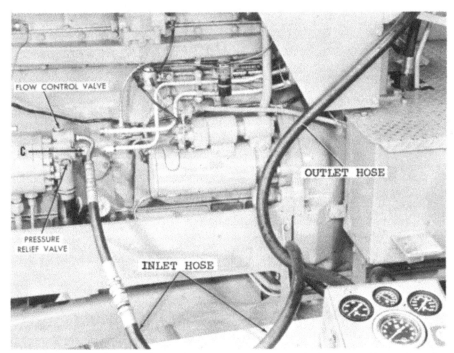

Fig. AC4A—View of OTC hydraulic tester connected to D-21 tractor for testing the power steering system. Both pressure and return tubes must be disconnected from pump and return connection must have cap (C) installed. Outlet hose must return oil to sump as shown.

FLOW CONTROL VALVE

C

OUTLET HOSE

PRESSURE RELIEF VALVE

INLET HOSE

210 and 220 Models

10B. The PTO clutch (on models so equipped) is operated by return or by-pass oil from the power steering system. PTO clutch operating pressure must be tested as outlined in paragraph 109 before checking power steering relief pressure.

Power steering relief valve pressure on all models can be checked by installing a high pressure (3000 psi) hydraulic gage in port (P–Fig. AC4B) after removing the plug. Turn the steering wheel briefly to one end of its travel and observe pressure. If tractor is **not** equipped with PTO, power steering relief pressure should be 1690-1810 psi. If tractor is equipped with PTO, relief pressure should be 1925-2075 psi. Power steering relief pressure can be adjusted at screw (S) after removing cap nut (C). The neutral by-pass system pressure can be checked as outlined in paragraph 10A.

The flow divider located in pump rear cover first supplies 4.25 GPM of

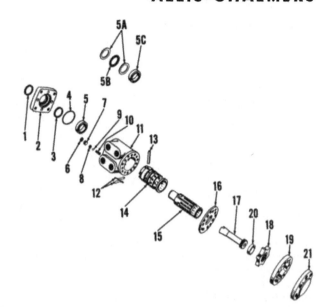

1. Oil seal
2. Mounting plate
3. Quad ring seal
4. "O" ring
5. Bushing
6. "O" ring
7. Plug
8. Check seat
9. Check valve
10. Valve spring
11. Control valve body
12. Centering springs (6)
13. Centering pin
14. Sleeve
15. Valve spool
16. Plate
17. Drive shaft
18. Rotor
19. Ring
20. Spacer
21. Cover (cap)

Fig. AC6—Exploded view of steering control valve assembly which is shown assembled in Fig. AC4. Centering springs (12) are installed in two groups of three springs with arch in each group back-to-back. Some models are equipped with needle thrust bearing (5A & 5B) and different type locator bushing (5C).

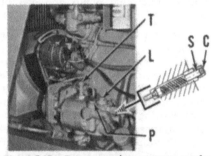

Fig. AC4B—Power steering pressure can be checked at port (P) on 210 and 220 tractors. Relief valve adjusting screw (S) and cap nut (C) are located on engine side of pump.

Fig. AC5 — View of steering control valve showing port locations. Pressure tube from pump connects to "IN" port; return tube to sump connects to "OUT" port; tube to rear end of cylinder connects to "L" (left turn) port; and tube to rod end of cylinder connects to "R" (right turn) port.

oil to the power steering system, with remainder of oil used for the power lift system. A hydraulic tester (such as OTC Y81-21) can be attached to port (P—Fig. AC4B) with outlet from tester routed to sump filler opening. With engine operating at approximately 1600 RPM, first check relief pressure. Refer to preceeding paragraph for recommended settings. Open tester valve until pressure is approximately 300 psi below relief pressure and observe volume of flow. Volume should be 4.25 GPM with engine speed between 1050 and 2200 RPM. The pump and/or flow divider valve should be serviced if volume of flow is incorrect.

POWER STEERING PUMP

11. A gear type pump driven from the engine camshaft gear is used as the power source for the steering system. On tractors equipped with a hydraulic system, the power steering pump is an integral part of the hydraulic system pump.

For service information on the power steering pump, on tractors both with and without a hydraulic system, refer to paragraph 139.

CONTROL VALVE

12. **REMOVE AND REINSTALL.** Remove the hood and disconnect the four hoses from control valve. Unbolt and remove the control valve assembly from the steering shaft housing.

Reinstall the control valve assembly by reversing the removal procedure. Refer to Fig. AC5 for proper hose

locations. Install new "O" ring seals on the hose fittings before connecting hoses to control valve and tighten fittings securely. Bleed trapped air from the power steering cylinder as outlined in paragraph 9 after assembly is completed.

13. **OVERHAUL CONTROL VALVE.** After removing the control valve assembly as outlined in paragraph 12, proceed as follows: Clean the valve thoroughly and remove paint from points of separation with a wire brush. Note: A clean work bench is necessary. Also, use only lint free paper shop towels for cleaning valve parts.

If oil leakage past seal (1—Fig. AC6) is the only difficulty, a new seal can be installed after removing the mounting plate (2) only. Remove plug (7) from valve body with a bent wire, install new "O" ring on plug, lubricate plug and "O" ring with motor oil and reinstall in valve body. Install new quad seal (3) and "O" ring (4) in mounting plate, reinstall mounting plate and tighten retaining cap screws equally to a torque of 250 inch-pounds.

To completely disassemble and overhaul the power steering control valve, refer to exploded view in Fig. AC6 and proceed as follows: Clamp valve mounting plate in vise with cap end of valve up. Remove end cap retaining cap screws. Remove end cap (21), gerotor set (18 & 19), plate (16) and drive shaft (17) from valve body as a unit. Remove valve from vise, place a clean wood block in vise throat and set valve assembly on block with mounting plate (2) end up. Lightly clamp vise against port face of valve

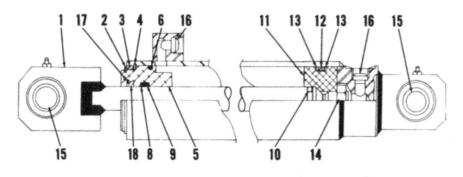

1. Cylinder rod	5. Support	10. "O" ring	14. Lock nut
2. Snap ring	6. "O" ring	11. Piston	15. Swivel
3. Spacer	8. Back-up ring	12. "O" ring	16. Tube seats
4. Snap ring	9. "O" ring	13. Back-up rings	17. Snap ring
			18. Scraper

body and remove mounting plate retaining screws. Hold spool assembly down against wood block while removing mounting plate. Remove valve body from vise and place on work bench with port face down. Carefully remove spool and sleeve assembly from 14-hole end of valve body. Use bent wire to remove check valve plug (7) and use $\frac{3}{16}$-inch Allen wrench to remove check valve seat (8). Then remove check valve (¼-inch steel ball) (9) and spring (10) from valve body.

Remove centering pin (13) from spool and sleeve assembly, hold sleeve and push spool, (splined end first) out of sleeve. Remove the six centering springs (12) from slot in spool.

Separate the end cap and plate from the gerotor set and remove the drive shaft, rotor and spacer.

Inspect all moving parts for scoring. Slightly scored parts can be cleaned by hand rubbing with 400 grit abrasive paper. To recondition gerotor section surfaces place a sheet of 600 grit paper on plate glass, lapping plate or other absolutely flat surface and remove sharp particles from paper with a flat piece of scrap steel. Stroke each surface of the gerotor section over the abrasive paper; any small bright areas indicate burrs that must be removed. Polish each part, rinse in clean solvent and air dry; keep these parts absolutely clean for reassembly.

Renew all parts that are excessively worn, scored or otherwise damaged. Install new seal kit when reassembling.

To reassemble valve, proceed as follows: Install check valve spring with small end out. Drop check valve ball on spring and install valve seat with counterbored side towards valve ball. Tighten seat to a torque of 150 inch-pounds.

Lubricate the valve spool and carefully insert spool in valve sleeve using a twisting motion; be sure that spring slots in spool and sleeve are at same end of assembly. Stand the assembly on end with spring slots up and aligned. Assemble the springs in two groups with extended edges of springs down. Place the two groups of springs back-to-back (arched sections together) and install the springs into the spring slot in sleeve and spool in this position. Use a small screwdriver to guide the springs through slots in opposite side of assembly. Center the springs in the sleeve with edges of springs flush with upper surface of sleeve. Insert centering pin (13) through the spool and sleeve assembly so that both ends of pin are below flush with outside of sleeve (14). On some models, nylon plugs are located at ends of centering pin to prevent pin from contacting inside diameter of valve body. Carefully insert spool and sleeve assembly, splined end first, in 14-hole end of valve body (11).

Set valve body on clean surface with 14-hole end down. Install new "O" ring (6) on check valve plug (7), lubricate plug and insert in check valve bore. Install locator bushing (5 or 5C) in valve bore with chamfered side up. On models so equipped, install thrust bearing assembly (5A and 5B) over valve spool (15). Install new quad seal (3) and "O" ring (4) in mounting plate, lubricate seal and install mounting plate over valve spool and locator bushing. Tighten the mounting plate retaining screws evenly to a torque of 250 inch-pounds. Clamp the mounting plate in a vise with 14-hole end of valve body up. Place plate (16) and gerotor outer ring (19) on valve body so that bolt holes align. Insert drive shaft (17) in gerotor inner rotor so that slot in shaft is aligned with valleys in rotor

and push shaft through rotor so that about ½ of length of the splines protrude. Holding the shaft and rotor in this position, insert them in valve housing so that notch in shaft engages centering pin in valve sleeve and spool. Install spacer (20) at end of drive shaft; if spacer does not drop down flush with rotor, shaft is not properly engaged with centering pin. Install end cap and tighten the retaining screws equally to a torque of 150 inch-pounds.

POWER STEERING CLYINDER

14. REMOVE AND REINSTALL. Thoroughly clean cylinder and hose fittings to avoid entry of dirt when lines are disconnected. Turn front wheels to right as far as possible. Remove the bolt at each end of cylinder. Disconnect hose fittings at cylinder and remove cylinder from tractor. Work the cylinder piston each way to end of stroke to expel all excess oil. Note: Only about 1 pint of oil will drain from hose when cylinder is disconnected.

When reinstalling cylinder, first attach hose fittings leaving the flare nuts loose. Install the two bolts retaining cylinder to crossbar and piston rod to steering arm with bolt heads up, then tighten the nuts and secure with cotter pins. Check to see that hoses are properly located, then tighten the flare nuts.

Bleed the power steering system as outlined in paragraph 9, check transmission oil level and add oil as necessary to bring oil level to full mark on dipstick.

15. OVERHAUL STEERING CYLINDER. After removing power steering cylinder as outlined in paragraph 14, refer to Fig. AC7 or AC8. The earlier type is shown in Fig. AC7.

On early models (Fig. AC7), remove snap ring (2) and withdraw spacer (3) from cylinder tube. Push support (5) into tube far enough to remove snap ring (4). Remove snap ring, then use piston rod to bump the support out of the cylinder tube. Remove nut (14) from inner end of piston rod, carefully remove the burr left from staking nut to rod, then remove piston (11). Remove "O" ring (10) and slide the support (5) off rod.

On later models (Fig. AC8), remove fitting (20) push support (5) into tube far enough to remove snap ring (4). Remove snap ring, then use piston rod to bump the support out of cylinder tube.

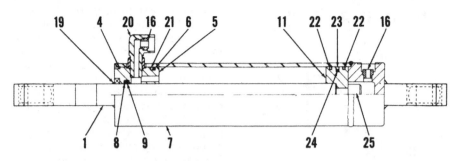

Fig. AC8—View of late power steering cylinder. Refer to Fig. AC7 for legend except the following.

7. Cylinder tube	21. Back-up ring	24. "O" ring
19. Scraper	22. Split wear ring	25. Nut
20. Fitting	23. Sealing ring	

The sharp edge of scrapers (18—Fig. AC7 and 19—Fig. AC8) should be toward outside. Refer to Figs. AC7 & AC8 for remainder of assembly. Stake the piston retaining nut to rod after tightening. Support (5) must be pushed into cylinder enough to install snap ring (4), then bump the support back out against snap ring by pulling on rod. On later models, make certain that hole in support is aligned with hole in cylinder, so that fitting (20—Fig. AC8) can be installed. On all models be sure that all parts are lubricated when assembling and carefully work seals onto rod and into cylinder.

ENGINE AND COMPONENTS

R&R ENGINE WITH CLUTCH ASSEMBLY

20. To remove the engine, first remove the cover(s) from the batteries and disconnect all battery ground straps. Note that each battery has a separate ground strap. Then, proceed as follows:

Drain oil pan if engine is to be disassembled. It is not necessary to drain the transmission as front end of pump suction tube is higher than transmission oil level. Remove the front support casting as outlined in paragraph 7. Unbolt and remove both side rails from tractor. Disconnect the starter cable from starting motor and disconnect all wiring to engine electrical units by separating the plug-in connectors or detaching from electrical units. Move wires out of the way. Disconnect power steering, hydraulic system and "Traction Booster" lines from pump, remove clamps and reposition lines away from engine. NOTE: Cap all openings as lines are disconnected. Shut off fuel supply and disconnect fuel lines. Disconnect speed control rod, fuel shut-off cable (or wires) and engine hour-meter cable.

NOTE: Depending on reason for engine removal, it may be easier to remove other components from engine at this time.

Attach hoist to lift hooks provided at front and rear end of cylinder head, then unbolt and remove the engine from torque housing. Refer to Fig. AC10.

Reinstall engine and clutch assembly by reversing removal procedures. Bleed the fuel system as outlined in paragraph 51 and, if necessary, bleed the power steering system as outlined in paragraph 9.

CYLINDER HEAD

21. **REMOVE AND REINSTALL.** Drain cooling system and remove air cleaner inlet cap and pipe, hood top and side panels. Disconnect upper radiator hose from thermostat housing and remove tube from air cleaner to intake manifold or turbocharger. Disconnect water by-pass tube from water pump and thermostat housing, disconnect temperature gage sender switch wire and remove thermostat housing from cylinder head. On turbocharged models, remove the turbocharger. On all models, remove exhaust manifold. Remove the intake manifold if so desired.

Disconnect breather tube from rocker arm cover, remove rocker arm

Fig. AC10—View of engine removed from D-21 tractor. Two-Ten, Two-Twenty and D-21 Series II engines are similar.

1. Lift hook	3. Oil filters	6. Fuel filter
2. Lift hook	4. TDC timing pin	7. Timing window
	5. Fuel primer pump	

cover, remove oil supply line to rocker arm assembly and equally loosen and remove the rocker arm assembly retaining cap screws. Lift off rocker arm assembly and remove the push rods. Disconnect fuel return line at front injector and cap all openings as each line is disconnected.

NOTE: Although not required for removal of the cylinder head, it is recommended that the fuel injector assemblies be removed from cylinder head at this time. The injector nozzle tips protrude through the flat bottom surface of the cylinder head and, if injectors are not removed, extreme care must be taken when removing and handling the removed cylinder head.

Remove the cylinder head retaining cap screws and carefully pry cylinder head up off of the dowel pins (DP—Fig. AC11) located at front and rear of cylinder block. Lift cylinder head from tractor.

When reinstalling cylinder head, be sure that gasket surfaces of cylinder head and block are clean and free of burrs. Check cylinder liner standout as outlined in paragraph 39. Place new gasket with side marked "THIS SIDE DOWN" against cylinder block and be sure gasket is properly located on the two dowel pins. Note: Do not apply grease or gasket sealer to gasket surfaces, block or cylinder head; new gasket is treated with phenolic sealer. Locate fire rings around cylinder sleeves then position cylinder head on the dowel pins. Lubricate the cylinder head retaining cap screws with motor oil and install the 20 short cap screws as indicated in View "A", Fig. AC12; tighten the short cap screws to a torque of 90-100 Ft.-Lbs. in sequence shown. Install the push rods and rocker arm assembly. Starting in the center and working toward each end, tighten the six long cylinder head retaining cap screws to a torque of 90-100 Ft.-Lbs. Retighten all 26 cylinder head retaining cap screws to a torque of 130-140 Ft.-Lbs. in se-

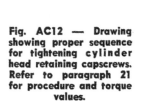

Fig. AC12 — Drawing showing proper sequence for tightening cylinder head retaining capscrews. Refer to paragraph 21 for procedure and torque values.

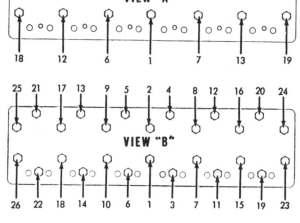

quence shown in View "B", Fig. AC12, and tighten the six ⅜-in rocker arm bracket cap screws to a torque of 28-33 Ft.-Lbs. NOTE: Adapter must be used to correctly tighten the screws located under rocker arm shaft. Adjust valve tappet gap cold as outlined in paragraph 28, start engine and bring to normal operating temperature. After engine has operated for one hour (preferably under load), retorque the cylinder head screws and readjust valve gap hot. After ten hours of normal tractor operation, retighten the cylinder head cap screws to a torque of 130-140 Ft.-Lbs. and readjust valve tappet gap to 0.015 hot.

NOTE: All torque values are given for clean threads lubricated with motor oil. Allis-Chalmers recommends that no other type of thread lubricant be used.

VALVES AND VALVE SEATS

Intake Valves

22. Intake valves have a face and seat angle of 30 degrees. A one degree interference angle may be used by machining the seat to 31 degrees. The seat width can be narrowed by using 15 and 60 degree stones to obtain the

desired seat width of $\frac{3}{32}$-inch. Intake valve stem diameter is 0.3715-0.3720.

Intake valves seat directly in the cylinder head although intake valve seats are available for service. When installing service intake valve seat inserts, machine counter bore in cylinder head to a diameter of 1.814-1.815 and to a depth of 0.4585-0.4605. Outside diameter of seat insert is 1.816-1.817 which provides a 0.001-0.003 interference fit.

Surface of intake valve head must be recessed at least 0.054 from cylinder head surface to provide proper clearance between valve head and the piston.

Exhaust Valves

23. Exhaust valves have a face and seat angle of 45 degrees. A one degree interference angle may be used by machining seat to 46 degrees. The seat width can be narrowed by using 30 and 60 degree stones to obtain desired seat width of $\frac{3}{32}$-inch. Exhaust valve stem diameter is 0.3705-0.3710.

Exhaust valves seat in renewable inserts. Seat insert diameter of 1.667-1.668 provides a 0.001-0.003 interference fit in the 1.665-1.666 diameter counterbore in cylinder head. If new exhaust valve seat insert fits too loosely, machine the counterbore diameter to 1.670-1.671 and install a 0.005 oversize insert. Depth of the counterbore should be 0.4735-0.4755 for both the standard size and 0.005 oversize inserts. Stake the seat insert at 3 points 120° apart after seat is installed. NOTE: Do not stake the seat near the injector nozzle hole.

Surface of exhaust valve head must be recessed at least 0.053 from cylinder head surface to provide proper clearance between valve head and the piston.

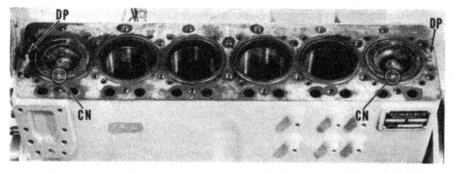

Fig. AC11 — View of engine showing cylinder head removed. Note the cylinder head locating dowel pin (DP) at each end of block. Cylinder numbers (CN) are stamped on side of piston towards the camshaft. Disregard number stamped on opposite side of piston.

VALVE GUIDES

24. Renew valve guides when stem to guide clearance exceeds 0.0035 on intake valves or 0.0055 on exhaust valves. Desired stem to guide clearance is 0.001-0.0015 for intake valves and 0.002-0.0025 for exhaust valves.

Intake and exhaust valve guides are interchangeable. When installing new guides, press new guides in from bottom of cylinder head. The top of intake valve guides should be 25/32-inch from counterbore around guide and bottom of guide should be approximately 1⅜-inches from cylinder head gasket surface. The top of exhaust valve guides should be $1\frac{3}{32}$-inches from counterbore around guide and bottom of guide should be approximately $1\frac{11}{16}$-inches from cylinder head gasket surface. Refer to Fig. AC13.

After installing guides, ream inside diameter of both intake and exhaust valve guides to 0.373 to provide proper stem to guide clearance.

VALVE SPRINGS

25. The interchangeable intake and exhaust valve springs should be renewed if rusted, distorted or if they vary more than 5% from following specifications:
Pounds pressure @
  2.237 inches 40-46
Pounds pressure @
  1.780 inches 105-115
Spring free length,
  approx. ....$2\frac{11}{32}$ inches
Spring are installed with flat steel dampener coil inside spring. Spring or dampener coil are not available separately.

CAM FOLLOWERS

26. The 0.748-0.7485 diameter mushroom type cam followers (valve lifters) ride directly in unbushed cylinder block bores and can be removed after removing the camshaft. Cam followers are available in standard size only and should be renewed if either end is chipped or worn or if they are loose in cylinder block bores. Desired follower to bore clearance is 0.001-0.0025.

Fig. AC13 — View showing method of measuring valve guide height above cylinder head counterbore when installing new valve guides. Refer to paragraph 24.

ROCKER ARMS

27. **R&R AND OVERHAUL.** Rocker arms and shaft assembly can be removed after removing hood top and rear side panels, disconnecting breather tube and removing rocker arm cover and oiling tube. Loosen and remove all retaining cap screws equally, then lift rocker arms and shaft assembly from the cylinder head.

The hollow rocker arm shaft is drilled for lubrication to each rocker arm. Lubricating oil to the oiling tube is supplied through a drilled passage in cylinder head and engine block.

To disassemble the rocker arms and shaft assembly, remove the hex plugs, spring washers and spacer washers from each end of shaft and remove locating cap screw and lock washer from No. 3 rocker arm support. Slide the rocker arms, supports and spacer springs from shaft.

Rocker arm shaft diameter is 0.999-1.000. Bore in rocker arm is 1.001-1.002 providing a clearance of 0.001-0.003 between shaft and rocker arms. If clearance is excessive, renew the shaft and/or rocker arms. All rocker arms are alike.

To reassemble the rocker arms and shaft assembly, proceed as follows: Place shaft on work bench with end nearest to the detent hole (DH—Fig. AC15) to left. Slide rocker arm support with locating capscrew (LC) on shaft with flat mounting surface down and large capscrew hole towards you. Tighten the locating capscrew into the hole in shaft to 28-33 Ft.-Lbs. torque. Complete the assembly of rocker arm shaft using Figs. AC14 and AC15 as a guide. End plugs (12—Fig. AC15) should be torqued to 40 Ft.-Lbs. The restrictor elbow (7) with a $\frac{1}{16}$-inch opening should be installed in rear plug and the inlet oil line (16) should be attached. The elbow at front is not a restricted type.

Before reinstalling the rocker arm assembly, loosen each tappet adjusting screw about two turns to prevent interference between valve heads and pistons due to interchanged components. Tighten the ⅜-inch support retaining capscrews to a torque of 28-33 Ft.-Lbs. and the $\frac{9}{16}$-inch (cylinder head) capscrews to a torque of 130-140 Ft.-Lbs. Adjust valve tappet clearance cold as outlined in paragraph 28. Complete the reassembly of tractor, then readjust tappet clearance after engine has reached normal operating temperature.

Fig. AC14—View of rocker arm assembly and related parts. Disconnect oiling tube (OT) from rear end of rocker arm shaft before removing the assembly. Note the position of shaft locating capscrew (LC). Oil return tube (RT) is located at front end of rocker arm shaft.

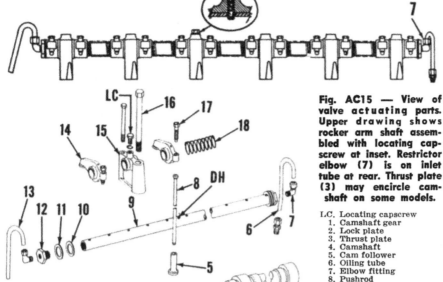

Fig. AC15 — View of valve actuating parts. Upper drawing shows rocker arm shaft assembled with locating capscrew at inset. Restrictor elbow (7) is on inlet tube at rear. Thrust plate (3) may encircle camshaft on some models.

LC. Locating capscrew
1. Camshaft gear
2. Lock plate
3. Thrust plate
4. Camshaft
5. Cam follower
6. Oiling tube
7. Elbow fitting
8. Pushrod
9. Rocker arm shaft
10. Washer
11. Wave washer
12. End plug
13. Oil return tube
14. Rocker arm
15. Support
16. Cylinder head capscrew
17. Self-locking screw
18. Spacer spring

VALVE CLEARANCE

28. Because of the limited clearance between the head of the valve and top of piston, valve clearance (Tappet Gap) should be checked and adjusted only with engine stopped. After any service to valve system, valve clearance should be adjusted cold before engine is started. Valve clearance should be rechecked with engine at operating temperature.

Clearance Cold—
 Inlet & Exhaust ....0.018 in.
Clearance Hot—
 Inlet & Exhaust ....0.015 in.

Two-position adjustment of all valves is possible as shown in Figs. AC15A & AC15B. To make the adjustment turn the crankshaft until No. 1 cylinder is at TDC on compression stroke. Mark on crankshaft pulley (Fig. AC43) indicates TDC and if clearances are nearly correct, both front rocker arms will be loose and both rear rocker arms will be tight with No. 1 cylinder on compression stroke. Adjust the six valves indicated in Fig. AC15A to the correct clearances. Turn the crankshaft one complete revolution until TDC marks are again aligned. This will position No. 6

cylinder at TDC on compression stroke. Adjust the remaining six valves shown in Fig. AC15B.

NOTE: When adjusting valve tappet clearance, be sure to note that the intake valve rocker arm is always to the right side of rocker arm support (as viewed from left side of engine) and exhaust valve rocker arm is to left side of support. There is a rocker arm support between the rocker arms for the valves of each cylinder. Refer to Fig. AC14.

TIMING GEAR COVER AND CRANKSHAFT FRONT OIL SEAL

30. REMOVE AND REINSTALL. To remove the timing gear cover, first remove the front support casting as outlined in paragraph 7, then proceed as follows:

Remove the fan blades, fan belts and the cap screw and washer from front end of crankshaft. Using a puller threaded into the tapped holes in crankshaft pulley hub as shown in Fig. AC16, remove pulley from crankshaft. CAUTION: Do not attempt to remove pulley by pulling on outer diameter as vibration dampener built into pulley will be damaged.

It is suggested to loosen all oil pan cap screws enough to lower the front of oil pan or preferably remove oil pan. Remove the plate from cover in front of the fuel injection pump drive gear and remove the thrust plunger if so equipped. Remove screws that attach timing gear cover to front of crankcase, and carefully pry cover forward to slide seal and cover forward off of crankshaft. NOTE: Some of the attaching screws enter from

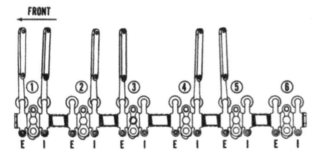

Fig. AC15A—With No. 1 cylinder at TDC on compression stroke, adjust tappet gap on valves shown. Refer to text for clearance.

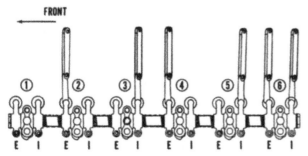

Fig. AC15B—With No. 6 cylinder at TDC on compression stroke and No. 1 cylinder at TDC of exhaust stroke, adjust valves shown. Refer to text for recommended clearance.

Fig. AC16—Do not pull crankshaft pulley by outer rim; thread pullers into hub of pulley as shown. Place correct size center plug over end of crankshaft after removing pulley retaining capscrew (CS) and washer.

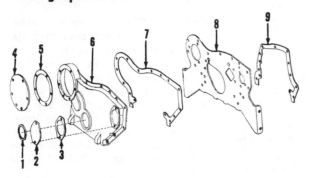

1. Crankshaft seal
2. Cover
3. Gasket
4. Cover
5. Gasket
6. Timing gear cover
7. Gasket
8. Engine front plate
9. Gasket

Fig. AC17—Drawing showing exploded view of engine front (timing gear) cover and related parts.

rear and timing gear cover is located on crankcase with two dowel pins.

Crankshaft front oil seal (1—Fig. AC17) can be renewed while timing gear cover is removed. Note: Inner diameter of seal grips crankshaft tightly and inner part of seal rotates with shaft.

When reinstalling timing gear cover, use new pan gasket front section (or complete gasket if pan was removed) and new timing gear cover gasket. Lubricate oil seals and carefully install timing gear cover over crankshaft and the two dowel pins. Tighten the socket head (grade 100) timing gear cover retaining screws to 45 Ft.-Lbs. and all other oil pan and timing gear cover cap screws to a torque of 28-33 Ft.-Lbs. Tighten the crankshaft pulley retaining cap screw to a torque of 200-220 Ft.-Lbs.

TIMING GEARS

31. VALVE TIMING. Valves are properly timed when punch marked tooth of the crankshaft gear is in register with the punch mark between two teeth on the camshaft gear as shown in Fig. AC18. Note: Ignore any timing marks found on fuel injection pump idler gears or injection pump drive gear; refer to paragraph 34 for timing procedure for injection pump gears.

32. R&R CRANKSHAFT GEAR. The crankshaft gear is keyed and press fitted to the crankshaft. The gear can be removed by using a suitable puller after first removing the timing gear cover as outlined in paragraph 30.

New gear can be installed by heating it in oil for fifteen minutes prior to installation and drifting it on the

crankshaft, or by pressing gear on crankshaft using crankshaft pulley retaining capscrew and suitable washers and spacers. Be sure timing marks on crankshaft gear and camshaft gear are aligned as outlined in paragraph 31 and retime fuel injection pump as outlined in paragraph 34.

33. R&R CAMSHAFT GEAR. The camshaft gear is keyed and press fitted to the camshaft. Gear can be removed and new gear installed in a press after camshaft is removed from engine as outlined in paragraph 36. The gear should be heated to approximately 300° F. before pressing onto camshaft. On some models, the camshaft thrust plate (3—Fig. AC15) encircles the camshaft and thrust plate must be installed before camshaft gear.

Camshaft end play is controlled by the thrust plate (3) that retains the camshaft and gear assembly in the cylinder block. Desired camshaft end play is 0.003-0.009. Renew the thrust plate if end play is excessive; if renewing thrust plate does not correct excessive camshaft end play, the camshaft and/or camshaft gear must be renewed.

DIESEL INJECTION PUMP DRIVE GEAR

34. REMOVE AND REINSTALL. The injection pump drive gear, idler gear and crankshaft gear are all provided with timing marks for correct assembling; however, because of the odd number of teeth on idler gear, the marks are aligned only every 70th revolution of the crankshaft. The injection pump drive gear can be repositioned without regard to marks on gear teeth to provide correct pump timing.

To remove the injection pump drive gear, drain cooling system, shut off fuel and remove the radiator assembly. Remove injection pump timing window (TW-Fig. AC41) and turn engine until pump timing marks are aligned as shown in Fig. AC43. (Number one piston should be 34° BTDC on compression stroke for D-21 and D-21 Series II models; 24° BTDC for Two-Ten and Two-Twenty models.) Remove cover (4—Fig. AC17) from timing gear cover and thrust plunger and spring from end of shaft. Remove nut from shaft and use a suitable puller to remove gear from tapered end of shaft.

CAUTION: If pump drive shaft is pulled out of pump, drive shaft seals (15—Fig. AC19) may be damaged. To install the drive shaft and seals, the injection pump must be

Fig. AC18—View of engine with timing gear cover removed. To time camshaft gear to crankshaft gear, align timing marks shown. Ignore any other timing marks that may be present on crankshaft gear, injection pump idler gears or injection pump drive gear. Note thrust plunger (10) in front end of injection pump driveshaft.

removed and seals installed after shaft is inserted through the adapter bushing.

To install injection pump drive gear, make certain that crankshaft is set at correct timing mark. (Number one piston should be 34° BTDC on compression stroke for D-21 and D-21 Series II models; 24° BTDC for 210 and 220 models.) Make certain that pump timing marks are aligned as shown in Fig. AC43 and install gear (8—Fig. AC19) on pump drive shaft. The key in tapered end of pump drive shaft must engage slot in gear hub. With gear in place, timing marks (Fig. AC43) must be aligned or closely aligned. Install lock washer and gear retaining nut. Nut should be tightened to 35-40 Ft.-Lbs. of torque. Check pump timing as outlined in paragraph 56.

INJECTION PUMP ADAPTER

34A. R&R AND OVERHAUL. To remove the injection pump drive adapter, first remove drive gear as outlined in paragraph 34 and injection pump in paragraph 56.

The bushing (3—Fig. AC20) should be a press fit in adapter housing (6). Journal of pump drive shaft should be 0.8735-0.8740 diameter and should have 0.001-0.0035 diametral clearance in bushing (3).

To assemble, install adapter assembly on engine front plate, making certain that oil hole (OH) is toward top, then insert pump drive shaft through adapter bushing. Install new drive shaft seals and injection pump as outlined in paragraph 57, then, install drive gear as in paragraph 34. Refer to paragraph 56 to check and adjust pump timing.

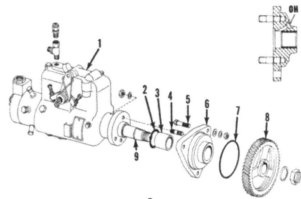

Fig. AC20 — Exploded view of fuel injection pump adapter, drive shaft and gear unit. Inset shows lubrication oil hole (OH) in adapter.

1. Injection pump
2. "O" ring
3. Bushing
4. Stud bolts
5. Bolts
6. Adapter
7. "O" ring
8. Drive gear
9. Pump drive shaft

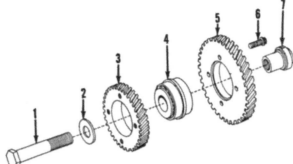

Fig. AC21 — Exploded view of fuel injection pump idler gear assembly. Idler shaft (7) is pressed into front face of crankcase as shown in Fig. AC22.

1. Capscrew
2. Retaining washer
3. Idler gear
4. Bearing assembly
5. Idler gear
6. Capscrews (4)
7. Idler shaft

DIESEL INJECTION PUMP IDLER GEARS

35. R&R AND OVERHAUL. To remove the fuel injection pump idler gears, first remove the timing gear cover as outlined in paragraph 30, remove the injection pump drive gear as outlined in paragraph 34, then remove the idler gear retaining cap screw (1—Fig. AC21) and washer (2). Pull the idler gear assembly from the idler shaft (7).

Idler shaft (7) is pressed into the cylinder block and can be removed by using a slide hammer threaded into front end of shaft. Thread in front of idler shaft is $\frac{9}{16}$ in.-18. Press or drive new idler shaft into cylinder block with flat on large diameter of shaft aligned with straight edge of engine front adapter plate as shown at (F—Fig. AC22). Oil hole in idler shaft must be aligned with oil passage (OP).

If any part of the bearing unit (4—Fig. AC21) is damaged beyond further use or the bearings are worn so that idler gear end play is excessive, the complete bearing unit must be renewed.

To reassemble, press the bearing cups into the idler gears with thick part of cup inward. Be sure that bearing cups are pressed tightly against spacer. Place a bearing cone on the idler shaft with taper forward and drive the cone back against shoulder on shaft. Align the single punch mark on camshaft and crankshaft gears, then install idler gears with large gear toward rear and the two marks on large idler gear aligned with the

two marks on crankshaft gear. Install idler gear front bearing cone, flat washer and retaining cap screw. Tighten the cap screw to a torque of 95-105 Ft.-Lbs.

Reinstall fuel injection pump drive gear, aligning the single mark on small idler gear and single mark on fuel pump drive gear. Refer to paragraph 34 and reinstall timing gear cover as outlined in paragraph 30.

HYDRAULIC PUMP DRIVE GEAR

Refer to paragraph 138 for information on the hydraulic pump drive gear.

Fig. AC19—The fuel injection pump drive shaft (9) and gear (8) unit can be withdrawn from front. "O" ring (14) fills groove left from machining the drive shaft and keeps governor from hanging up in the groove. Oil seals (15) should be renewed whenever shaft is removed.

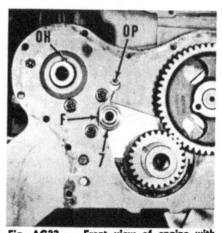

Fig. AC22 — Front view of engine with timing gear cover, injection pump drive shaft and gear unit and the injection pump idler gears removed. Pump adapter is installed with oil hole (OH) up and idler gear shaft (7) is installed with flat (F) against straight edge of engine front plate.

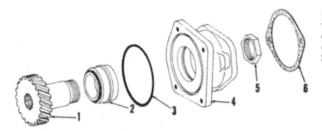

Fig. AC24 — Exploded view of hydraulic pump drive unit. Refer to paragraph 138 in Hydraulic Section for service information.

1. Gear and shaft
2. Bearing assembly
3. "O" ring
4. Adapter
5. Nut
6. Gasket

CAMSHAFT AND BUSHINGS

36. R&R CAMSHAFT. To remove the camshaft, first remove oil pan, oil pump, timing gear cover and rocker arm assembly; then, proceed as follows:

Working through holes in timing gear, bend down tabs on locking plate and remove the cap screws, locking plate and end thrust plate. Cam followers can be held in up position to facilitate camshaft removal as follows: Wooden dowels 5/8-inch diameter and 16 inches long can be used to hold cam followers up. Be sure that dowels are seated in followers, then install rubber bands over each pair of dowels to hold side pressure on cam followers. Push the cam followers up out of the way and remove camshaft. Press camshaft out of camshaft timing gear.

Backlash should be 0.0015-0.009 between camshaft gear and either mating gear. Renew parts as necessary if backlash exceeds 0.015 inch. Camshaft journal diameter (all six journals) is 2.130-2.131. Desired camshaft to bushing clearance is 0.002-0.006. Camshaft bushings should be renewed as outlined in paragraph 37 if clearance exceeds 0.008. Bushings are available in 0.010 undersize as well as standard size. If camshaft journals are excessively worn or scored and camshaft is otherwise serviceable, journals may be ground to 2.120-2.121 and the 0.010 undersize bushings installed. Cam lift is 0.318 for intake valves and 0.285 for exhaust valves. Rocker arm ratio is 1.5:1.

Camshaft end play is controlled by the 0.205-0.206 thick thrust plate (3—Fig. AC15). End play should be 0.003-0.009 inch. Renew the thrust plate if worn so that camshaft end play is excessive.

37. CAMSHAFT BUSHINGS. The camshaft is supported in six bushings. The camshaft should have 0.002-0.006 clearance in the bushings. If clearance is excessive, the camshaft and/or bushings should be renewed or the camshaft journals ground to 2.120-2.121 and 0.010 undersize bushings installed.

To renew the camshaft bushings, the engine must be removed from the tractor and the flywheel and rear adapter plate must be removed from the engine. Press old bushings out and new bushings in with suitable bushing tools. Front bushing is 1⅜-inches wide and the five rear bushings are one inch wide. Bushings have a 0.003-0.006 interference fit in bores in cylinder block. Bushings are pre-sized and should not require reaming if carefully installed. However, they should be checked after installation for localized high spots. Standard bushing diameter after installation should be 2.133-2.136; 0.010 undersize bushing inside diameter should be 2.123-2.126. Be sure that bushings are installed so that oil holes in bushings are aligned with oil passages in cylinder block. Front bushing should be installed flush or slightly below flush with front face of cylinder block. Rear bushing should be installed flush with front face of cylinder block bore.

ROD AND PISTON UNITS

38. Piston and connecting rod units are removed from above after removing cylinder head, oil pan and rod bearing caps.

Connecting rods beginning at engine serial number 3D-09649 are different than earlier type. The earlier type connecting rod can be easily identified by the serrated mating surface of rod and cap; the mating surface on later rod is smooth and flat. The new type connecting rod assembly (including cap and special screws) can be installed in earlier engines individually or as complete sets. **The cap retaining screws must be the correct type according to the rod assembly.** Early connecting rods use cap screws with six point head and a washer is installed below head of each screw. Threads of early type screws must be cleaned, then lubricated with "Molykote G" before installing. The later type connecting rods use special socket head screws with no washer and engine lubricating oil should be used to lubricate the late (socket head) screws. The tightening torque

is increased for the late type connecting rod and screws.

On all models, each rod and cap bear an assembly number in addition to the cylinder number. The cylinder number is stamped on the left (camshaft) side of connecting rod and cap, and rods are numbered 1 through 6 from front to rear. Disregard the assembly numbers.

When reinstalling rod and piston units, be sure that cylinder numbers are in register and face toward the camshaft; both bearing insert tangs must be toward same side of rod.

The six point cap screws used with early connecting rods must be thoroughly cleaned and dried; then, lubricate threads and under the head with "Molykote G" before installing. Make certain that the special hardened washer is installed under head of each screw and torque both screws in each rod evenly to 65-70 Ft.-Lbs.

The special socket head screws used with late connecting rods should be lubricated with engine oil before installing. The special screws require a male ⅜ inch hex wrench. The following procedure should be used when tightening. Tighten both screws evenly to 8-12 Ft.-Lbs. torque, then tap the lower balance pad with a plastic hammer toward front then toward rear to align cap. Tighten both screws to 40-42 Ft.-Lbs., then evenly to the final torque of 80-85 Ft.-Lbs. Side play on crankpin should be 0.005-0.010 after both screws are tightened.

PISTONS, SLEEVES (CYLINDER LINERS) AND RINGS

39. The cam ground aluminum alloy pistons are fitted with three compression rings and one oil control ring. Top compression ring is chrome plated. All rings are located above the piston pin. Pistons and rings are available in standard size only.

Install compression rings with side of ring marked "TOP" or "T" toward top of piston. When installing oil control ring, be sure that ends of expander are butted together and not overlapped.

With piston and connecting rod removed, use suitable puller to remove the wet type sleeve. Thoroughly clean all sealing and mating surfaces of block and cylinder sleeve. Insert sleeve in bore **without** sealing rings; sleeves should be free enough to be pushed into place and then be rotated by hand pressure. If sleeve cannot be inserted and turned by hand, more thorough cleaning is necessary. When sleeve is removed, check the thrust surface of the counterbore in cylinder block. If eroded or otherwise damaged, counterbore must be reseated.

Cylinder sleeve standout must be maintained at 0.002-0.005 above the cylinder block surface. To check standout, measure thickness of sleeve flange with micrometer (do not include the "fire wall" on top end of sleeve) and depth of counterbore in block with a micrometer depth gage. Then, use special shims as necessary to provide the proper 0.002-0.005 standout. Shims are available in thicknesses of 0.005, 0.010, 0.015 and 0.020. Note: Do not attempt to measure standout with straight edge and feeler gage after installing sleeve as rough edges on shims could give incorrect standout measurement.

When sleeve and bore in block are properly cleaned, install new sealing rings **without** lubricant of any kind on **dry** cylinder sleeve. On sleeves with three grooves for sealing rings, install a black packing ring in the top seal groove and red sealing rings in the two lower seal ring grooves. Note: All three sealing ring grooves are at bottom end of cylinder sleeve. Sleeves for later models have two seal grooves and two red seal rings are used. An all models, a seal ring pro-

tector should be used when installing sleeve to prevent damaging seal rings. Prior to installing sleeve, brush the block bore, then the sealing rings with light motor oil; then, **immediately** install the sleeve. The sealing ring material expands rapidly upon contact with oil. Be extremely careful that the soft rubber sealing rings are not cut on the sharp edges of cylinder block.

Check the pistons, rings and sleeves against the following values.

D-21 MODEL

Ring Side Clearance in Groove—
Top Ring 0.0065-0.0080
2nd and 3rd Rings 0.0030-0.0050
Oil Control Ring 0.0025-0.0085
Ring End Gap—
Top Ring 0.030-0.047
2nd and 3rd Rings 0.015-0.030
Oil Ring Side Rails 0.013-0.058
Piston Skirt to Sleeve
Clearance 0.0065-0.0090

D-21 Series II, 210 and 220 Models

Ring Side Clearance in Groove—
Top Ring 0.004-0.006
2nd and 3rd Rings 0.003-0.005
Oil Control Ring 0.0015-0.0035
Ring End Gap—
Top 3 (Compression)
Rings 0.013-0.033
Oil Control Ring 0.008-0.028
Piston Skirt to Sleeve
Clearance 0.0025-0.0050

PISTON PINS

40. Piston pins are retained in pistons by snap rings. Piston pin should be a tight fit in piston at room temperature. Remove and install piston pin in a pin press or by heating piston to 180° F. in hot water.

Desired clearance between piston pin and connecting rod bushing is 0.0014-0.0021. Renew the pin and/or bushing in connecting rod if clearance between pin and bushing is excessive. After installing new bushing, finish ream or hone the bushing inside diameter to provide correct clearance.

CONNECTING RODS AND BEARINGS

41. Connecting rod bearings are of the renewable precision slip-in type. The bearing inserts can be renewed after removing oil pan and bearing caps. When installing new bearing inserts (liners), be sure that liner projections (tangs) engage the milled slot in connecting rod and bearing cap and that the cylinder numbers on rod and cap are in register and face towards the camshaft side of engine. Bearing

inserts are available in standard size and undersizes of 0.002, 0.010, 0.020, 0.030 and 0.040. Check the crankshaft crankpin journals and connecting rod bearings against the following values:

Crankpin diameter,
(Std.) 2.7470-2.7485
Rod bearing clearance ..0.001-0.004
Rod side play 0.005-0.010

Refer to paragraph 38 for recommended tightening torques, installation procedures and cautions regarding the different types of connecting rods.

CRANKSHAFT AND MAIN BEARINGS

42. The crankshaft is supported in seven precision slip-in type main bearings. Main bearing inserts (liners) can be renewed after removing the oil pan, oil pump, oil tube and main bearing caps. Desired crankshaft end play is 0.007-0.013. Maximum allowable end play is 0.021. Crankshaft end play is controlled by renewable thrust flanges at each side of the center main bearing. Thrust flanges are available in standard size and oversizes of 0.005, 0.010 and 0.015. The thrust flanges are retained from turning by pins located in the center main bearing cap.

To remove the crankshaft, first remove the engine as outlined in paragraph 20. Then remove clutch, flywheel, engine rear adapter plate, crankshaft pulley, timing gear cover, injection pump drive gear and shaft, injection pump drive idler gears, oil pan, oil pump, oil tube, camshaft, engine front plate and rod and main bearing caps. Lift crankshaft from engine. Remove crankshaft gear with suitable puller.

Check crankshaft and main bearing inserts against the following specifications:

Crankpin dia., (Std.) 2.7470-2.7485
Main journal dia.,
(Std.) 3.2465-3.2480
Main bearing clearance,
desired 0.0019-0.0051
Main bearing bolt torque,
(Ft.-Lbs.) 170-190

Main bearing inserts are available in standard size and undersizes of 0.002, 0.010, 0.020, 0.030 and 0.040. Center main bearing inserts are available in a kit only containing upper and lower bearing inserts and a set of standard thickness thrust flanges and pins. The thrust flanges are also available separately from the kit. All main bearing inserts except for center main bearing are available as separate up-

Fig. AC25 — View of early type cylinder sleeve. Seal rings are placed in three grooves at bottom of sleeve. Head gasket seals top end of sleeve. Later cylinder sleeves have two grooves for seal rings.

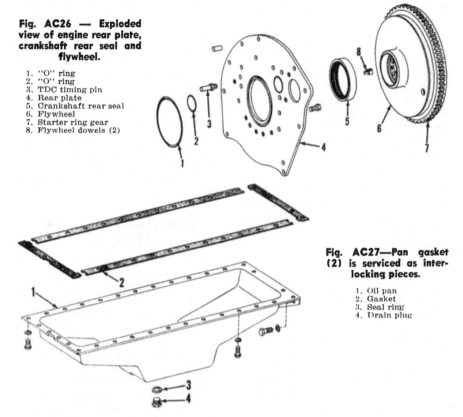

Fig. AC26 — Exploded view of engine rear plate, crankshaft rear seal and flywheel.

1. "O" ring
2. "O" ring
3. TDC timing pin
4. Rear plate
5. Crankshaft rear seal
6. Flywheel
7. Starter ring gear
8. Flywheel dowels (2)

Fig. AC27—Pan gasket (2) is serviced as interlocking pieces.

1. Oil pan
2. Gasket
3. Seal ring
4. Drain plug

OIL PAN (SUMP)

46. REMOVE AND REINSTALL. To remove the oil pan, it is first necessary to remove the front axle assembly as follows: Support the tractor under torque housing so that there is no weight on the front axle pivot points. Detach the power steering cylinder from steering arm and front axle main member; it is not necessary to disconnect the hose from the cylinder. Remove the front axle pivot pin and pivot bolt, raise front end of tractor to clear the axle assembly and roll the axle and front wheels forward out of the way. Place a safety support under the front support casting; then, drain the oil pan and unbolt and remove pan from engine.

The oil pan is sealed at the rear by a large square-cut sealing ring (1—Fig. AC26) that is located between the adapter plate and rear end of the engine. Care should be taken not to damage this sealing ring when removing and reinstalling the oil pan as removal of the clutch, flywheel and rear adapter plate is required to renew the sealing ring. Reinstall pan using new gasket. Install all pan retaining capscrews loosely, turn the pan to engine cap screws in just far enough to bring pan, gasket and cylinder block surfaces into contact. Then, tighten the pan to rear adapter plate cap screws to a torque of 28-33 Ft.-Lbs. After tightening the pan to rear adapter plate cap screws, tighten the pan to engine cap screws to a torque of 28-33 Ft.-Lbs.

per or lower half. Bearing lower halves do not have oil holes. Semifinished main bearing caps are available; after installing service bearing cap, they must be finished with line boring bar. Diameter of bearing bore without inserts is 3.5607-3.5614. Tighten bearing cap bolts to a torque of 170-190 Ft.-Lbs.

adapter plate. Tighten the adapter plate to cylinder block cap screws to a torque of 68-73 Ft.-Lbs. and the oil pan to adapter plate screws to 28-33 Ft.-Lbs. torque. Reinstall flywheel and tighten flywheel retaining cap screws to a torque of 95-105 Ft.-Lbs. Complete the reassembly of tractor by reversing disassembly procedure.

CRANKSHAFT OIL SEALS

43. FRONT SEAL. The crankshaft front oil seal is located in the timing gear cover and can be renewed as outlined in paragraph 30.

44. REAR SEAL. The crankshaft rear oil seal is installed in the adapter plate at rear of engine. To renew the seal, first remove engine from tractor as outlined in paragraph 20, then proceed as follows: Remove clutch, flywheel and engine rear adapter plate. Remove old seal from the adapter plate and install new seal with lip of seal to front. Oil seal sleeve on hub of flywheel should be renewed if nicked or worn. Beveled outside diameter of wear sleeve should be toward front (inside) of engine with flywheel installed. New sleeve should be pressed on flywheel flush with edge of flywheel hub. NOTE: Do not lubricate or heat wear sleeve when installing. Renew the two "O" rings located between the adapter plate and engine, lubricate the oil seal and reinstall the

FLYWHEEL

45. REMOVE AND REINSTALL. To remove the flywheel, first remove engine clutch as outlined in paragraph 73. The flywheel is retained to the engine crankshaft with six cap screws and two dowel pins. One of the cap screw holes is offset so that installation of flywheel is possible in one position only. The oil seal wear surface on flywheel is renewable. To renew flywheel ring gear, flywheel must be removed from crankshaft. Inspect clutch friction surface and crankshaft rear oil seal surface of flywheel. Remove ring gear after grinding notch at one of the tooth valleys. Heat new ring gear evenly to approximately 300°-325° F. and install ring gear with tooth bevel to front. Clutch pilot bearing in flywheel is a sealed unit.

When reinstalling flywheel, tighten the retaining cap screws to a torque of 95-105 Ft.-Lbs. Complete the reassembly of tractor by reversing disassembly procedure.

OIL PUMP AND RELIEF VALVES

All Models Except 210

47. R&R AND OVERHAUL PUMP. Removal procedure will be evident after removal of oil pan as outlined in paragraph 46.

To disassemble the removed pump, refer to Fig. AC28 or AC28A and proceed as follows: Remove screen (18), cover (16) and idler gear (12). Remove pin (9) from pump so equipped, then press shaft (13) out of gear. If necessary to renew idler shaft (11), press shaft out of pump body. Remove pin (4), retainer (1), spring (2) and 70-90 psi relief valve (3).

Check pump body, gears and shafts against the following specifications and renew any part excessively worn, scored or damaged.

Clearance, drive shaft
  to pump body 0.0015-0.0030
Clearance, drive shaft
  to bushing 0.0010-0.0035
Clearance, drive shaft to pump
  cover* 0.002-0.003
*Late pumps only

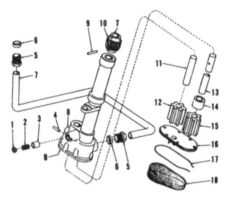

Fig. AC28—Exploded view of early type engine oil pump assembly and pressure tube. Shafts (11 & 13) do not extend into cover (16) and surge pressure relief valve (1, 2 & 3) is different than later types shown in Fig. AC28A.

1. Spring retainer	10. Drive gear
2. Spring	11. Idler shaft
3. Relief valve	12. Idler gear
4. Pin	13. Drive shaft
5. Packing nuts	14. Bushing
6. Packing seals	15. Driver gear
7. Oil pressure tube	16. Cover
8. Pump body	17. Wire
9. Pin	18. Oil screen

Clearance, idler gear
　to idler shaft 0.0005-0.0030
End clearance,
　both gears 0.0020-0.0045
Radial clearance, both gears,
　to body bore 0.00075-0.00175

If renewing bushing (14), finish bore or ream the bushing to 0.6235-0.6255 after bushing is pressed into pump body. Lower end of bushing should be flush to 0.004 recessed from surface of bore in pump body for gear (15).

To reassemble pump, press idler shaft (11) into pump body (8) and driver gear (15) onto drive shaft (13). If pump cover (16) has holes for pump shafts, idler shaft should protrude ¾-inch from surface (B) of body and lower end of drive shaft should protrude ¾-inch from lower surface of gear (15). On early pumps, cover (16) does not have holes for shafts (11 & 13) and the shafts must not hold cover away from body (8). On early models, idler shaft should be pressed into body until end is flush with surface (B) and gear (15) should be pressed onto shaft (13) until lower surface of gear is flush with end of shaft. Slight clearance between ends of shafts and cover is permissable, but make certain that shafts do not hold cover out.

On early models, gear (10) is retained by pin (9). On later models, gear is retained by a tight press fit and no pin or key is used. When assembling early type pump, press the drive gear onto shaft making certain that holes for pin are aligned, then install pin. Service shafts are not pre-drilled. Press gear onto shaft until distance from top (T) of gear to lower surface

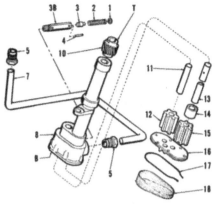

Fig. AC28A—Exploded view of late type engine oil pump. Shafts (11 & 13) extend into cover (16) and surge pressure relief valve (1 thru 4) is different than early type.

(B) of pump body is 13⅜-inches. Use hole in gear as a guide, drill a 0.125 hole through shaft and insert pin (9). On later type pump without drive pin, heat gear (10) slightly (not more than 300° F.), support end of shaft (13) and press gear onto shaft until top (T) of gear is 13⅜-inches from lower surface (B) of pump body.

Install idler gear and pump cover. Tighten cover retaining cap screws to 9-11 Ft.-Lbs. Turn the pump drive shaft and check pump for any binding condition and correct if necessary. Install screen (18), relief valve (3), spring (2), retainer (1) and pin (4).

When reinstalling the oil pump, be sure the slot in oil pump drive gear engages the drive pin for tachometer (operation meter) drive shaft on models so equipped. NOTE: On later models tachometer is driven from rear of alternator. Tighten the pump retaining cap screw to a torque of 44-49 Ft.-Lbs. Install oil tube.

210 Model

47A. R&R AND OVERHAUL PUMP. The oil pump can be removed after first removing the oil pan as outlined in paragraph 46. The pump is located on lower surface of block with dowel pins and is attached with two cap screws. Oil line brackets must be de-

tached and oil line must be detached from block before removing pump. Refer to Fig. AC28B for exploded view of pump. Drive gear (10) is pressed onto drive shaft. Pump gears and shafts (12 & 15) are available only with shafts. Relief valve (1 thru 4) is for surge pressure only.

All Models

48. RELIEF VALVES. Two oil pressure relief valves are used. The relief valve (3—Fig. AC28, AC28A or AC-28B) in the oil pump body is non-adjustable and should by-pass oil at 82-98 psi. As this is well above normal oil pressure of 40-45 psi, valve (3) opens only due to surge pressure when engine oil is cold.

The relief valve (Fig. AC29) located on the right hand side of the engine just to the rear of the engine oil filters is adjustable to control normal engine oil pressure of 40-45 psi. To adjust the relief valve, loosen lock nut (25) and turn adjusting screw (26) in or out until oil pressure is 40-45 psi with engine running at 2200 RPM and with oil at normal operating temperature (180°-200° F.); oil pressure at slow idle speed of 700 RPM should then be approximately 20 psi. Tractor is equipped with an oil pressure

Fig. AC29—View showing location of oil pressure adjusting screw (26). Loosen nut (25) and turn screw in to increase pressure.

Fig. AC28B — Exploded view of engine oil pump used on Two-Ten model tractors.

1. Spring retainer	
2. Spring	
3. Surge relief valve	
4. Pin	
5. Packing nut	
6. "O" rings	
7. Pressure tube	
8. Pump body	
10. Drive gear	
12. Idler gear and shaft	
15. Driver gear and shaft	
16. Cover	
18. Screen and inlet tube	

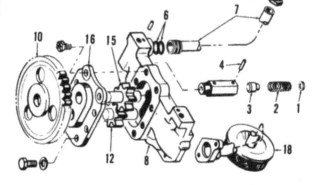

warning light instead of a pressure gage; therefore, a master pressure gage must be installed in place of the oil pressure sending switch when checking or adjusting system pressure. Adjust relief valve only when engine is running at normal operating temperature and speed. Do not attempt to adjust relief valve to compensate for worn oil pump or engine bearings.

OIL COOLER

The turbocharged engines used are equipped with an engine oil cooler. It is important for the oil cooler to operate correctly for proper cooling. Overheating of the oil will usually be indicated by low oil pressure.

49. R&R AND OVERHAUL. Drain cooling system and remove plug (1—Fig. AC31) from bottom of oil cooler. Remove fuel lines (2 & 3), then remove complete fuel filter assembly (4) including filter base. Remove front oil filter element (5) and turbocharger oil drain tube. Oil cooler can now be unbolted and removed from side of cylinder block. Disassembly can be accomplished by removing the cap screws that attach the parts together. The water tube (7) should be withdrawn for cleaning and renewal of "O" rings.

The oil passages can be soaked in cleaning solvent; however, it is **difficult** to determine when passages are clean. If badly clogged, the oil cooler core should be renewed. If cooler is cleaned with solvent, be sure to flush thoroughly before installing.

Two steel plates can be locally manufactured for pressure testing oil cooler. One of the test plates (P1—Fig. AC32) must have air hose connected for pressurizing the oil passage of cooler. Clamp test plates (P1 & P2) over oil passages, attach air hose (H) and check for leaks with approximately 200 psi air pressure. Make certain that "O" rings (5—Fig.

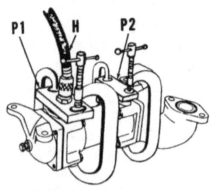

Fig. AC32—Oil cooler can be checked for leaks using plates (P1 & P2). One of the plates must have connector for attachment of air hose (H).

AC33) are installed under test plates and that air connections are not leaking. Continue to test with unit submerged in hot water until the cooler is heated to approximately 150° F. Cooler should be renewed if core is leaking.

When assembling, notch (N) or word "TOP" on the core (4) should be toward top. Bridge of gasket (6), header (7) and core (4) should be aligned, then install the four attaching screws loosely. If the unit can be assembled on a surface plate, make certain that surfaces (A, B, C & D) are flat against surface plate when screws are tightened to 28-33 Ft.-Lbs. torque. Install oil cooler bonnet (2) and gasket (3). Flanges (E) should be flat against surface plate when surfaces (A, B, C & D) are against plate and bonnet attaching screws are tightened to 28-33 Ft.-Lbs. torque. When assembled correctly, all of the machined surfaces (A, B, C, D & E) should be on same plane within 0.005 inch. Leakage or damage may result if not correctly aligned.

If surface plate is not available, tighten the eight oil cooler assembly

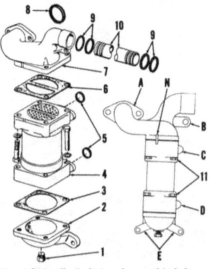

Fig. AC33—Exploded and assembled drawing of oil cooler. Machined surfaces (A, B, C, D & E) must be on same plane when unit is assembled. Notch (N) or word "TOP" indicates correct assembly of core (4).

1. Coolant drain plug	6. Gasket
2. Bonnet	7. Header
3. Gasket	8. "O" ring
4. Core	9. "O" rings
5. "O" rings	10. Tube
	11. Assembly screws

screws (11) just enough to hold assembly together. Clean mounting surfaces on cylinder block and install oil cooler on cylinder block without "O" rings (5 & 8) and tube (10). Tighten the five oil cooler mounting screws, then tap cooler core (4) toward block until the machined surfaces (C & D) are flat against block. Loosen the two lower mounting screws that attach bonnet (2) to cylinder block, then tighten the oil cooler assembly screws (11) to 28-33 Ft.-Lbs. torque. Remove oil cooler and reinstall using new "O" rings (5, 8 & 9). Recommended torque for the oil cooler to cylinder block screws is 44-49 Ft.-Lbs. for $\frac{7}{16}$-inch screws; 68-73 Ft.-Lbs. for ½-inch screws.

Fig. AC31—View showing location of oil cooler on turbocharged engines. Refer to text.

DIESEL FUEL SYSTEM

The diesel fuel system consists of three basic units; the fuel filter, injection pump and injection nozzles. When servicing any unit associated with the fuel system, the maintenance of absolute cleanliness is of utmost importance.

Probably the most important precaution that servicing personnel can impart to owners of diesel powered tractors is to urge them to use an approved fuel that is absolutely clean and free from foreign material. Extra precaution should be taken to make certain that no water enters the fuel storage tanks. This last precaution is based on the fact that all diesel fuels contain some sulphur. When water is mixed with sulphur, an acid is formed which will quickly erode

the closely fitting parts of the injection pump and fuel injection nozzles.

FUEL TANK

50. R&R FUEL TANK. To remove fuel tank, the tank must first be drained and all fuel lines disconnected. On D-21 and D-21 Series II, remove cover from fuel gage sender switch and disconnect gage wire from sender switch terminal. On all models, unbolt and remove tank from tractor. (In some instances, mechanic may prefer to remove operator's seat assembly from tank before removing tank from tractor.)

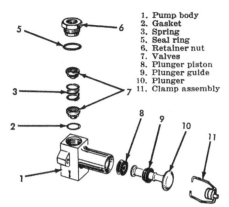

1. Pump body
2. Gasket
3. Spring
5. Seal ring
6. Retainer nut
7. Valves
8. Plunger piston
9. Plunger guide
10. Plunger
11. Clamp assembly

Fig. AC35—Exploded view of the Roosa-Master fuel primer pump used. Refer to Fig. AC10 for location of pump on engine.

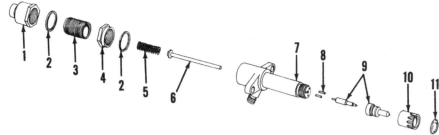

Fig. AC36—Exploded view of typical fuel injector nozzle and holder assembly.

1. Cap	4. Lock nut	7. Holder body	9. Nozzle
2. Copper washer	5. Spring	8. Dowel pins	10. Nut
3. Adjusting screw	6. Spindle		11. Copper washer

NOTE: On D-21 tractors equipped with a 3-point hitch, it may be necessary to remove right rear wheel guard and the "Traction-Booster" valve from control valve assembly. Then unbolt "Traction-Booster" and position control lever bracket from hydraulic manifold to gain clearance for loosening right front tank retaining nut.

FILTERS AND BLEEDING

The fuel filtering system consists of a fuel strainer and sediment bowl and a single unit throw-away type filter.

51. BLEEDING. Each time the filter element is renewed, or the fuel lines are disconnected, it will be necessary to bleed air from the system as follows:

To bleed the filter, remove the pipe plug from filter base, open the fuel shut-off valve and operate the primer pump (See Fig. AC35) until fuel flows freely and all air has escaped. Then, reinstall the pipe plug.

Normally, the injection pump is self-bleeding. However, it may be necessary to bleed the pump and injector pressure lines as follows: Loosen the inlet line at pump and operate the primer pump until stream of fuel from loosened connection is free of air bubbles. Then, tighten the connection and continue to operate the primer pump for a while.

Loosen the high pressure fuel line connections at all injectors and crank the engine with starting motor until fuel appears at all injectors. Tighten the fuel line connections and start the engine.

52. FILTERS. The fuel sediment bowl at the fuel shut-off valve should be checked daily or after each 10 hours of operation and the bowl and fuel strainer screen cleaned as necessary.

The throw-away type filtering element should be renewed every 500 hours of operation. Poor fuel handling and storage facilities will decrease the effective life of filter elements; conversely, clean fuel will increase the life of filter elements. Filter elements should never remain in the fuel filtering system until a decrease in power or engine speed is noticed because some dirt may enter the pump and/or fuel injector nozzles and result in severe damage.

INJECTION NOZZLES

A direct type fuel injection system is used. The injectors each have 4 spray orifices.

WARNING: Fuel leaves the injection nozzles with sufficient force to penetrate the skin which could cause blood poisoning. When testing nozzles, keep clear of the nozzle spray.

53. TESTING AND LOCATING A FAULTY NOZZLE. If rough or uneven engine operation, or misfiring, indicates a faulty injector, the defective unit can usually be located as follows:

With engine running at low idle speed, loosen the high pressure connection at each injector in turn. As in checking spark plugs, the faulty unit is the one which, when its line is loosened, least affects the running of the engine.

If a faulty nozzle is found and considerable time has elapsed since the injectors have been serviced it, is recommended that all injectors be removed and serviced or that new or reconditioned units be installed.

Remove the suspected nozzle as outlined in paragraph 54, place nozzle in test stand. Check the nozzle against the following specifications:

D-21 With 3400 Engine
Opening pressure ....2750 psi
Nozzle should not leak at 200 psi less than opening pressure.

D-21 Series II with turbocharged 3500 Engine
Opening pressure ....2900 psi
Nozzle should not leak at 200 psi less than opening pressure.

Two-Ten and Two-Twenty with 3500 Engine.
Opening pressure ....3100-3150 psi
Nozzle should not leak at 200 psi less than opening pressure.

Operate the tester handle at approximately 100 strokes per minute and observe the spray pattern. All four sprays must be similar and well atomized. If spray pattern is not as described, overhaul the nozzle as outlined in paragraph 55.

54. REMOVE AND INSTALL INJECTORS. Before loosening any lines, wash the nozzle holder and connections with clean diesel fuel or kerosene. After disconnecting the high pressure and leakoff lines, cover open ends of connections with composition caps to prevent the entrance of dirt or other foreign material. Remove the nozzle holder stud nuts and carefully withdraw the nozzles from cylinder head, being careful not to strike the tip end of nozzle against any hard surface.

Thoroughly clean the nozzle recess in the cylinder head before reinstalling the nozzle and holder assembly. Use only wood or brass cleaning tools to avoid damage to seating surfaces and make sure the recess is free of all dirt and carbon. Even a small particle could cause the unit to be cocked and result in a compression loss and improper cooling of the fuel injector.

When reinstalling the injector, always renew the copper gasket (11—Fig. AC36). Install washer with cup side up. Torque each of the two nozzle holder stud nuts in 2 Ft.-Lb. increments until each reaches the final torque of 9-12 Ft.-Lbs. This method of tightening will prevent holder from being cocked in the bore.

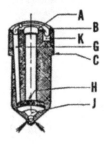
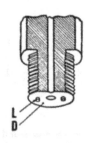

Fig. AC37—Drawing showing text reference points on injector nozzle body and holder body.

A. Pressure surface
B. Pressure surface
C. Shoulder
D. Pressure surface
G. Fuel passage
H. Fuel gallery
J. Needle seat
K. Hole for dowel
L. Dowel

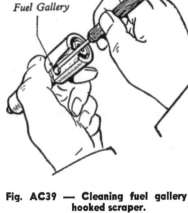

Fig. AC39 — Cleaning fuel gallery with hooked scraper.

Fig. AC38 — Cleaning fuel passage in nozzle body.

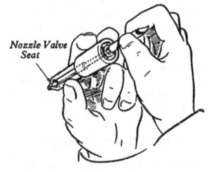

Fig. AC40 — Scraping carbon from needle valve seat.

55. OVERHAUL. Hard or sharp tools, emery cloth, crocus cloth, grinding compounds or abrasives of any kind should NEVER be used in the cleaning of nozzles.

Wipe all dirt and loose carbon from the injector assembly with a clean, lint free cloth. Carefully clamp injector assembly in a soft jawed vice or injector fixture and remove the protecting cap (1—Fig. AC36). Loosen the jam nut (4) and back off the adjusting screw (3) enough to relieve the load from spring (5). Remove the nozzle cap nut (10) and nozzle assembly (9). Normally, the nozzle valve needle can easily be withdrawn from the nozzle body. It it cannot, soak the assembly in suitable solvent to facilitate removal. Be careful not to permit the valve or body to come in contact with any hard surface.

If more than one injector is being serviced, keep the component parts of each injector separate from the others by placing them in a clean compartmented pan covered with fuel oil or solvent. Examine the nozzle body and remove any carbon deposits from exterior surfaces using a brass wire brush. The nozzle body must be in good condition and not blued due to overheating.

All polished surfaces should be relatively bright without scratches or dull patches. Pressure surfaces (A, B and D—Fig. AC37) must be absolutely clean and free from nicks, scratches or foreign material as these surfaces must register together to form a high pressure joint.

Clean out the small fuel feed channels (G), using a small diameter wire as shown in Fig. AC38. Insert the special groove scraper into nozzle body until nose of scraper locates in the fuel gallery. Press nose of scraper hard against side of cavity and rotate scraper to clean all carbon deposits from the gallery as shown in Fig. AC39. Using seat scraper, clean all carbon from valve seat (J—Fig. AC-37) by rotating and pressing on the scraper as shown in Fig. AC40.

Using a pin vise with the correct size cleaning wire, thoroughly clean the four spray holes in the nozzle body end. Cleaning wire should be slightly smaller than spray holes.

Examine the stem and seat end of the nozzle valve and remove any carbon deposit using a clean, lint free cloth. Use extreme care, however, as any burr or small scratch may cause valve leakage or spray pattern distortion. If valve seat has a dull circumferential ring indicating wear or pit-

ting, or if valve is blued, the valve and body should be turned over to an authorized diesel service station for possible overhaul.

Before reassembling, thoroughly rinse all parts in clean diesel fuel and make certain that all carbon is removed from the nozzle holder nut. Assemble parts while still wet with diesel fuel. Install nozzle assembly and cap nut making certain that the valve stem is located in the hole of the holder body and the two dowel pins (8—Fig. AC36) enter holes in nozzle body. Tighten the holder nut to a torque of 40-60 Ft.-Lbs.

Install the spindle (6), spring (5), adjusting screw (3) and lock nut (4) using a new copper washer (2). Connect the injector to a nozzle tester and adjust opening pressure to 2750 psi for D-21 models with 3400 engines, 2900 psi for D-21 Series II with 3500 engine or 3125 psi for 210 and 220 models. Use new copper gasket and install cap nut (1) to torque of 75-90 Ft.-Lbs. Recheck nozzle opening pressure to be sure adjustment was not changed by tightening the lock nut and cap nut.

Retest the injector and renew the nozzle and needle assembly (9) if still faulty.

INJECTION PUMP

56. PUMP TIMING. To check or adjust pump timing, proceed as follows: Shut off the fuel supply and remove timing window cover (See Fig. AC41) from injection pump. Turn engine slowly in normal direction of travel until timing mark on governor weight assembly appears at bottom of timing window in pump housing. Then, continue turning engine slowly until the correct timing mark on crankshaft pulley is aligned with pointer on engine front cover as shown in Fig. AC42. **Correct timing is 34° BTDC for D-21 and D-21 Series II; 24° BTDC for 210 and 220 models.** Timing marks on governor weight re-

Fig. AC41 — Timing window (TW) cover location on fuel injection pump.

tainer and cam in injection pump should then be aligned as shown in Fig. AC43. If not, loosen the injection pump mounting stud nuts, turn pump housing so that pump timing marks are aligned and retighten the stud nuts. Recheck timing by backing the engine up ⅛ to ¼ turn, then turning the engine slowly to the correct timing mark. Check to see that timing marks are aligned in pump timing window and readjust timing if they are not aligned.

Note: Timing procedure as outlined assumes normal timing check and adjustment on engine that is in operative condition; if engine will not start and improper installation of fuel injection pump is suspected, check pump timing as follows: Remove rocker arm

Fig. AC42 — View showing location of timing pointer (TP) on engine front (timing gear) cover. Align proper degree mark on crankshaft pulley (CP) with pointer.

Fig. AC43 — Align timing marks on fuel injection pump governor weight retainer and cam as shown. Marks are visible after removing timing window cover (TW—Fig. AC41) and rotating pump to bring mark on weight retainer in alignment with mark on cam.

cover or loosen the No. 1 cylinder injector. Turn engine slowly until both the intake and exhaust valves are closed on No. 1 cylinder or corpression leak occurs around the loosened injector. Then, continue to turn engine slowly until the correct timing mark on crankshaft pulley is aligned with pointer on engine front cover. Shut off the fuel supply valve and remove timing window cover from fuel injection pump. The pump timing marks should then be aligned as shown in Fig. AC43. If both timing marks are visible, but not aligned, loosen the pump mounting stud nuts and turn pump body to bring marks into alignment. If timing mark on governor weight retainer is not visible in timing window, or pump cannot be rotated far enough in mounting slots to bring timing marks into alignment, refer to paragraphs 34 and 57 for proper installation and timing of fuel injection pump and pump drive gear.

57. R&R FUEL INJECTION PUMP. Remove the lower side panel from right side of hood and on models so equipped, remove the engine oil cooler as outlined in paragraph 49. Before loosening any connections, thoroughly clean the pump fuel lines and fittings with diesel fuel or petroleum solvent; then, proceed as follows: Shut off the fuel supply, remove timing window

cover from fuel injection pump and turn engine so that the correct timing mark (34° BTDC for D-21 and D-21 Series II; 24° BTDC for 210 and 220 models) on crankshaft pulley is aligned with pointer on engine front cover and injection pump timing marks are aligned in timing window as shown in Fig. AC43. Disconnect the fuel supply line, throttle rod, shut off wire (or cable) and fuel return line from pump. Remove the two clamps from the high pressure (nozzle) lines; then, remove the line one at a time from pump and nozzles in order that they were retained in the front clamp. Remove the pump mounting stud nuts and slide pump from pump drive shaft.

Before reinstalling pump, check to be sure that engine has not been turned while pump was removed; the correct timing mark on crankshaft pulley should be aligned with pointer on engine front cover and slot in rear end of pump drive shaft (See Fig. AC45) should be in horizontal position and pointing away from engine. Install new lip seals and "O" ring (See Fig AC46) on injection pump drive shaft and apply lubricant such as Lubriplate to seals. Remove the timing window cover from outside (side with nameplate) of pump housing and turn the pump so that timing marks are aligned. Carefully work the pump

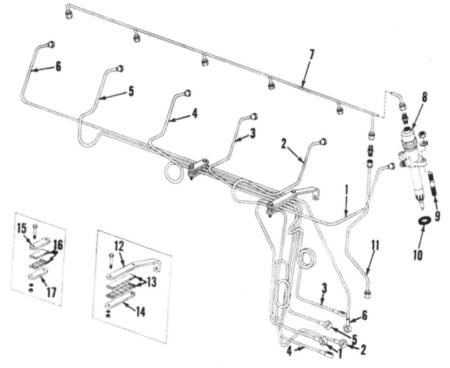

Fig. AC44 — Drawing showing layout of fuel injection lines. Pump to injector pressure lines are numbered 1 through 6 corresponding to engine cylinder number.

7. Injector return line
8. Fuel injector
9. Injector stud bolts (12)
10. Copper washer

11. Pump return line
12. Bracket
13. Cushions

14. Clamp
15. Clamp
16. Cushinons
17. Clamp

on shaft over the seals and attach pump to adapter with mounting studs and nuts so that the timing marks are in alignment. CAUTION: Be sure that lip of rear seal on pump shaft is not rolled or bent back as pump is worked onto shaft. If this occurs, remove pump and renew seal as the seal will have been damaged and early failure may result. Installation of pump on drive shaft can be simplified by using special seal compressor available from Roosa-Master.

Reconnect the high pressure lines in reverse order from which they were removed leaving the fittings loose both at nozzle and at pump. Connect the throttle rod, fuel supply line, fuel return line and the shut-off wire (or cable).

On D-21 and D-21 Series II, turn key switch on and off to be sure solenoid is working; a click should be heard from within the pump as the key switch is actuated.

On 210 and 220 models, adjust the fuel shut-off cable to provide complete operation of the shut-off lever on pump.

On all models, open the fuel shut-off valve and prime the system with primer pump until fuel can be heard re-entering the fuel tank through the fuel return line. Crank engine until fuel is ejected from the loose fittings at pump, tighten line fittings at pump and crank engine until fuel is ejected from the loose fittings at nozzles. Note: On D-21 and D-21 Series II, key switch must be in "ON" position to energize the fuel shut-off solenoid. On 210 and 220 models, fuel shut-off knob must be pulled out to operate. After tightening the fittings at nozzles, engine should start. Reinstall the two clamps on high pressure (nozzle) lines as shown in Fig. AC44. Check throttle lever travel at pump and equalize lever overtravel by repositioning throttle rod in trunnion on injection pump lever if necessary. Throttle lever on pump is spring loaded.

58. SPEED ADJUSTMENTS. Actuate the throttle hand lever and observe over-travel of the spring loaded throttle arm after both the idle adjustment and high speed adjustment screws contact the stops. Over-travel in each direction should be approximately equal; if not, loosen the screw (14—Fig. AC48) in pump lever trunion and readjust position of the throttle rod to obtain equal overtravel. Then, check and adjust speed adjustments as follows:

Start the engine and move hand throttle lever to low idle speed position; engine low idle speed should be 650-700 RPM. Loosen the locknut on idle speed adjustment screw (L—Fig. AC47) and turn screw in to increase or out to decrease idle speed. Move the hand throttle lever to high idle speed position and adjust the high idle speed screw (H) to obtain correct high idle speed. High idle RPM should be 2375-2425 for D-21 and D-21 Series II; 2390-2420 for 210 and 220 models.

NOTE: If tractor is not equipped with an hourmeter, engine speed can be checked at PTO shaft. PTO speed at 2400 engine speed is 1089 RPM; PTO speed at 700 RPM engine speed is 318 RPM. If tractor is not equipped with either an hourmeter or PTO, a stroboscopic tachometer must be used or the optional equipment hourmeter be installed.

Fig. AC47 — View of fuel injection pump showing location of torque screw (T), low idle speed screw (L) and high idle speed screw (H).

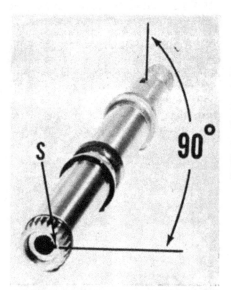

Fig. AC45 — Slot (S) in rear end of pump drive shaft must be aligned with tang on pin in pump rotor so that splines on shaft will engage splines in pump rotor. Slot should be at 90° to keyway at front end of shaft.

Fig. AC46 — View of injection pump drive shaft and gear unit showing proper installation of seals (15) and "O" ring (14).

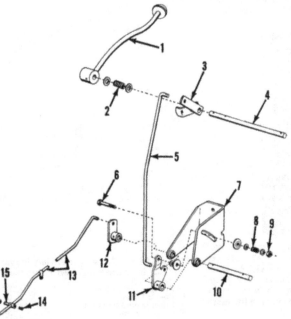

Fig. AC48 — Exploded view of typical throttle linkage. Adjust tension on spring (8) by holding bolt (6) and turning nut (9) so that throttle lever (1) will stay in set position. Adjust position of trunnion (15) on throttle rod (13) so that over-travel of spring loaded throttle lever on fuel injection pump is equal in each direction.

1. Throttle lever
2. Spring
3. Upper lever
4. Upper shaft
5. Lever to bellcrank rod
6. Friction bolt
7. Bracket
8. Friction spring
9. Adjusting nut
10. Lower shaft
11. Bellcrank
12. Lower lever
13. Throttle rod
14. Set screw
15. Trunnion

TURBOCHARGER

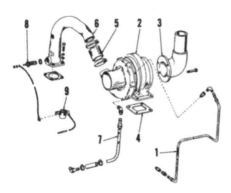

Fig. AC49 — View of Thompson turbocharger unit showing position of pressure line (1) and return line (7).

1. Lubrication pressure line
2. Turbocharger
3. Exhaust elbow
4. Gasket
5. Hose
6. Tube to inlet manifold
7. Oil return line
8. Air heater element
9. Heater solenoid

Engine model 3500 used in D-21 Series II tractors is equipped with a Thompson exhaust driven turbocharger. Engine model 3500 used in Two-Twenty tractors is usually equipped with an AiResearch exhaust driven turbocharger; however, some early models may be equipped with the Thompson unit. All Two-Ten models ar equipped with AiResearch type. External differences will be readily apparent.

TURBOCHARGER UNIT

All Models So Equipped

60. **REMOVE AND REINSTALL.** Remove hood, air inlet tube (from air cleaner), air outlet tube (6—Fig. AC49 or AC49A) and exhaust outlet elbow (3). Disconnect oil inlet and return lines (1 & 7), then unbolt and remove unit from exhaust manifold.

To reinstall, reverse removal procedure. Leave oil return line (7) disconnected and crank engine with starter until oil begins to flow from return port. NOTE: Do not start engine until it is certain that turbocharger is receiving lubricating oil. Tighten stud nuts attaching turbocharger to exhaust manifold to 18-21 Ft.-Lbs. torque. Exhaust elbow mounting screws should be tightened to 28-33 Ft.-Lbs. torque.

Thompson Turbocharger

61. **OVERHAUL.** When servicing turbocharger as originally installed on D-21 Series II tractors before engine serial number 3D-05886, it is advisable to use special tools available from

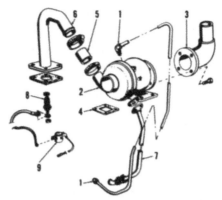

Fig. AC49A—View of AiResearch turbocharger unit showing position of pressure line (1) and return line (7).

Kent-Moore Organization, Inc. Some of the early turbochargers have been changed by installing later type parts and the special tools are not required. Attention will be called to difference in service procedure.

Remove turbocharger unit as outlined in paragraph 60. Scribe a mark across turbine housing (22—Fig. AC50), clamp (20), bearing housing (13) and compressor housing (1) to aid reassembly and proceed as follows:

Remove nut (19) and clamp (20), then remove turbine housing (22). Remove cap screws attaching bearing housing (13) to compressor housing (1) and remove compressor housing. Hold blades of turbine wheel (18) with a shop towel, remove nut (3) and washer (4) from turbine shaft (L.H. threads). On early models, impeller (5E) is a tight fit on turbine shaft (18) and shaft must be pressed out of impeller as shown in Fig. 50A. On later model, impeller (5—Fig. AC 50) will slide easily off shaft.

CAUTION: On all models, do not allow turbine wheel and shaft to drop when impeller is removed.

Remove snap ring (6), then remove oil retainer plate (7 or 7E). On early type, retainer plate (7E) must be pressed out of bore using the Kent-Moore special tool. On later type (7), a ¾-inch driver can be used as shown in Fig. AC 50B to remove seal and retainer plate (7—Fig. AC50), mating piece (10) and bearing (12).

It is recommended that late type seal (7) and mating ring (10) be renewed each time turbocharger is dis-

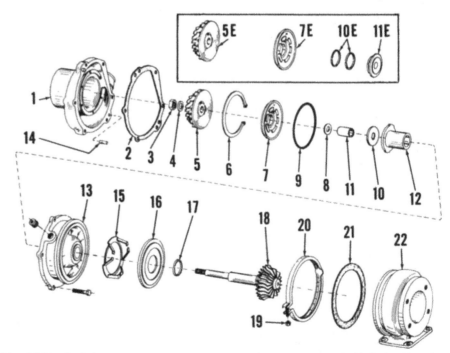

Fig. AC50—Exploded view of Thompson turbocharger used on late D-21 Series II and early Two-Twenty models. Inset shows parts that are different on earlier unit.

1. Compressor housing
2. Gasket
5 & 5E. Impeller
6. Snap ring
7. Oil seal & retaining plate
7E. Oil retaining plate
8. Shims
9. "O" ring
10. Mating ring
10E. Turbine shaft seals
11. Shaft sleeve
11E. Turbine shaft spacer
12. Bearing
13. Bearing housing
14. Groove pins
15. Spring ring
16. Turbine wheel shield
17. Turbine shaft seal
18. Turbine wheel & shaft
20. Clamp
21. Gasket
22. Turbine housing

assembled for service. The early type turbocharger can be changed to later type by installing new type parts (5, 7, 10 & 11).

Clean all parts except late type oil seal (7) by washing in kerosene or diesel fuel. A nylon bristle brush may be used to clean carbon from parts. CAUTION: Do not use wire brush, caustic cleaners, etc., on turbocharger parts. Inspect all parts for burring, eroding, nicks, breaks, scoring, excessive carbon build-up or other defects and renew all questionable parts.

On early type turbocharger, compressor impeller (5E) should be press fit on turbine shaft (18). Thrust surfaces of spacer (11E) must not be rough. Inspect grooves in spacer (11E) for excessive wear and make certain carbon is removed from bottom of grooves. Check O.D. of retainer plate (7E) for nicks that would damage "O" ring (9). Inspect bore in retainer plate for grooving caused by seal rings (10E) sticking and turning with shaft. Clean the bore chamfer to ease installation of seals (10E).

Fig. AC50A—View showing method of removing early type impeller using special Kent-Moore tool.

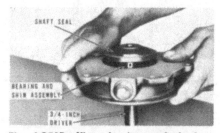

Fig. AC50B—View showing method of removing oil seal and retainer from late type turbocharger. Parts should move out with only a small amount of pressure.

On late type turbocharger, inspect seal (7) carefully if the seal is to be reinstalled. The carbon face insert should move freely in and out and must not be scored or excessively worn. Mating ring (10) should not show evidence of wear or scuffing on either side. Remove any carbon from seal contact side. Examine sleeve (11) for burrs, scoring and wear. The sleeve is precision ground and any defect will distort the mating ring and cause seal leakage.

On all models, pay particular attention to blades of turbine wheel (18) and compressor impeller (5 or 5E) and to bushing (12). If bushing (12) and/or turbine shaft is worn excessively, the blades of turbine wheel and compressor impeller may have rubbed against housings. Shaft clearance in bushing should be measured at compressor impeller end of shaft using a dial indicator with unit assembled. If indicated clearance exceeds 0.022, bushing (12) and/or shaft (18) should be renewed. Inspect bore in bearing housing (13) for evidence of stuck seal ring (17). If seal ring was sticking in shaft groove, the bore in housing will be grooved and should be renewed. Make certain that seal bore chamfer is clean and smooth to ease installation of shaft and seal ring.

To reassemble turbocharger, proceed as follows: Assemble bearing (12), mating ring (10) and sleeve (11) or bearing (12), and spacer (11E) on turbine shaft (18). Hold sleeve and mating ring (or spacer) tight against shoulder of turbine shaft and check end play of bearing with feeler gage as shown in Fig. AC 50C. End play should be 0.004-0.006. If clearance exceeds 0.006, bearing (12—Fig. AC50) should be renewed. **Record amount of end play for use in later step in reassembly.** Remove parts from turbine shaft.

Lubricate bearing (12) with motor oil and position in bearing housing bore with tabs on bearing between lugs in housing. Lubricate "O" ring (9) with silicone lubricant (or light grease) and position in bottom of retainer bore in bearing housing (13).

On early type, lubricate grooves in spacer (11E) and install new seal rings (10E) with end gaps opposite each other. Carefully slide spacer (11E) with seal rings (10E) into bore of retainer (7E). Install retainer and seal assembly in bearing retainer bore with one lug centered over each of the two tabs on bearing (12). Press retainer lightly into "O" ring (9) and

install snap ring (6) with tapered side out.

On later type, lubricate mating ring (10) and position over hole in bearing (12). Install oil seal and retainer (7) in bearing retainer bore, press oil seal and retainer down lightly into "O" ring (9) and install snap ring (6) with tapered side out. When correctly assembled, lugs on inside of retainer will hold mating ring (10) in position. CAUTION: Do not damage mating ring, "O" ring, seal and retainer by attempting to force installation with parts misaligned.

Check clearance between bearing and retainer as follows: Insert turbine shaft (18—Fig. AC50) through bearing (12) and mating ring (10) or bearing and spacer (11E). NOTE: Seal ring (17), spring ring (15) and shield (16) should not be installed for this check. On late type, install sleeve (11) over end of shaft. On all models, attach a dial indicator as shown in Fig. AC50D. Measure and record the total amount of end play measured at sleeve (11—Fig. AC50) or spacer (11E). If total end play is not within limits of 0.005-0.009, check for incorrect assembly. The end play of shaft in bearing (previously measured Fig. AC50C)

Fig. AC50C—Measure turbine shaft end play with feeler gage. Early type is shown.

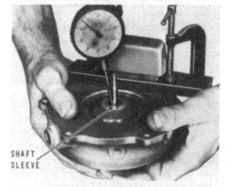

Fig. AC50D — Total end play of turbine shaft in bearing and bearing housing can be measured as shown. Late type turbocharger is shown but early type is similarly measured. It is usually necessary to push sleeve or spacer down, then push up on turbine shaft when measuring.

should be 0.004-0.006 and end play of bearing between retainer and housing should be 0.001-0.003. If end play measurement is again within limits, **record the total shaft and bearing end play for use when selecting shims (8).** Remove the turbine shaft (18) and proceed as follows:

Position impeller (5) over oil seal and retainer (7 or 7E). On late models, make certain that sleeve (11) is installed. On all models, install compressor housing (1) using a new gasket and three of the attaching screws torqued to 80 inch-pounds. Check end play of impeller in housing as shown in Fig. AC50E. Subtract total shaft and bearing end play (found in preceding paragraph) from impeller end play. In final assembly, add shims (8—Fig. AC50) of total thickness of 0.015-0.020 less than this determined value. As an example, if impeller to housing clearance is 0.049 and total shaft and bearing end play is 0.006; Subtract as follows:

Impeller end clearance		0.049
Shaft and bearing end play		−0.006
		0.043
Determined value	0.043	0.043
Desired impeller clearance	−0.015	−0.020
Shim thickness required	0.028	0.023

Shims (8) are available in thicknesses of 0.010 and 0.015. The addition of one 0.010 and one 0.015 thick shim will be within the range of the shim thickness required in the preceding example.

Remove compressor housing (1) and impeller (5). Lubricate groove in tur-

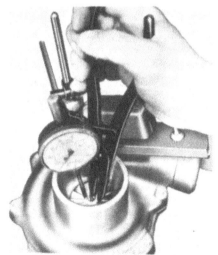

Fig. AC50E—Measure impeller clearance in housing as shown. Turbine shaft is not installed when checking this clearance.

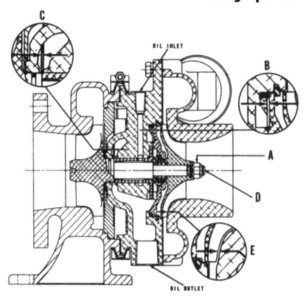

Fig. AC50G—Cross section of Thompson turbocharger. Shaft side play in bearing, measured at (A) should not exceed 0.022. Shaft end play in bearing (C) is measured as shown in Fig. AC50C. Total shaft end play (D) is measured as shown in Fig. AC50D. End clearance of bearing in housing (B) is difference of end play (D & C). Impeller to compressor housing clearance (E) is measured as shown in Fig. AC50E.

bine shaft (18) with motor oil and install new seal ring (17) in groove. Place plastic seal compressor (furnished with seal ring) over seal ring. Install spring ring (15) in bearing housing and place shield (16) over spring with projections on shield against the flat sections of the spring. Use two small "C" clamps to compress spring and hold shield in place. Insert turbine shaft (18) through hole in shield and bore of bearing. Plastic band used to compress seal ring (17) will be pushed off of seal ring as shaft is inserted in bearing and plastic will disintegrate from heat as soon as engine is started. Rotate turbine shaft to be sure it is a free fit in bearing. Place shim pack (8) (of thickness determined in previous step) on turbine shaft; then, install impeller onto turbine shaft. On early type, impeller must be pressed onto turbine shaft. On all models, install washer (4) and nut (3) on turbine shaft (L.H. threads) and tighten nut to a torque of 80-100 inch-pounds while holding turbine with shop towel. Check turbine shaft assembly for free rotation.

Reinstall compressor housing (1) and gasket (2) to bearing housing and tighten cap screws to a torque of 80-100 inch-pounds. Install turbine housing and gasket (21). Tighten clamp screw nut to a torque of 15-20 inch-pounds. Remove plug from bearing housing and fill reservoir with same type oil as used in engine. Protect all openings of turbocharger until unit is installed on tractor.

AiResearch Turbocharger

62. **OVERHAUL.** Remove turbocharger unit as outlined in paragraph 60. Mark across compressor housing (1—Fig. AC51), center housing (14) and turbine housing (21) to aid alignment when assembling.

CAUTION: Do not rest weight of any parts on impeller or turbine blades. Weight of only the turbocharger unit is enough to damage the blades.

Remove clamp (3), compressor housing (1) and diffuser (2). Remove screws (T), lock plates (19L) and clamp plates (19C); then, remove turbine housing (21). Hold turbine shaft from turning using the appropriate type of wrench at center of turbine wheel (20) and remove locknut (4). NOTE: Use a "T" handle to remove locknut in order to prevent bending turbine shaft. On some turbine shafts, an allen wrench must be used at turbine end while others are equipped with a hex and can be held with a standard socket. Lift compressor impeller (5) off, then remove center housing from turbine shaft while holding shroud (18) onto center housing. Remove back plate retaining screws (C), then remove back plate (6), thrust bearing (11), thrust collar (8) and spring (9). Carefully remove bearing retainers (12) from ends and withdraw bearings (13 & 15).

CAUTION: Be careful not to damage bearings or surface of center housing when removing retainers. The center two retainers do not have to be removed unless damaged or unseated. Always renew bearing retainers if removed from grooves in housing.

Clean all parts in a cleaning solution which is not harmful to aluminum. A stiff brush and plastic or wood scraper should be used after deposits have softened. When cleaning, use extreme caution to prevent parts

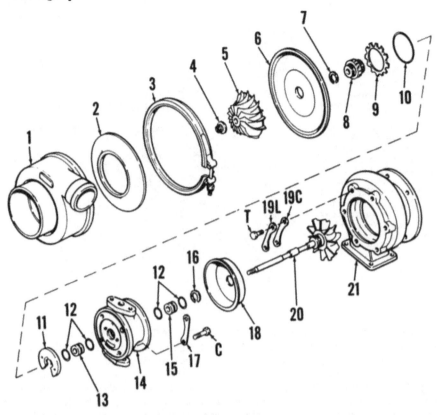

Fig. AC51—Exploded view of AiResearch turbocharger. Use extreme care to prevent damage to parts.

C. Backplate screws	4. Locknut	10. Seal ring	17. Lock plates
T. Turbine housing screws	5. Compressor impeller	11. Thrust bearing	18. Turbine shroud
1. Compressor housing	6. Backplate	12. Bearing retainers	19C. Clamp plates
2. Diffuser	7. Seal ring	13. Bearing	19L. Lock plates
3. Clamp	8. Thrust collar	14. Center housing	20. Turbine wheel and shaft
	9. Spring	15. Bearing	21. Turbine housing
		16. Seal ring	

Fig. AC51A — View showing method of checking turbine shaft end play. Shaft end play should be checked after unit is cleaned to prevent false reading caused by carbon build-up.

from being nicked, scratched or bent.

Inspect bearing bores in center housing (14—Fig. AC51) for scored surfaces, out of round or excessive wear. Bearing bore diameter must not exceed 0.6228 and maximum permissable out of round is 0.0003. Make certain bore in center housing is not grooved in area where seal (16) rides. Inside diameter of bearings (13 & 15) must not be more than 0.4019 and outside diameter must not be less than 0.6182. Thrust bearing (11) should be measured at three locations around collar bore. Thickness should not be more than 0.1720 or less than 0.1711. Inside diameter of bore in backplate (6) must not exceed 0.5015 and seal contact area must be clean and smooth. Compressor impeller (5) must not show signs of rubbing with either the compressor housing (1) or the back plate (6). Impeller should have 0.0002 tight to 0.0004 loose fit on turbine shaft. Make certain that impeller blades are not bent, chipped, cracked or eroded. Oil passages in thrust collar (8) must be clean and thrust faces must not be warped or scored. Ring groove shoulders must not have step wear. Bearing area width must not

exceed 0.1758 and width of groove for seal ring (7) must not exceed 0.0665. Clearance between thrust bearing (11) and groove in collar (8) must be 0.001-0.004, when checked at three locations. Inspect turbine shroud (18) for evidence of turbine wheel rubbing. Turbine wheel (20) should not show evidence of rubbing and vanes must not be bent, cracked, nicked, or eroded. Turbine wheel shaft must not show signs of scoring, scratching or overheating. Diameter of shaft journals must not be less than 0.3992 and out of round must not exceed 0.0003. Groove in shaft for seal ring (16) must not be stepped and diameter of hub near seal ring should be 0.682-0.683. Check shaft end play and radial clearance when assembling.

If the bearing inner retainers (12) were removed, install new retainers using special Kent-Moore tools (JD-274). CAUTION: Bore in housing may be damaged if special retainer installing tool is not used. Oil bearings (13 & 15) and install outer retainers using the special tool. Position the shroud (18) on turbine shaft (20) and install seal ring (16) in groove. Apply a light, even coat of engine oil to shaft

journals, compress seal ring (16) and install center housing (14). Install new seal ring (7) in groove of thrust collar (8), then install thrust bearing so that smooth side of bearing (11) is toward seal ring (7) end of collar. Install thrust bearing and collar assembly over shaft, making certain that pins in center housing engage holes in thrust bearing. Install new rubber seal ring (10), make certain that spring (9) is positioned in back plate (6), then install backplate making certain that seal ring (7) is not damaged. Install lock plates (17) and screws (C), tightening screws to 40-60 inch-pounds torque. Install compressor impeller (5) and make certain that impeller is compeltely seated against thrust collar (8). Install lock nut (4) to 18-20 inch-pounds torque, then use a "T" handle to turn lock nut an additional 90°. CAUTION: If "T" handle is not used, shaft may be bent when tightening nut (4). Install turbine housing (21) with clamp plates (19C) next to housing, tighten screws (T) to 100-130 In.-Lbs., then bend lock plates (19L) up around screw heads.

Check shaft end play and radial play at this point of assembly. If shaft end play (Fig. AC51A) exceeds 0.0042, thrust collar (8—Fig. AC51) and/or thrust bearing (11) is worn excessively. End play of less than 0.001 indicates incomplete cleaning (carbon

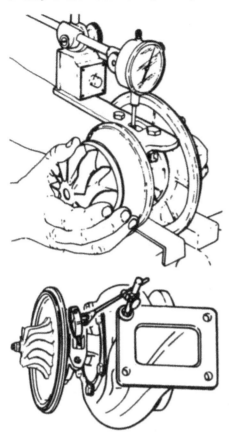

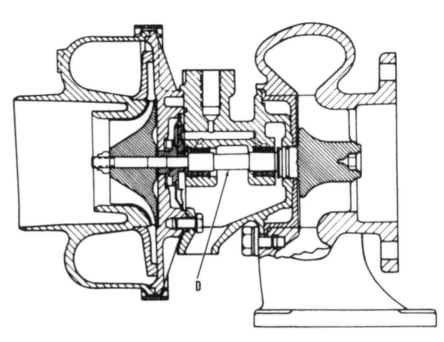

Fig. AC51C—Cross sectional drawing of AiResearch turbocharger. Shaft radial play is checked through oil outlet hole as shown at (D).

Fig. AC51B—Turbine shaft radial play is checked with dial indicator through the oil outlet hole and touching shaft at (D—Fig. AC51C). Two methods of attaching dial indicator to center housing are shown above.

not all removed) or dirty assembly and unit should be disassembled and cleaned. Refer to Figs. AC51B & AC-51C. If turbine shaft radial play exceeds 0.007, unit should be disassembled and bearings, shaft and/or center housing should be renewed. Maximum permissable limits of all of these parts may result in radial play which is not acceptable.

Make certain that legs on diffuser (2—Fig. AC51) are aligned with spot faces on backplate (6) and install diffuser. Install compressor housing (1) and tighten nut of clamp (3) to 40-80 inch-pounds torque. Fill reservoir with engine oil and protect all openings of turbocharger until unit is installed on tractor.

COOLING SYSTEM

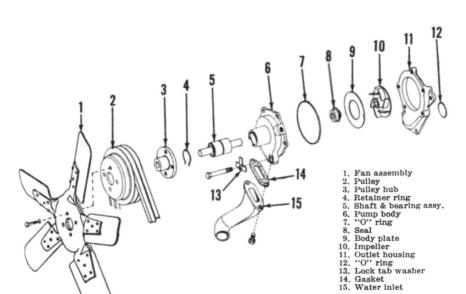

1. Fan assembly
2. Pulley
3. Pulley hub
4. Retainer ring
5. Shaft & bearing assy.
6. Pump body
7. "O" ring
8. Seal
9. Body plate
10. Impeller
11. Outlet housing
12. "O" ring
13. Lock tab washer
14. Gasket
15. Water inlet

Fig. AC54—View of water pump assembly. Slinger (S) shown in drawing at top should be pressed onto rear of shaft until distance (D) is 1-15/16 inches.

RADIATOR

67. REMOVE AND REINSTALL. To remove the radiator, proceed as follows: Remove grille and drain cooling system. Remove hood top panel and disconnect the hose to air cleaner. Unbolt and remove the hood supports and air cleaner. Disconnect radiator hoses and remove radiator from tractor.

Reinstall radiator by reversing removal procedure. Proper coolant level is 2 inches below radiator neck.

WATER PUMP

68. REMOVE AND REINSTALL. Remove radiator, loosen alternator mounting screws, then remove drive belts, fan, spacer and pulley. Remove hoses and (on early models) tube from water pump. Unbolt and remove water pump.

If outlet housing (11—Fig. AC54) is removed, renew "O" ring (12) and "O" ring on tube to oil cooler on models so equipped. Outlet housing (11) to cylinder block screws, water pump (6) to outlet housing and inlet housing (15) to water pump screws should all be tightened to 28-33 Ft.-Lbs. torque. Fan and pulley to shaft hub screws should be tightened to 30-35 Ft.-Lbs. torque.

69. OVERHAUL WATER PUMP. Refer to the exploded view of water pump in Fig. AC54. Using a suitable puller, remove pulley hub (3). Remove retaining ring (4), then press shaft and bearing assembly (5) forward out of pump body (6) and impeller (10). Unbolt and remove water pump outlet housing (11) from pump body and remove the body plate (9). Press seal (8) out towards rear of body.

Renew the seal (8) and other parts as necessary. Apply gasket sealer to outer rim of seal and press the seal into pump body. Press slinger (S) onto rear of shaft until distance (D) is $1\frac{15}{16}$ inches. Press shaft and bearing assembly (5) into front of pump body so that retainer (4) can be installed. Position the plate (9) into recess in pump body and press impeller onto rear end of shaft so that there is a maximum clearance of 0.015 inch between impeller vanes and body plate. Press pulley hub onto front end of shaft with flange side towards pump body. Front edge of hub flange should be 4 inches from rear surface of pump body (6). Install new "O" ring (7) into groove in pump body and install outlet housing (11). Tighten outlet housing retaining cap screws to a torque of 28-33 Ft.-Lbs.

Fig. AC56—View of clutch pedal linkage used on D-21 and D-21 Series II models. Shafts (2 & 10) are located in cast iron rear hood support.

1. Foot pedal
2. Upper shaft
3. Spring links
4. Return spring
5. Lock nuts
6. Turnbuckle
7. Needle bearings
7S. Seals
8. Release lever rod
9. Bellcrank
10. Lower shaft

ELECTRICAL SYSTEM

CAUTION: An alternator (A.C. generator) is used to supply charging current for the system. Due to the fact that certain components of the alternator can be seriously damaged by procedures that would not affect a D. C. generator, the following precautions must be observed:

1. Always be sure that when installing batteries or connecting a booster battery, the negative post of all batteries is grounded.

2. Never short across any of the alternator or regulator terminals.

3. Never attempt to polarize the alternator.

4. Always disconnect all battery ground straps before removing or replacing any electrical unit.

5. Never operate the alternator on an open circuit; be sure that all leads are properly connected and tightened before starting the engine.

70. ALTERNATOR. A Delco-Remy alternator (A.C. generator) and voltage regulator are used. Specification data for each unit follows:

Alternator—1100630
Field Current @ 12 volts
(at 80° F.): 1.9-2.3
Cold Output
Amps @ 2000 RPM 21
Amps. @ 5000 RPM 30
Hot Output:
Rated .... 32 Amps.

Alternator—1100735
Field Current @ 12 volts
(at 80° F.) 2.2-2.6
Cold Output
Amps @ 2000 RPM 21
Amps @ 5000 RPM 30

Hot Output:
Rated .... 32 Amps

Regulator
Relay Unit:
Air gap 0.015
Point opening 0.030
Closing voltage 3.8-7.2
Regulator:
Air gap
(lower points closed) ......0.067[1]
Upper point opening
(lower points closed) 0.014
Voltage setting:

Temp., °F.[2]	Volts[3]
65	13.9-15.0
85	13.8-14.8
105	13.7-14.6
125	13.5-14.4
145	13.4-14.2
165	13.2-14.0
185	13.1-13.9

[1]. Air gap setting of 0.067 is only a starting point; correct air gap is obtained by adjusting unit for proper voltage regulation.

[2]. Ambient temperature measured ¼-inch away from voltage regulator cover; adjustment should be made only when at normal operating temperature.

[3]. Regulated voltage when regulator is working on upper set of points; when regulator is working on lower set of points, voltage should be 0.1 to 0.4 volts less than given in table. Voltage setting may be increased up to 0.3 volts to correct chronic battery undercharge or decreased up to 0.3 volts to correct battery over-charging condition.

71. STARTING MOTOR. A Delco-Remy starting motor is used. Specifications data follows:
No load test:
Volts 9.0
Amperes 120-190*
RPM 4000-7000
*Includes solenoid

ENGINE CLUTCH

72. ADJUSTMENT. Clutch pedal free travel should be 2⅛-inches on D-21 and D-21 Series II models and 1½-inches on 210 and 220 tractors. On all models, pedal free travel is measured at pedal pad. Pedal free play is adjusted at turnbuckle (6—Fig. AC56) or clevis (6—Fig. AC56A). On D-21 and D-21 Series II, loosen locknuts (5—Fig. AC56) and lengthen or shorten link rod as necessary. On 210 and 220, loosen lock nut (5—Fig. AC56A), remove pin (P) and lengthen or shorten rod as necessary. Adjust position of clip (11) on rod so that safety starter switch is depressed 0.050-0.180 when clutch pedal is all the way down. Clip must be correctly positioned to allow starting with clutch pedal all the way down, but switch may be damaged if pressed in too far.

73. R&R ENGINE CLUTCH. The engine clutch may be unbolted and removed from the flywheel after removing the engine as outlined in paragraph 20.

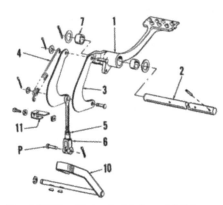

Fig. AC56A—View of clutch pedal linkage used on 210 and 220 tractors. Shaft (2) is located in rear hood support; shaft (10) is located in torque housing.

P. Pin	6. Clevis
1. Foot pedal	7. Bushing
2. Pedal shaft	10. Release lever
3. Link plates	shaft
4. Return spring	11. Starting switch
5. Lock nut	clip

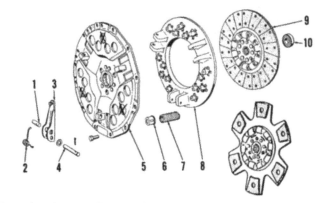

1. Lever pin
2. Anti-rattle spring
3. Lever assembly
4. Lever pin
5. Clutch cover
6. Spring cup
7. Clutch spring
8. Pressure plate
9. Lined disc
10. Pilot bearing

Fig. AC57—Exploded view of typical clutch assembly. On some models, springs are not located in holes marked (X) above. Pilot bearing (10) is installed in flywheel with identification marking out.

Reverse removal procedures to reinstall the clutch assembly. Use a suitable pilot tool to align clutch disc hub with the pilot bearing in flywheel. Tighten the pressure plate retaining cap screws to a torque of 22-27 Ft.-Lbs.

74. OVERHAUL CLUTCH UNIT. Three different clutch discs and pressure plate assemblies have been used and must be correctly matched. D-21 tractors before tractor serial number 1964 were originally equipped with fifteen clutch springs (7—Fig. AC57). Later D-21 and D-21 Series II tractors are equipped with a ceramic clutch disc and only twelve clutch springs are used. If the later clutch disc is installed in early tractors, three springs (7) and cups (6) should be removed from positions (X). On 210 and 220 tractors, all fifteen springs are used.

Overhaul of clutch cover assembly is conventional, Refer to Fig. AC57 for exploded view of unit. Following specifications will apply to the pressure springs:

Pressure spring
free length ....$3\frac{5}{32}$ inches
Lbs. pressure @ $1\frac{13}{16}$ in. 130

Reassemble and install clutch on flywheel using a NEW lined disc. Do not attempt to adjust the release lever adjusting screws with a used lined disc. Distance from surface of cover plate (5—Fig. AC57) to release bearing contacting surface of each adjusting screw in arm (3) should be equal. After release level adjusting screws are properly positioned, the cover assembly can be removed and a used but still serviceable lined disc can be installed if so desired.

75. RELEASE BEARING. After removing engine from tractor as outlined in paragraph 20, the release bearing can be removed as follows:

On D-21 and D-21 Series II, remove pivot pin (5—Fig. AC58) and slide release bearing and hub (shifter) off. On 210 and 220 models, disconnect clevis (C—Fig. AC58A) from release lever shaft (10) and remove snap ring and washer from right end of shaft on outside of torque housing. Pull shaft and lever out left side of torque housing, removing woodruff keys as soon as they are clear of clutch shift fork. Disconnect spring from hub and slide release bearing and shifter hub off.

On all models, bearing can be pressed off and new bearing pressed onto shifter hub.

On D-21 and D-21 Series II tractors, if clutch release lever (3—Fig. AC58) has been disconnected from control rod, reinstall lever with lettered side facing out. Check lever for any binding condition before reconnecting con-

trol rod; reverse the position of lever on pivot pin (5) if binding is noted.

On 210 and 220 tractors, straight side of shift fork (3—Fig. AC58A) should be toward rear. Slide the release lever shaft (10) in left side and install woodruff keys as keyways pass the bore in left side of torque housing. Install washer and snap ring on right end of shaft outside of torque housing. Make certain to reconnect return spring.

76. CLUTCH SHAFT. To remove the engine clutch shaft, the transmission must be split from torque housing. Refer to paragraph 81. Clutch shaft is integral with four-speed transmission input shaft and is serviced as outlined in paragraph 84.

77. CLUTCH (TORQUE) HOUSING. For normal service requirements, components within the torque housing can be removed as outlined in paragraph 113 after engine is removed as outlined in paragraph 20 and the transmission lubricant is drained. Procedure for renewing the torque housing will depend upon whether the tractor is assembled, or if the engine or transmission assemblies have been

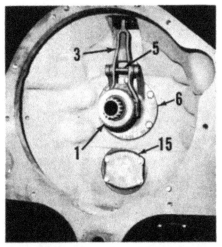

Fig. AC58 — Front view of torque housing showing clutch release lever and release bearing used on D-21 and D-21 Series II tractors.

1. Release bearing	6. Retainer & sleeve
3. Release lever	assembly
5. Pivot pin	15. Bearing retainer

Fig. AC58A—Front view of torque housing used on 210 and 220 tractors showing clutch release bearing and associated parts.

	6. Retainer & sleeve
C. Clevis	10. Release lever &
1. Release fork	shaft
3. Release fork	15. Bearing retainer

detached from the torque housing. For procedure to remove torque housing from assembled tractor or when engine is detached, refer to paragraph 78; if transmission has been detached from torque housing, refer to paragraph 79. Remove components as outlined in paragraph 113.

78. If necessary to renew torque housing in assembled tractor, first remove engine as outlined in paragraph 20, drain transmission lubricant and proceed as follows:

Disconnect all control linkage, wiring and instrument connections from components installed on the hood rear support casting. Attach hoist to support casting, unbolt and remove support casting from torque housing. Remove the platform and transmission top (shifter) cover as outlined in paragraph 80. Attach hoist to torque housing, then unbolt and remove torque housing from front end of transmission housing.

79. If necessary to renew torque housing and transmission has been detached from torque housing, proceed as follows:

Securely chain the front support casting to the front axle center member under sufficient tension so that front support cannot possibly slide from front pivot pin. Remove the hood top panel. Block up under front support, rear of axle center member, rear of engine and rear ends of side rails so that no movement of these components can occur when torque housing is unbolted. Hoist can be used to support engine by hooking into engine lift hooks. Disconnect all controls, wires and connections to instruments located on hood rear support casting. Attach hoist to support casting and unbolt and remove support from torque housing. Attach hoist to torque housing, remove front axle rear pivot bolt, unbolt torque housing from side rails and engine rear adapter plate and remove torque housing from tractor.

FRONT (4-SPEED) TRANSMISSION

NOTE: The PTO gears in torque housing, front (four-speed) transmission, rear (dual range) transmission, differential, final drive gears, PTO clutch, hydraulic system and power steering system all share a common oil supply. Fluid capacity on tractor equipped with PTO and hydraulic system is approximately 21 gallons for D-21 and D-21 Series II; 27 gallons for 210 and 220 tractors. "Allis-Chalmers Transmission and Hydraulic Power Fluid 821" or equivalent should be used. Oil should be drained and the system refilled with new oil once each year. Removal of all drain plugs (one plug in bottom plate on torque housing, one in each final drive compartment in rear axle housing and one or two in the center of rear axle center housing) is required to completely drain the system. Oil levels should be maintained between marks on dipstick (Fig. AC59).

SHIFTER ASSEMBLY

80. **R&R AND OVERHAUL.** The front four-speed transmission shifter assembly is removed with the transmission cover. To remove cover, first remove the center section of operator's platform, shift transmission to neutral position, then unbolt and remove the transmission cover and shifter assembly.

NOTE: Prior to disassembly of shifter mechanism, location and number of overshift limiting washers (28—Fig. AC60) should be noted so that they can be reinstalled in same position. If shift cover, shift

rail (6) or fork (5) is renewed, it may be necessary to add or remove washers (28).

To remove gear shift lever, remove dust cover (11), snap ring (15) and lever pivot washer (16). Then, lift lever from cover. Remove the two lever pivot pins (12). Pivot insert (20) for shift lever in cover can be renewed if necessary; press old insert out towards top of cover and press new insert into place.

To remove the reverse shifter fork (7), remove lock screw, rotate shift rail ¼-turn, and catch detent ball and spring (9) while sliding rail forward out of cover. With the reverse shift rail removed, long interlock plunger (23) and interlock pin (26)

can be removed. Reverse shift lug (24), latch plunger (21) and spring (22) can be removed from rail if necessary.

Remove lock screw from first and second shifter fork (5), rotate rail ¼-turn and catch detent ball and spring as rail is removed from cover. Keep the spacer (25) and washer(s) (28) with shift fork (5). Remove shifter lug (27) from rail if necessary.

Rotate third and fourth gear shift fork (2) and rail (4) ¼-turn and catch detent ball and spring while removing rail from cover. Remove the short interlock plunger (3) if not removed from cover after removing first and second gear shift rail.

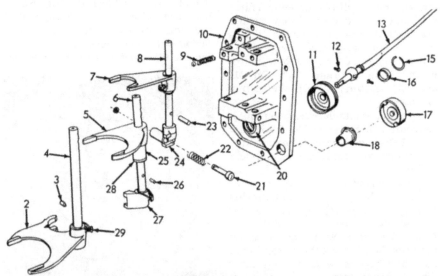

Fig. AC60 — Exploded view of transmission cover and shifter assembly. Refer to Fig. AC61 for cross-sectional view of cover (10) showing proper installation of interlock plungers and pin.

Fig. AC59 — Transmission and hydraulic system dipstick and filler cap are located below covers on platform.

2. 3rd & 4th shift fork	10. Transmission cover	21. Latch plunger
3. Short interlock plunger	11. Dust cover	22. Plunger spring
4. 3rd & 4th shift rail	12. Pivot pins	23. Long interlock plunger
5. 1st & 2nd shift fork	13. Gearshift lever	24. Reverse shift lug
6. 1st & 2nd shift rail	15. Snap ring	25. Spacer
7. Reverse shift fork	16. Pivot washer	26. Interlock pin
8. Reverse shift rail	17. Filler cap	27. 1st & 2nd shift lug
9. Detent spring & ball	18. Filler tube	28. Shift stop washers
	20. Insert	29. Lock screws

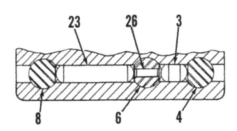

Fig. AC61 — Cross-sectional view of transmission cover showing placement of the interlock pin and plungers. Refer to Fig. AC60 for legend.

Renew bent, sprung or worn shifter rails, forks or lugs. Wear on ends of interlock plungers or pin will allow more than one shift rail to be moved at a time; check length of plungers and pin against the following values: Long interlock plunger, 1.691; short interlock plunger, 0.566; and interlock pin, 0.551. Refer to Fig. AC61 for cutaway view of transmission cover showing location of interlock plungers and pin.

To reassemble cover and shifter mechanism, reverse the disassembly procedure. Face of the reverse latch plunger (21—Fig. AC60) should be flush with reverse shifter lug (24); insert cotter pin through castellated nut and plunger after adjusting plunger position. Insert detent spring and ball into cover and depress detent ball with punch while sliding shift rail into place. Place shift lever in neutral position and measure distance between shifter lug (27) to stop surface of transmission cover and distance between spacer (25) or washer(s) (28) and cover on first and second gear shift rail. Both distances should measure from 0.550 to 0.630. Add or remove flat washers (28) as necessary to provide these measurements.

Tighten lug and shifter fork lock screws to a torque of 15-20 Ft.-Lbs. and secure the screws with wires. Place all shifter couplings in neutral position and install cover using a new gasket. Be sure that shifter forks entered grooves in the shift couplings; then tighten the cover retaining cap screws to a torque of 35-40 Ft.-Lbs. Note: Two of the cover retaining cap screws are used as locating dowels and are not threaded the full length. The dowel type screws must be installed in the left front and right rear holes in cover. Reinstall the center section of the operator's platform.

R&R TRANSMISSION ASSEMBLY

80A. To remove the four-speed transmission from the tractor, first perform the split between the transmission housing and torque housing as outlined in the following paragraph 81; then, remove transmission assembly from the rear section as outlined in paragraph 83. Refer to Fig. AC65.

81. **TRACTOR SPLIT, TRANSMISSION FROM TORQUE HOUSING.** First, drain all oil from the torque housing, transmission and rear axle center housing by removing the drain plugs from torque housing and rear axle center housing.

Remove operator's platform, batteries and battery boxes from tractor. Disconnect all interfering wiring and hydraulic lines.. Shut off the fuel supply valve and disconnect the fuel supply and return lines. Remove brake control rods and dual range transmission control rod.

Drive wood 2 X 4's in between front axle support casting and axle center (main) member to hold front end of tractor upright. Block up securely under drawbar **and** under front end of rear axle center housing. Place a rolling floor jack from front of tractor under the front axle rear pixot pin or remove the hood top panel and attach moveable hoist to engine rear lift hook.

Remove the four-speed transmission cover and shifter assembly as outlined in paragraph 80. Disconnect the PTO shifter rod. Remove the cap screws and bolts retaining transmission housing to torque housing (two capscrews are located inside transmission under top cover), split transmission from torque housing and roll front section of tractor away from transmission.

Reassemble tractor by reversing procedure followed to perform split. Refill transmission and recheck oil level after engine has been started and the power steering and hydraulic systems bled.

82. **TRACTOR SPLIT, TRANSMISSION FROM REAR AXLE CENTER HOUSING.** If performing a split between transmission and rear axle center housing to service differential or range transmission, and transmission does not need to be removed, follow the same general procedures as outlined in paragraph 81. However, unbolt the transmission from the rear axle center housing instead of the torque housing. It is not necessary to remove the four-speed transmission shifter cover or to disconnect the PTO control rod to perform the split at rear of transmission.

82A. NOTE: The PTO drive shaft will usually be moved forward with transmission assembly. This is not a problem on D-21 and D-21 Series II tractors; however, the drive shaft will probably pull the PTO clutch hub forward on early Two-Twenty tractors allowing PTO clutch discs to slide down behind hub. On Two-Twenty tractors before serial number 220-1560D, the lift shaft housing and PTO clutch should be removed as outlined in paragraph 133. Install lift shaft housing after transmission housing is attached to rear axle center housing.

On later 220 models (Ser. No. P 1563 and up), and all 210 models, the PTO clutch shield will usually hold clutch hub in position. If difficulty is encountered when assembling, lift shaft housing should be removed to prevent possible damage to the PTO clutch.

83. If performing the split between transmission and rear axle center housing as part of procedure to remove the four-speed transmission assembly, first perform all steps outlined in paragraph 81. Then, attach a hoist to the four-speed transmission housing, unbolt the transmission housing from the rear axle center housing

Fig. AC65 — Removing the front (four-speed) transmission from D-21 tractor. Other models are similar.

and remove the assembly. Refer to NOTE paragraph 82A.

OVERHAUL TRANSMISSION

84. INPUT (CLUTCH) SHAFT. The four-speed transmission input shaft can be removed, after splitting the transmission from torque housing as outlined in paragraph 81, by proceeding as follows:

Remove the capscrews securing the bearing retainer (2—Fig. AC67) to the transmission housing and removing the retainer and shims (3) being careful not to lose or damage the shims. Withdraw the input shaft and ball bearing (5). Remove snap ring (4) and press ball bearing from shaft. The roller type pilot bearing (7) floats in the rear end of input shaft and on the pilot of transmission mainshaft (25).

Install the ball bearing (5) on input shaft with shielded side of bearing towards the rear (gear) end of shaft. Install snap ring (4) at front of ball bearing. Place roller bearing (7) on pilot end of transmission mainshaft, then insert input shaft and bearing assembly so that snap ring on outer race of ball bearing contacts transmission case. Install bearing retainer without shims and tighten the cap screws evenly and snugly. Using a feeler gage, measure the clearance between retainer and housing in several places, make certain the measurements are equal and record the measurement. Remove the bearing retainer. Reinstall the bearing retainer, using a shim pack thickness equal to the measured clearance plus 0.000 to 0.005 to establish the desired clearance between the ball bearing race

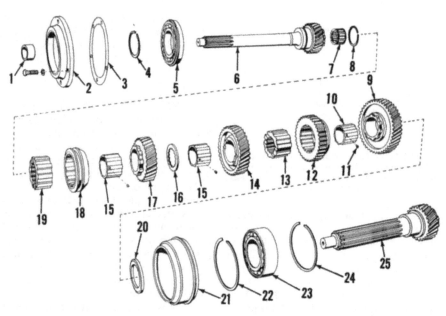

Fig. AC67 — Exploded view of four-speed transmission input shaft and mainshaft assemblies. Sleeve (1) is used on tractors without PTO only. Refer to Fig. AC66.

1. Sleeve	8. Snap ring	14. 2nd gear	20. Thrust washer
2. Retainer	9. 1st gear	15. Bushings	21. Bearing sleeve
3. Shims	10. Bushing	16. Splined washer	22. Snap ring
4. Snap ring	11. Lock pins	17. 3rd gear	23. Ball bearing
5. Ball bearing	12. Reverse gear	18. Coupling	24. Snap ring
6. Input shaft	13. Splined collar	19. Splined collar	25. Mainshaft
7. Roller bearing			

and bearing retainer. Tighten the retaining cap screws to a torque of 35-40 Ft.-Lbs. Shims are available in thickness of 0.003 and 0.010.

85. REVERSE IDLER. The reverse idler assembly can be removed from the four-speed transmission after removing transmission from tractor as outlined in paragraphs 81 and 83, or when the transmission top (shifter) cover has been removed and the transmission split from rear axle center (differential) housing as outlined in paragraph 82.

Remove the capscrew, lockwasher and flat retainer plate (3—Fig. AC68) from rear face of transmission housing and bump or pull the idler shaft (2) out towards rear. Withdraw the idler gear (1) from top opening of transmission.

The reverse idler gear and bushing are serviced as an assembly only. Shaft to bushing running clearance should be 0.002-0.004; renew shaft and/or gear and bushing assembly if the clearance is excessive.

86. TRANSMISSION MAINSHAFT. To remove the transmission mainshaft, first remove transmission assembly from tractor as outlined in paragraphs 81 and 83. Although not required, removal of the reverse idler gear and shaft as outlined in paragraph 85 will facilitate reinstalling the mainshaft assembly.

Remove the input (clutch) shaft from transmission as outlined in paragraph 84 and withdraw the pilot bearing (7—Fig. AC67) if it remained on mainshaft pilot. Remove the snap ring (8) from front end of mainshaft and withdraw the splined collar (19) and coupling (18). Press or bump the mainshaft rearward and remove the gears, bushings, splined collar and washer from top of transmission housing. Be careful not to lose the lock pins (11) from bushings.

Remove snap ring (22) and press bearing sleeve (21) from bearing; then, press ball bearing off of the mainshaft.

To reinstall the mainshaft, reverse the removal and disassembly procedure. Place the rear thrust washer (20) on mainshaft next to ball bearing and while inserting shaft through rear opening of transmission, install parts in following order:

Install the 1.845 wide bushing (10) on shaft with lock pins forward; pins are inserted in bushing from inside and can be held in place with heavy grease or by peening outer end. Install first speed gear (9) (38 teeth) over bushing with clutch teeth forward. Install 1.875 wide splined collar and install reverse gear (12) (spur teeth) over collar with shifter groove

Fig. AC66 — View showing front end of four-speed transmission. Sleeve (1 — Fig. AC67) is pressed on input shaft at (S) on tractors not equipped with PTO. Seal (8—Fig. AC91) contacts sleeve or, on tractors with PTO, contacts PTO drive shaft shown below input shaft.

rearward. Install 1.593 wide bushing over shaft with lock pins forward and install second speed gear (14) (34 teeth) over bushing with clutch teeth rearward. Install splined thrust washer (16), then install the second 1.593 wide bushing on shaft with lock pins forward. Install third speed gear (17) (28 teeth) over bushing with clutch teeth forward. Press shaft forward into housing until snap ring on outer diameter of bearing sleeve (21) contacts transmission housing. Install shifter coupling (18) over shaft with groove forward; then, install 1.375 wide splined collar over shaft and into shifter coupling. Install the thickest snap ring (8) that will seat into the groove in shaft in front of the splined collar. Snap rings are available in thicknesses of 0.091 to 0.121 in steps of 0.006.

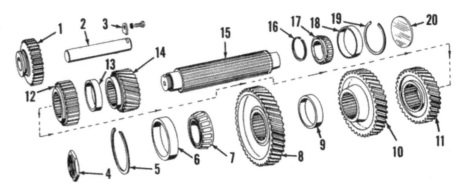

Fig. AC68—Exploded view of reverse idler and transmission countershaft assemblies.

1. Reverse idler	6. Front bearing cup	11. 2nd gear	16. Snap ring
2. Idler shaft	7. Front bearing cone	12. Reverse gear	17. Rear bearing cone
3. Retainer plate	8. Driven gear	13. Spacer	18. Rear bearing cup
4. Nut	9. Spacer	14. 1st gear	19. Snap ring
5. Snap ring	10. 3rd gear	15. Countershaft	20. Plug

87. TRANSMISSION COUNTERSHAFT. To remove the transmission countershaft, first remove the mainshaft as outlined in paragraph 86; then, proceed as follows:

Carefully unstake and remove the nut (4—Fig. AC68) from front end of countershaft. Remove the plug (20) and snap ring (19) at rear end of countershaft; then, push or drive the shaft to rear. Rear bearing cup (18) will be pushed from housing with the shaft. Front bearing cone (7) will be pushed from front end of shaft by the countershaft driven gear (8). Remove the shaft from rear of housing and remove the gears, spacers and front bearing cone from top opening of transmission housing. Remove the front bearing cup (6) and snap ring (5) from transmission housing and remove rear bearing cone (17) and snap ring (16) from countershaft.

To reassemble and reinstall the countershaft, proceed as follows: Install snap ring (16) on rear end of shaft and press bearing cone (17) on rear end of shaft tightly against the shoulder. Install front bearing cup (6) and a 0.105 thick snap ring (5) in front bore of transmission housing. Set transmission housing on its front face and place cone (7) in cup. Stack gears and spacers in housing in following order: Countershaft driven gear (8) with long hub up. 0.599 wide spacer (9); third gear (10) with long hub down; second gear (11) with long hub up; reverse gear (12) (spur teeth) with chamfer on gear teeth up; 0.599 wide spacer (13); and first gear (14) with long hub down. Then, insert the countershaft down through

the stack of gears and spacers into the front bearing cone and turn transmission right side up. Press or bump the countershaft forward until the shoulder on shaft is tight against the front bearing cone and install rear bearing cup, driving the cup forward until all end play is removed. Then, install nut (4) on front end of shaft, tighten the nut to a torque of 80-100 Ft.-Lbs. and stake nut securely to slot in shaft. Select a rear snap ring (19) that will provide a 0.002-0.005 end play of the countershaft assembly in the bearing cups. Bump the shaft assembly forward, then to rear to seat the bearing cups against the snap rings before checking end play. Snap rings are available in thicknesses of 0.095 to 0.125 in steps of 0.002. If correct bearing adjustment is impossible, install a different thickness snap ring (5) at front. Install new plug (20) at rear of countershaft.

REAR (DUAL RANGE) TRANSMISSION

SHIFTER ASSEMBLY

90. R&R AND OVERHAUL. Refer to Figs. AC69, AC69A and AC69B for an exploded view of the dual range transmission shifter assembly and shift linkage. Shifter cover assembly

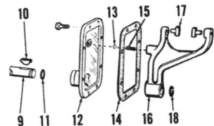

Fig. AC69—View of dual range shift cover and fork. Shifter lever (9) is shown in Fig. AC69A or AC69B.

9. Shifter lever	14. Gasket
10. Key	15. Detent spring
11. "O" ring	16. Shift fork
12. Shifter cover	17. Inserts
13. Detent ball	18. Snap ring

can be removed after draining the lubricating oil from transmission and, if tractor is equipped with four batteries, removing the right hand battery box and batteries. Disconnect the adjustable shift link (8—Fig. AC69A or AC69B) from lever (9); then, unbolt and remove the shifter cover assembly from the rear axle center housing. Remove inserts (17—Fig. AC-69) from bottom of compartment if they fell out of fork during removal of the shifter assembly.

NOTE: It is necessary to remove only the drain plug from the bottom of the dual range transmission compartment to remove enough oil to permit removal of shifter cover; however, to completely drain the transmission lubricating oil, it is necessary to remove all drain plugs.

To disassemble the shifter unit, remove the snap ring (18) and catch

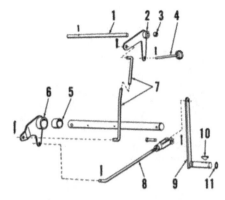

Fig. AC69A—View of dual range shift linkage used on D-21 and D-21 Series II tractors. Refer to Fig. AC69B for linkage used on 210 and 220 models.

1. Upper shaft	6. Bellcrank
2. Bellcrank	7. Linkage rod
3. Bushing	8. Linkage rod
4. Shifter rod	9. Shifter lever
5. Bushing	10. Key
	11. "O" ring

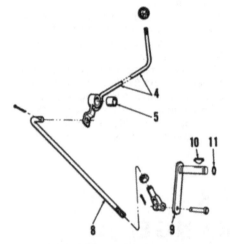

Fig. AC69B—View of dual range shift linkage used on 210 and 220 tractors.

4. Handle	9. Shifter lever
5. Bushing	10. Key
8. Linkage rod	11. "O" ring

the detent ball (13) while pulling shifter fork (16) from lever shaft. Remove the woodruff key (10) from shaft and withdraw lever shaft from cover assembly. Remove detent spring (15) from bore in shifter fork and remove "O" ring (11) from lever shaft.

Reassemble unit using new "O" ring (11). Reinstall assembly using new gasket (12). Stick the inserts (17) into shifter fork with heavy grease and be sure the inserts enter the groove in shift collar. Tighten the retaining cap screws to a torque of 35-40 Ft.-Lbs. Note: Two of the cover retaining screws are used as dowel bolts and are not threaded the full length. Be sure these two screws are inserted in the holes in cover surrounded by a raised boss. If necessary, adjust length of link (8—Fig. AC69A

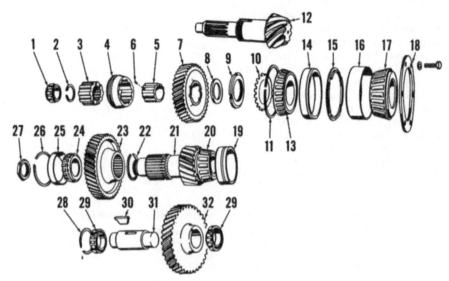

Fig. AC70 — Exploded view of dual range transmission mainshaft (bevel pinion shaft), countershaft and gears. Idler shaft and gear (31 & 32) are for Two-Twenty tractors with F.W.A. only.

1. Pilot bearing	12. Bevel pinion shaft	22. Snap ring
2. Snap ring	13. Narrow bearing cone	23. Driven gear
3. Splined collar	14. Narrow bearing cup	24. Front bearing cone
4. Coupling	15. Spacer	25. Front bearing cup
5. Bushing	16. Wide bearing cup	26. Snap ring
6. Lock pins	17. Wide bearing cone	27. Nut
7. Lo range gear	18. Retainer plate	28. Snap ring
8. Thrust washer	19. Rear bearing cup	29. Bearings
9. Nut	20. Rear bearing cone	30. Key
10. Lock tab washer	21. Countershaft	31. Idler shaft
11. Snap ring		32. Idler gear

or AC69B) to synchronize travel of push-pull shifter rod (4—Fig. AC69A) or lever (4—Fig. AC69B) with detent positions of shifter fork (16—Fig. AC-69).

DUAL RANGE TRANSMISSION OVERHAUL

91. To remove either the dual range transmission mainshaft (bevel pinion shaft) or countershaft, major tractor units must first be removed in following order (reference paragraphs in parentheses); lift shaft housing (133), both brakes (105A or 106A), both axle shafts (103 or 104A), both bull pinions (99), both bull gears (100) and differential (96). Block up securely under the torque housing, attach hoist to rear axle center housing and unbolt and separate center housing from rear end of four speed transmission.

92. **MAIN (BEVEL PINION) SHAFT.** Refer to paragraph 91, then proceed as follows:

Pull bearing (1—Fig. AC70) from front end of shaft. Remove snap ring (2), then remove shifter collar (4) and splined coupling (3) from shaft. Remove capscrews and lockwashers securing bearing retainer (18) to front side of the differential compartment and remove retainer. Press or bump pinion shaft rearward until free of housing bore and remove gear (7),

bushing (5) and thrust washer (8) from side opening in housing while withdrawing shaft and bearing assembly from rear.

To remove bearings from pinion shaft, unlock tab washer (10) from retaining nut (9) and remove nut and washer. Press bevel pinion shaft out of the bearings. The bevel pinion shaft is renewable only in a matched set with the ring gear.

The mesh position (center distance) of the bevel gears is controlled by the thickness of the snap ring (11) which is available in thicknesses of 0.178 to 0.190 in steps of 0.002. If necessary to renew the snap ring only, be sure to use replacement of exactly the same thickness as original. However, if the bevel pinion shaft and differential ring gear, bevel pinion shaft bearings or rear axle center housing is renewed, it will be necessary to follow the procedure outlined in paragraph 93 to select the proper thickness of the snap ring (11).

To reassemble and reinstall bevel pinion shaft after the thickness of snap ring (11) has been determined, proceed as follows; Press wide bearing cone (17) onto pinion shaft tightly against the gear and install wide bearing cup (16), spacer (15), narrow bearing cup (14) and press narrow bearing cone (13) onto shaft so that there is no end play of the bearing cups and spacer. Install the locktab washer (10) and nut (9) and tighten the nut so that 5 to 20 inch pounds torque is required to rotate the shaft in the bearings. Bend a washer tab into a notch in the nut.

Install snap ring (11) of correct thickness into groove in housing bore. Insert bevel pinion shaft assembly into bore from differential compartment and install the thrust washer (8), and gear (7) through side opening in housing as bevel pinion shaft is pressed or bumped forward. Press or bump shaft forward until front (narrow) bearing cup is tight against the snap ring (11). Insert the two pins (6) into holes in bushing (5) and retain with heavy grease or by peening outer ends of the pins. Install the bushing, with the pins to front, on the pinion shaft and into the hub of the gear (7). Position coupling (4) on shaft with radius (or long splines) toward front, then install collar (3) on shaft and in coupling with long splines forward. Install thickest snap ring (2) that can be installed in the groove in shaft. Snap rings (2) are available in nominal thicknesses of 0.097, 0.103 and 0.109; however, the snap ring thickness may vary up to 0.004 from the nominal thickness. Press the roller bearing (1) onto front end of bevel pinion shaft.

93. BEVEL PINION BEARING SNAP RING SELECTION.
The mesh position of the bevel gears is controlled by the thickness of the snap ring (11—Fig. AC70) located in front of the bearing cup (14) in bearing bore of rear axle center housing. If bevel pinion (12) and ring gear set, bearing (13, 14, 16 and 17), spacer (15) and/or rear axle center housing are renewed, the following procedure must be observed in selecting the thickness of a new snap ring.

Assemble the bearing cones, cups and the bearing spacer on the bevel

pinion shaft as shown in Fig. AC71 and adjust bearing preload so that 5 to 20 inch-pounds of torque is required to rotate pinion shaft in bearings and lock the adjustment nut in this position. Then, measure distance (D—Fig. AC71) from rear face of rear bearing cone inner race to front edge of front bearing cup. Add to this measurement the figure etched on rear face of the pinion shaft (See Fig. AC72). Subtract this sum from the figure stamped on the lower right hand corner of the rear face of rear axle center housing (See Fig. AC73). The result should be the thickness of the snap ring (11—Fig. AC70) to be used in reassembly of the dual range transmission. Snap rings are available in thicknesses of 0.178 to 0.190 in steps of 0.002.

As an example of above procedure, assume measurement (D—Fig. AC71) be 2.906. Add this to the figure etched on rear face of pinion gear (See Fig. AC72) which is, on the pinion shown, 6.720. This gives a sum of 9.626; subtract this sum from the figure stamped

Fig. AC72 — Pinion measurement (PM) etched on rear face of bevel pinion is used in determining thickness of snap ring (11—Fig. AC70) used in reassembly of dual range transmission.

on rear axle center housing (See Fig. AC73) which, on the housing shown, is 9.810. The result is 0.184 which should be the thickness of the snap ring used in reassembly of dual range transmission. If determined thickness for snap ring is not the same as a snap ring available for service (falls in between two available thicknesses), use snap ring with thickness closest to determined value.

94. COUNTERSHAFT ASSEMBLY.
To remove the dual range transmission countershaft, first remove the main (bevel pinion) shaft as outlined in paragraph 92; then, proceed as follows:

Carefully unstake nut (27—Fig. AC70) from slot in front end of countershaft and remove nut. Working through opening in left side of housing, disengage snap ring (22) from countershaft at rear side of gear (23) and slide snap ring rearward. Remove snap ring (26) from front of bore in housing and bump countershaft forward, removing bearing cup (25). Pull bearing cone (24) from front end of shaft, turn rear end of shaft towards opening in side of housing and withdraw the countershaft from gear (23) and housing. Pull bearing cup (19) from rear bore in housing and bearing cone (20) from rear end of countershaft.

Reverse removal procedure to reinstall countershaft assembly. Tighten nut (27) to a torque of 80-100 Ft.-Lbs. and stake nut into slot machined in shaft. Install a snap ring (26) of thickness that will provide 0.003-0.005 end play of shaft assembly in bearing cups. Be sure to bump shaft forward to seat bearing cup against snap ring before checking end play with dial indicator. Snap rings are available in thicknesses of 0.099 to 0.129 in steps of 0.002.

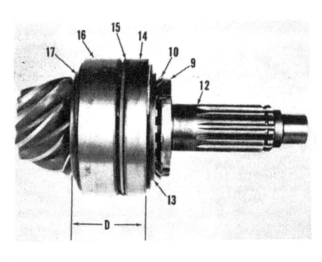

Fig. AC71 — View of bevel pinion shaft and bearing assembly. Distance "D", measured after bearings are assembled on shaft and properly adjusted, is used in determining thickness of snap ring (11—Fig. AC70) when reinstalling bevel pinion assembly. Refer to Fig. AC70 for legend.

Fig. AC73 — Measurement stamped on rear face of rear axle (differential) center housing is used along with bearing dimension ("D"—Fig. AC71) and measurement (PM—Fig. AC72) in determining thickness of snap ring (11—Fig. AC70) to be used.

DIFFERENTIAL, FINAL DRIVE AND REAR AXLE

DIFFERENTIAL

95. BEARING ADJUSTMENT. Adjustment of the differential carrier bearings is controlled by shims (8—Fig. AC74) located between each bearing carrier (10) and the rear axle center housing. Transferring shims from under one bearing carrier to under the opposite bearing carrier will adjust backlash of the bevel ring gear and bevel pinion gear.

To adjust the bearings, first remove the differential assembly as outlined in paragraph 96. Remove nuts retaining bevel ring gear to differential housing and remove bevel ring gear. Then, using spacers if necessary, reinstall the nuts and tighten to a torque of 90-100 Ft.-Lbs. Reinstall differential housing and bearing carriers without the ring gear using the previously removed shims (8). Tighten bearing carrier retaining capscrews securely. Wrap a cord around the differential housing, attach a spring pull scale to the cord and note amount of pull required to steadily rotate differential assembly in carrier bearings. Adjustment of carrier bearings is correct when a steady pull of 1-3 pounds is required to rotate the as-

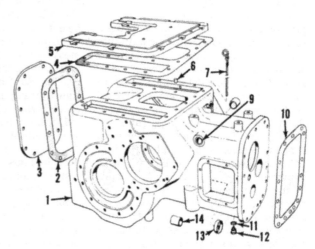

Fig. AC75 — Rear axle center (differential) housing and related parts. Plug (13) is used to fill PTO shaft bore through wall between differential and dual range transmission compartments on tractors without PTO.

1. Center housing
2. Gasket
3. Cover (w/o PTO)
4. Gasket
5. Top cover
6. Dowel pins
7. Oil dipstick
8. Brake shaft bushings
10. Gasket
11. Seal washer
12. Drain plug
13. Plug
14. Hydraulic oil outlet

sembly in the bearings. Add shims under the right bearing carrier or remove shims from under the left bearing carrier to obtain desired bearing adjustment.

After correct bearing adjustment is obtained, remove differential assembly from rear axle center housing being careful **not** to lose or inter-mix the shims (8). Reinstall bevel ring gear on differential assembly and tighten retaining nuts to a torque of 90-100 Ft.-Lbs. Reinstall ring gear and differential assembly in rear axle center housing using the previously

determined shim stack thickness under each bearing carrier. Using a dial indicator, check ring gear to bevel pinion gear backlash at several points around the ring gear. Minimum backlash should be 0.010; maximum backlash should be 0.013. If difference between minimum and maximum backlash exceeds 0.003, run-out of bevel ring gear is excessive and should be corrected before proceeding further. If run-out is less than 0.003, but backlash is less than 0.010 or more than 0.013, adjust backlash by transferring shims from under one bearing carrier to under opposite bearing carrier. NOTE: Only transfer shims; do not add or remove shims or carrier bearing adjustment will be changed. Transferring shims from under right bearing carrier to under left bearing carrier will decrease backlash; transferring shims from under left bearing carrier to under right bearing carrier will increase backlash. Shims are available in thicknesses of 0.003, 0.010 and 0.020.

NOTE: Adjustment of differential carrier bearings and backlash of bevel ring gear to bevel pinion gear will affect bull pinion shaft end play. Be sure to check and adjust bull pinion shaft end play as outlined in paragraph 99 when reassembling tractor.

96. R&R AND OVERHAUL. To remove differential assembly, major tractor units must first be removed in following order (reference paragraphs in parentheses); lift shaft housing (133), both brakes (105A or 106A), both axle shafts (103 or 104A), both bull pinions (99) and both bull gears (100). Unbolt and remove the two

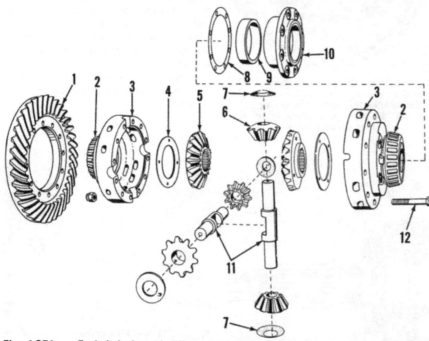

Fig. AC74 — Exploded view of differential and bevel ring gear assembly. Differential housing bolts (12) also retain bevel ring gear to differential housing.

1. Bevel ring gear
2. Carrier bearing cone
3. Differential housing
4. Thrust washer
5. Pinion shaft gears
6. Differential pinions
7. Thrust washers
8. Shims
9. Carrier bearing cup
10. Bearing retainer
11. Pinion shafts
12. Assembly bolts

differential bearing carriers, then remove differential assembly from rear opening in rear axle center housing.

To disassemble the differential unit, first be sure that the two housing halves (3—Fig. AC74) are correlation marked; then, remove the twelve locknuts and bolts (12) retaining ring gear to differential housing and holding the two halves of differential housing together. The unit can then be disassembled as shown in exploded view in Fig. AC74. Renew all damaged or questionably worn parts.

If backlash between teeth of side (pinion shaft) gears (5) and differential pinions (6) is excessive, renew the thrust washers (4 & 7). If backlash is still excessive, it may be necessary to renew the side gears and differential pinions. Differential gears are available separately. When reassembling, tighten the twelve self-locking nuts to a torque of 90-100 Ft.-Lbs. The cap screws attaching differential bearing carriers should be tightened to 70-75 Ft.-Lbs. torque.

MAIN DRIVE BEVEL GEAR

97. MESH AND BACKLASH ADJUSTMENT. The mesh position of the bevel pinion gear is controlled by the thickness of the snap ring (11—Fig. AC70) which is fitted to a particular assembly of rear axle center housing, bevel gear set and the assembled thickness of the bevel pinion bearings. Refer to paragraph 93.

To adjust the backlash between bevel pinion and bevel ring gear, refer to paragraph 95.

98. RENEW BEVEL GEARS. The bevel pinion is an integral part of the rear (dual range) transmission

Fig. AC76—When reinstalling bull gear in rear axle center housing, be sure that side of hub with raised boss (B) is placed towards outside of housing.

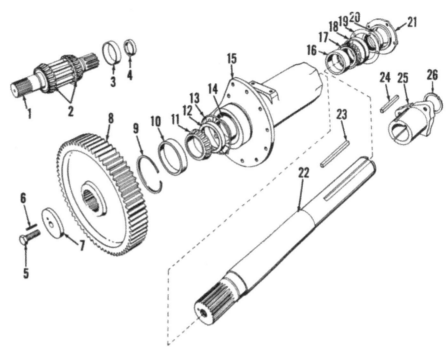

Fig. AC77—Exploded view of D-21 and D-21 Series II rear axle and final drive unit. Bearing cup (3) is located in brake housing.

1. Bull pinion shaft	8. Bull gear	14. Sleeve	20. Grease seal
2. Bearing cones	9. Snap ring	15. Axle housing	21. Retainer
3. Bearing cup	10. Bearing cup	16. Sleeve	22. Axle shaft
4. Oil seal	11. Bearing cone	17. Bearing cone	23. Key
5. Capscrew	12. Oil seal	18. Bearing cup	24. Key
6. Roll pin	13. "O" ring	19. Shims	25. Wheel hub
7. Washer			26. Snap ring

mainshaft; refer to paragraph 92 for service procedures.

The bevel ring gear is renewable when the differential assembly is removed as outlined in paragraph 96. The ring gear is retained to the differential housing by the same bolts and self-locking nuts that hold the two halves of the differential housing together. Tighten the self-locking nuts to a torque of 90-100 Ft.-Lbs. when installing new gear.

Main drive bevel pinion and bevel ring gear are available in a matched set only.

FINAL DRIVE GEARS

99. BULL PINION SHAFTS. Bull pinion gear shafts are carried in tapered roller bearings located in the differential bearing carriers and in the brake housing. Number of shims used between differential bearing carriers and rear axle center housing to adjust differential carrier bearings and bevel gear backlash will affect bull pinion shaft end play; shims are used between brake housing and rear axle center housing to control bull pinion shaft end play.

To remove the bull pinion shafts, proceed as follows: On D-21 and D-21 Series II, remove rear axle assemblies

as outlined in paragraph 103. On 210 and 220 tractors, remove axle shaft as outlined in paragraph 104A. On all models, remove the brake housings as outlined in paragraph 105A or 106A. With the bull gears resting on bottom of housing, the bull pinion shafts can be withdrawn.

Reassemble by reversing disassembly procedure. Use proper number of shims between brake housing and rear axle center housing to provide 0.002-0.005 end play of bull pinion shafts. Shims are available in thicknesses of 0.005, 0.007 and 0.020. Tighten the brake housing retaining capscrews to a torque of 90-100 Ft.-Lbs. when checking shaft end play.

100. BULL GEARS. To remove the bull gears, first remove the bull pinion shafts as outlined in paragraph 99 and remove top cover from rear axle center housing. The bull gears can then be lifted from the housing.

CAUTION: When reinstalling bull gears in housing on D-21 and D-21 Series II tractors, be sure that side of gear hub having raised boss (B—Fig. AC76) is towards outside of housing. Flat side of hub must be towards the retainer washer (7—Fig. AC77). On 210 and 220 tractors, bearing cup and cone (10 & 11—Fig. AC77A) must be

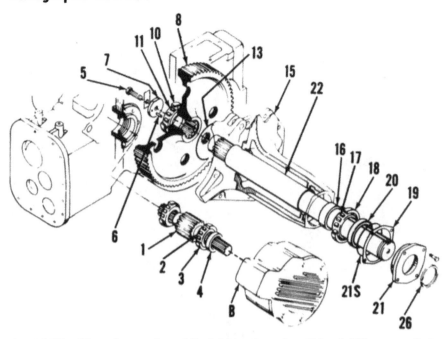

Fig. AC77A—View of rear axle and final drive unit used on 210 and 220 tractors. Brake assembly (B) is located at outside end of bull pinion shaft (1).

1. Bull pinion shaft	7. Washer	15. Axle housing
2. Bearing cones	8. Bull gear	16. Sleeve
3. Bearing cups	10. Bearing cup	17. Bearing cone
4. Oil seal	11. Bearing cone	18. Bearing cup
5. Cap screw	13. "O" ring	19. Shims
6. Roll pin		
20. Seal		
21. Retainer		
21S. "O" ring		
22. Axle shaft		
26. Snap ring		

center housing then completely remove the shaft and housing.

When reinstalling the axle and housing assembly, renew "O" ring at inner end of axle shaft housing. Align splines on axle shafts with splines in bull gear and push into gear until housing reaches the retaining studs; then, lift the bull gear with the axle and housing assembly until the axle housing will enter the bore in center housing. Work the axle and housing into place and tighten the retaining nuts to a torque of 180-190 Ft.-Lbs. Install the pinned washer and capscrew at inner end of axle shaft and tighten the capscrew to a torque of 300-315 Ft.-Lbs.

104. RENEW REAR AXLE, AXLE BEARINGS, INNER OIL SEAL AND/ OR AXLE SHAFT HOUSING. On 210 and 220 tractors, refer to paragraph 104A. On D-21 and D-21 Series II, remove axle shaft and housing assembly as outlined in paragraph 103, then disassemble as follows:

Remove seal and bearing retainer (21—Fig. AC77) and shims (19). Press axle shaft outward until outer bearing cup (18) is free of housing and remove the cup. Remove snap ring (9) at inner end of housing and press shaft inward until inner bearing cup (10) is free of housing and remove the cup. Continue pressing shaft inward until outer bearing cone (17) is free of shaft and remove the cone. Pull

in place before positioning bull gear in housing cavity. Bull gear must be installed with side of hub with splines recessed toward outside (wheel).

101. BEARING ADJUSTMENT. Bearing adjustment of each rear axle shaft is controlled by shims (19—Fig. AC77 or AC77A) between axle housing (15) and retainer (21). To adjust the bearings, remove retainer (21) and add enough shims (19) to provide axle shaft with a measurable amount of end play. Pull axle out to seat bearing cup (18) against retainer (21) and measure axle shaft end play. Remove retainer and vary the number of shims (19) to provide axle bearing with 0.005-0.008 inch pre-load. Shims are available in thicknesses of 0.003 and 0.010. Screws attaching retainer to axle housing should be torqued to 70-75 Ft.-Lbs. for D-21 and D-21 Series II; 90-100 Ft.-Lbs. for 210 and 220 tractors.

102. RENEW AXLE OUTER SEAL. To renew grease seal at outer end of axle shaft, remove wheel guard (fender), support rear end of tractor and remove rear wheel as shown in Fig. AC78. The seal retainer (21—Fig. AC77 or AC77A) can then be removed. Install new seal with lip to inner side of retainer. Reinstall retainer with same number of shims (19) as were removed. Tighten retainer cap screws to a torque of 70-75 Ft.-Lbs. for D-21 and D-21 Series II

tractors; 90-100 Ft.-Lbs. for 210 and 220 tractors.

NOTE: On D-21 and D-21 Series II tractors, axle outer bearing is lubricated by grease through a lubrication fitting (zerk) in rear axle housing. Any evidence of transmission lubricating oil inside rear axle housing indicates inner seal (12—Fig. AC77) is leaking and that the seal should be renewed as outlined in paragraph 104.

103. R&R REAR AXLE SHAFT AND HOUSING. On 210 and 220 tractors, axle and housing can be removed separately. Refer to paragraph 104A. Most service work requiring removal of axle can be accomplished while leaving housing installed.

On D-21 and D-21 Series II tractors, to remove the rear axle and housing as an assembly, proceed as follows:

Drain oil from rear axle center housing. Remove wheel guard (fender) from side of tractor being serviced, support tractor under center housing and remove rear wheel as shown in Fig. AC78. Remove rear cover or PTO/lift shaft housing from rear face of center housing as in paragraph 133. Remove capscrew and pinned washer from inner end of axle shaft. Attach hoist to axle housing, unbolt housing from center housing and withdraw axle shaft and housing from bull gear and center housing. When axle housing is clear of the retaining studs, lower the housing until the bull gear is resting on bottom of

Fig. AC78 — Removing the tapered rear wheel hub. Refer to Fig. AC79 for special capscrews used to remove wheel from hub.

inner bearing cone (11) from shaft, remove sleeve (14), then remove shaft from housing. Remove oil seal (12) from inner end of housing.

To reassemble the unit, proceed as follows: Install seal (12) with lip of seal facing toward inner end of housing. Place the sleeves (14 and 16) on axle shaft with chamfer on sleeves toward ends of shaft. Slide the shaft through housing and inner oil seal and install inner bearing cone (11) on shaft. Drive inner bearing cup (10) into housing past the snap ring groove and install snap ring (9). Bump or press shaft inward to seat the inner bearing against snap ring. Pack outer bearing cone (17) with grease and install cone against spacer on outer end of shaft. Drive outer bearing cup (18) in far enough to remove all shaft end play. Install seal (20) in retainer (21) with seal lip to inside of retainer. Install retainer with proper thickness of shims (19) to provide a shaft rolling torque of 3-7 Ft.-Lbs. when retainer capscrews are tightened to a torque of 70-75 Ft.-Lbs. Note: Bump or press shaft outward to seat bearing against retainer before checking shaft rolling torque (bearing preload). Shims are available in thicknesses of 0.003 and 0.010.

104A. On 210 and 220 tractors, the axle and axle housing are removed separately and if so desired the other unit may remain in place.

To remove the axle shaft (22—Fig. AC77A), drain oil from axle center housing and leave plugs out of final drive compartments. Support tractor under center housing and remove rear wheel as shown in Fig. AC78. Remove rear cover or PTO/lift shaft housing from rear face of center housing as in paragraph 133. Remove bearing retainer (21—Fig. AC-77A) from outer end of axle housing and capscrew (5) and pinned washer (6) from inner end of axle shaft.

Press inner end of axle shaft out until axle shaft is out of inner bearing cone (11) and outer bearing cup (18) is out of axle housing bore. Pull axle shaft out of bull gear (8) and axle housing. Axle shaft, outer bearing (17 & 18) and seal may be serviced at this time. The preceding will also allow removal of pinion shaft (1); which is necessary for removal of some other components.

To renew inner bearing (10 & 11), it is necessary to remove bull pinion shaft (1) as outlined in paragraph 99 and the final drive bull gear (8) as outlined in paragraph 100. The axle housing (22) can be unbolted and removed if required.

If inner bearing (10 & 11) was removed, drive bearing cup into bore until thick side of cup is seated against shoulder of bore. Use grease to hold bearing cone (11) in cup while installing bull gear (8). Bull gear should be installed with recessed

spline toward outside (wheel). Raise bull gear enough to slip axle shaft into hub of bull gear. Continue to press axle into bearing cone (11). When correctly seated, shoulder on axle shaft will be against gear hub and inside of gear hub will be against bearing cone inner race. Install pinned washer (6) and tighten capscrew (5) to 300-315 Ft.-Lbs. torque. Install outer bearing (17 & 18) and adjust axle bearings as outlined in paragraph 101.

To remove only the axle housing, drain center housing, removing wheel and support tractor under axle center housing. Remove retainer (21), unbolt axle housing (22) from center housing, then use a puller to pull axle housing off axle. NOTE: A wall in axle housing will contact outer bearing cone (17) and cone will be pulled from axle as housing is removed. To prevent damage, keep axle housing centered while removing.

BRAKES

D-21 and D-21 Series II tractors use band-disc type brakes as shown in Fig. AC82. Two-Ten and Two-Twenty models are equipped with disc type brakes (Fig. AC82A) which have three discs. Refer to the appropriate following paragraphs.

D-21 and D-21 Series II

105. **ADJUSTMENT.** To adjust either brake, turn capscrew (4—Fig. AC82) in or out as required to provide 3 inches of pedal free travel. Capscrew has self-locking threads. Equalize pedal height by adjusting turnbuckles

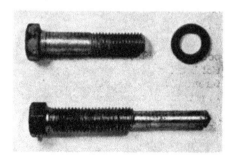

Fig. AC79 — To push rear wheel from tapered hub, remove the three retaining capscrews and washers (top) and insert the three special capscrews. After wheel is pushed from hub, hub can be removed from axle shaft as shown in Fig. AC78. The three special capscrews are provided with each tractor.

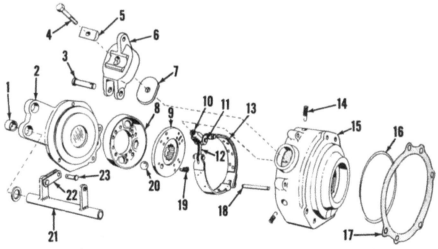

Fig. AC82 — Exploded view of brake unit used on D-21 and D-21 Series II tractors. Outer bearing cup and seal for bull pinion shaft are located in brake housing (15). Shims (17) control end play of bull pinion shaft bearings.

1. Bushing	7. Dust seal	13. Brake band	19. Disc return springs
2. Brake cover	8. Brake drum	14. Band return springs	20. Steel balls
3. Pin	9. Brake disc	15. Brake housing	21. Control shaft
4. Adjusting screw	10. Yoke	16. "O" ring	22. Links
5. Bar seat	11. Pin	17. Shims	23. Pin
6. Actuating lever	12. Link	18. Pivot pin	

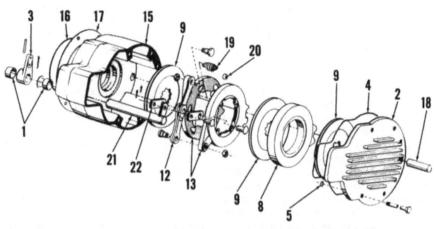

Fig. AC82A—Exploded view of brake assembly used on 210 and 220 tractors. Outer bearing cup and seal for bull pinion shaft are located in brake housing (15). Shims (17) control end play of bull pinion shaft bearings.

1. Bushings	8. Intermediate disc	15. Brake housing	19. Return springs
2. Cover	9. Lined discs	16. "O" ring	20. Steel balls
3. Lever	12. Links	17. Shims	21. Control shaft
4. Spacer wire	13. Actuating disc	18. Pin	22. Links
5. Spacer washers			

on brake control rods (3—Fig. AC83).

105A. R&R AND OVERHAUL. Unless rear wheels are set at or near minimum tread width, the brake covers can be unbolted and removed and and the disc/drum assembly withdrawn. To remove contracting band, use punch or screwdriver to disengage band return springs (14—Fig. AC82) from band (13); then, remove adjusting screw (4) and withdraw the band assembly from pivot pin (18).

After removing the disc/drum assembly and contracting band, the brake housing (15) can be unbolted and removed from rear axle center housing. When reinstalling brake

housing, renew the sealing "O" ring (16) and install correct number of shims (17) for proper bull pinion shaft bearing adjustment as outlined in paragraph 99.

To disassemble the removed disc/-drum unit, insert punch or screwdriver through disc return springs (19) and extend the springs only far enough to unhook them. Remove disc (9) and the four steel balls (20) from drum (8).

Linings are not available separately from disc, drum or band. If necessary to renew linings, renew the complete disc, drum or band assembly. Inspect friction surfaces of brake housing (15) and cover (2) for scoring, scuffing or warpage and renew either part if friction surface is not suitable for further use.

Condition of return springs (14 and 19) is of utmost importance when servicing band/disc brakes. Renew any springs if coils do not fit tightly together and be careful not to stretch the springs any farther than necessary when reassembling brakes. In-

sufficient spring tension will allow the brakes to drag as they will not be returned to proper position when pedals are released.

Reverse removal procedures to reinstall brakes. Tighten the brake cover retaining capscrews to a torque of 70-75 Ft.-Lbs. Adjust brakes as outlined in paragraph 105.

Two-Ten and Two-Twenty

106. **ADJUSTMENT.** Equalize pedal height by removing operator's platform and adjusting length of brake rods (21—Fig. AC83A) by turning the turnbuckles. Adjustment to compensate for lining wear is limited to removal of spacer wire (4—Fig. AC82A) and spacers (5) around capscrews. The center lined disc can be expected to wear faster than the other two discs.

106A. **R&R AND OVERHAUL.** Unbolt and remove cover (2—Fig. AC-82A), then withdraw outer lined disc (9), intermediate disc (8) and the center lined disc. Remove pin connecting links (12 & 22), then remove actuating disc (13) as an assembly with balls (20) and springs (19). Remainder of disassembly and inspection will be self-evident. Intermediate disc (8) is 1.370 inch thick.

If required, the brake housing (15) can be removed after disconnecting linkage and removing the two retaining screws. When reinstalling, renew "O" ring (16) and install correct thickness of shims (17) for proper bull pinion shaft bearing adjustment as outlined in paragraph 99. The two brake housing retaining screws should be tightened to 90-100 Ft.-Lbs. torque.

When installing actuating disc (13) and intermediate disc (8), make certain that notch is engaged with pin (18). Spacer washers (5) are 0.148-0.150 thick and should be used with spacer wire (4) when new lined discs (9) are installed. Tighten the cover retaining screws to 70-75 Ft.-Lbs. torque.

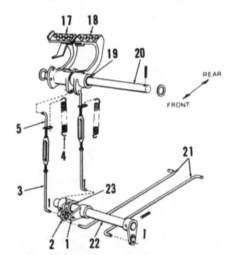

Fig. AC83—View of brake linkage used on D-21 and D-21 Series II tractors. Brake rods (21) connect to control shafts (21—Fig. AC82) which are mounted on the rear axle center housing.

1. Lever	18. Pedal, L.H.
2. Bellcrank	19. Bushings
3. Rod, lower	20. Upper shaft
4. Return springs	21. Brake rods
5. Rod, upper	22. Lower shaft
17. Pedal, R.H.	23. Bushings

Fig AC83A — View of brake linkage used on 210 and 220 tractors. Refer to Fig. AC83 for legend.

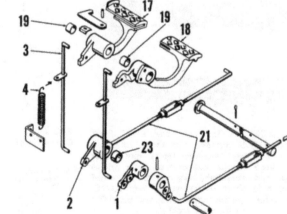

POWER TAKE-OFF

All models are available with an independent 1000 RPM power take-off. Power is transmitted to the PTO reduction gears in the torque housing via a hollow shaft that is splined into the engine clutch cover. On D-21 and D-21 Series II tractors, a shifter unit to engage or disengage the PTO drive is located in the four speed transmission housing to rear of reduction gears but in front of the PTO clutch. The shifter coupling can be engaged only when the engine is NOT running.

On D-21 and D-21 Series II tractors, the PTO clutch is a dual multiple disc type with over-center linkage and is submerged in transmission oil. When the three-position clutch lever is in forward position, the rear section of the clutch is engaged to drive the PTO output shaft. When lever is in rear position, the front section of the dual clutch is engaged which acts as a PTO output shaft brake. With lever in center position, both sections of the dual clutch are disengaged.

On 210 and 220 tractors, the PTO clutch and brake are actuated by hydraulic pressure. With lever forward, clutch is engaged and with lever to rear, brake is engaged.

TEST AND ADJUST

D-21 and D-21 Series II

107. **ADJUST PTO CLUTCH.** To adjust the PTO clutch, the lift shaft housing and clutch assembly must be removed from the tractor as shown in Fig. AC89.

Clutch plate pressure is applied through a spring (Belleville) washer (8—Fig. AC92A) that is located between the preload plate (7) and pressure plate (9) of each clutch pack.

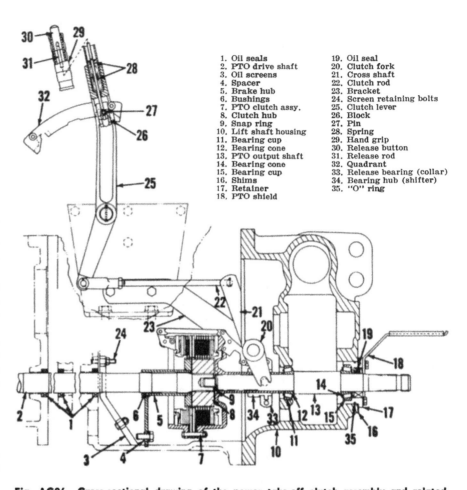

1. Oil seals
2. PTO drive shaft
3. Oil screens
4. Spacer
5. Brake hub
6. Bushings
7. PTO clutch assy.
8. Clutch hub
9. Snap ring
10. Lift shaft housing
11. Bearing cup
12. Bearing cone
13. PTO output shaft
14. Bearing cone
15. Bearing cup
16. Shims
17. Retainer
18. PTO shield
19. Oil seal
20. Clutch fork
21. Cross shaft
22. Clutch rod
23. Bracket
24. Screen retaining bolts
25. Clutch lever
26. Block
27. Pin
28. Spring
29. Hand grip
30. Release button
31. Release rod
32. Quadrant
33. Release bearing (collar)
34. Bearing hub (shifter)
35. "O" ring

Fig. AC86—Cross-sectional drawing of the power take-off clutch assembly and related parts. The D-21 has a 1000 RPM PTO output shaft (13) only. Oil seals (1) on PTO drive shaft (2) help retain oil level in each compartment during up or down-hill operation, but have no function when tractor is operating on level ground.

Fig. AC85—Clutch adjusting shims on D-21 and D-21 Series II models are located at A, B and C. Measure distance at D and E with clutch both in engaged and disengaged positions.

The spring washer must be compressed 0.042-0.046 inch when clutch pack is engaged. If compression is less, slippage of clutch will result. If compression of spring washer is greater than 0.046, clutch pack will not release properly. Adjustment is provided with three shim stacks (A, B & C—Fig. AC85) placed between the clutch housings and adjoining center plates and between the center plates.

To check clutch adjustment, use an inside hole gage of 2/10 to 3/10-inch capacity and a micrometer to measure clearance at (D & E) between the preload plate and pressure plate of each clutch pack, first with clutch pack engaged, then in the disengaged position. Measurements should be made at each of the three openings around the clutch housings and an average of these dimensions used. Subtract the average engaged dimension from the average disengaged dimension for each clutch pack. If the

difference between the two averaged dimensions for each clutch pack is between 0.042 and 0.046, no adjustment of the clutch is necessary. If the difference between the averaged engaged and disengaged dimensions of either pack is less than 0.042, remove sufficient shim thickness from between that clutch pack housing and adjacent center plate to increase spring compression to 0.042-0.046 and add this same thickness to the shim stacks between the two center plates. For example, if the difference between the average engaged and disengaged dimension (E) of the front clutch (brake) pack was 0.035, removing a 0.010 thick shim at each of the three stacks at (C) and adding a 0.010 thick shim at each of the three stacks at (B) would increase compression of the spring washer to 0.045 which is within limits of proper adjustment.

If the difference between the average engaged and disengaged dimen-

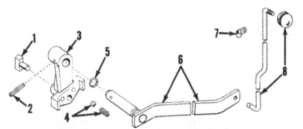

1. Insert
2. Roll pin
3. Shifter arm
4. Detent ball & spring
5. "O" ring
6. Shift lever & shaft
7. Rod guide
8. Shifter rod & knob

Fig. AC88—Exploded view of PTO shifter assembly and linkage on D-21 and D-21 Series II tractors. Shifter assembly is located in four-speed transmission housing.

sions is more than 0.046, add sufficient shim thickness between that clutch pack housing and adjacent center plate and remove the same thickness from between the two center plates. For example, if the difference between the average engaged and disengaged dimension (D) of the rear clutch pack was 0.050, adding 0.005 thickness of shims at (A) and removing 0.005 thickness of shims at (B) would decrease compression of spring washer to 0.045.

NOTE: As 0.005 thick shims are not provided for use between clutch housings and the center plates, add a 0.010 shim and remove a 0.015 shim at each of the three shim stacks (A or C) to reduce shim stack height by 0.005; or, add a 0.015 thick shim and remove a 0.010 shim at each of the three shim stacks to add 0.005 to shim stack height. Shims of 0.005 thickness are provided for service use between the two center plates; however, no 0.005 shims are used in original assembly of the clutch unit.

108. **ADJUST PTO CLUTCH LEVER.** Refer to Fig. AC86 and proceed as follows: Move control lever forward until clutch is engaged (linkage snaps over-center). Lever detent should then be centered in front notch of inner section of the quadrant (32) with control rod (22) free. If not, disconnect and adjust length of con-

trol rod (22) so that rod can be easily reconnected with lever detent in forward notch and the clutch engaged.

Then, move lever rearward until brake section of clutch unit is engaged. Lever detent should then be centered in notch in outer section of the quadrant (32). If not, loosen quadrant mounting bolts and readjust position of outer section of quadrant.

Two-Ten and Two-Twenty

109. **RELIEF PRESSURE.** System pressure can be checked by installing a suitable pressure gage in the two ports located in the right rear flange of lift housing. The upper port is in the clutch system and lower port is for PTO brake. Normally pressure of brake system will be slightly higher than clutch system pressure; however, if difference is excessive, the cause should be determined before changing relief valve setting.

With pressure gage installed in upper test port, allow oil to reach 100° F., then operate engine at high idle RPM. Pressure should increase gradually after PTO control handle is moved to the forward (engage) detent position until pressure reaches 225-275 psi. If pressure immediately reaches relief setting or if pressure does not exceed 50 psi, check modulating piston (4—Fig. AC90) for dam-

age and/or contaminated fluid (modulating piston sticking). If relief valve setting is incorrect add or remove shims (9) as required. Addition of one 0.030 thick shim will increase pressure approximately 13 psi. No more than six shims should ever be installed. If relief valve setting is changed, refer to paragraph 10B and check power steering relief pressure.

CONTROLS

D-21 and D-21 Series II

110. **SHIFTER COUPLING.** To service the PTO shifter coupling (23—Fig. AC91) and linkage (see Fig. AC-88), remove four-speed transmission from tractor as outlined in paragraphs 81 and 83, and remove mainshaft and countershaft as outlined in paragraphs 86 and 87. Then, proceed as follows:

Slide the PTO rear drive shaft out of transmission if not already removed and extract shifter coupling from bottom of transmission. Remove insert (1—Fig. AC88) from shifter arm (3). Drive roll pin out of arm and shift lever shaft, remove the arm, detent ball, spring and lever shaft from transmission housing.

To reassemble, reverse the disassembly procedure. Install new "O" ring (5) on shift lever shaft and lubricate the "O" ring before inserting shaft through bore in housing.

111. **PTO CLUTCH FORK AND CROSS SHAFT.** To remove the PTO clutch fork and cross shaft, first remove the lift shaft housing as outlined in paragraph 133. Using a pin punch, drive shaft retaining pin from hole (P—Fig. AC89) in the flange at right side of housing. Slide the cross shaft out far enough to remove the fork retaining keys (K), then completely withdraw the cross shaft and remove the fork.

Remove "Traction-Booster" control shaft (23—Fig. AC115) from PTO cross shaft on tractors with 3-point hitch. Bushings (24) in each end of the hollow control shaft are renewable.

Renew the cross shaft sealing "O" ring before reinstalling shaft. Lubricate the "O" ring and be sure there are no burrs on shaft that would damage the "O" ring. Drive the shaft retaining pin (P—Fig. AC89) in until flush with front face of lift shaft housing.

Two-Ten and Two-Twenty

112. **CONTROL VALVE.** The PTO control valve is located behind panel below the control lever. The control

Fig. AC89 — View of hydraulic lift housing removed from D-21 tractor showing power take-off clutch assembly. Cross shaft (21) is retained in housing by pin (P).

K. Woodruff keys
P. Pin
7. Clutch assembly
20. Clutch fork
21. Cross shaft
33. Release bearing

valve should be removed for all service except adjusting relief pressure.

The accumulator valve plug (13—Fig. AC90) can be withdrawn using a ¼-28 UNF screw in tapped hole, after removing snap ring (12). The modulating valve piston (4) can be removed by using a hooked wire caught in the 0.120 diameter cross drilled hole. To remove control valve spool without damage, first remove detent plug (25), the two outside detent springs (21 & 22) and the outside detent plunger (23). Remove end cap (30), screw (29), spring seats (28 & 33), spacer (32), spring (31) and plate (26). Push control valve spool (20) up toward lever end just far enough to remove "O" ring (27) from groove in valve body, then remove valve spool from bottom (cap) end of valve body. NOTE: Do not lose the inner detent plunger (23) or springs (21 & 22).

Reassemble in reverse of disassembly procedure, using new "O" rings. Install control valve spool (20) from lower end of valve body before installing "O" ring (27). Use a punch to compress the inner detent plunger (23) when installing spool. Push control valve spool up far enough to install "O" ring (27), then lower the

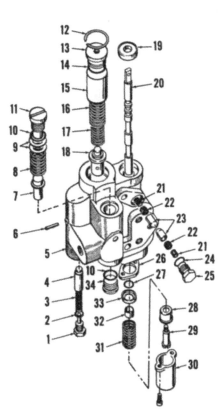

Fig. AC90—Exploded view of PTO control valve. "O" ring seal (27) is located in groove at lower end of valve spool bore in body (5).

1. Modulating valve plug	18. Pressure reducing valve
2. "O" ring	19. Oil seal
3. Spring	20. Valve spool
4. Modulating valve piston	21. Inner detent springs
5. Valve body	22. Outer detent springs
6. Taper pin	23. Detent plungers
7. Relief valve piston	24. "O" ring
8. Spring	25. Detent cap
9. Shims	26. Seal plate
10. "O" rings	27. "O" ring
11. Relief valve plug	28. Spring seat
12. Snap ring	29. Screw
13. Accumulator plug	30. Cap
14. "O" ring	31. Centering spring
15. Accumulator piston	32. Spacer
16. Outer spring	33. Spring seat
17. Inner spring	34. Plug

spool. The hole in end of modulating valve piston (4) is 0.040 and the cross drilled hole is 0.120.

REDUCTION GEARS AND SHAFTS

All Models

113. **R&R AND OVERHAUL.** On D-21 and D-21 Series II models, PTO reduction gears and shafts can be serviced after transmission oil is drained and engine is removed as outlined in paragraph 20. The input shaft can be removed from 210 and 220 tractors after engine is removed, but torque housing must be removed as outlined in paragraph 78 then coupling (23—Fig. AC91A) and retaining ring (23R) must be removed before driven gear and shaft. On all models, bearing cup (13—Fig. AC91 or AC91A) is installed in transmission input shaft bearing retainer (2—Fig. AC67) and torque housing must be separated from transmission housing before removing this bearing cup. Refer to Fig. AC91 or AC91A and proceed as follows.

Remove clutch fork and release bearing as outlined in paragraph 75. Unbolt bearing retainer (6) from torque housing and remove the retainer and shims (10). PTO drive shaft and gear (11) can then be withdrawn

Remove cover (15) and shims (16) from front of torque housing and remove plate (25) from bottom of housing. On D-21 and D-21 Series II, be sure that PTO shift coupling is in the engaged position. On all models, thread a slide hammer adapter into front of shaft (19) and bump the shaft forward out of housing. Bearing cup (17) will be withdrawn with shaft; catch gear (20) and remove the gear and bearing cone (21) from bottom

Fig. AC89A—View of rear opening in D-21 rear axle center housing with lift shaft housing and PTO clutch assembly removed.

2. PTO drive shaft
3. Oil screen
5. Brake hub

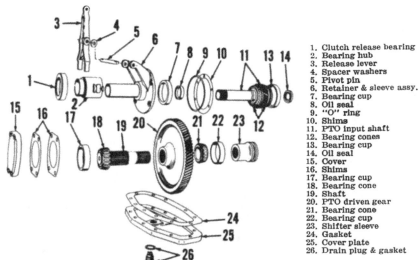

1. Clutch release bearing
2. Bearing hub
3. Release lever
4. Spacer washers
5. Pivot pin
6. Retainer & sleeve assy.
7. Bearing cup
8. Oil seal
9. "O" ring
10. Shims
11. PTO input shaft
12. Bearing cones
13. Bearing cup
14. Oil seal
15. Cover
16. Shims
17. Bearing cup
18. Bearing cone
19. Shaft
20. PTO driven gear
21. Bearing cone
22. Bearing cup
23. Shifter sleeve
24. Gasket
25. Cover plate
26. Drain plug & gasket

Fig. AC91—Exploded view of PTO reduction gears and related parts used on D-21 and D-21 Series II models. PTO shifter sleeve (23) is controlled by shifter fork and linkage shown in Fig. AC88.

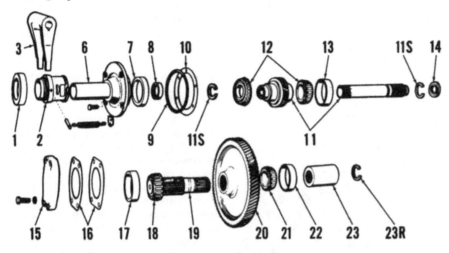

Fig. AC91A—Exploded view of PTO reduction gears and related parts used on 210 and 220 tractors. Retaining ring (23R) holds coupling (23) to shaft (19). Snap rings (11S) hold gear to shaft (11). Refer to Fig. AC91 for legend.

opening. Bearing cup (22) can be pulled forward out of bore with cup puller attachment on slide hammer.

Reinstall PTO reduction gears and shafts as outlined in paragraph 113A or in paragraph 113 B.

113A. If torque housing has been detached from transmission, reassemble as follows:

Install bearing cone (18—Fig. AC91 or AC91A) on front end of shaft (19). Install bearing cup (22) in rear bore of housing and stick cone (21) in cup with heavy grease. Insert gear (20) through bottom opening of housing with long hub of gear forward. Insert shaft (19) through gear and bearing cone and bump the shaft to rear until there is no end play of gear between the two bearing cones. Install bearing cup (17) in front bore of housing. Install cover (15) with correct thickness of shims (16) to provide 0.001-0.005 end play of shaft and gear as-

sembly in the bearing cups. Note: Alternately place the 0.006 thick paper and 0.010 thick metal shims for proper sealing. Bump the shaft forward to seat bearing cup against cover (15) before checking end play.

Install bearing cup (13) in transmission input shaft bearing retainer (2—Fig. AC67). On 210 and 220 models, position retaining ring (23R—Fig. AC91A) in groove at rear of shaft (19), then slide coupling (23) onto shaft until retaining ring enters groove in center of coupling. On all models, re-attach torque housing to transmission. Tighten the ½-inch capscrews to a torque of 70-75 Ft.-Lbs. and the ⅝-inch capscrews to a torque of 130-140 Ft.-Lbs.

Install bearing cones (12—Fig. AC91 or 91A) on PTO input shaft and install new oil seal (14) inside shaft with lip of seal to rear. Lubricate seal and slide PTO drive shaft over the transmission input shaft. Install bear-

ing cup (7) and new oil seal (8) in front bearing retainer (6) with lip of seal to rear. Lubricate the seal and install retainer with correct thickness of shims (10) to provide 0.003-0.005 end play of PTO drive shaft in bearings. Shims are available in thicknesses of 0.003 and 0.010. Tighten the retaining cap screws to a torque of 35-40 Ft.-Lbs. Reinstall clutch fork and release bearing and reinstall engine in tractor.

113B. If servicing PTO reduction gears with torque housing attached to transmission, reassemble unit as follows:

Install bearing cup (22) in rear bore of torque housing and stick cone (21) in cup with heavy grease. Install cone (18) on front end of shaft (19), place gear (20) in housing, with long hub forward, and insert shaft through gear and rear bearing and into shift coupling. Bump the shaft to rear with hollow driver that will contact only the inner race of bearing cone (18) until there is no end play of gear on shaft between the two bearings. Install front bearing cup (17) part of the way in bore; then, install the cover (15) and shims (16). Tighten the cover retaining cap screws to draw the bearing cup into place, then remove the cover and check end play of the shaft carefully to avoid moving front bearing cup while end play is being checked. Note: A spare cover (15) can be reworked to check shaft end play by drilling a hole in center of cover and checking end play through hole. If end play is greater than 0.005, remove sufficient shim thickness to bring end play within limits of 0.001-0.005. Alternately place the 0.006 paper and 0.010 thick metal shims for proper sealing. If end play is less than 0.001, bump the bearing cup out a short distance, then reinstall the cover with additional shims to bring end play within limits.

Reinstall the PTO drive shaft (11) and retainer as outlined in paragraph 113A.

CLUTCH AND BRAKE

D-21 and D-21 Series II

114. R&R AND OVERHAUL. The PTO multiple disc clutch assembly is removed with the hydraulic lift shaft cover. Refer to Fig. AC89, and to paragraph 133 for removal procedure.

To remove the PTO clutch assembly from the lift shaft cover, proceed as follows: Extract the clutch hub (8 — Fig. AC92) from clutch assembly and remove snap ring (9) from front end of PTO output shaft (13). The

Fig. AC92 — To remove assembly (7) from output shaft (13), extract clutch hub (8) from clutch assembly and remove snap ring (9) from front end of shaft. It is not necessary to remove the cross shaft and fork.

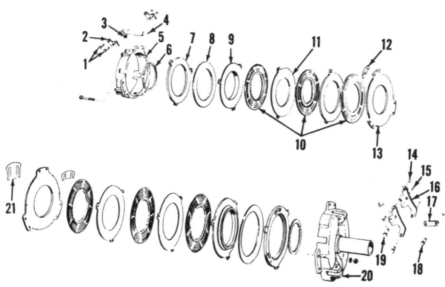

Fig. AC92A—Exploded view of the PTO clutch assembly. Three bronze discs (10) and two steel discs (11) are used in the brake (front) section of the assembly. Seven bronze discs and six steel discs are used in the clutch (rear) section. Refer to Fig. AC85 for assembled view of clutch. Note: Not all of the above parts are used and some may differ from that shown.

1. Pins	6. Snap ring	11. Steel discs	17. Links
2. Release levers	7. Pre-load plate	12. Shims	18. Pins
3. Rollers	8. Belleville washer	13. Center plate	19. Pins
4. Links	9. Pressure plate	14. Snap rings	20. Rear housing
5. Front housing	10. Bronze discs	15. Snap rings	21. Shims
		16. Release levers	

clutch assembly can then be withdrawn from the output shaft by pivoting the cross shaft and clutch fork forward.

To disassemble the PTO clutch assembly, refer to Fig. AC92A and proceed as follows: Mark the front and rear housings (5 & 20) prior to disassembly as the housings are a balanced assembly. Disconnect the three links (4) from release levers, loosen the six bolts retaining the housings (5 & 20) together and remove the shim stacks at each of the three bolting points. Unbolt and separate the clutch housings, discs and center plates. Compress the preload plate (7), spring washer (8) and pressure plate assembly (9) to remove the snap rings (6).

Inspect the clutch discs and renew any that are excessively worn or have damaged notches for the clutch hub splines. Inspect all other parts and renew any that are questionable. The free height of the spring washer in each clutch pack should be 0.270-0.302. Clutch plates with internal notches for clutch hub splines should measure 0.117-0.123 thick. Steel plates should be renewed if scored or showing signs of being overheated. All plates should be flat within 0.009.

To reassemble pre-load plate, spring washer and pressure plate, place parts in clutch housing to keep drive tangs aligned, compress spring and install snap ring. Reassemble clutch packs and install three 0.110 thick shim stacks at (A—Fig. AC85), three 0.080

thick shim stacks at (B) and three 0.130 thick stacks at (C). The 0.110 thick stacks should contain four 0.015 thick shims and five 0.010 thick shims; the 0.080 thick shim stacks should contain four 0.015 thick shims and

two 0.010 thick shims; and the 0.130 shim stacks should contain six 0.015 and four 0.010 thick shims. Shims of 0.005 thickness are available for service installation at (B).

All pins should be installed with heads in direction of clutch rotation to prevent failure of the retaining snap rings.

Adjust clutch as outlined in paragraph 108 before reinstalling lift shaft housing to tractor.

Two-Ten and Two-Twenty

115. R&R AND OVERHAUL. Remove the lift shaft housing as outlined in paragraph 133, unbolt and remove shield (19—Fig. AC93), then slide clutch hub (18) out of clutch. Remove thrust washers (17) and snap ring (15), then lift clutch assembly off. Remove snap ring (1—Fig. AC93A) and withdraw end plate (2), lined discs (3) and driven plate (4). Remainder of clutch disassembly will be self evident.

The Teflon sealing rings (9 & 11) are easily broken when cold. Install steel expanders (8 & 10) in grooves, heat Teflon sealing rings in water or oil to 150°-200° F., then work Teflon sealing rings into grooves using a blunt tool with no sharp edges. Make certain that all seal contacting areas

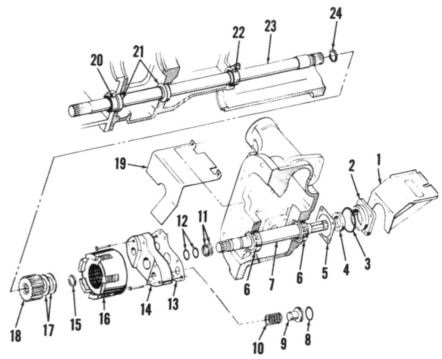

Fig. AC93—View of PTO rear and center shafts and associated parts. Clutch (16) is shown in Fig. AC93A. Two brake pistons (9) are used.

1. Guard	7. Rear shaft	13. Gasket	19. Clutch shield
2. Bearing retainer	8. "O" rings	14. Oil manifold	20. Bushing
3. "O" ring	9. Brake piston	15. Snap ring	21. Oil seal
4. Oil seal	10. Spring	16. Clutch assembly	22. Shaft bushing
5. Shims	11. Cast iron seal	17. Thrust washers	23. Center shaft
6. Bearings	rings	18. Clutch hub	24. Snap ring
	12. "O" rings		

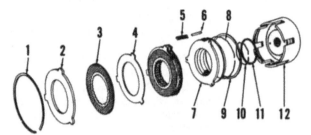

1. Snap ring
2. End plate
3. Lined discs (7 used)
4. Driven plates (6 used)
5. Return spring (3 used)
6. Pins (3 used)
7. Clutch piston
8. Steel expander
9. Teflon seal ring
10. Steel expander
11. Teflon seal ring
12. Clutch drum

Fig. AC93A—Exploded view of PTO clutch assembly. Refer to Fig. AC93 for associated parts including the clutch hub.

of clutch drum (12) and piston (7) are smooth and well lubricated, then slide piston into clutch drum. Install the seven lined discs (3) and six driven plates (4) alternately, starting and ending with a lined disc. Locate the three pins (6) in holes in piston and install springs (5) around pins. Position end plate (2) and compress against springs (5), making sure that pins (6) engage holes in end plate, then install snap ring (1).

The oil manifold (14—Fig. AC93) can be unbolted and removed from front of lift shaft housing after clutch is removed. Make certain that brake bores in manifold are not nicked or worn excessively. Ends of cast iron seal rings must be locked together before installing manifold with brake pistons (9) and springs (10). The eight cap screws attaching manifold to lift shaft housing should be tightened to 35-40 Ft.-Lbs.

Renew "O" rings (12), lubricate shaft and bore in clutch drum, then slide clutch assembly onto PTO rear shaft (7). Install snap ring (15) and use grease to hold thrust washers (17) in place. Install hub (18), making certain that hub is completely into clutch and is correctly engaging all seven lined discs, then install shield (19).

Install lift shaft housing, making certain that hub (18) is not pulled out of clutch disc while assembling. If lift shaft housing will not slide completely against rear face of axle center housing, check to make certain that hub has not moved forward and allowed one (or more) of the clutch lined discs to fall down behind the hub. CAUTION: The PTO clutch will be damaged if lift housing is attached to axle center housing with a lined disc behind hub. A tab on shield (19) is provided on models after tractor serial number 220-1560D (transmission

serial number P1563) which should be bent toward PTO clutch hub to prevent hub from moving forward. Clearance between tab and clutch hub should be ⅛-inch.

PTO OUTPUT SHAFT & BEARINGS

D-21 and D-21 Series II

116. **R&R AND OVERHAUL.** To remove the PTO output shaft, first remove the PTO clutch unit from front end of shaft as outlined in paragraph 114. The shaft and bearing cones can then be withdrawn fom rear of housing after removing the rear bearing retainer and shims as shown in Fig. AC94. Remove the front bearing cup from lift shaft housing and the rear bearing cup and seal from bearing retainer.

When reassembling shaft in housing, reinstall bearing retainer with proper number of shims to provide 0.003-0.005 end play of shaft in bearings. Shims are available in thicknesses of 0.003 and 0.005. Install new seal (19—Fig. AC86) in retainer (17) with lip of seal towards front (inside) of housing.

Two-Ten and Two-Twenty

117. **R&R AND OVERHAUL.** The center PTO shaft can be withdrawn from rear after the lift shaft housing is removed as outlined in paragraph 133. Bushing (20—Fig. AC93) is located in rear of transmission housing. To renew bushing (20) it is necessary to split tractor between transmission and rear axle center housing as outlined in paragraph 82.

To remove the PTO rear shaft (7), remove clutch assembly and manifold (14) as outlined in paragraph 115. Unbolt and remove retainer (2) and guard (1), then pull shaft (7) out toward rear. The cup for rear bearing will be removed when shaft is removed.

Shims (5) are available in thicknesses of 0.003 and 0.005 and should be selected to provide shaft end play of 0.002-0.005 in bearings. Cap screws attaching guard (1) and retainer to lift shaft center housing should be tightened to 30-35 Ft.-Lbs. torque. Renew cast iron seal rings (11), install manifold (14) and clutch assembly (16) as outlined in paragraph 115 after PTO rear shaft is installed and end play is adjusted.

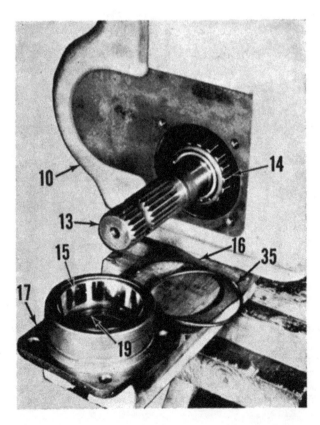

Fig. AC94 — View showing bearing retainer (17) removed from rear face of lift shaft housing.

10. Lift shaft housing
13. PTO output shaft
14. Bearing cone
15. Bearing cup
16. Shims
17. Retainer
19. Oil seal
35. "O" ring

HYDRAULIC SYSTEM

Tractors are available with a hydraulic system designed for control of remote cylinders only, or with a Category III 3-point hydraulic lift system having provision for the operation of remote hydraulic cylinders. Also, the model D-21 is available without a hydraulic lift system. All models are equipped with a hydraulic power steering system.

On D-21 tractors having a remote control type hydraulic system, either a one-spool or a three-spool remote control valve is used.

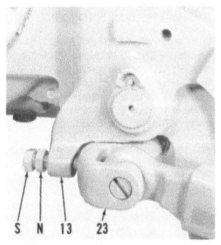

Fig. AC95 — View of torsion bar adjusting screw (S) and lock nut (N). Refer to paragraph 121 for adjustment procedure.

11. Torsion bar
13. Bracket, L.H.
23. Torsion bar tube

On 210 and 220 tractors without 3-point hitch, two remote control valves are used. All tractors having a 3-point hitch are equipped with three-spool control valves; one spool is used to operate the 3-point hitch and the other two spools can be used to control remote hydraulic cylinders. In addition to the three-spool valve, a separate one-spool valve is used for draft control (Traction Booster) action and for position control of the 3-point hitch.

All tractors are equipped with a one, two or three section gear type pump depending upon type of hydraulic equipment. A one section pump is used for the power steering system on D-21 and D-21 Series II tractors without other hydraulic equipment. A two section pump is used for D-21 and D-21 Series II tractors without Traction Booster. A three section pump is used on D-21 and D-21 Series II tractors with Traction Booster. A one section pump is used on 210 and 220 tractors and flow is divided to provide approximately 4.25 GPM to the power steering (and PTO clutch if so equipped) and 17.2 GPM to the lift system. The two section pump used on 210 and 220 tractors equipped with Traction Booster is similar

to the one section pump with an additional (smaller) section for the "Traction Booster".

Oil from the transmission and final drive sump is used as fluid for the hydraulic system and power steering. Refer to note preceding paragraph 80.

CHECKS AND ADJUSTMENTS

D-21 and D-21 Series II

121. **TORSION BAR ADJUST-MENT.** Remove any weight or implement attached to the three point hitch. Loosen lock nut (N—Fig. AC95) and back adjusting screw (S) out until torsion bar tube (23) is free to turn in the support brackets. Then, turn adjusting screw in just far enough to eliminate all free movement of the torsion bar tube and tighten the lock nut while holding the screw in this position.

121A. **TRACTION BOOSTER (DRAFT) ADJUSTMENT.** Remove any weight or implement attached to three-point hitch and/or drawbar, then adjust the torsion bar pre-load as outlined in paragraph 121. Move lift arm lever (1—Fig. AC95B) to Trac-

Fig. AC95A — View of three-point hitch linkage on D-21 tractor, showing Traction Booster (TB) and position control (PC) adjustment points. Traction Booster link pivot should be installed in outer hole (H) of torsion bar tube when using semi-mounted implements.

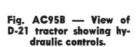

Fig. AC95B — View of D-21 tractor showing hydraulic controls.

A. Lower position for 1
B. Traction Booster position for 1
C. Hold position for 1
D. Raise position for 1
1. 3-point hitch control lever
2. Remote control levers
3. Traction Booster control lever
4. Position control lever
5. Friction adjustment nut
6. Traction Booster valve
7. High volume bleed-off screw
8. Relief valve plug
9. Control valve assy.
10. Control valve manifold
11. Traction Booster gage line

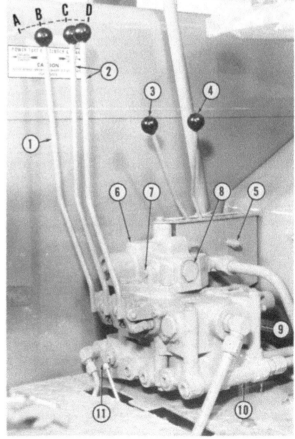

tion Booster position (B), move position control lever (4) all the way forward and Traction Booster lever (3) all the way to the rear. Observe position of lift arms with engine running at 1000 RPM. If lift arms are not horizontal (straight back), turn the adjusting nut (TB—Fig. AC95A) as required until the lift arms are as near horizontal as possible.

NOTE: Traction Booster link pivot should be installed in outer hole (H) of torsion bar tube when using semi-mounted implements.

121B. POSITION CONTROL ADJUSTMENT. With engine running at slow idle speed, proceed as follows: Move Traction Booster control lever (3—Fig. AC95B) to full forward position. Lower the lift arms, then move 3-point hitch control lever (1) to the Traction Booster detent position (B). Move position control lever (4) fully rearward and back-off the position control adjusting nut (PC—Fig. AC95A) until the lift arms reach top of stroke and Traction Booster relief valve is by-passing oil which can be detected by a squealing or buzzing sound. Then tighten the position control nut until relief valve seats (Traction Booster gage pressure drops).

121C. LEVER FRICTION ADJUSTMENT. With the engine stopped, completely lower the lift arms. Move the Traction Booster control lever (3—Fig. AC95B) and the position control lever (4) to full rearward position. If the levers will not stay in this position, tighten the friction adjusting nut (5).

121D. HIGH VOLUME BLEED-OFF ADJUSTMENT. The bleed-off adjusting screw (7—Fig. AC95B) should normally be set four turns out from seated position. To provide faster rate of lift for Traction Booster and position control operation, turn screw (7) in slightly. To slow the rate of lift, turn screw out. When setting the bleed-off screw at other than normal (four turns open) position, move screw ¼-turn and check operation.

122. LOWERING RATE ADJUSTMENT. To adjust the lowering rate of the three-point hitch, disconnect linkage between control lever (1—Fig. AC95B) and 3-point hitch control valve spool. Hold spool from turning with wrench on flats at front end of valve spool sleeve and turn the spool end in to decrease rate of lower or out to increase rate of lower. The normal rate of lowering can be obtained by turning the spool end all the way

in, then backing it out three to three and one-half turns.

122A. SYSTEM RELIEF PRESSURE. To check hydraulic system relief pressure, install a 3000 psi gage in a remote cylinder port and pressurize that port. Note: Control valve lever must be held in position when checking pressure. Gage pressure should be 2250-2350 psi with engine running at 2000 RPM. Relief pressure can be adjusted by removing the cap (11—Fig. AC117) and turning adjusting screw (10) in to increase or out to decrease pressure.

122B. CONTROL LEVER RELEASE PRESSURE. When engine is running at normal operating speeds, the remote cylinder control levers should automatically return to neutral position when remote cylinder reaches end of stroke or the 3-point hitch control lever should return to neutral position from raising position when lift arms reach fully raised position. If not, or if lever returns to neutral prematurely, remove plug (14—Fig. AC117) from spring end cap of affected valve spool and turn the screw (16) in to increase or out to decrease the pressure at which the lever will return to neutral.

122C. TRACTION BOOSTER RELIEF VALVE. Pressure in the Traction Booster and position control circuit is controlled by a relief valve, located within the single spool valve housing. To check relief pressure, install a 3000 psi gage at connection where Traction Booster gage line is installed on control valve manifold. With engine running at 2000 RPM, actuate the Traction Booster valve spool with screwdriver and observe gage reading. If not between 2000-2200 psi, remove plug (8—Fig. AC95B) and adjust relief valve by adding or removing shims. The Traction Booster relief valve adjusting shims are shown at (13—Fig. AC116).

122D. PRESSURE AND VOLUME TESTS. Relief valve settings and volume of flow from lift system and Traction Booster sections of pump can be checked using an OTC tester (Y81-21) or equivalent as follows: Disconnect Traction Booster gage tube from port (11—Fig. AC95B) in oil manifold and attach tester inlet hose. Attach return hose from tester to the sump filler opening. Remove plug (14—Fig. AC117) from end of valve and turn adjusting screw (16) in enough to prevent lever (1—Fig. AC95B) from automatically returning to hold position.

Open tester valve and operate engine at 2200 RPM. Move lever (1) to lift position (rear) and slowly close the tester until presure is 2000 psi. Volume of flow should be approximately 16.5 GPM when pressure is 2000 psi and engine speed is 2200 RPM. Volume of flow should be checked at 200 psi less than relief valve setting. Close tester until volume of flow is zero and note the relief pressure. Lift system relief pressure should be 2250-2350 psi at 2000 RPM engine speed. Refer to paragraph 122A for setting lift system relief valve.

The Traction Booster relief valve and pump volume can be checked with same connections. Move control lever (1—Fig. AC95B) to TRACTION BOOSTER position and the Traction Booster lever (3) all the way to rear. Adjust Traction Booster rod at nut (TB—Fig. AC95A) so that sensing valve plunger is pushed all the way forward when Traction Booster lever (3—Fig. AC95B) is moved to rear. Turn the high volume bleed-off screw (7) in until seated, then back screw out 7 turns. Operate engine at 2000 RPM and slowly close tester until pressure is 1800 psi. Volume of flow should be 2.5 GPM at 1800 psi and engine speed at 2000 RPM. Volume of flow should be checked at 200 psi less than relief setting. Close tester until volume of flow is zero and note the relief pressure. Traction Booster relief pressure should be 2000-2200 psi. Refer to paragraph 122C for setting Traction Booster relief valve.

After tests are completed, it is necessary to adjust length of Traction Booster rod as outlined in paragraph 121A and control valve release pressure as outlined in paragraph 122B. Reset the high volume bleed-off screw (7). Normal setting for the bleed-off screw is a 4-turns out from seated position.

Two-Ten and Two-Twenty

123. TORSION BAR ADJUSTMENT. The procedure for adjusting the torsion bar is the same as for D-21 models. Refer to Fig. AC95 and paragraph 121.

123A. TRACTION BOOSTER (DRAFT) ADJUSTMENT. Remove any weight or implement attached to three-point hitch and/or drawbar, then adjust the torsion bar pre-load as outlined in paragraph 121. Move the lift arm lever (1—Fig. AC96A) to TRACTION BOOSTER position, move position control lever (4) all the way forward, and move Traction Booster lever (3) all the way to the rear. Observe position of lift arms with engine operating at 1000 RPM. If lift arms are not horizontal (straight back), turn the adjusting nut (TB—Fig. AC96) as required until the lift arms are as near horizontal as possible.

123B. POSITION CONTROL ADJUSTMENT. First make certain that position control upper rod (Fig. AC-96B) is $5\frac{9}{32}$-inches long, measured between centers of rod attachment points. Move lift arm lever (1—Fig. AC96A) to TRACTION BOOSTER position, Traction Booster lever (3) to full forward location and the position control lever (4) all the way to the rear. With engine running at 1000 RPM, turn the self locking adjusting nut (PC—Fig. AC96) as required until lift arms are within one inch from fully raised position.

123C. LEVER FRICTION ADJUSTMENT. If the position control lever and/or Traction Booster control lever move from their set position when operating normally, the lever friction

should be increased. To adjust, remove the side panel as shown in Fig. AC96B and lower the lift arms completely. Tighten the adjusting nut against spring until the position control lever will stay when moved fully to the rear. The engine should be stopped when checking and adjusting.

123D. HIGH VOLUME BLEED-OFF ADJUSTMENT. The high volume bleed-off screw shown in Fig. AC96C is normally set 4-turns open from seated position. If Traction Booster system does not lift fast enough under heavy draft loads, turn screw in slightly. At normal setting (4-turns open), Traction Booster system will be controlled by volume of Traction Booster pump only. With screw turned in from this setting, part of the lift system oil will be used by the Traction Booster system.

124. LOWERING RATE ADJUSTMENT. The rate of lowering is adjusted at screw shown in Fig. AC96C. Turn screw in to increase speed of lowering. Normal setting is accomplished by turning needle in until it seats, then backing screw out ¾-turn. The adjusting needle is located at bottom of lift arm valve body, just ahead of the lift arm ram outlet connection. NOTE: The high volume bleed-off adjustment screw (also shown in Fig. AC96C) should **NOT** be mistaken for the rate of lowering adjustment screw. Normal setting for the high volume bleed-off screw is 4-turns open.

124A. LIFT SYSTEM RELIEF PRESSURE. To check hydraulic system relief pressure, install a 3000 psi gage in a remote cylinder port and pressurize that port. NOTE: Control valve lever must be held in position when checking pressure. Gage pressure should be 2250-2350 psi with engine running at 2200 RPM. To adjust relief pressure, remove cap (15—Fig. AC118), loosen lock nut (17), then turn screw (18) in to increase or out to decrease pressure.

124B. CONTROL RELEASE PRESSURE. When engine is operating at normal speed, the remote cylinder control levers should automatically return to neutral (hold) position when

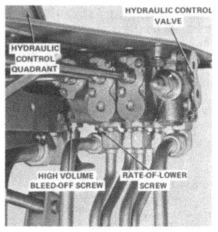

Fig. AC96C — View of hydraulic control valves showing location of high volume bleed-off screw and rate-of-lower screw. Refer to text for adjustment procedure.

Fig. AC96 — View of three-point hitch linkage on 210 and 220 tractors showing Traction Booster (TB) and position control (PC) adjustment points.

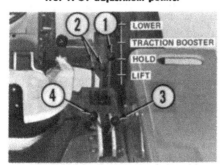

Fig. AC96A—View showing hydraulic control levers used on 210 and 220 tractors.

1. Three-point hitch (lift arm) control lever
2. Remote cylinder control levers
3. Traction Booster control lever
4. Position control lever

Fig. AC96B—View of hydraulic control linkage with side panel removed. Refer to text for adjustment.

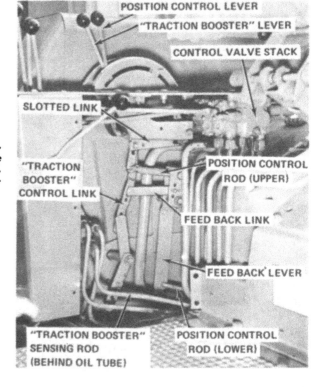

remote cylinder reaches end of stroke. The 3-point hitch control lever should also return from raising psition to neutral (hold) position when lift arms reach fully raised position. If control levers do not return to neutral or if control levers return prematurely, remove plug (53 or 74—Fig. AC118) from spring end cap of affected valve spool and turn screw (51 or 72) in to increase or out to decrease the pressure at which the lever will return to neutral.

124C. TRACTION BOOSTER RELIEF VALVE. Pressure in the Traction Booster and position control circuit is controlled by relief valve (15 thru 33—Fig. AC118). To check relief pressure, attach a 3000 psi gage to the "T" fitting (T—Fig. AC4B) located at top of pump. With engine running at 2200 RPM, actuate the Traction Booster sensing valve and observe gage pressure. If pressure is not within range of 2125-2225 psi, remove cap nut (15—Fig. AC118), loosen lock nut (17) and turn adjusting screw (18) as required.

124D. PRESSURE AND VOLUME TESTS. Relief valve settings and volume of flow can be checked using an OTC tester (Y81-21) or equivalent as follows: Remove test plug (L—Fig. AC4B) and attach tester inlet hose. Attach return hose from tester to the sump filler opening. Gage pressure with zero flow (relief pressure) should be 2250-2350 psi with engine speed at 2200 RPM. Refer to paragraph 124A for setting relief pressure. Volume of flow should be approximately 17.2 GPM when engine speed is 2200 RPM and pressure is 2000 psi (at least 200 psi less than relief pressure).

To test pressure and volume of flow for the Traction Booster system attach the tester inlet hose to "T" fitting after removing cap (T—Fig. AC4B). Before checking, make certain that high volume bleed-off screw (Fig. AC96C) is set exactly 4-turns out from seated position. Gage pressure with zero flow should be 2125-2225 psi with engine speed at 2200 RPM. Refer to paragraph 124C for setting Traction Booster relief pressure. Volume of flow should be approximately 3.2 GPM when engine speed is 2200 RPM and pressure is 2000 psi.

Power steering system pressure and volume are tested at port (P—Fig. AC4B). Refer to paragraph 10B for testing.

3-POINT HYDRAULIC LIFT SYSTEM
All Models

129. R & R LIFT CYLINDERS (RAMS). To remove the 3-point hitch lift cylinders, first move the hitch to fully lowered position and block up under rear ends of lower (draft) links to take weight off of the lift cylinders. Disconnect hydraulic lines from cylinders and remove the cylinder attaching pins. Snap rings are used to retain the pins on some models.

130. OVERHAUL LIFT CYLINDERS. Refer to paragraph 130A for overhauling lift cylinders used on D-21 and D-21 Series II tractors or to paragraph 130B for overhauling lift cylinders used on 210 and 220 models.

130A. On D-21 and D-21 Series II tractors, unscrew piston rod bearing nut (8—Fig. AC98) with a pin type spanner wrench, then the piston rod, nut, bearing and piston assembly can be removed from cylinder tube as shown.

Using two pin type spanner wrenches, hold rear side (6) of piston and unscrew head end (3) of piston from piston rod (9). Remove packing (5) from piston and unscrew remaining part of piston from rod. Withdraw piston rod from bearing (7) and retaining nut.

Inspect cylinder tube (1) for wear or scoring and hone or renew cylinder tube if necessary. Clean or renew the metallic breather screen (2) in vent hole near open end of cylinder tube.

Install new scraper and seal in piston rod bearing retaining nut. Scraper must be towards outer side of nut with sharp edge out. Lip of seal is towards scraper. Install retaining nut on piston rod, outer side first, and slide bearing on rod with ridge on outer diameter towards nut. Screw rear part of piston on rod and install new seals (5) and guide rings (4). Lips of the chevron type seals must be towards head end (3) of piston. Install and securely tighten head end of piston and stake end of piston rod with center punch.

Lubricate cylinder tube and piston, then carefully install piston and rod assembly. Securely tighten bearing retaining nut with spanner wrench.

130B. On 210 and 220 models, disassemble the lift arm cylinders as follows: Align retaining ring (1—Fig. AC99) with cylinder port with end of ring slightly to one side of port opening as shown in Fig. AC99A. Insert screwdriver or similar tool through port opening and work retaining ring into the deep second groove in piston rod (Fig. AC99B and Fig. AC99C). Piston can now be withdrawn.

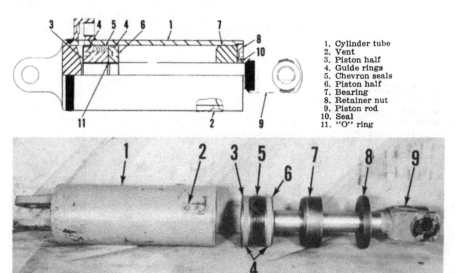

1. Cylinder tube
2. Vent
3. Piston half
4. Guide rings
5. Chevron seals
6. Piston half
7. Bearing
8. Retainer nut
9. Piston rod
10. Seal
11. "O" ring

Fig. AC98—Partly disassembled and cross sectional views of three-point hitch lift arm cylinder used on D-21 and D-21 Series II tractors. Vent (2) is on lower side of cylinder tube when installed.

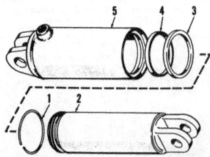

Fig. AC99—Exploded view of three-point hitch lift arm cylinder used on 210 and 220 models. Refer to Figs. AC99A through AC99D for removal and installation of piston rod.

1. Retaining ring
2. Piston rod
3. Wiper
4. Seal
5. Cylinder

Install seal (4—Fig. AC99) with thick portion toward inside of cylinder (5). Install wiper (3) and position the retaining ring (1) in the deep second groove as shown in Fig. AC-99C. Lubricate seal, wiper and piston, then insert piston into cylinder. Pull

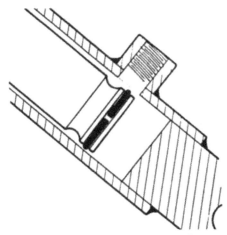

Fig. AC99A—When disassembling lift arm cylinder, align retaining ring with hose port as shown.

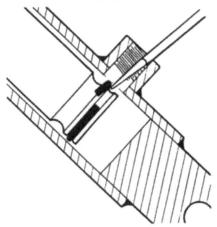

Fig. AC99B—Working through port opening, move retaining ring into the deep second groove.

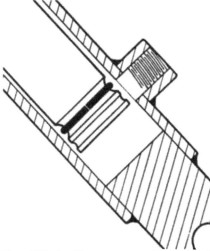

Fig. AC99C—Piston rod can be removed and installed with retaining ring in the deep second groove.

the retaining ring out of deep groove and into groove at end as shown in Fig. AC99D. Pull piston against stop after retaining ring is repositioned, to make sure that ring is fully seated.

133. **R&R LIFT SHAFT HOUSING ASSEMBLY.** To remove the hydraulic lift shaft housing assembly, proceed as follows: Move the 3-point hitch to fully lowered position. Detach draft arm leveling (winging) screw trunions by removing the linch pins (21—Fig. AC103) or spirol pins. Unbolt torsion bar supports (13 & 17) from rear face of rear axle center housing and remove the torsion bar and draft arm assembly.

On D-21 and D-21 Series II tractors, disconnect the PTO clutch control rod and remove the support at outer (right) end of clutch cross shaft. Disconnect Traction Booster and position control linkage. Remove cap screw that retains fuel sediment bowl bracket at top of lift shaft housing and spring the bracket up far enough to clear the housing flange.

On 210 and 220 models, disconnect PTO hydraulic lines (if so equipped),

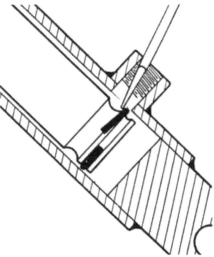

Fig. AC99D—When assembling, work retaining ring from deep groove back into end groove.

Traction Booster linkage, position control linkage and remote cylinder outlets and brackets. Move the disconnected lines and linkage out of the way. It may be necessary to tie hoses up to prevent interference when removing the housing.

On all models, drain all oil from rear axle center housing. Remove pins connecting rod ends of lift cylinders to lift arms. Support lift shaft housing, unbolt and carefully pry lift shaft housing from rear face of rear axle center housing, then remove assembly from tractor. On models with PTO, the multiple disc clutch assembly and PTO output shaft will be removed with the lift shaft housing.

To reinstall lift shaft housing, reverse removal procedure. On tractors equipped with PTO, alignment studs threaded into rear face of rear axle center housing will facilitate alignment. Tighten the housing retaining ⅝-inch cap screws to 180-190 Ft.-Lbs. torque, ¾-inch stud nuts to 310-320 Ft.-Lbs. torque.

NOTE: On 210 and 220 models with PTO, make certain that hub (18—Fig. AC93) is not pulled out of clutch disc while installing lift shaft housing. If lift shaft housing will not slide completely against rear face of axle center housing, check to make certain that hub has not moved forward and allowed one (or more) of the clutch lined discs to fall down behind the hub. PTO clutch will be damaged if lift housing is attached to axle center housing with a lined disc behind hub. Refer to paragraph 115.

134. **LIFT SHAFT AND LIFT ARMS.** Remove snap rings (1—Fig. AC102) retaining lift arms on shaft and remove the lift arms. Washers (2) are used between the lift arms and lift shaft housing to control shaft end play. Remove any washers present and remove lift shaft from housing.

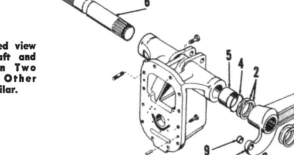

Fig. 102—Exploded view of lift arms, shaft and housing used on Two Twenty tractors. Other models are similar.

1. Snap ring
2. Washers
3. Lift arm
4. "O" ring
5. Bushing
6. Shaft
7. Snap rings
8. Cylinder pivot pin
9. Bushing

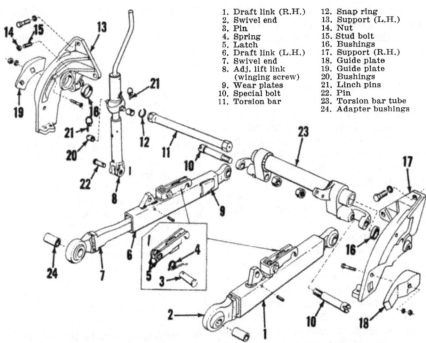

1. Draft link (R.H.)
2. Swivel end
3. Pin
4. Spring
5. Latch
6. Draft link (L.H.)
7. Swivel end
8. Adj. lift link
   (winging screw)
9. Wear plates
10. Special bolt
11. Torsion bar
12. Snap ring
13. Support (L.H.)
14. Nut
15. Stud bolt
16. Bushings
17. Support (R.H.)
18. Guide plate
19. Guide plate
20. Bushings
21. Linch pins
22. Pin
23. Torsion bar tube
24. Adapter bushings

Fig. AC103—Exploded view of three point hitch draft links and torsion bar unit typical of all models. Wear plates (9) on draft links (1 & 6) are renewable. Side sway of draft links is controlled by renewable guide plates (18 & 19) attached to torsion bar supports (13 & 17). Bushings (24) are used to convert the Category III hitch to use Category II implements.

Renew bushings in lift shaft housing if excessively worn. Bushings are pre-sized and should not require reaming. Renew "O" rings when re-installing shaft. Install washers (2) as required to remove shaft end play; use equal number of washers at each side of housing where possible.

PUMP

On D-21 and D-21 Series II tractors, a one section gear type pump is used if only equipped wtih power steering, an additional pumping section is used if equipped with hydraulic lift system and a third section is added if equipped with Traction Booster (3-point hitch). Oil flow from the power steering pump, or from the power steering section of a dual or three stage pump, is limited to 5.25 GPM at 1500 PSI and at 2200 engine RPM by a flow control valve and a pressure relief valve in the pump manifold. The hydraulic lift pump section delivers 16.5 GPM to the hydraulic control valve at 2000 engine RPM. Hydraulic system pressure is limited to 2000 PSI by a relief valve located in the control valve body. The Traction Booster section of the pump delivers 2.5 GPM at 2200 engine RPM. Refer to paragraph 122D for testing pressure and volume.

On 210 and 220 tractors, a one section gear type pump is used if not equipped with Traction Booster (3-point hitch) and a second section is added if equipped with Traction Booster. Oil flow from the main pump section is routed through a flow divider valve which directs 4.25 GPM to the power steering system and the remainder to the hydraulic lift system. Oil from the power steering system is used to actuate the PTO clutch and brake on models so equipped. Oil flows out of PTO valve (or power steering valve if not equipped with PTO), through a renewable filter and back to the sump. The oil available to the hydraulic lift system should be approximately 17.2 GPM when engine speed is 2200 RPM and pressure is 2000 psi (relief pressure is 2250-2350 psi). The Traction Booster section of pump, on models so equipped, should deliver 3.2 GPM when engine speed is 2200 RPM and pressure is 2000 psi (relief pressure is 2125-2225 psi). Refer to paragraph 124D for testing pressure and volume.

137. R&R HYDRAULIC PUMP. Before removing the hydraulic pump, be sure to clean the pump and connections thoroughly. If inlet tube is not to be removed, it will not be necessary to drain the transmission. Disconnect the hydraulic lines from the

pump (and inlet manifold on D-21 and D-21 Series II models); then, unbolt and remove the pump from the pump drive adapter. Be sure to cap or cover all hydraulic connections and openings after removing pump from tractor.

When reinstalling pump, use a new gasket between pump and pump drive adapter and tighten the two pump retaining cap screws to a torque of 28-33 Ft.-Lbs.

NOTE. On D-21 and D-21 Series II, the manifold unit can be removed for service without removing the pump.

138. R&R AND OVERHAUL PUMP DRIVE ADAPTER. Refer to exploded view in Fig. AC106. First, remove the hydraulic pump assembly as outlined in paragraph 137; then, proceed as follows:

Remove the pump drive shaft cover from the front side of the engine timing gear cover. Unstake the nut (5) from pump drive shaft and gear (1) and remove nut from shaft. Push the shaft and gear forward out through opening in timing gear cover; the front bearing cone and roller assembly and inner spacer from bearing unit (2) will be removed with the shaft and gear. Remove the rear bearing cone and roller assembly from rear of adapter housing. Unbolt and remove the adapter housing from the engine front plate. Note: Two adapter housing retaining cap screws are located inside the timing gear cover. To remove these cap screws with timing gear cover in place, re-shape the handle and grind the head of a $\frac{9}{16}$-inch box end wrench and remove the cap screws by working through opening in front of timing gear cover.

Remove the front bearing cone and roller assembly and the inner spacer from the pump drive shaft and gear and remove the two bearing cups and outer spacer from the adapter housing if necessary. If any of the bearing cups or cone and roller assemblies are worn or damaged beyond further use, it will be necessary to renew both bearing cups, both cone and roller assemblies and the two spacers as a

Fig. AC106 — Exploded view of hydraulic pump drive adapter.

1. Gear and shaft
2. Bearing assembly
3. "O" ring
4. Adapter
5. Nut
6. Gasket

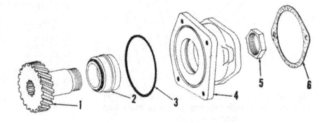

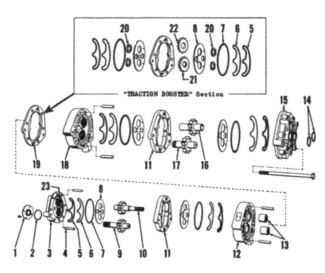

1. Seal body
2. "O" ring
3. Body
4. Dowel pins
   (6 used)
5. Sealing rings
   (6 used)
6. Back-up rings
   (6 used)
7. Seal ring
   (6 used)
8. Wear plates
   (6 used)
9. Lift pump gear
10. Lift pump gear
11. Gear plates
    (2 used)
12. Support plate
13. Bearings
    (8 used)
14. "O" rings
    (2 used)
15. Cover plate
16. Power steering
    pump gear
17. Power steering
    pump gear
18. Support plate
19. Gear plate
20. Back-up washers
    (4 used)
21. Traction Booster
    pump gear
22. Traction Booster
    pump gear
23. "O" ring
    (4 used)

Fig. AC107—Exploded view of three section gear type pump used on D-21 and D-21 Series II tractors. The inlet manifold used with this pump is shown in Fig. AC108.

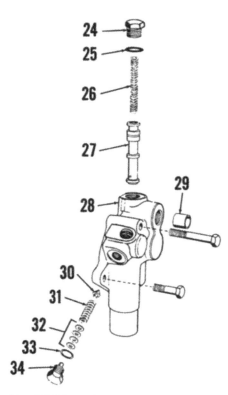

Fig. AC108 — The hydraulic pump inlet manifold, on D-21 and D-21 Series II tractors, contains power steering flow control and pressure relief valves.

24. Plug	29. Bushing
25. "O" ring	30. Power steering
26. Spring	relief valve
27. Flow control	31. Spring
valve spool	32. Shims
28. Manifold	33. Seal
	34. Plug

complete unit. When the pump drive adapter is properly assembled with a new bearing unit, the pump drive shaft end play should be 0.004-0.006.

To reassemble and reinstall the pump drive adapter, proceed as follows: Insert outer bearing spacer in the groove in bore of adapter housing. Be sure that there are no burrs on the outer spacer or the adapter housing bore; then, press the two bearing cups into the housing tightly against the spacer with the thick side of both cups towards the spacer. Place a new "O" ring (3) in the groove in adapter housing and install housing on engine front plate. Tighten the retaining cap screws to a torque of 28-33 Ft.-Lbs. Special seal washers should be used on screws inside timing gear cover. Install a bearing cone and roller assembly on pump drive shaft tightly against the gear and place the bearing inner spacer against bearing cone. Insert the shaft through opening in front of timing gear cover and install the rear bearing cone and roller assembly on the shaft. Install the retaining nut (5) and tighten the nut to a torque of 70-80 Ft.-Lbs. for D-21 and D-21 Series II models; 240-260 Ft.-Lbs. for 210 and 220 models. On all models, stake the nut to the shaft. The shaft should then have 0.004-0.006 end play; bump the shaft forward, then to rear before checking end play.

NOTE: If the bearing unit is being re-used, the bearing cone and roller assemblies should be reinstalled with their mating cups. Bearing cups fit in adapter housing with a zero to 0.002 interference fit; bearing cones fit on pump drive shaft with a 0.0005 interference to 0.001 clearance fit.

Reinstall cover on front side on timing gear cover and tighten the retaining cap screws to a torque of 28-33 Ft.-Lbs.

139. OVERHAUL HYDRAULIC PUMP. The pump will be a one, two or three section gear type pump depending upon type of hydraulic system used on tractor. Service instructions which follow are for the three section pump used on D-21 tractors equipped with a 3-point hitch. However, service procedures are similar for the single and two section pumps. Differences will be noted where applicable.

140. DRIVE SHAFT SEAL. To service the pump drive shaft seal, first remove pump assembly as outlined in paragraph 137; then, proceed as follows:

On D-21 and D-21 Series II tractors, remove the four screws retaining seal plate (1—Fig. AC107) to pump front cover and remove the seal plate and "O" ring (2). Shaft seal is serviced as an assembly with the seal plate. Install new "O" ring in pump body, lubricate the seal and shaft and carefully install seal plate with flat side out.

On 210 and 220 models, seal (1—Fig. AC109) can be removed and installed after removing the pump. Spring loaded lip of seal should be toward inside and seal should be pressed into bore until seal bottoms in bore.

141. POWER STEERING FLOW CONTROL (OR FLOW DIVIDER) AND RELIEF VALVES. On D-21 and D-21 Series II tractors, the inlet manifold (Fig. AC108) containing the power steering flow control and relief valves can be removed from the pump without removing pump from tractor. The flow control valve (27) should limit the volume of oil for power steering system to 5.25 GPM at 1200 psi and at engine between 1050-2200 RPM. Refer to paragraph 10A for checking pressure and volume. Disassembly procedure of inlet manifold is evident.

On 210 and 220 models, the pump should be removed before removing power steering relief valve (30 thru 37—Fig. AC109) and flow divider valve (24 thru 27). Pump must be disassembled to service check valve assembly (38 thru 41). The flow divider valve (27) first supplies 4.25 GPM to the power steering system, with remainder of oil used for the power lift system. Refer to paragraph 10B for checking power steering pressure and volume and to paragraph 124D for checking lift system pressure and volume.

On all models, be sure that valve spool slides freely in its bore. Small burrs can be removed with crocus

1. Seal
2. Gasket
3. Body
4. Dowel pins
5. Sealing rings
6. Back-up rings
7. Seal rings
8. Wear plates
9. Drive shaft and pump gear
10. Idler shaft and pump gear
11. Gear plate
12. Support plate
13. Bearings
15. Cover plate
19. Gear plate
20. Back-up washers
21. Traction Booster pump gear
22. Traction Booster pump idler gear
24. Plugs
25. "O" rings
26. Spring
27. Flow divider valve spool
30. Power steering relief valve
31. Spring
32. Adjusting screw
33. "O" ring
34. Plug
35. Guide
36. Gasket
37. Lock nut
38. Guide pin
39. Spring
40. Check valve ball
41. Check valve seat

Fig. AC109—Exploded view of two section pump used on 210 and 220 models. Models not equipped with Traction Booster are similar except parts (11, 19, 21 & 22) are not used.

sage and power steering pressure passage is closed. Suction will cause check valve to open when steering manually because of the volume difference in the two ends of steering cylinder.

142. **PUMP BODY AND GEARS.** After removing pump as outlined in paragraph 137, thoroughly clean outside of pump and mark each section of pump for correct assembly. Remove the screws attaching pump sections together; then bump the drive shaft against work bench or wood block to separate the pump sections. Do not pry sections apart as this may damage the sealing surfaces. Refer to Fig. AC107 for pump used on D-21 and D-21 Series II tractors; Fig. AC109 for pump used on 210 and 220 models.

Check the wear plates, gear plates, gears and bearing surfaces for excessive wear or scoring. Renew needle bearings if needles are loose or scored. Renew the pump as an assembly if excessive defects render renewing individual parts impractical.

If renewing needle bearings, drive or press on lettered end of bearing cage only. Be very careful to keep the needle bearing assemblies clean; it is very difficult to remove dirt or foreign material from within the bearing.

cloth. Renew valve spool if excessively worn or deeply scored. Renew spring if not approximately equal in length to new spring or if worn, cracked or distorted in any way. Use new sealing "O" ring when installing valve plug.

Inspect relief valve (30—Fig. AC-108 or AC109) and the valve seat for nicks, excessive wear or grooving. Renew relief valve and/or manifold (pump rear cover on 210 and 220 models) if serious defects are noted. Check spring free length against that of new spring; renew the spring if short, worn, distorted or cracked. On D-21 and D-21 Series II tractors, reinstall relief valve with same number of shims as were removed and recheck the pressure as outlined in paragraph 10A before adding shims.

The ball check valve (38 thru 41—Fig. AC109) on 210 and 220 models, is used only when manual steering is required (without engine running). In normal operation, hydraulic pressure plus pressure of spring (39) keeps the ¼-inch ball (40) seated and passage between pump suction pas-

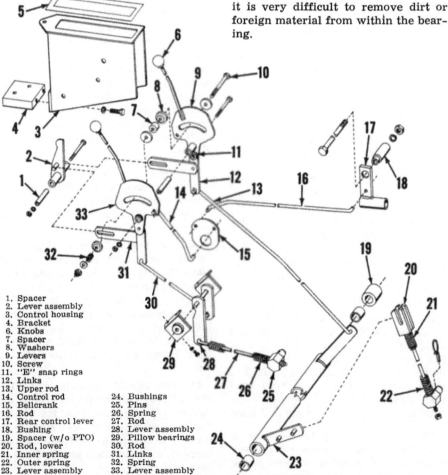

1. Spacer
2. Lever assembly
3. Control housing
4. Bracket
6. Knobs
7. Spacer
8. Washers
9. Levers
10. Screw
11. "E" snap rings
12. Links
13. Upper rod
14. Control rod
15. Bellcrank
16. Rod
17. Rear control lever
18. Bushing
19. Spacer (w/o PTO)
20. Rod, lower
21. Inner spring
22. Outer spring
23. Lever assembly
24. Bushings
25. Pins
26. Spring
27. Rod
28. Lever assembly
29. Pillow bearings
30. Rod
31. Links
32. Spring
33. Lever assembly

Fig. AC110—Pumps used on D-21 and D-21 Series II should be assembled with 1/8-inch holes (H) aligned on suction side of pumps.

Fig. AC115—Exploded view of three point hitch "Traction-Booster" and position control linkage used on D-21 and D-21 Series II tractors. To operate three point hitch with either lever or for "Traction-Booster" (draft control) action, the three point hitch lift control lever (1—Fig. AC97) must be in "Taction-Booster" detent position (B).

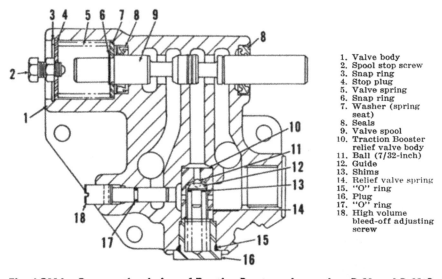

1. Valve body
2. Spool stop screw
3. Snap ring
4. Stop plug
5. Valve spring
6. Snap ring
7. Washer (spring seat)
8. Seals
9. Valve spool
10. Traction Booster relief valve body
11. Ball (7/32-inch)
12. Guide
13. Shims
14. Relief valve spring
15. "O" ring
16. Plug
17. "O" ring
18. High volume bleed-off adjusting screw

Fig. AC116—Cross sectional view of Traction Booster valve used on D-21 and D-21 Series II tractors.

143. When reassembling pump, refer to Fig. AC107 or Fig. AC109 and proceed as follows: Install the "E" shaped neoprene rubber sealing ring (5) in groove with flat side down. Install nylon or steel back-up (6) for "E" ring in groove at outer side of rubber ring with flat side of back-up ring up. Install wear plate sealing ring (7) in large oval groove with flat side down and on D-21 models, place small "O" ring (23—Fig. AC107) in recess of plate. On all models, brass side of wear plates should be toward gears. If wear plates have diamond shaped opening, this opening should be toward inlet side of pump. If wear plates have two small (⅛-inch) holes, this side should be toward "E" rings (5 & 6).

On D-21 and D-21 Series II tractors, install gear plates (11 & 13—Fig. AC-107) so that the small ⅛-inch holes (H—Fig. AC110) are all on the suction side of pump and aligned with similar holes in bearing plates.

On all models, shellac gear plate sealing surfaces lightly before assembling. All sections of the pump are similarly assembled.

On D-21 and D-21 Series II, tighten the screws that attach pump sections together to 33-37 Ft.-Lbs. torque and the oil inlet manifold attaching screws to 18-22 Ft.-Lbs. torque.

On 210 and 220 models, tighten the screws that attach pump sections together to 31-36 Ft.-Lbs. torque.

On all models, screws should be tightened evenly while shaft is turned

to check for binding. Refer to paragraph 140 for installation of drive shaft seal.

CONTROL VALVE

Several different control valves are used depending upon tractor model and type of equipment. Refer to the appropriate following paragraphs for service.

D-21 and D-21 Series II

150. **R&R AND OVERHAUL "TRACTION-BOOSTER" VALVE.** To remove the "Traction-Booster" valve (Fig. AC116), first disconnect the two hydraulic lines attached to valve; then, unbolt and remove valve from the top of the three-spool control valve. Renew valve spool and/or housing if scored or valve is sticking and renew seals. Make system adjustments as outlined in paragraphs 121 through 122C.

152. **R&R CONTROL VALVE ASSEMBLY.** To remove the single-spool control valve assembly, disconnect the pump pressure and sump return lines from the valve, then unbolt and remove the valve assembly from manifold.

To remove the three spool control valve, first remove fuel tank. Disconnect oil sump return tube, remote cylinder tubes, oil supply (pressure) tubes and on models with 3-point

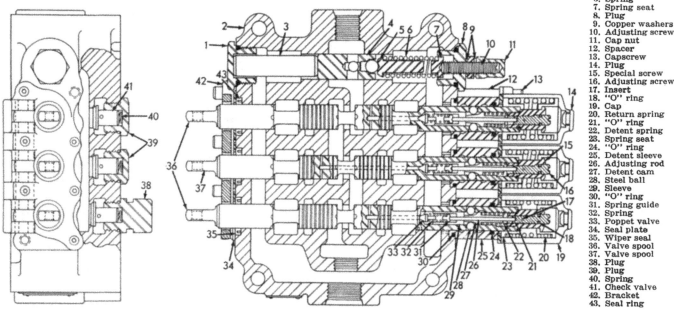

1. Plug
2. Valve body
3. Spacer
4. Relief valve seat
5. Spring guide
6. Spring
7. Spring seat
8. Plug
9. Copper washers
10. Adjusting screw
11. Cap nut
12. Spacer
13. Capscrew
14. Plug
15. Special screw
16. Adjusting screw
17. Insert
18. "O" ring
19. Cap
20. Return spring
21. "O" ring
22. Detent spring
23. Spring seat
24. "O" ring
25. Detent sleeve
26. Adjusting rod
27. Detent cam
28. Steel ball
29. Sleeve
30. "O" ring
31. Spring guide
32. Spring
33. Poppet valve
34. Seal plate
35. Wiper seal
36. Valve spool
37. Valve spool
38. Plug
39. Plug
40. Spring
41. Check valve
42. Bracket
43. Seal ring

Fig. AC117—Cross-sectional drawings of three spool control valve used on D-21 and D-21 Series II tractors not equipped with three point hitch. Single spool remote control valve is similarly constructed. On D-21 and D-21 Series II tractors with three point hitch, valve spool and bore adjacent to relief valve is ported for "Traction-Booster" circuit.

hitch, disconnect "Traction-Booster" gage tube and tube to lift cylinders. Unbolt and remove the hydraulic manifold and control valve from tractor; then, unbolt and remove the control valve assembly from manifold.

Reverse removal procedure to reinstall the control valve assembly. On models with 3-point hitch, check adjustment of the "Traction-Booster" and postition control levers as outlined in paragraphs 121A and 121B.

153. OVERHAUL CONTROL VALVE. Refer to Fig. AC117 for cross-sectional view of the three spool control valve used on tractors without a 3-point hitch. The single spool valve and the three-spool control valve used on tractors equipped with 3-point hitch are similarly constructed.

Seals on front ends of valve spools can be renewed with valve installed on tractor after removing the control levers and lever brackets from front of valve assembly.

Pressure relief valve and related parts can be removed and reinstalled with valve installed on tractor. Renew valve spring if distorted or cracked. Renew relief valve (steel ball) and/or valve seat if valve is rough or seat is peened. Readjust relief pressure as outlined in paragraph 122A after valve is reassembled.

For other service, the valve must be removed from tractor as outlined in paragraph 152. Valve spools, centering springs, detent mechanism and related parts are all serviced separately. However, the valve body (housing) is serviced in a complete control valve assembly only. Use the cross-sectional view in Fig. AC117 as a guide in disassembly and reassembly of the valve. Always renew the

sealing "O" rings when reassembling valve. Readjust control lever release pressure as outlined in paragraph 122B and on tractors equipped with 3-point hitch, readjust hitch lowering rate as outlined in paragraph 122.

Two-Ten and Two-Twenty

154. R&R AND OVERHAUL. The control valve assembly (Fig. AC118) can be removed from bracket after disconnecting hydraulic lines and control rods.

Individual sections of the control valve assembly can be overhauled. Valve spools and housings (36 & 37; 57 & 58) are not available separately and if damaged, the complete section of valve must be renewed. Refer to paragraphs 123 through 124D for system checks and adjustments.

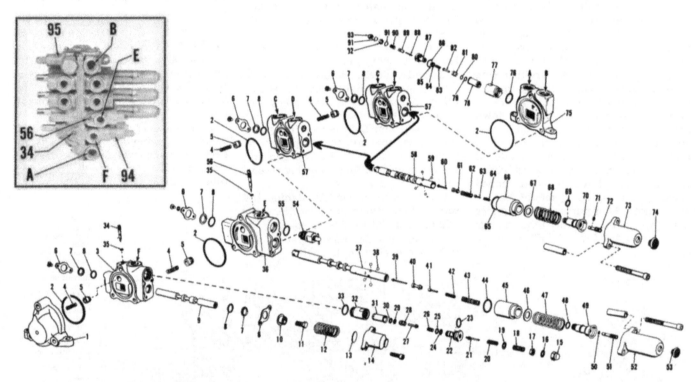

Fig. AC118—Exploded view of Traction Booster and power lift control valves. Outlet port (A) is located in the outlet housing (1) as shown at inset.

A. Outlet port to sump	12. Valve spring	33. "O" ring
B. Inlet from lift pump	13. Spacer	34. High volume bleed-off adjusting screw
C & D. Double acting remote cylinder ports	14. Cover	35. "O" rings
	15. Cap nut	36. "Traction Booster"—lift arm valve housing
E. Port to lift arm rams	16. Copper washer	37. Valve spool
F. Inlet from "Traction Booster" pump	17. Lock nut	38. Steel balls
	18. Adjusting screw ("Traction Booster" relief valve)	39. Poppet
1. Outlet housing	19. Copper washer	40. Cam
2. "O" rings	20. Spring	41. Spring guide
3. "Traction Booster" valve housing	21. Plunger	42. Spring
4. Check valve springs	22. Plug	43. Detent spring
5. Check valves	23. "O" ring	44. "O" ring
6. Seal plates	24. "O" ring	45. Sleeve
7. Seal wiper rings	25. Back-up ring	46. Washer
8. "O" rings	26. Spring	47. Plunger spring
9. "Traction Booster" sensing valve	27. Relief valve ("Traction Booster")	48. "O" ring
10. Spring seat	28. Piston	49. Spring seat
11. Screw	29. "O" ring	50. "O" ring
	30. Back-up ring	51. Adjusting screw (detent release)
	31. Valve sleeve	52. Cover
	32. "Traction Booster" relief valve cap	53. Rubber plug

54. Shut-off valve	76. "O" ring
55. "O" ring	77. Lift system relief valve cap
56. Lift arm rate of lowering adjusting screw	78. Valve sleeve
	79. Back-up ring
57. Remote cylinder control housing	80. "O" ring
58. Valve spool	81. Piston
59. Steel balls	82. Relief valve (hydraulic lift system)
60. Poppet	83. Spring
61. Cam	84. Back-up washer
62. Detent spring	85. "O" ring
63. Spring guide	86. "O" ring
64. Spring	87. Plug
65. "O" ring	88. Plunger
66. Sleeve	89. Spring
67. Washer	90. Adjusting screw (hydraulic lift system relief valve)
68. Plunger spring	91. Copper washers
69. "O" ring	92. Lock-nut
70. Spring seat	93. Cap nut
71. "O" ring	94. Lift system relief valve
72. Adjusting screw (detent release)	95. "Traction-Booster" relief valve
73. Cover	
74. Rubber plug	
75. Inlet housing	

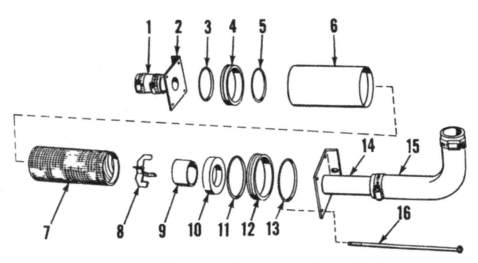

Fig. AC120—Exploded view of magnetic filter used on D-21 and D-21 Series II tractors. Filter is located in pump suction line at inner side of engine left side rail.

1. Hose	5. "O" ring	9. Spacer	13. "O" ring
2. Rear end assembly	6. Filter cover	10. Magnet	14. Front end assembly
3. "O" ring	7. Filter screen	11. "O" ring	15. Hose
4. Metal ring	8. Support baffle	12. Metal ring	16. Clamp bolts (4)

Fig. AC121—Cross sectional view showing the inlet filter screen used on 210 and 220 tractors.

1. Drain plugs
2. Suction tube to hydraulic pump
3. Cover
4. Gasket
5. "O" ring
6. Filter screen
7. Filter support

HYDRAULIC OIL FILTERS

All Models

The transmission, differential and final drive lubricating oil is used as the working fluid for the hydraulic power steering system and the hydraulic lift system. The pump and other components, therefore, must be protected from any abrasive particles in the oil.

155. On D-21 and D-21 Series II tractors, a screen filter with magnetic trap (Fig. AC120) is located inside the engine left side rail in the pump suction line. A throw away type filter is located at rear of the lift arm housing in the sump return line. The magnetic screen filter should be removed, disassembled, cleaned and reinstalled each 200 hours of operation. The throw away filter located at rear of lift arm housing should be renewed and all transmission, final drive and hydraulic fluid drained and refilled with new oil each 500 hours or once each year.

On 210 and 220 tractors, a screen type filter (Fig. AC121) is located at the bottom of the rear axle center housing in the pump suction line. A throw away filter is located on right side of axle center housing under the platform. The throw away filter should be renewed each 150 hours of operation and can be removed from below. All transmission, final drive and hydraulic fluid should be drained and refilled with new fluid every 600 hours or once each year. The inlet screen filter should be cleaned whenever any service is performed on components which could contaminate the fluid and once each year.

FRONT WHEEL ASSIST

Two-Twenty tractors are available with a front drive axle which is driven by a gear located on the range transmission countershaft (21—Fig. AC133) via idler gear (32), drive adapter (Fig. AC131) and a drive shaft. The shifting mechanism in the drive adapter allows connecting or disconnecting power to front wheels as required.

LUBRICATION

164. The planetary drive units are lubricated by SAE-90EP oil. Units are drained by rotating wheel until drain plug (Fig. AC130) is at bottom, then removing plug. To fill, rotate wheel until filler plug is in the 3 o'clock position and add the necessary amount of oil until fluid is at edge of filler plug opening.

The front axle universal joints are lubricated at grease fitting (G—Fig. AC134). Remove the plug (Fig. AC130A) and rotate wheel until grease fitting is accessible. Reinstall the opening plug after universal joint is lubricated.

Fig. AC130 — View of front wheel showing location of drain and filler plugs for the planetary drive units. When draining, drain plug should be at bottom. Filler plug should be at 3 o'clock position when filling.

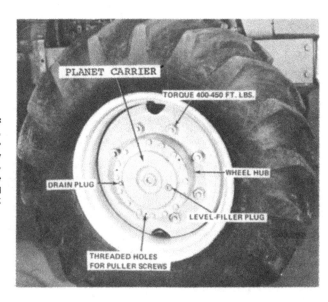

PLANET CARRIER

TORQUE 400-450 FT. LBS.

DRAIN PLUG

WHEEL HUB

LEVEL-FILLER PLUG

THREADED HOLES FOR PULLER SCREWS

Each of the four pin caps (23 & 35 —Fig. AC134) is equipped with grease fitting (F). Grease will exit inside the steering knuckle flange (25).

The inner axle bushings and the differential carrier assembly is lubricated by SAE-90EP oil. The drain plug is located at (D—Fig. AC130A) and filler plug at (F). Oil should be maintained at level of filler plug opening.

DRIVE ADAPTER

165. REMOVE AND REINSTALL. Drain transmission oil, disconnect rear of drive shaft from universal joint yoke (10—Fig. AC131) and disconnect lower end of handle (2) from bellcrank (3). Remove attaching cap screws and pull drive adapter away from axle center housing. Two dowel pins are used to align adapter housing with axle center housing.

Before installing the drive adapter, thoroughly clean and dry the mating surfaces of axle center housing and adapter housing. No gasket is used so that mesh position of gears can be maintained, but surfaces should be coated with sealer to prevent leakage. Make certain that surface of axle center housing remains dry until drive adapter is attached, to prevent leakage. Coat threads of the front two attaching screws with sealer. Tighten the special attaching screws to 90-100 Ft.-Lbs. torque. Reconnect drive shaft and shift handle, then fill transmission with oil.

166. OVERHAUL. Refer to paragraph 165 for removal procedure. Remove nut (8—Fig. AC131), washer (9), yoke (10) and spacer (11), then unbolt and remove retainer (7). Be careful not to lose or damage shims (12). Remove pin (29) from clevis, then unbolt and remove bellcrank and bracket (3). Remove plug (31) from bore in rear of housing and remove

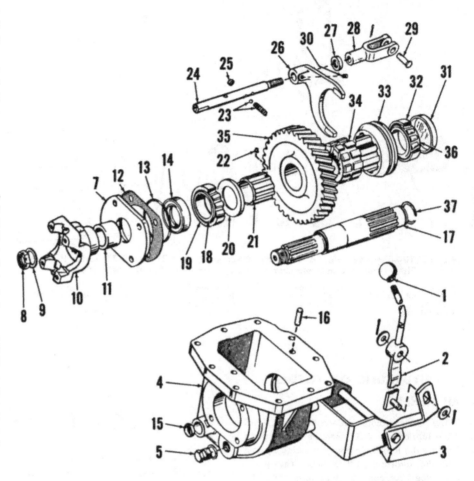

Fig. AC131—Exploded view of drive adapter used on Front Wheel Assist tractors.

1. Knob	11. Spacer	20. Spacer	29. Pin
2. Handle	12. Shims	21. Bushing	30. Set screw
3. Bellcrank and bracket	13. "O" rings	22. Pins (2 used)	31. Plug
4. Housing	14. Seal	23. Detent ball and spring	32. Bearing cup
5. Drain plug	15. Plug	24. Shift rod	33. Coupling
6.	16. Dowel (2 used)	25. Plug	34. Collar
7. Retainer	17. Shaft	26. Shift fork	35. Gear
8. Nut	18. Front bearing cone	27. Seal	36. Rear bearing cone
9. Washer	19. Bearing cup	28. Clevis	37. Snap ring
10. Yoke			

snap ring (37) from groove in rear of shaft. Press shaft (17) out toward front of housing. If necessary to remove shift rod (24), plug (25) must be removed to install the detent ball and spring (23). Remainder of disassembly will be self-evident.

Shift rod oil seal (27) should be installed with lip toward inside. If removed, the self locking set screw (30) must engage relief in shaft and should be tightened to 25-30 Ft.-Lbs. torque. Bearing cup (32) must be bottomed in housing bore. Press front bearing cone (18) on shaft (17) so that front of inner race is approximately flush with step on shaft (at rear of oil seal surface). Slide spacer (20) onto shaft behind the front bearing cone. Use grease to hold pins (22) inside bushing (21) while assembling, then slide bushing onto shaft with pins toward rear engaging the splines. Position gear (35) in housing with shifter teeth toward rear and start shaft into housing and gear. Assemble shift coupling (33) and collar (34), then position bearing cone (36) in rear bearing cup and the coupling and collar assembly between gear and bearing. NOTE: Shift groove in coupling (33) must engage fork (26) and be

Fig. AC130A — View of front axle showing differential carrier lubrication filler plug (F) and drain plug (D). The plug indicated must be removed for greasing axle universal joint.

REMOVE PLUG TO LUBRICATE UNIVERSAL JOINT

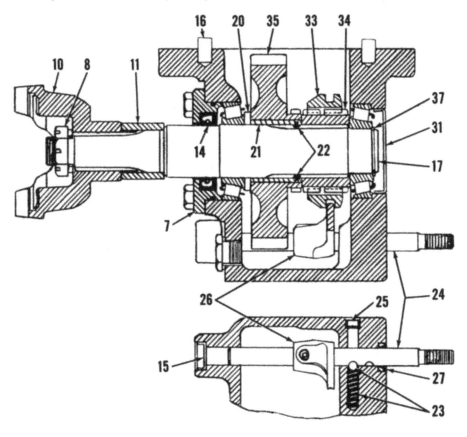

Fig. AC132—Cross sectional view of drive adapter used on Front Wheel Assist equipped tractors. Refer to Fig. AC131 for legend.

Lbs. (minimum) before installing cotter pin.

IDLER GEAR, SHAFT AND BEARINGS

167. **R&R AND OVERHAUL.** The idler gear, shaft and bearings can be removed for service after removing the drive adapter assembly as outlined in paragraph 165 and separating rear axle center housing from rear face of the four speed transmission housing as outlined in paragraph 82. Damage to the idler assembly may also damage the range transmission countershaft. Refer to paragraph 94.

Idler shaft (31—Fig. AC133) and gear (32) can be withdrawn after removing snap ring (28). Snap ring (28) is available in thicknesses from 0.068 to 0.092 in steps of 0.003. Select the thickest snap ring that can be installed after front bearing cup is installed. Shaft end play should be within limits of 0.003-0.005 after front bearing cup is seated against snap ring.

FRONT AXLE ASSEMBLY

168. **PLANETARY DRIVE UNITS.** The planetary drive units can be disassembled with tractor weight supported by front wheels.

Drain oil from the planetary drive unit as outlined in paragraph 164. Unbolt planet carrier (16—Fig. AC134) from hub, then use three ⅜-16 screws in the threaded holes to push planet carrier away from hub. Remove snap ring (10) from end of axle (37) and remove sun gear (9). After sun gear is removed, ring gear (7) can be withdrawn from spindle (29). Planet gears (18) and thrust washers (19) can be removed after snap ring (21) is removed from ends of shafts (20). Planet gear shafts (20) can be pressed from bores in carrier (16) after removing set screws (17). Shafts (20) should be a press fit in bores. When assembling, make certain that holes in shafts (20) are aligned with set screw holes in planet carrier (16).

Assemble ring gear (7) over splines and pin (P). NOTE: Refer to paragraph 169 if adjustment of bearings (1 & 5) is required or if nut (8) has been moved. Install sun gear (9) over splines on axle (37), then install snap ring (10). Position "O" ring (6) in groove of carrier (16), then install planet carrier assembly. NOTE: It may be necessary to turn the drive shaft slightly, while installing the planet carrier asesmbly in order to align

toward rear of housing. The 45 degree chamfer on coupling (33) is toward front. Slide shaft through gear, collar and bearing cone. Make certain that bushing (21) does not catch on edge of gear (35). When shaft is seated, install snap ring (37) at rear and front bearing cup (19) in housing bore. Lip of seal (14) should be toward inside. Install retainer (7), using a new "O" ring (13) and correct thickness of shims (12) to provide shaft end play of 0.002-0.005. Screws attaching retainer (7) to the housing should be tightened to 70-75 Ft.-Lbs. torque. Tighten nut (8) to a torque of 350 Ft.-

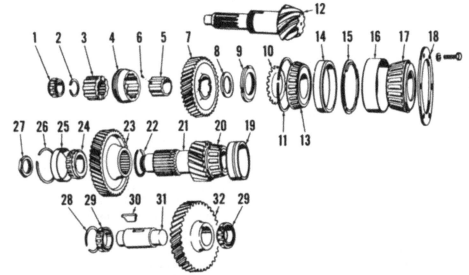

Fig. AC133—The idler gear (32) drives the Front Wheel Assist drive adapter shown in Figs. AC131 and AC132. Driven gear (23) drives the idler gear.

1. Pilot bearing	10. Tab washer	17. Wide bearing cone	25. Front bearing cup
2. Snap ring	11. Snap ring	18. Retainer plate	26. Snap ring
3. Collar	12. Bevel pinion shaft	19. Rear bearing cup	27. Nut
4. Coupling	13. Narrow bearing cone	20. Rear bearing cone	28. Snap ring
5. Bushing	14. Narrow bearing cup	21. Countershaft	29. Bearings
6. Lock pins	15. Spacer	22. Snap ring	30. Key
7. Lo range gear	16. Wide bearing cup	23. Driven gear	31. Idler shaft
8. Thrust washer		24. Front bearing cone	32. Idler gear
9. Nut			

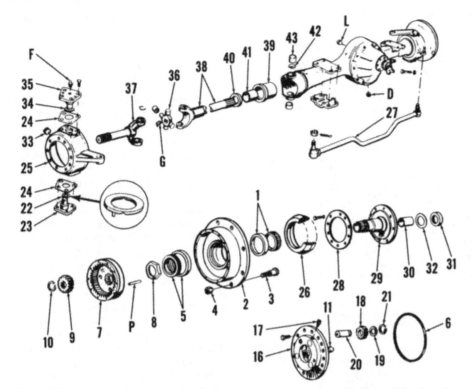

Fig. AC134—Exploded view of the front axle used on tractors with Front Wheel Assist. Refer to Fig. AC135 for exploded view of the differential and carrier assembly. Number used indicates quantity for each side.

1. Inner wheel bearing	11. Thrust plunger	24. Shims	35. Top (left) pin cap
2. Hub	16. Planet carrier	25. Steering knuckle flange	36. Universal joint
3. Wheel stud (8 used)	17. Set screw (3 used)	26. Oil seal guard	37. Outer axle
4. Nut (8 used)	18. Planet gear (3 used)	27. Tie rod	38. Inner axle
5. Outer wheel bearing	19. Thrust washer (3 used)	28. Oil seal	39. Retainer
6. "O" ring	20. Planet shaft (3 used)	29. Hollow spindle	40. Oil seal
7. Ring gear	21. Snap ring	30. Bushing	41. Bushing
8. Adjusting nut and lock pin	22. Lower (pinned) thrust bearing	31. Oil seal	42. Plug (2 used)
9. Sun gear	23. Lower pin cap	32. Washer	43. Bushing (2 used)
10. Snap ring		33. Grease opening plug	D. Drain plug
		34. Top thrust bearing	F. Grease fitting
			G. Grease fitting
			L. Filler plug

teeth of gears (7, 9, & 18). Cap screws attaching carrier (16) to hub (2) should be tightened to 80-90 Ft.-Lbs. torque. Refer to paragraph 164 for lubrication requirements.

169. WHEEL HUB AND BEARINGS. Refer to paragraph 168 and remove the planetary drive unit. Support tractor and remove the tire and wheel unit. Remove nut (8—Fig. AC134), then withdraw hub (2) and bearings (1 & 5). Method of renewing bearing cups is conventional. Bearings are lubricated by oil contained in planetary drive unit and seal (28) should not leak. Seal (28) is attached by same screws that attach hollow spindle (29) to steering knuckle (25). CAUTION: If spindle is withdrawn, be careful not to damage seal (31) with axle splines.

When assembling, the spindle (and oil seal) retaining screws should be tightened to 120-130 Ft.-Lbs. torque. Make certain that bearing cups are seated in hub and that inner bearing cone is seated on spindle. Install hub (2) and bearings (1 & 5) and tighten nut (8). To set bearing adjustment,

nut (8) should be tightened until rolling torque of hub is 3.5 Ft.-Lbs. This setting can be checked by attaching a cord to one of the planet carrier attaching screws, wrapping cord around hub and pulling cord with a spring scale. Make certain that reading is rolling torque not starting effort. With hub radius of 6-inches, spring scale should indicate within limits of 6-10 pounds. With nut (8) set to provide correct bearing adjustment, slide ring gear (7) on splines so that hole in back of ring gear is closely aligned with pin (P) in nut. Nut (8) can be tightened or loosened slightly to align pin with hole in gear. Refer to paragraph 168 for assembly of planetary drive. Wheel retaining stud nuts should be tightened to 400-450 Ft.-Lbs. torque.

170. SPINDLES AND STEERING KNUCKLES. The hollow spindle (29 —Fig. AC134) can be removed as outlined in paragraph 169. Bushing (30) can be renewed if required. Press new bushing in until flush with counterbore. Install washer (32), then press

seal (31) into counterbore until past flush with surface of spindle flange. Lip of seal (31) should be toward bushing (30). Leakage of seal (31) can be indicated by reduction of fluid level in planetary drive and presence of oil in steering knuckle (25). Leakage of seal (40) will allow oil from axle housing to enter steering knuckle also.

After the hollow spindle is removed, the steering knuckle (25) can be removed as follows: Disconnect tie rod (27) from steering knuckle and, if knuckle on right side is to be removed, detach steering cylinder. Remove the upper and lower pin caps (23 & 35) and lift steering knuckle off. Jack screws ($\frac{7}{16}$-14) can be used in the threaded holes in pin caps to push pin caps out of bore.

NOTE: Do not lose or damage shims (24) or thrust washers (22 & 34). If axle (36, 37 & 38) is withdrawn, it will be necessary to drain axle housing. Be careful not to damage seal (40) with splines at inner end of shaft.

Bushings (43) should be pressed into bores until slightly past flush with flat surface. Make certain that plug (42) is not damaged and that one plug is located in each of the four bearing bores before pressing bushing into position. Bearings are presized, but fit should be checked with pin caps (23 & 35) before assembling. Thrust washers (22) located on bottom pin caps are pinned to caps (23). The upper thrust washers (34) are not pinned.

Install steering knuckles using original shims (24). Tighten screws attaching pin caps (23 & 35) to recommended torque and check end clearance (vertical play) of steering knuckle. If end clearance is not within limits of 0.005-0.013, add or remove shims (24) as required. Cap screws attaching the upper right side pin cap (with steering arm) should be tightened to 180-195 Ft.-Lbs. torque. Screws attaching the other three pin caps (23 & 35) should be tightened to 120-130 Ft.-Lbs. torque. Thickness of shims (24) should be the same under top pin cap (35) as under bottom pin cap (23) in order to maintain axle alignment.

171. AXLE SHAFTS. The axle shafts can be withdrawn after removing the hollow spindles as outlined in paragraph 170. Oil should be drained from axle center housing before withdrawing axle shaft. Removal and installation of seal (40—Fig. AC134), bushing (41) and retainer (39) is more

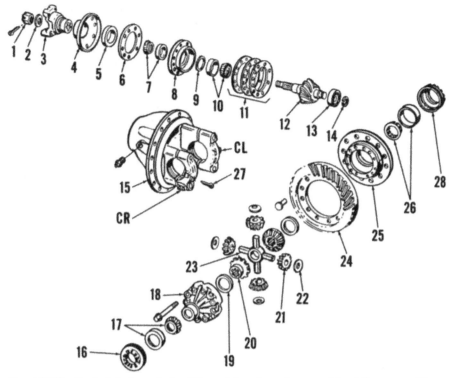

Fig. AC135—Exploded view of the differential and carrier assembly. Adjustment of bearings (7 & 10) is set by thickness of spacer (9). Carrier bearings (17 & 26) and backlash are adjusted at adjusting rings (16 & 28). Mesh position of ring gear and pinion is adjusted by varying thickness of shims (11).

CL. Carrier bearing cap (Left)	7. Pinion outer thrust bearing	14. Snap ring	21. Spider gears
CR. Carrier bearing cap (Right)	8. Retainer	15. Carrier housing	22. Thrust washers
1. Nut	9. Spacer	16. Adjusting ring	23. Spider
2. Washer	10. Pinion inner thrust bearing	17. Right carrier bearing	24. Ring gear
3. Yoke	11. Shims	18. Differential case half	25. Differential case half
4. Cover	12. Pinion shaft	19. Thrust washers	26. Left carrier bearing
5. Seal	13. Pinion radial load bearing	20. Side gears	27. Cotter (lock) pin
6. Gasket			28. Adjusting ring

easily accomplished with steering knuckle (25) removed.

The universal joint bearings and cross (36) can be pressed from yokes after removing the retaining snap rings. The retainer (39) is removed and installed with seal (40) and bushing (41) installed. Bushing should be flush with small end of retainer and should have 0.007-0.014 clearance on the 1.876-1.879 diameter journal of axle shaft (38). Bushing (41) is pre-sized and should not need reaming if

Fig. AC136 — Differential case halves should be marked for correct assembly as shown. Safety wire ends of screws as shown.

carefully installed. Lip of oil seal (40) should be toward bushing (41). Leakage of seal can be indicated by reduction of fluid level in axle center housing and presence of oil in steering knuckle (25). Leakage of seal (31) will allow oil from planetary drive to enter steering knuckle also. Press retainer (39) into axle center housing until seated. Seal and bushing must be assembled in retainer before retainer is installed. Be extremely careful to prevent damage to seal (40) when installing axle shaft. Refer to paragraph 170 for installation of steering knuckle and hollow spindle.

172. DIFFERENTIAL AND CARRIER. The differential and carrier assembly can be removed after blocking up front of tractor and removing the axle shafts from both sides as outlined in paragraph 171.

Before disassembling, punch mark the carrier bearing caps (CL & CR—Fig. AC135) and matching bosses of carrier housing (15) so caps can be reinstalled in original positions. Remove cotter (lock) pins (27), cut safety wire and remove the carrier bearing caps. Lift differential from

carrier and keep carrier bearing cups identified with their bearing cones. NOTE: Carrier bearings (17 & 26) are NOT interchangeable.

Differential case halves (18 & 25) can be unbolted and separated. NOTE: Case halves should be punch marked for correct assembly (Fig. AC136) before separating. Side gears (20—Fig. AC135), thrust washers (19), spider gears (21) and thrust washers (22) should be renewed in sets. Do not mix old and new gears or old and new thrust washers when reassembling. If ring gear (24) is removed from case half (25), center punch rivets and drill out using $\frac{15}{32}$-inch bit. CAUTION: Case half (25) will be damaged if rivet heads are chiseled off. Ring gear (24) and pinion (12) are available only as matched set.

When assembling, the cold annealed steel rivets attaching ring gear to case half will require approximately 22-45 tons of pressure to correctly shape the formed end. When correctly shaped, the formed end will be approximately the same height as the original rivet head and diameter will be approximately ⅛-inch larger than rivet hole. After assembly, the ring gear should be checked for eccentricity and run-out which must not exceed 0.008. Cap screws attaching differential case halves together, should be tightened to 92-118 Ft.-Lbs. torque, then route safety wire through ends of all screws. Refer to paragraph 174 for setting backlash and mesh position of bevel gears.

173. The pinion shaft (12—Fig. AC-135) is removed as follows: Remove cotter pin, nut (1) and washer (2). Use a suitable puller to remove yoke (3), then unbolt and remove cover (4). Seal (5) is pressed into cover with lip toward inside of carrier and should be bottomed in bore. Pull the pinion assembly from carrier, using jack screws as shown in Fig. AC137. CAUTION: Shims (11—Fig. AC135) are used to adjust ring gear and pinion mesh position.

Remainder of disassembly procedure will be evident. Bearing (13) is a press fit on pinion shaft and is retained with snap ring (14). NOTE: Bearing cups and cones (7 & 10) are not interchangeable.

When assembling, adjustment of bearings (7 & 10) should be checked. Bearing cups should be pressed into retainer (8) until seated firmly against shoulder. Press the inner bearing cone onto pinion shaft until race is firmly seated against shoulder. Position original spacer (9) on shaft and insert pinion shaft through retainer (8) and bearing cups. Install outside bearing cone and position the assembly in a press as shown in Fig. AC138, with a sleeve around the pinion shaft and pressing against the bearing inner race. When correctly positioned apply pressure to hold both pinion bearing inner races and spacer tightly together and against shoulder of pinion. Attach cord to one of the retainer mounting holes, wrap cord around retainer and check rolling torque of bearings as shown in Fig. AC138 with a spring scale. Rolling torque (not starting torque) should be within limits of 5-15 inch-pounds. If rolling torque is too low, install a thinner spacer (9—Fig. AC135). If torque is too high install a thicker spacer (9). Spacer is available in thicknesses of 0.172, 0.173, 0.174, 0.175, 0.176, 0.177, 0.183, 0.189, 0.195, and 0.201. After correct thickness of spacer is selected, remaining parts of pinion can be assembled, including yoke (3) and nut (1). Nut (1) should be torqued to 300-400 Ft.-Lbs.

Do not back nut off to install cotter pin. Refer to paragraph 174 for installation of differential and pinion assemblies in carrier housing and for adjustment of bevel gear backlash and mesh position.

174. Before assembling and adjusting, install adjusting rings (16 & 28), caps (CL & CR) and cups for carrier bearings (17 & 26). Tighten the bearing cap retaining screws to 160 Ft.-Lbs. torque, then move the bearing cups in bores. If bearing cups are not a hand push fit, carefully sand the bores in carrier housing. Bearing cup must be able to move freely when adjusting. After checking fit of bearing cups, remove the bearing caps (CL & CR), adjusting rings (16 & 28) and bearing cups. Position the pinion assembly (1 thru 14) in carrier housing (15) using the original shims (11). At least two of the attaching screws should be started before retainer (8)

is bumped into carrier bore to maintain alignment of the holes. Make certain that retainer is seated tightly against shims and carrier, then tighten the retainer attaching screws. NOTE: Shims (11) may need to be changed, but screws should be tightened enough to hold pinion assembly firmly against carrier while checking backlash and mesh position. Recommended torque for the pinion retaining cap screws is 30-35 Ft.-Lbs.

Position the differential assembly (17 thru 26) and adjusting rings (16 & 28) in carrier, then install carrier bearing caps (CL & CR). Install the carrier bearing cap retaining screws finger tight. Maintain some clearance between pinion and ring gear teeth and tighten the adjusting rings (16 & 28) until carrier bearing cups are seated and all end play is eliminated. After all end play is removed, tighten each adjusting ring (16 & 28) one additional notch to preload the bearings.

Mount a dial indicator as shown in Fig. AC139 and measure backlash of the ring gear. If backlash is not within limits 0.005-0.015, shift the differential as required by loosening one adjusting ring (16 or 28) and tightening the other adjusting ring an equal amount.

After the differential carrier bearing preload and bevel gear backlash have been set, the mesh position of the bevel gears should be checked and adjusted if necessary. To check the mesh position of the bevel drive gears, paint the teeth of ring gear with red lead or prussian blue, then maintain a slight drag on ring gear and rotate the bevel pinion in direction of forward travel. If mesh position is

Fig. AC137—Remove retainer and pinion shaft using jack screws as shown. Shims for adjusting mesh position of ring gear and pinion are between retainer and carrier housing.

Fig. AC139 — View of dial indicator mounted for checking ring gear to pinion backlash.

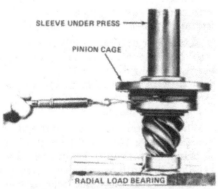

Fig. AC138 — Pinion assembly should be mounted in press as shown when checking pinion shaft bearing adjustment. Refer to text.

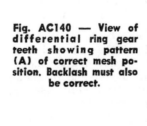

Fig. AC140 — View of differential ring gear teeth showing pattern (A) of correct mesh position. Backlash must also be correct.

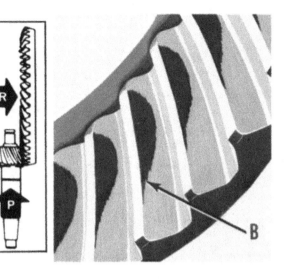

Fig. AC141—View of differential ring gear teeth showing pattern (B) obtained when pinion is too far out. Arrow (P) shows direction to move pinion to correct mesh position. It will be necessary to move differential and ring gear in direction (R) in order to correct backlash of gears.

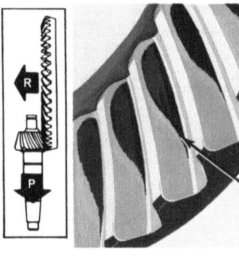

Fig. AC142—View of differential ring gear teeth showing pattern (C) obtained when pinion is too far in. Arrow (P) shows direction to move pinion to correct mesh position. It will be necessary to move differential and ring gear in direction (R) in order to correct backlash of gears.

correct, the coating will be removed from ring gear as shown at (A—Fig. AC140). The tooth contact pattern should be well toward the toe of gear under no-load condition. However, as load is applied, the contact area will extend toward heel of gear to give full tooth contact.

If tooth pattern such as shown at (B—Fig. AC141) is obtained, it indicates that bevel pinion is set too far out. Remove part of shims (11—Fig. AC135) as required between retainer (8) and carrier housing (15). Recheck backlash as previously outlined. It may be necessary to move ring gear in direction shown at arrow (R—Fig. AC141) in order to obtain correct backlash of bevel gears. Recheck mesh position.

If tooth pattern such as shown at (C—Fig. AC142) is obtained, it indicates that bevel pinion is set too far in. Add shims (11—Fig. AC135) as required between retainer (8) and carrier housing (15). Recheck backlash as previously outlined. It may be necessary to move ring gear in direction of arrow (R—Fig. AC142) in order to obtain correct backlash of bevel gears. Recheck mesh position.

After adjustment is complete, the pinion retaining cap screws should be tightened to 30-35 Ft.-Lbs. torque and the carrier bearing cap retaining screws should be torqued to 160-205 Ft.-Lbs. Bearing cap retaining screws should be locked with safety wire through heads of screws.

NOTE: It must be remembered that when the backlash of bevel gears has been set, any subsequent movement of the bevel pinion shaft either fore or aft will change backlash setting and backlash must be reset.

175. Install carrier, using a new gasket between carrier and axle housing. Make certain that lock washers and stud nuts are started under housing offsets before the carrier is drawn completely against axle housing. All carrier to axle housing stud nuts should be tightened to 53-67 Ft.-Lbs. torque.

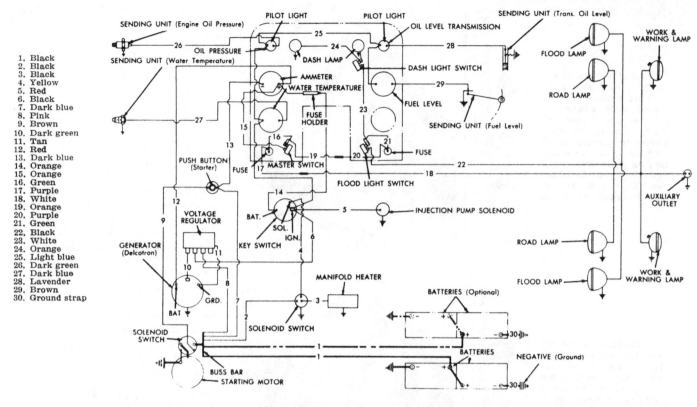

1. Black
2. Black
3. Black
4. Yellow
5. Red
6. Black
7. Dark blue
8. Pink
9. Brown
10. Dark green
11. Tan
12. Red
13. Dark blue
14. Orange
15. Orange
16. Green
17. Purple
18. White
19. Orange
20. Purple
21. Green
22. Black
23. White
24. Orange
25. Light blue
26. Dark green
27. Dark blue
28. Lavender
29. Brown
30. Ground strap

Fig. AC54—Wiring diagram for the Allis-Chalmers D-21 tractor. Wiring is color coded as shown at left.

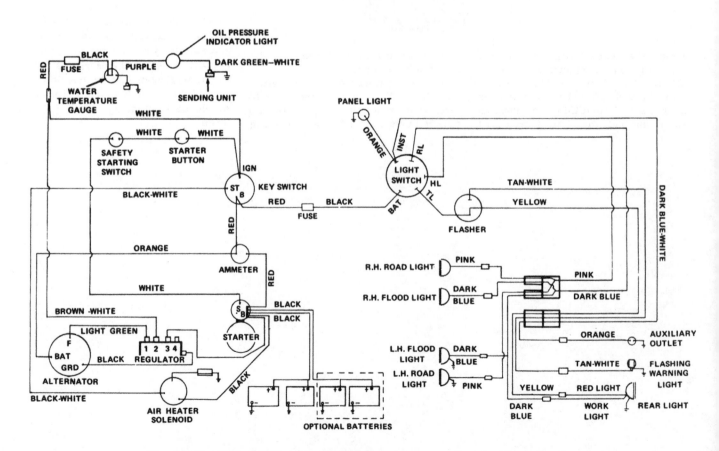

Fig. AC151—Wiring diagram for Allis-Chalmers two-ten and two-twenty tractors.

ALLIS-CHALMERS

Models ■ 7010 ■ 7020 ■ 7030 ■ 7040 ■ 7045
■ 7050 ■ 7060 ■ 7080

Previously contained in I&T Shop Service Manual No. AC-33

SHOP MANUAL
ALLIS-CHALMERS

MODELS
7010-7020-7030-7040-7045-7050-7060-7080

Tractor serial number is stamped on rear side of differential housing above power take-off shield. Engine serial number plate is on upper left side of engine block.

INDEX (By Starting Paragraph)

INDEX (Cont.)

DUAL DIMENSIONS

This service manual provides specifications in both the U.S. Customary and Metric (SI) systems of measurement. The first specification is given in the measuring system perceived by us to be the preferred system when servicing a particular component, while the second specification (given in parenthesis) is the converted measurement. For instance, a specification of "0.011 inch (0.28 mm)" would indicate that we feel the preferred measurement, in this instance, is the U.S. system of measurement and the metric equivalent of 0.011 inch is 0.28 mm.

CONDENSED SERVICE DATA
Models 7010-7020-7030-7040

	7010	7020	7030	7040
GENERAL				
Engine Make	Own	Own	Own	Own
Engine Model	649T	649I	3500 MARK II	3500 MARK II
Number of Cylinders	6	6	6	6
Bore	3.875 in. (98.43 mm)	3.875 in. (98.43 mm)	4.250 in. (107.95 mm)	4.250 in. (107.95 mm)
Stroke	4.250 in. (107.95 mm)	4.250 in. (107.95 mm)	5.000 in. (127.0 mm)	5.000 in. (127.0 mm)
Displacement	301 cu. in. (4909 cc)	301 cu. in. (4909 cc)	426 cu. in. (6982 cc)	426 cu. in. (6982 cc)
Main Bearings, Number of	7	7	7	7
Cylinder Sleeves	Wet	Wet	Wet	Wet
Alternator & Starter Make	DELCO-REMY*	DELCO-REMY*	DELCO-REMY*	DELCO-REMY*

*Some models are equipped with Niehoff.

	7010	7020	7030	7040
TUNE-UP				
Firing Order	1-5-3-6-2-4	1-5-3-6-2-4	1-5-3-6-2-4	1-5-3-6-2-4
Valve Tappet Gap (Hot) Intake & Exhaust	0.015 in. (0.38 mm)	0.015 in. (0.38 mm)	0.015 in. (0.38 mm)	0.015 in. (0.38 mm)
Valve Seat Angle Inlet & Exhaust	30°	30°	30°	30°
Injection Timing	18°BTDC	18°BTDC	24°BTDC	16°BTDC
Injection Pump Make	ROOSA-MASTER			
Timing Mark Location	CRANKSHAFT PULLEY			
Battery Terminal, Ground	NEG.	NEG.	NEG.	NEG.
Engine Low Idle Rpm	750-800	750-800	700-750	700-750
Engine High Idle Rpm, No Load	2480-2580	2480-2580	2500-2550	2500-2645
Engine Full Load Rpm	2300	2300	2300	2300

SIZES-CAPACITIES-CLEARANCES

	7010 / 7020	7030 / 7040
Crankshaft Main Journal Diameter	2.7465-2.7480 in. (69.761-69.799 mm)	3.2465-3.248 in. (82.461-82.499 mm)
Crankpin Diameter	2.3720-2.3735 in. (60.248-60.286 mm)	2.7470-2.7485 in. (69.773-69.811 mm)
Camshaft Journal Diameter, All	2.130-2.131 in. (54.10-54.13 mm)	2.130-2.131 in. (54.10-54.13 mm)
Piston Pin Diameter	1.2515-1.2517 in. (31.78-31.79 mm)	1.5011-1.5013 in. (38.127-38.133 mm)
Valve Stem Diameter, Inlet	0.3715-0.3720 in. (9.436-9.448 mm)	0.3715-0.3720 in. (9.436-9.448 mm)
Exhaust	0.3705-0.3710 in. (9.410-9.423 mm)	0.3705-0.3710 in. (9.410-9.423 mm)
Main Bearing Diametral Clearance	0.0016-0.0048 in. (0.04-0.12 mm)	0.0019-0.0051 in. (0.048-0.129 mm)
Rod Bearing Diametral Clearance	0.0009-0.0039 in. (0.02-0.10 mm)	0.001-0.004 in. (0.02-0.10 mm)
Piston Skirt Diametral Clearance	0.0045-0.0070 in. (0.11-0.18 mm)	0.0025-0.0050 in. (0.063-0.127 mm)
Crankshaft End Play	0.004-0.010 in. (0.10-0.25 mm)	0.007-0.013 in. (0.18-0.33 mm)
Camshaft Bearings Diametral Clearance	0.002-0.005 in. (0.05-0.13 mm)	0.002-0.006 in. (0.050-0.15 mm)

CONDENSED SERVICE DATA (Cont.)

	7010	7020	7030	7040
SIZES-CAPACITIES-CLEARANCES (Cont.)				
Camshaft End Play	————0.001-0.011in.————		————0.0027-0.0083 in.————	
	(0.03-0.28 mm)		(0.068-0.210 mm)	
Cooling System Capacity*	26 qts.	26 qts.	32. qts.	32 qts.
	(24.5 L)	(24.5 L)	(30.2 L)	(30.2 L)
Crankcase Oil*	16 qts.	16 qts.	19 qts.	19 qts.
	(15.1 L)	(15.1 L)	(17.9 L)	(17.9 L)
Transmission*	7.4 gal.	7.4 gal.	7.4 gal.	7.4 gal.
	(28.0 L)	(28.0 L)	(28.0 L)	(28.0 L)
Differential*	16.7 gal.	16.7 gal.	16.7 gal.	16.7 gal.
	(63.46 L)	(63.46 L)	(63.46 L)	(63.46 L)

*Approximate capacity

TIGHTENING TORQUES

General Recommendations	See End of Shop Manual			
Rod Bearing Cap Screws	————See Paragraph 30————			
Cylinder Head Cap Screws	————165.ft.-lbs.————		————150 ft.-lbs.————	
	(224.0 N•m)		(203.2 N•m)	
Flywheel Cap Screws	————135 ft.-lbs. (182.9 N•m)————			
Injection Nozzle Nuts	————40-60 ft.-lbs. (54.2-81.3 N•m)————			
Main Bearing Screws	————135 ft.-lbs.————		————170-190 ft.-lbs.————	
	(183.0 N•m)		(230.3-257.4 N•m)	

Models 7045-7050-7060-7080

	7045	7050	7060	7080
GENERAL				
Engine Make	Own	Own	Own	Own
Engine Model	670T	3700	3700	3750 MARK II
Number of Cylinders	6	6	6	6
Bore	4.250 in.	4.250 in.	4.250 in.	4.250 in.
	(107.95 mm)	(107.95 mm)	(107.95 mm)	(107.95 mm)
Stroke	5.000 in.	5.000 in.	5.000 in.	5.000 in.
	(127.0 mm)	(127.0 mm)	(127.0 mm)	(127.0 mm)
Displacement	426 cu. in.	426 cu. in.	426 cu. in.	426 cu. in
	(6982 cc)	(6982 cc)	(6982 cc)	(6982 cc)
Main Bearings, Number of	7	7	7	7
Cylinder Sleeves	Wet	Wet	Wet	Wet
Alternator & Starter Make	DELCO-REMY*	DELCO-REMY*	DELCO-REMY*	DELCO-REMY*

*Some models are equipped with Niehoff.

	7045	7050	7060	7080
TUNE-UP				
Firing Order	1-5-3-6-2-4	1-5-3-6-2-4	1-5-3-6-2-4	1-5-3-6-2-4
Valve Tappet Gap (Hot)				
Intake & Exhaust	0.015 in.	0.015 in.	0.015 in.	0.015 in.
	(0.38 mm)	(0.38 mm)	(0.38 mm)	(0.38 mm)
Valve Seat Angle				
Inlet & Exhaust	30°	30°	30°	30°
Injection Timing	16°BTDC	26°BTDC	18°BTDC	22°BTDC
Injection Pump Make	———————— ROOSA-MASTER ————————			
Timing Mark Location	———————— CRANKSHAFT PULLEY ————————			
Battery Terminal, Ground	NEG.	NEG.	NEG.	NEG.
Engine Low Idle Rpm	700-750	700-750	700-750	725-775
Engine High Idle Rpm,				
No Load	2500-2550	2500-2550	2500-2645	2800-2850
Engine Full Load Rpm	2300	2300	2300	2550

SIZES-CAPACITIES-CLEARANCES

Crankshaft Main Journal				
Diameter	————3.2465-3.248 in. (82.461-82.499 mm)————			
Crankpin Diameter	————2.7470-2.7485 in. (69.773-69.811 mm)————			
Camshaft Journal				
Diameter, All	————2.130-2.131 in. (54.10-54.13 mm)————			

CONDENSED SERVICE DATA (Cont.)

	7045	7050	7060	7080

SIZES-CAPACITIES-CLEARANCES (Cont.)

Piston Pin Diameter	———————1.5011-1.5013 in.(38.127-38.133 mm)———————	
Valve Stem Diameter		
Inlet	————————0.3715-0.3720 in.(9.436-9.448 mm)————————	
Exhaust	————————0.3705-0.3710 in (9.410-9.423 mm)————————	
Main Bearing Diametral		
Clearance	———————0.0019-0.0051 in.(0.048-0.129 mm)————————	
Rod Bearing Diametral		
Clearance	————————0.001-0.004 in (0.02-0.10 mm)————————	
Piston Skirt Diametral		
Clearance	———————0.0025-0.0050 in.(0.063-0.127 mm)————————	
Crankshaft End Play	————————0.007-0.013 in.(0.18-0.33 mm)————————	
Camshaft Bearings		
Diametral Clearance	————————0.002-0.006 in.(0.05-0.15 mm)————————	

	7045	7050	7060	7080
Camshaft End Play	0.001-0.008 in. (0.02-0.20 mm)		0.0027-0.0083 in. (0.068-0.210 mm)	
Cooling System Capacity*	32 qts. (30.2 L)	32 qts. (30.2 L)	32 qts. (30.2 L)	36 qts. (34.1 L)
Crankcase Oil*	19 qts. (17.9 L)	19 qts. (17.9 L)	19 qts. (17.9 L)	19 qts. (17.9 L)
Transmission*	7.4 gal. (28.0 L)	7.4 gal. (28.0 L)	7.4 gal. (28.0 L)	7.4 gal. (28.0 L)
Differential*	16.7 gal. (63.46 L)	16.7 gal. (63.46 L)	16.7 gal. (63.46 L)	17.1 gal. (64.98 L)

*Approximate capacity

TIGHTENING TORQUES

General Recommendations	——————See End of Shop Manual——————
Rod Bearing Cap Screws	————————See Paragraph 30————————
Cylinder Head Cap Screws	——————150 ft.-lbs. (203.2 N•m)——————
Flywheel Cap Screws	——————135 ft.-lbs. (182.9 N•m)——————
Injection Nozzle Nuts	————40-60 ft.-lbs. (54.2-81.3 N•m)————
Main Bearing Screws	————170-190 ft.-lbs. (230.3-257.4 N•m)————

FRONT AXLE SYSTEM

SPINDLES AND BUSHINGS

1. **R&R SPINDLES.** To remove front spindle (15–Fig. 1), support front of tractor, remove front wheel and proceed as follows: Remove snap ring (10) and pull steering arm (9) from spindle. Removal of steering arm will probably require it to be cut off using a suitable torch due to extreme press fit. Remove key (12) and withdraw spindle from bottom of axle extension (11). Remove thrust washers (16) from spindle.

Install two thrust washers (16) on spindle (15) and install it in axle extension (11) from bottom. Install key (12). Heat steering arm (9) to 600°F (315°C) and press it on spindle so maximum shaft end play is 0.030 inch (0.76 mm). Spindle should rotate freely between stops and steering arm must be clear of snap ring groove in spindle. Any adjustment of steering arm on spindle must be made prior to steering arm cooling to below 300°F (148°C). Reseating or removal of steering arm after it has cool-

ed below 300°F (148°C) will probably require it to be cut off with a torch due to extreme press fit.

2. **R&R SPINDLE BUSHINGS.** Spindle bushings can be renewed after removing spindle as outlined in paragraph 1. Remove bushings (13 and 14–Fig. 1) using a suitable bushing driver or drift punch. New bushings are presized and should not require reaming if carefully installed. Install upper and lower bushings in axle extension (11), using a suitable press or shoulder punch,

Fig. 1—Exploded view of adjustable front axle assembly.

1. Seal
2. Wear sleeve
3. Bearing assy. (inner)
4. Hub
5. Bearing assy. (outer)
6. Washer
7. Nut
8. Cap
9. Arm
10. Snap ring
11. Axle extension
12. Key
13. Bushing (upper)
14. Bushing (lower)
15. Spindle
16. Thrust washers
17. Washers
18. Pivot pin
19. Axle main member

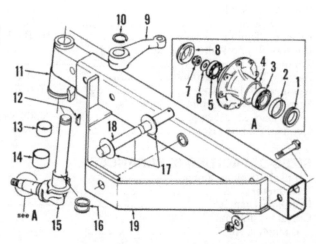

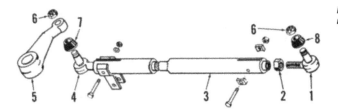

Fig. 2—Exploded view of tie rod and components. Refer to text.

1. Tie rod end
2. Locknut
3. Tube
4. Tie rod end
5. Arm
6. Nuts
7. Seal
8. Seal

until they are flush to 0.030 inch (0.76 mm) below end of bore. Reinstall spindle as outlined in paragraph 1.

TIE RODS AND TOE-IN

3. Toe-in of front wheels should be 3/32 to 7/16 inch (2.4 to 11.1 mm) measured from OD of tire at spindle height. To adjust toe-in remove nut (6–Fig. 2) and disconnect tie rod end (1) from spindle. Loosen jam nut (2) and turn tie rod end in or out as necessary to correct toe-in.

AXLE EXTENSIONS

4. To renew axle extension (11–Fig. 1), remove spindle as outlined in paragraph 1, then remove tread width adjusting bolts and withdraw axle extension from main member (19).

AXLE CENTER (MAIN)MEMBER AND PIVOT PIN

5. The axle center (main) member is a welded one-piece assembly (19–Fig. 1).

The center member pivots on one long pin. The pin pivots in renewable bushings in front support casting.

To renew the pivot pin, proceed as follows: Support tractor under torque housing so that no weight is carried on front axle. Then, remove the retaining bolt and drive out pivot pin. Then, raise front of tractor until front support is clear of axle and drive bushings out of front support castings with suitable driver. New bushings are presized and should not require reaming if carefully installed. Install pivot pin (18) with one thrust washer (17) at rear. Install enough thrust washers (17) at front to establish an end play of 0.00-0.17 inch (0.0-4.3 mm).

To renew axle center (main) member, remove both axle extensions as outlined in paragraph 4, disconnect power steering cylinder at both ends and remove front axle pivot pin.

POWER STEERING SYSTEM

All models are equipped with hydrostatic power steering system that has no mechanical linkage between steering wheel and front steering cylinder. Refer to Fig. 3 for drawing showing the steering system.

There are three pumps located together under the range transmission. Pumps are driven by a shaft from the pto gear train. Pumps are bolted together in one housing but are of different type and supply oil to three separate circuits. A brief description of the pumps from front to rear follows. Refer to Fig. 4.

Front pump is an axial piston pump which delivers oil to the following:
1. 3-point hitch.
2. Remote control valve.
3. Power take-off valve.
4. Differential lock valve.
5. Brakes control valve.

Middle pump is a gear type with a flow divider. Flow divider splits output into priority and secondary flow and supplies oil for the following functions:
A. Priority flow:
1. Power steering.
2. Cooling oil in rear axle.
3. Filtering oil in rear axle.
B. Secondary flow:
1. Lubricating and cooling oil for brakes.
2. Lubricating and cooling oil for pto clutch.

Rear pump is a gerotor type and is recessed into range transmission housing. It is the only pump that is not visible

from the outside of tractor. This pump supplies oil to the following:
1. Power director clutch or power shift clutch.
2. Power director or power shift clutches, lubrication and cooling.
3. Lubrication, cooling and filtering of standard and range transmissions.

The control valve unit (8–Fig. 3) contains a rotary metering motor, a commutator feed valve sleeve and a selector valve spool. In event of engine or hydraulic power failure, the metering motor becomes a rotary hand pump to actuate the power steering cylinder

when steering wheel is turned. A check valve within the gear pump housing allows recirculation of fluid within the control valve and steering cylinder during manual operation.

NOTE: The maintenance and absolute cleanliness of all parts is of utmost importance in the operation and servicing of the hydraulic power steering system. Of equal importance is the avoidance of nicks or burns on any of the working parts. Do not use cloth shop towels in cleaning internal parts; use only lint-free shop towels.

Fig. 3—A schematic showing components of hydraulic power steering system. No mechanical linkage is used between steering wheel and tractor front wheels.

1. Filters
2. Tube (intake)
3. Pump
4. Tube (return)
5. Tube (to inlet hose)
6. Hose (outlet)
7. Hose (inlet)
8. Steering control valve
9. Tube
10. Hose
11. Oil cooler
12. Hose
13. Tube
14. Steering cylinder
15. Hose (valve to cylinder)
16. Hose (valve to cylinder)

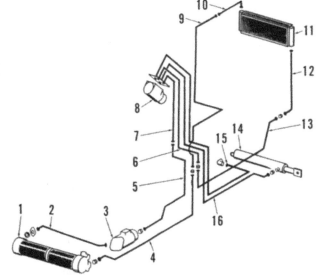

TROUBLESHOOTING

6. Before attempting to adjust or repair the power steering system, the cause of any malfunction should be located. Refer to the following paragraphs for possible causes of power steering system malfunction:

Irregular or "Sticky" Steering. If irregular or "sticky" feeling is noted when turning steering wheel with forward motion of tractor stopped and with engine running at rated speed, or if steering wheel continues to rotate after being turned and released, foreign material in the power steering fluid (transmission oil) is probable cause. Clean or renew all hydraulic system filters. It will be necessary to also drain transmission oil and refill with clean oil. Also check for air leaks in gear pump suction line which will cause pump to cavitate. If trouble is not corrected, the power steering control valve assembly should be removed and serviced; refer to paragraph 10.

Steering Cylinder "Hesitates". If steering cylinder appears to pause in travel when steering wheel is being turned steadily, probable cause of trouble is air trapped in the power steering cylinder. Bleed cylinder as outlined in paragraph 7.

Slow Steering. Slow steering may be caused by low oil flow from pump. Check time required for full stroke travel of power steering cylinder; first, with tractor weight on the front wheels; then, with front end of tractor supported by a jack. If time between the two checks varies considerably, overhaul power steering pump as outlined in paragraph 151.

Loss of Power. Loss of steering power may be caused by system relief setting being too low. Check and adjust relief valve setting as outlined in paragraph 11.

Left, rear fluid filter (9 – Fig. 5) plugged or wrong size. Filter should be 6.10-6.16 inches (154.9-156.4 mm) long.

Orifice disc (8 – Fig. 6) in gear pump flow divider plugged. Remove spool (4)

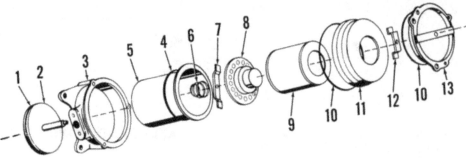

Fig. 5 — Filter on left side of rear housing is filter for gear pump.

1. Seal	5. Housing	8. Valve	11. Cap
2. Check valve	6. Spring	9. Filter	12. Strap
3. Head	7. Bar	10. "O" rings	13. Cover
4. Seal			

from gear pump and clean orifice plate.

Relief valve (8 – Fig. 9) in steering control valve stuck open. Steering control valve assembly should be removed and serviced; refer to paragraph 10.

System Overheats. Overheating of system may be caused by neutral (bypass) pressure being too high. If neutral pressure is excessive, check for obstructions in system such as kinked tube, etc. If neutral pressure is erratic, check for binding of control valve shaft, spool and/or sleeve.

Too much oil in rear housing sump. Shut off engine and allow tractor to stand for 10 minutes, check oil level through sight glass in rear housing. Drain as necessary.

LUBRICATION AND BLEEDING

7. Differential and final drive oil is utilized as the power steering fluid. The filter, shown in Fig. 5 is located in lower front part of differential housing on left-hand side. This filter is for the gear pump which supplies the power steering, and should be renewed after the first 50 hours and thereafter not to exceed over 1000 hours. The power steering system is usually self-bleeding. With engine running at high idle speed, cycle system through several full strokes of the cylinder. In some cases, it may be necessary to loosen connections at cylinder to bleed trapped air.

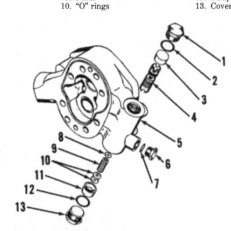

Fig. 6 — Exploded view of flow divider.

1. Plug	
2. "O" ring	8. Orifice disc
3. End cap	9. Spring
4. Spool	10. Shims
5. Gear pump housing	11. End cap
6. Test port plug	12. "O" ring
7. "O" ring	13. Plug

SYSTEM OPERATING PRESSURE AND RELIEF VALVE

8. Power steering relief pressure can be checked by connecting a Flo-Rater into test port of gear pump. Refer to Fig. 4 showing test port location and to Fig. 7 showing Flo-Rater connected. Connect

Fig. 4 — View of the three hydraulic pumps used. Test port (T) is used for diagnosis of gear pump hydraulic system.

Fig. 7 — View of hydraulic tester connected for testing power steering system. Refer to text.

outlet from Flo-Rater to remote outlet on tractor. MAKE SURE REMOTE VALVE LEVER FOR OUTLET THAT IS CONNECTED TO FLO-RATER IS IN FLOAT POSITION. This is to assure sufficient oil to charge the piston pump. Start engine and operate at 1500 rpm. Turn steering wheel to one end of its travel and hold in that position. Turn restrictor valve on Flo-Rater until 1000-1500 psi (6.8-10.3 MPa) reading is obtained. Operate in this manner until fluid reaches a temperature of 130°-140°F (54.4°-60.0°C). After temperature has been reached, increase engine speed to 2300 rpm and check flow at 1000 psi (6.96 MPa) with steering wheel held at end of travel. Flow should be 5.25 to 5.75 gpm (19.8-21.7 L/min.). Then, reduce engine speed to 1100 rpm at which point flow should not be less than 3.0 gpm (11.3 L/min.) for Model 7010, 7020 and 7045 tractors, or 3.5-4.5 gpm (13.2-17.0 L/min.) for all other models. Set engine speed at 2300 rpm and close restrictor valve to check steering relief valve pressure. This pressure should be 1800-2000 psi (12.4-15.88 MPa) at 2300 rpm. If relief valve pressure is found to be incorrect, the upper end of orbitrol motor will have to be removed to adjust the relief valve. If flow is not as stated, flow divider could be defective or pump may be worn and not producing enough volume.

To check pump volume proceed as follows: Remove flow divider cap (3 – Fig. 6), spool (4), orifice disc (8) and spring (9). Disconnect priority line to steering valve and plug fitting at flow divider. Disconnect secondary line to differential lock valve and plug fitting at flow divider.

CAUTION: Make sure restrictor valve on flow-tester is in open position as there is no relief valve protection.

Fig. 8 – *View of steering control valve showing port locations. Pressure tube from pump connects to "IN" port; return tube to sump connects to "OUT" port; tube to rear end of cylinder connects to "L" (left turn) port; and tube to rod end of cylinder connects to "R" (right turn) port.*

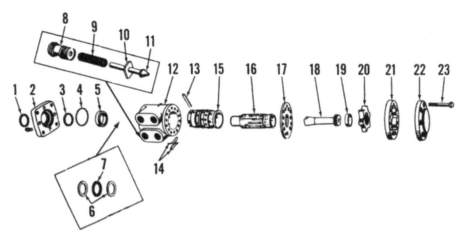

Fig. 9 – *Exploded view of steering control assembly. Centering springs (14) are installed in two groups of three springs with arch in each group back-to-back.*

1. Oil seal	7. Thrust bearing
2. Mounting plate	8. Relief valve plug
3. Quad ring	9. Spring
4. "O" ring	10. Guide
5. Locator	11. Spool
6. Bearing races	12. Control valve body

13. Centering pin	18. Drive shaft
14. Centering springs (6)	19. Spacer
	20. Rotor
15. Sleeve	21. Ring
16. Valve spool	22. Cover (cap)
17. Plate	23. Cap screw

Start engine, run at 2300 rpm and adjust restrictor on Flo-Rater until there is a pressure reading of 2000 psi (13.79 MPa). At this time, flow should be 9 gpm (34.0 L/min.) minimum. If correct pressure and flow are found, but there is still difficulty with the steering, steering valve or steering cylinder could be the trouble. Before removing steering valve, check steering cylinder as follows: Remove both hoses from cylinder and connect cylinder hoses together. Attempt to rotate steering wheel. If steering wheel will not rotate, steering valve is okay and trouble is in steering cylinder. If steering wheel will rotate, trouble is in steering valve.

POWER STEERING PUMP

9. The gear type pump is the center of three pumps under the range transmission where they are driven by a shaft from pto gear train.

For service information on all three pumps refer to paragraph 149.

CONTROL VALVE

10. **REMOVE AND REINSTALL.** Remove hood and disconnect the four hoses from control valve. Unbolt and remove control valve assembly from steering shaft housing.

Reinstall control valve by reversing removal procedure. Refer to Fig. 8 for proper hose locations. Install new "O" ring seals on hose fittings before connecting hoses to control valve and tighten fittings securely. Bleed trapped air from power steering cylinder as outlined in paragraph 7 after assembly is completed.

11. **OVERHAUL CONTROL VALVE.** After removing control valve as outlined in paragraph 10, proceed as follows: Clean valve thoroughly and remove paint from points of separation with a wire brush.

NOTE: A clean work bench is necessary. Also, use only lint-free paper shop towels for cleaning valve parts.

If oil leakage past seal (1 – Fig. 9) is the only difficulty, a new seal can be installed after removing mounting plate (2). Install new quad ring (3) and "O" ring (4) in mounting plate; then reinstall mounting plate and tighten retaining cap screws equally to a torque of 21 ft.-lbs. (28.35 N·m).

To completely disassemble and overhaul, refer to exploded view in Fig. 9 and proceed as follows: Clamp valve mounting plate in vise with cap end up. Remove retaining cap screws (23). Remove end cap (22), gerotor set (20 and 21), plate (17) and drive shaft (18) from valve body as a unit. Remove valve from vise, place a clean wood block in vise throat and set valve assembly on block with mounting plate (2) end up. Lightly clamp vise against port face of valve body and remove mounting plate retaining screws. Hold spool assembly down against wood block while removing mounting plate. Remove valve body from vise and place on work bench with port face down. Carefully remove spool and sleeve assembly (15 and 16) from 14-hole end of valve body. Using a screwdriver, remove relief valve plug (8) from housing; then, remove spool, guide and spring.

Remove centering pin (13) from spool and sleeve assembly, hold sleeve and

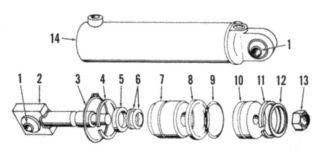

Fig. 10 — Exploded view of power steering cylinder.

1. Swivel
2. Cylinder rod
3. Snap ring
4. Retainer
5. Wiper
6. Seal
7. Head
8. Back-up ring
9. "O" ring
10. Piston
11. Piston seal
12. "O" ring
13. Locknut
14. Cylinder

push spool, (splined end first) out of sleeve. Remove the six centering springs (14) from slot in spool.

Separate end cap and plate from gerotor set and remove drive shaft, rotor and spacer.

Inspect all moving parts for scoring. Slightly scored parts can be cleaned by hand, rubbing with 400 grit abrasive paper. To recondition gerotor section surfaces, place a sheet of 600 grit paper on plate glass, lapping plate or other absolutely flat surface and remove sharp particles from paper with a flat piece of scrap steel. Stroke each surface of the gerotor section over the abrasive paper; any small bright areas indicate burrs that must be removed. Polish each part, rinse in clean solvent and air dry; keep these parts absolutely clean for reassembly.

Renew all parts that are excessively worn, scored or otherwise damaged. Install new seal kit when reassembling.

To reassemble valve, proceed as follows: Install relief valve guide on spool and install in relief valve hole. Then, install spring and plug with new "O" ring. Turn plug in until it is at least flush with housing surface. The exact relief valve setting should be checked on a test stand after valve is assembled.

Lubricate valve spool and carefully insert spool in valve sleeve using a twisting motion. Be sure that spring slots in spool and sleeve are at same end of assembly. Stand assembly on end with spring slots up and aligned. Assemble the springs in two groups with ex-

tended edges of springs down. Place the two groups of springs back-to-back (arched sections together) and install springs into spring slot in sleeve and spool in this position. Use a small screwdriver to guide springs through slots in opposite side of assembly. Center springs in sleeve with edges flush with upper surface of sleeve. Insert centering pin (13) through spool and sleeve assembly so that both ends of pin are below flush with outside of sleeve (15). On some models, nylon plugs are located at ends of centering pin to prevent pin from contacting inside diameter of valve body. Carefully insert spool and sleeve assembly, splined end first, in 14-hole end of valve body (12).

Set valve body on clean surface with 14-hole end down. Install locator bushing (5) in valve bore with chamfered side up. Install thrust bearing assembly (6 and 7) over valve spool (16). Install new quad ring (3) and "O" ring (4) in mounting plate; lubricate seal and install mounting plate over valve spool and locator bushing. Tighten mounting plate retaining screws evenly to a torque of 21 ft.-lbs. (28.35 N·m). Clamp mounting plate in a vise with 14-hole end of valve body up. Place plate (17) and gerotor outer ring (21) on valve body so bolt holes align. Insert drive shaft (18) in gerotor inner rotor so slot in shaft is aligned with valleys in rotor and push shaft through rotor so that about ½ of the splines protrude. Holding shaft and rotor in this position, insert them in valve housing so notch in shaft engages

centering pin in valve sleeve and spool. Install spacer (19) at end of drive shaft. If spacer does not drop down flush with rotor, shaft is not properly engaged with centering pin. Install end cap and tighten retaining screws equally to a torque of 20 ft.-lbs. (27.0 N·m).

If a source of hydraulic power for testing is available, relief valve may be checked as follows: Refer to Fig. 8 and plug "R" and "L" ports. Connect a pressure gage in pressure line from pump and to "IN" port. Connect a line from "OUT" port to sump. Support valve in a vise and turn shaft until relief valve opens. Relief valve should open at 1750-1800 psi (11.44-12.43 MPa). If relief pressure is incorrect, remove end plate (2 – Fig. 9) and adjust relief valve until correct reading is obtained.

If no bench testing is available then adjust as outlined in paragraph 8.

POWER STEERING CYLINDER

12. **R&R AND OVERHAUL.** Thoroughly clean cylinder and hose fitting to avoid entry of dirt when lines are disconnected. After hoses are disconnected remove bolts at each end of cylinder and remove cylinder. Work cylinder piston each way to end of stroke to expel oil from cylinder.

13. With cylinder removed as outlined in paragraph 12, refer to Fig. 10 and proceed as follows: Remove retaining ring (3); then push head toward piston end of cylinder and remove snap ring (4). Slide rod, piston and head out of cylinder tube. Remove nut (13), piston (10) and head (7) from rod.

Inspect inside of cylinder tube and rod assembly for excess scratching, scoring or pitting. Inspect seals for nicks and cuts.

Reassembly is reverse of disassembly, making sure that back-up washer (8) goes to rod end of cylinder and head (7) is installed with tapered end toward piston end of rod. Torque locking nut (13) to 220-250 ft.-lbs. (297-337.5 N·m).

ENGINE AND COMPONENTS

Model 7010 tractor uses the 649T engine while Model 7020 tractor uses the 649I engine. Both engines are turbocharged with the 649I engine also being intercooled.

Models 7030 and 7040 use the 3700 engine with turbocharger and intercooler.

Model 7045 tractors use the turbocharged 670T engine.

Models 7050 and 7060 use the 3700

engine with turbocharger and intercooler.

Model 7080 tractor is equipped with the 3750 engine with turbocharger, intercooling and counterbalanced crankshaft.

R&R ENGINE ASSEMBLY

14. Removal of engine is best accomplished by removing radiator, front

support, front axle and wheels assembly. Then, unbolt and remove engine and side rails from transmission housing as follows:

Drain cooling system, disconnect battery cables and remove right and left hood assemblies. Remove panel at front of console, and remove the top hood assembly. Disconnect radiator hoses and air cleaner inlet hose. Identify and disconnect power steering lines, power

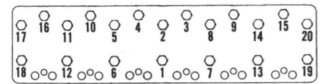

Fig. 11—On all models except 7010 and 7020 use sequence shown to tighten short cylinder head bolts. See text for complete outline of cylinder head installation procedure.

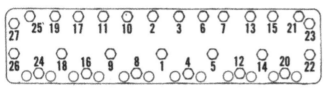

Fig. 12—Use sequence shown when tightening all cylinder head bolts after rocker arm assembly has been installed on Model 7010 and 7020 tractors. See text for complete outline of cylinder head installation procedure.

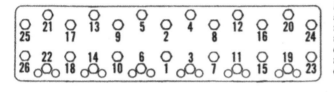

Fig. 13—Use sequence shown when performing final cylinder head bolt tightening on all models except 7010 and 7020. See text for complete outline of cylinder head installation procedure.

steering oil cooler in front of radiator and transmission cooler lines at bottom of radiator. Unbolt and remove radiator. Drain oil pan if engine is to be disassembled. Disconnect all electrical wiring to engine and move to rear.

NOTE: Cap all openings as lines are disconnected.

Remove platform panel, step and brace from left side of tractor. Disconnect vent line and fuel return line from top of fuel tank. Shut of fuel and disconnect line from main fuel tank (large tank). Remove the cross-over fuel line between main tank (left side) and auxiliary tank (right side) if so equipped. Plug auxiliary fuel line with plastic cap if tractor is so equipped. Loosen main fuel tank support straps at trunnion on inner side of tank. Support tank on floor jack. Disconnect battery wire from fuel level sending unit and disconnect ground wire from bolt on platform support. Remove fuel tank support straps, lower fuel tank to floor and move from under tractor. Disconnect speed control cable, fuel shut-off cable and engine hour-meter cable.

Support tractor under transmission housing, attach hoist to front support, unbolt front support from side rails and roll complete front assembly forward from tractor.

NOTE: If additional clearance is needed for removal of front support from side rails, loosen the engine front mounting cap screws.

Attach hoist to engine, then unbolt and remove engine and side rails from transmission housing.

Reinstall engine and torque limiter clutch assembly by reversing removal procedures. Bleed diesel fuel system as outlined in paragraph 44 and, if necessary, bleed power steering system as outlined in paragraph 7.

CYLINDER HEAD

15. REMOVE AND REINSTALL. Drain cooling system and remove hood top and side panels. Disconnect upper radiator hose from thermostat housing and remove tube from air cleaner to turbocharger. Disconnect oil lines and remove turbocharger. Disconnect water by-pass tube from water pump and thermostat housing, disconnect temperature gage sender switch wire and remove thermostat housing from cylinder head. If so equipped, remove cap screws securing coolant filter bracket to intake manifold allowing coolant filter with hoses intact to hang to one side. Remove exhaust and intake manifolds.

Disconnect breather tube from rocker arm cover and remove rocker arm cover. Remove oil supply line to rocker arm assembly and equally loosen and remove rocker arm assembly retaining cap screws. Lift off rocker arm assembly and remove push rods. Disconnect and remove the fuel return manifold, disconnect fuel lines from injectors and cap all openings to prevent entrance of dirt or dust.

NOTE: Although not required for removal of cylinder head, it is recommended that fuel injector assemblies be removed at this time. Injector nozzle tips protrude through the flat bottom surface of

cylinder head and, if injectors are not removed, extreme care must be taken when removing and handling the cylinder head.

Remove cylinder head retaining cap screws and carefully pry cylinder head up off dowel pins located at front and rear of cylinder block. Lift cylinder head from tractor.

When reinstalling cylinder head, be sure that gasket surfaces of cylinder head and block are clean and free of burrs. Check cylinder liner standout as outlined in paragraph 31. Place new gasket with side marked "THIS SIDE DOWN" against cylinder block and be sure gasket is properly located on the two dowel pins.

NOTE: Do not apply grease or gasket sealer to gasket surfaces, block or cylinder head; new gasket is treated with phenolic sealer.

If applicable locate fire rings around cylinder sleeves then position cylinder head on dowel pins. Lubricate cylinder head retaining cap screws with motor oil and install the 20 short cap screws. On Models 7010 and 7020 tighten short cap screws evenly from center of cylinder head towards each end to 63-73 ft.-lbs. (85.0-98.5 N·m). On all other models tighten the 20 short cap screws to a torque of 90-110 ft.-lbs. (121.9-149.1 N·m) in sequence shown in Fig. 11.

Install push rods and rocker arm assembly. Starting in center and working toward each end, tighten the six long rocker arm shaft/cylinder head retaining cap screws to a torque of 63-73 ft.-lbs. (85.0-98.5 N·m) for Models 7010 and 7020, and 90-110 ft.-lbs. (121.9-149.1 N·m) for all other models.

On Models 7010 and 7020 retighten all cylinder head retaining cap screws to a torque of 150 ft.-lbs. (203.2 N·m) in sequence shown in Fig. 12. For all other models retighten all cylinder head retaining cap screws to a torque of 165 ft.-lbs. (223.5 N·m) in sequence shown in Fig. 13; an adapter must be used to correctly tighten cap screws located under rocker arm shaft.

On all models, adjust valve tappet gap cold to 0.018 inch (0.45 mm) as outlined in paragraph 21. Install intake manifold and tighten retaining cap screws to 18-21 ft.-lbs. (24.3-28.3 N·m). Install exhaust manifold and tighten retaining cap screws to 46 ft.-lbs. (62.1 N·m).

Start and run engine for approximately one hour (preferably under a load) until it reaches normal operating temperature of approximately 160°F (71.1°C), then retorque cylinder head, intake and exhaust manifold retaining cap screws. Readjust valve tappet gap hot to 0.015 inch (0.38 mm). After ten hours of normal tractor operation, retor-

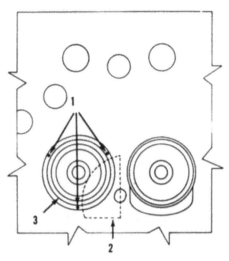

Fig. 14—Stake exhaust valve seat insert at points (1), but do not stake in area (2).

que cylinder head retaining cap screws and readjust valve tappet gap.

NOTE: All torque values are given for clean threads lubricated with motor oil. Allis-Chalmers recommends that no other thread lubricant be used.

VALVES AND SEATS

16. Intake and exhaust valves have a face and seat angle of 30 degrees. A one degree interference angle may be used by machining the seat to 31 degrees. Seat width should be 3/32 inch (2.38 mm) and can be narrowed by using 15 and 60 degree stones. Intake and exhaust valves seat in renewable valve seat inserts with standard and 0.005 inch (0.12 mm) oversize inserts available. Diameter of counterbore for standard intake valve seat insert should be 1.6035-1.6045 inch (40.728-40.754 mm) for Models 7010 and 7020, and 1.809-1.810 inch (45.94-45.97 mm) for all other models. Depth of intake valve seat counterbore should be 0.4735-0.4755 inch (12.026-12.077 mm) for all models except 7010 and 7020 which should be 0.4465-0.4485 inch (11.341-11.391 mm).

Diameter of counterbore for standard exhaust valve seat inserts should be 1.4805-1.4815 inch (37.604-37.630 mm) for Models 7010 and 7020, and 1.665-1.666 inch (42.29-42.31 mm) for all other models. Depth of exhaust valve seat counterbore should be 0.4735-0.4755 inch (12.026-12.077 mm) for all models except 7010 and 7020 which should be 0.4465-0.4485 inch (11.341-11.391 mm).

All inserts should have an interference fit of 0.001-0.003 inch (0.02-0.07 mm) in counterbore. If standard size insert is loose it will be necessary to install a 0.005 inch (0.12 mm) oversize insert. Us-

ing a center punch and hammer, stake exhaust valve seat insert into cylinder head at three points 1/32 inch (0.75 mm) in from outside edge of insert and 120 degrees apart as shown in Fig. 14. Be careful not to stake insert in area of injection nozzle and do not use old stake points. Head of intake valve should be recessed 0.0345 inch (0.874 mm) below gasket surface of cylinder head for Models 7010 and 7020, and 0.0505-0.0695 inch (1.282-1.765 mm) for all other models. Head of exhaust valve should be recessed 0.046 inch (1.16 mm) below gasket surface of cylinder head for Models 7010 and 7020, and 0.054-0.068 inch (1.37-1.73 mm) for all other models. Intake valve stem diameter should be 0.3715-0.3720 inch (9.436-9.448 mm) and exhaust valve stem diameter should be 0.3705-0.3710 inch (9.410-9.423 mm) for all models.

VALVE GUIDES

17. Intake valve stem-to-guide clearance should be 0.0010-0.0015 inch (0.025-0.038 mm) and must not exceed 0.0035 inch (0.088 mm) for all models except 7045. Intake valve stem-to-guide clearance for Model 7045 tractors should be 0.0015-0.0027 inch (0.038-0.068 mm) and must not exceed 0.0035 inch (0.088 mm). Exhaust valve stem-to-guide clearance should be 0.0020-0.0025 inch (0.050-0.063 mm) and must not exceed 0.0055 inch (0.140 mm) for all models except 7045. Model 7045 exhaust valve stem-to-guide clearance should be 0.0025-0.0037 inch (0.063-0.093 mm) and must not exceed 0.0055 inch (0.139 mm).

Intake and exhaust valve guides are

interchangeable. If necessary to renew valve guides, remove old guide from combustion chamber side of cylinder head using a suitable press and shoulder punch. Lubricate OD of new guide with a mixture of white lead and engine oil. Press guide into place, chamfered end first, from combustion chamber side of cylinder head using a suitable press and shoulder punch. When measured as shown in Fig. 15 valve guide protrusion should be as follows:

Models 7010 and 7020
Intake   21/32-inch (16.67 mm)
Exhaust   7/8-inch (22.22 mm)
All Other Models
Intake   25/32-inch (19.84 mm)
Exhaust   1-3/32-inch (27.78 mm)

After installation it will be necessary to ream inside diameter of guides to establish proper valve stem-to-guide clearance.

VALVE SPRINGS

18. On engines using a two-piece valve and damper spring assembly, the intake and exhaust valve spring assemblies are interchangeable and should be renewed as unit assemblies if rusted, distorted or cracked, or if they vary more than 5 percent from the following specifications:

Spring Height	Spring Pressure
With Damper Spring	
2.237 inches	40-60 lbs
(56.82 mm)	(177.9-266.9 N)
Without Damper Spring	
2.237 inches	38-42 lbs.
(56.82 mm)	(169.0-186.8 N)

Fig. 15—View showing method of measuring valve guide height above cylinder head counterbore. Refer to paragraph 17 for outline of procedure.

With Damper Spring
1.730 inches	105-115 lbs.
(43.94 mm)	(467.0-511.5 N)

Without Damper Spring
1.730 inches	95-105 lbs.
(43.94 mm)	(422.5-467.0 N)

On engines equipped with single valve springs (no damper spring), the intake and exhaust valve springs are interchangeable and should be renewed if rusted, cracked or distorted, or if they vary more than 5 percent from the following specifications:

Model 7010 and 7020
Pressure @ 2.223 in.	41-45 lbs.
(56.46 mm)	(182.4-200.2 N)
Pressure @ 1.836 in.	73-81 lbs.
(46.63 mm)	(324.7-360.3 N)

Model 7080
Pressure @ 2.231 in.	57-63 lbs.
(56.67 mm)	(253.5-280.2 N)
Pressure @ 1.780 in.	121-133 lbs.
(45.21 mm)	(538.2-591.5 N)

All other Models
Pressure @ 2.237 in.	40-46 lbs.
(56.82 mm)	(177.9-204.6 N)
Pressure @ 1.780 in.	105-115 lbs.
(45.21 mm)	(467.0-510.5 N)

All valve springs should be installed with closed coil end seated in cylinder head.

VALVE LIFTERS

19. The 0.7480-0.7485 inch (18.999-19.011 mm) diameter valve lifters ride directly in cylinder block bores and can only be renewed after removing the camshaft. Cylinder block valve lifter bore ID should be 0.7495-0.7505 inch (19.037-19.062 mm). Valve lifters are available in standard size only and should be renewed if either end is chipped or worn or if lifter-to-bore clearance exceeds 0.0035 inch (0.088 mm). Desired lifter-to-bore clearance is 0.0010-0.0025 inch (0.025-0.063 mm).

ROCKER ARMS

20. **R&R AND OVERHAUL.** Rocker arms and shaft assembly can be removed after removing hood, disconnecting breather tube and removing rocker arm cover and oiling tube. Loosen and remove all retaining cap screws equally, then lift rocker arms and shaft assembly from cylinder head.

Hollow rocker arm shaft is drilled for lubrication to each rocker arm. Lubrication to oiling tube is supplied through a drilled passage in cylinder head and engine block.

To disassemble rocker arms and shaft assembly, remove hex plugs, spring

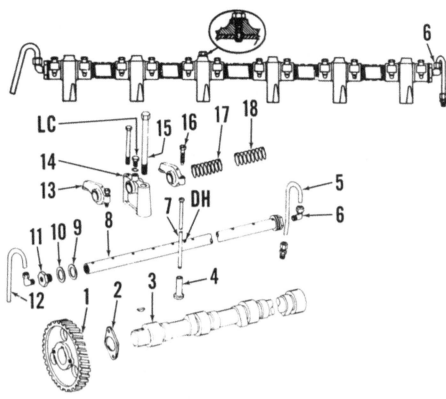

Fig. 16—View of valve actuating parts. Upper drawing shows rocker arm shaft assembled with locating cap screw at inset. Restrictor elbow (6) is on inlet tube at rear.

DH. Detent hole		12. Return tube
LC. Locating cap screw	6. Elbow	13. Rocker arm
1. Camshaft gear	7. Push rod	14. Support
2. Thrust plate	8. Rocker arm shaft	15. Cylinder head cap screw
3. Camshaft	9. Washer	16. Self-locking screw
4. Cam follower	10. Wave washer	17. Spacer spring (inner)
5. Inlet tube	11. End plug	18. Spacer spring (outer ends)

washers and spacer washers from each end of shaft and remove locating cap screws and lockwasher from number 3 rocker arm support. Slide rocker arms, supports and spacer spring from shaft.

NOTE: The two outer springs (18 – Fig. 16) are different than springs (17).

Rocker arm shaft diameter is 0.999-1.000 inch (25.37-25.40 mm). Bore in rocker arm is 1.001-1.002 inch (25.42-25.45 mm) providing a clearance of 0.001-0.003 inch (0.02-0.07 mm) between shaft and rocker arms. If clearance is excessive, renew shaft and/or rocker arms. All rocker arms are alike.

On Models 7010 and 7020, rocker arm shaft is stamped "CAM SIDE" between third and fourth shaft stands. Make sure that rocker arm adjusting screws are positioned towards this side of rocker arm shaft otherwise excessive oil could be forced out of holes in top of rocker arms. To assemble rocker arms on shaft of all other models place shaft on work bench with end nearest to detent hole (DH – Fig. 16) to left. Slide rocker arm support with locating cap screw (LC) on shaft with flat mounting surface down

and large cap screw hole toward you. Tighten locating cap screw into hole in shaft to 28-33 ft.-lbs. (37.9-44.7 N·m) torque. Complete assembly of rocker arm shaft using Fig. 16 as a guide.

On all models, end plugs (11 – Fig. 16) should be torqued to 40 ft.-lbs. (54.3 N·m). Restrictor elbow (6) with a 1/16-inch (1.5 mm) opening should be installed in rear plug and inlet oil line (5) should be attached. Elbow at front is not a restricted type.

Before installing rocker arm assembly, loosen each tappet adjusting screw about two turns to prevent interference between valve heads and pistons due to interchanged components. Tighten 3/8-inch support retaining cap screws to a torque of 28-33 ft.-lbs. (37.9-44.7 N·m) and 9/16-inch (cylinder head) cap screws to a torque of 130-140 ft.-lbs. (176.1-189.7 N·m). Adjust valve tappet clearance cold as outlined in paragraph 21. Complete assembly of tractor, then readjust tappet clearance after tractor engine has been run for one hour, preferably under a load, and minimum coolant temperature is 160°F (71.0°C). Retorque all rocker arm shaft stand bolts to the above specifications.

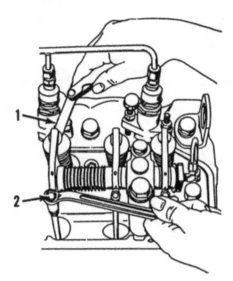

Fig. 17—Adjust valve clearance as outlined in paragraph 21 by using a suitable feeler gage (1) and turning adjustment screw (2).

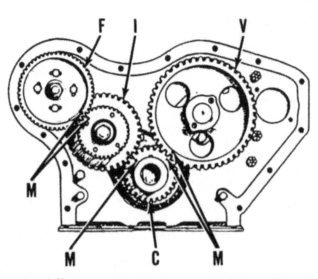

Fig. 19—Timing gears are correctly timed when punch marks (M) are aligned as shown.

C. Crankshaft gear
F. Injection pump gear
I. Idler gear
V. Camshaft gear

VALVE CLEARANCE

21. Because of the limited clearance between head of valve and top of piston, valve clearance should only be checked and adjusted with engine stopped. After any mechanical work has been done which would disturb valve clearance adjustment, clearance should be set at 0.018 inch (0.45 mm), engine started and allowed to warm up to 170°F (76.6°C). After engine has reached normal operating temperature clearance for both intake and exhaust valves should be reset to 0.015 inch (0.38 mm).

Crank engine until both valves of cylinder to be adjusted are fully closed and push rods are at their lowest position. Check clearance between valve stem and rocker arm using a suitable feeler gage (1–Fig. 17). Feeler gage should pass between valve stem and rocker arm with a slight drag when

Fig. 18—Do not pull crankshaft pulley by outer rim; thread pullers into hub of pulley as shown. Place correct size center plug over end of crankshaft after removing pulley retaining cap screw (CS) and washer.

valve clearance is correct. Adjust clearance by turning self-locking adjustment screw (2).

Two crankshaft revolution adjustment of all valves is possible knowing that number 1 and 6 pistons move up and down in their respective cylinders together, and that when one piston is on its compression stroke the other is on its exhaust stroke. This relationship is the same for number 2 and 5 pistons, and number 3 and 4 pistons. Also remember that it is necessary for crankshaft to rotate only 120 degrees to position next piston in firing order to TDC. Adjust valve clearance on cylinder listed in following table after positioning rocker arms of opposite paired cylinder (1 and 6, 2 and 5, 3 and 4) so exhaust valve is just closing and intake valve is just opening:

Adjust Valves On Cylinder	Ex. Valve Closing, In. Valve Opening On Cylinder
Number 1	Number 6
Number 5	Number 2
Number 3	Number 4
Number 6	Number 1
Number 2	Number 5
Number 4	Number 3

TIMING GEAR COVER AND CRANKSHAFT FRONT OIL SEAL

22. **REMOVE AND REINSTALL.** To remove the timing gear cover, first remove axle main member as outlined in paragraph 5, then proceed as follows: Drain cooling system then remove air cleaner, front support and radiator. Remove fan blade, water pump and alternator drive belts, and front pulley retaining bolt (CS–Fig. 18). Using a suitable puller, threaded into front pulley as shown in Fig. 18, remove pulley from crankshaft.

CAUTION: Do not attempt to remove pulley by pulling on outer circumference

as vibration dampener built into pulley will be damaged.

Remove bolts retaining oil pan to timing gear cover. It is possible to loosen remaining oil pan retaining bolts enough to clear timing case cover for its removal; however, it is recommended by manufacturer that oil pan be removed completely from engine and all gaskets be replaced at time of installation.

Remove fuel injection pump drive gear inspection cover from front of timing gear cover and withdraw button and spring from end of fuel injection pump drive shaft. Remove remaining timing gear cover retaining bolts and using a soft mallet tap cover from locating dowels and lift cover from engine.

NOTE: Some of the timing gear cover retaining bolts enter cover from the rear.

After timing gear cover has been removed place it on a flat surface and, using a suitable seal driver, remove crankshaft front oil seal from cover. Outside diameter of replacement seal is precoated with sealant. Install seal in opposite order of removal using a seal driver that contacts only the outer edge of the new seal. After seal has been installed check that inner circumference of seal rotates freely.

When installing timing gear cover, use a new oil pan gasket front section if oil pan was not removed. If oil pan was removed from engine install new gaskets. Install new timing gear cover gaskets. Lubricate oil seals and carefully install timing gear cover over crankshaft and the two dowel pins. Tighten timing gear cover and oil pan retaining cap screws to 28-33 ft.-lbs. (37.9-44.7 N·m). Install crankshaft pulley and tighten retaining cap screw to 180 ft.-lbs. (243.9 N·m) on Models 7010 and 7020 or to 220 ft.-lbs (298.1 N·m) on all other models. Install spring and button in end of fuel injection pump drive shaft and install inspection cover.

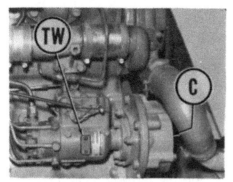

Fig. 20—View of right side of diesel engine showing location of timing window (TW) and pump drive gear cover (C).

TIMING GEARS

23. **GEAR TIMING.** Gear train is pressure lubricated through a 1/16-inch (1.5 mm) hole in front oil gallery plug which directs oil onto idler gear (I – Fig. 19) which in turn splashes oil on remaining gears in train. Correct backlash between crankshaft gear (C) and idler gear or camshaft gear (V) is 0.0015-0.0085 inch (0.038-0.215 mm). Correct backlash between idler gear and injection pump drive gear (F) is 0.002-0.010 inch (0.05-0.25 mm). New gears should be installed when backlash between any two mating gears exceeds 0.015 inch (0.38 mm). Gears are correctly timed when punch marks (M) on gears are aligned as shown in Fig. 19.

24. **R&R CRANKSHAFT GEAR.** The crankshaft gear is keyed and press fitted to the crankshaft. Use a suitable puller to remove crankshaft gear after first removing timing gear cover as outlined in paragraph 22.

Install crankshaft gear by first heating it in oil to 300°F (149°C) and then driving it onto crankshaft. Gear may also be pressed onto crankshaft by using front pulley retaining bolt and

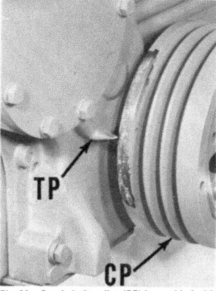

Fig. 22—Crankshaft pulley (CP) is provided with timing marks in degrees. When adjusting valve clearance or injection pump timing correct degree mark should be aligned with pointer (TP) on timing gear cover.

washers as spacers. Be sure timing marks on crankshaft, camshaft and idler gears are aligned as shown in Fig. 19.

25. **R&R CAMSHAFT GEAR.** Camshaft gear is keyed and press fitted to camshaft. It is also held in place by a retaining bolt and washer. To remove camshaft gear first remove camshaft from engine as outlined in paragraph 28, then remove gear from shaft by using a suitable press and press plate. Carefully inspect bore of gear for wear, gear bore should be 1.373-1.374 inch (34.87-34.89 mm) in diameter. Correct interference fit between camshaft and gear is 0.0025-0.004 inch (0.063-0.101 mm). Inspect thrust plate for wear and roughness, thrust plate thickness should be 0.204-0.206 inch (5.18-5.23 mm). Renew thrust plate and gear as necessary.

To install camshaft gear first place camshaft in a suitable press using a press plate located beneath first bearing journal. Place thrust plate on camshaft and install key in slot in end of camshaft. Heat camshaft gear in oil to 350°-400°F (177°-204°C). Press gear onto camshaft

until gear hub is flush with end of camshaft, or clearance between thrust plate and gear hub is 0.003-0.009 inch (0.07-0.22 mm). Install retaining bolt and washer and tighten bolt to 73 ft.-lbs. (98.9 N·m) torque.

26. **R&R FUEL INJECTION PUMP DRIVE GEAR.** Injection pump drive gear, idler gear and crankshaft gear are all provided with timing marks for correct assembling; however, because of the odd number of teeth on idler gear, marks are aligned only every 70th revolution of crankshaft. Injection pump drive gear can be repositioned without regard to marks on gear teeth to provide correct pump timing.

To remove pump drive gear, disconnect throttle and shut-off linkage. Shut off fuel supply. Remove injection pump timing window (TW – Fig. 20) and turn engine until pump timing marks are aligned as shown in Fig. 21. Number one piston should be on compression and timing mark on crankshaft pulley as follows:

Model	Injection Timing
7010	18° BTDC
7020	18° BTDC
7030	24° BTDC
7040	16° BTDC
7045	16° BTDC
7050	26° BTDC
7060	18° BTDC
7080	22° BTDC

Remove cover (C – Fig. 20) from front of timing gear cover. Remove four cap screws securing pump drive gear to pump drive hub and remove gear.

To install injection pump drive gear, make certain that No. 1 piston is on compression stroke and crankshaft is set at correct timing mark. Rotate pump drive hub clockwise until timing lines are in register as shown in Fig. 21. Install gear and tighten retaining cap screws to 27-29 ft.-lbs. (36.7-39.4 N·m).

27. **R&R IDLER GEAR AND SHAFT.** To remove idler gear (I – Fig. 19) and/or shaft, first remove timing gear cover as outlined in paragraph 22. Turn crankshaft in normal direction of rotation until number 1 piston is at TDC

Fig. 21—View of properly aligned diesel injection pump timing marks.

Fig. 23—Exlploded view of idler gear and support shaft assembly.

0. Oil supply hole
1. Bolt
2. Washer
3. Support bearings
3A. Spacer
4. Idler gear
5. Support shaft

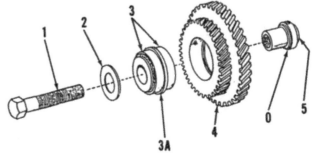

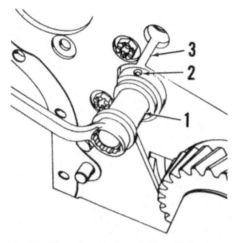

Fig. 24—View showing installation of idler gear shaft using cap screw and washer drawing shaft in place. Oil hole (2) is lined with slot (3).

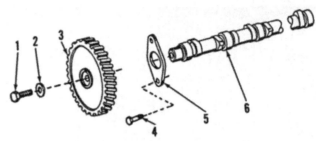

Fig. 25—Exploded view of camshaft and gear assembly.

1. Gear retaining bolt
2. Washer
3. Gear
4. Bolt
5. Retainer plate
6. Camshaft

on compression and timing marks (M) are aligned as shown in Fig. 19. Remove injection pump drive gear as outlined in paragraph 26. Remove idler gear retaining bolt (1 – Fig. 23) and washer (2) then withdraw gear (4) and bearings (3) as an assembly from shaft (5).

Inside diameter of bearings (3 and 3A – Fig. 23) should be 1.000-1.008 inch (25.4-25.6 mm) and OD of bearing outer races should be 1.980-1.981 inch (50.29-50.31 mm). Inside diameter of idler gear (4) should be 1.9785-1.9795 inch (50.253-50.279 mm). If gear or bearing assembly is worn, drive outer races out of gear using a suitable bearing driver and press in new races until they contact newly installed center spacer.

NOTE: Early model engines used a two-piece idler gear held together by a cap screw fitted from engine side of larger gear. When assembling this type of gear be sure to line up timing marks on smaller outer gear with timing marks in larger inner gear.

Idler gear shaft (5 – Fig. 23) is pressed into cylinder block and, if necessary, should be removed using a suitable slide hammer threaded into shaft. Outside diameter of both ends of idler shaft should be 0.998-0.999 inch (25.34-25.36 mm). To install idler gear shaft into cylinder block, position idler shaft so that oil hole (2 – Fig. 24) in shaft (1) lines up with slot (3) in engine front plate. Install retaining bolt (1 – Fig. 23) and washer (2) through shaft and tighten retaining bolt until idler shaft bottoms in cylinder block. Apply a light coat of engine oil to idler shaft. Install idler gear and bearing assembly on shaft making sure that timing marks on idler gear, crankshaft gear and injection pump gear are aligned as shown in Fig. 19. Install idler gear retaining bolt (1 – Fig. 23) and washer (2). Tighten re-

taining bolt to 95-105 ft.-lbs. (121.9-142.2 N·m). After installation idler gear end play should be 0.002-0.011 inch (0.05-0.28 mm).

CAMSHAFT AND BUSHINGS

28. R&R CAMSHAFT. To remove camshaft, first remove oil pan, oil pump, timing gear cover and rocker arm assembly; then, proceed as follows:

Working through holes in timing gear, bend down tabs on locking plate and end thrust plate. Cam followers can be held in up position to facilitate camshaft removal as follows: Wooden dowels 5/8-inch (15.87 mm) diameter and 16 inches (406 mm) long can be used to hold cam followers up. Be sure dowels are seated in followers, then install rubber bands over each pair of dowels to hold side pressure on cam followers. Push cam followers up out of the way and remove camshaft. Press camshaft out of camshaft timing gear.

Backlash should be 0.0015-0.009 inch (0.038-0.228 mm) between gear and either mating gear. Renew parts as necessary if backlash exceeds 0.015 inch (0.38 mm). Camshaft journal diameter (all six journals) is 2.130-2.131 inch (54.100-54.125 mm). Desired camshaft to bushing clearance is 0.002-0.006 inch (0.05-0.15 mm). Camshaft bushing should be renewed as outlined in paragraph 29, if clearance exceeds 0.008 inch (0.20 mm). Bushings are available in 0.010 inch (0.25 mm) undersize as well as standard size. If camshaft journal is excessively worn or scored and camshaft is otherwise serviceable, journal may be ground to 2.120-2.121 inch (53.85-53.87 mm) and 0.010 inch undersize bushings installed. Cam lift is 0.318 inch (8.08 mm) for intake valves and 0.285 inch (7.24 mm) for exhaust valves.

Camshaft end play is controlled by the 0.204-0.206 inch (5.18-5.23 mm) thick thrust plate (5 – Fig. 25). End play should be 0.003-0.009 inch (0.08-0.23 mm). Renew thrust plate if worn so that camshaft end play is excessive. If end play is still excessive after renewing thrust plate, camshaft gear should be renewed.

Install camshaft in opposite order of removal tightening gear retaining bolt

(1 – Fig. 22) to 69-74 ft.-lbs. (93.5-99.5 N·m) and thrust plate retaining bolts (4) to 18-20 ft.-lbs. (24.4-27.1 N·m).

29. CAMSHAFT BUSHINGS. Camshaft is supported in six bushings. Camshaft should have 0.002-0.006 inch (0.05-0.15 mm) clearance in bushings. If clearance is excessive, camshaft and/or bushing should be renewed or camshaft journals ground to 2.120-2.121 inch (53.85-53.87 mm) and 0.010 inch undersize bushings installed.

To renew camshaft bushings, engine must be removed from tractor and flywheel and rear adapter plate must be removed from engine. Press old bushings out and new bushings in with suitable bushing tools. Front bushing is 1-3/8 inches (34.93 mm) wide and the five other bushings are one inch (25.4 mm) wide. Bushings have a 0.003-0.006 inch (0.08-0.15 mm) interference fit in bores in cylinder block. Bushings are presized and should not require reaming if carefully installed. However, they should be checked after installation for localized high spots. Standard bushing diameter after installation should be 2.133-2.136 inch (54.18-54.25 mm); 0.010 inch undersize bushing inside diameter should be 2.123-2.126 inch (53.92-54.00 mm). Be sure bushings are installed so oil holes in bushings are aligned with oil passages in cylinder block. Front bushing should be installed flush or slightly below flush with front face of cylinder block. Rear bushing should be installed flush with front face of cylinder block bore.

ROD AND PISTON UNITS

30. Piston and connecting rod units are removed from above after removing cylinder head, oil pan, oil pump, rod bearing caps and piston ring travel ridge from cylinder sleeve.

There are two types of connecting rods (early and late). Earlier type connecting rod can be easily identified by serrated mating surface of rod and cap. Mating surface on later rod is smooth and flat. New type connecting rod assembly (including cap and special screws) can be installed in earlier engine individually or as complete sets. Cap re-

Fig. 26—View showing correct method of measuring cylinder sleeve flange thickness. Be sure not to measure thickness of fire wall at top inside edge of sleeve.

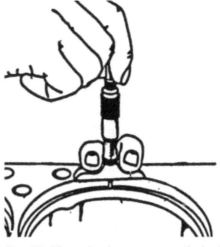

Fig. 27—View showing correct method of measuring cylinder block counterbore depth.

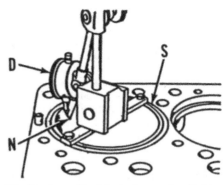

Fig. 28—As outlined in text, mount a dial indicator (D) on a new cylinder sleeve (S) that has had a notch (N) cut out of its upper flange. This procedure is used to check concentricity of cylinder block counterbore to cylinder block sleeve bore.

taining screws must be correct type according to rod assembly. Early connecting rods use cap screws with six-point head (early Model 7045 uses twelve-point cap screws) and a washer is installed below head of each screw. Threads of early type screws must be cleaned, then lubricated with "Molykote G" before installing. Later type connecting rods use special socket head screws with no washer and engine lubricating oil should be used to lubricate threads. Tightening torque is increased for late type connecting rod and screws.

On all models each rod and cap bear an assembly number in addition to cylinder number. Cylinder number is stamped on left (camshaft) side of connecting rod and cap, and rods are numbered 1 through 6 from front to rear. Disregard assembly numbers.

When reinstalling rod and piston units, be sure that cylinder numbers on rod and cap are in register and both bearing insert tangs are on same side of rod. On engines with slotted rod bearing insert, connecting rod must be installed with slot toward camshaft.

Six-or twelve-point cap screws used with early connecting rods must be thoroughly cleaned and dried; then, lubricate threads and under the head with "Molykote G" before installing. Make certain that special hardened washer is installed under head of each screw and torque both screws in each rod evenly to 65-70 ft.-lbs. (88.0-94.8 N•m).

On Models 7010 and 7020 lubricate rod screws with engine oil before installing. Tap lower balance pad with a plastic hammer toward front then toward rear to align cap on rod. Tighten screws to 40-45 ft.-lbs. (54.2-60.9 N•m). Side play on crankpin should be 0.005-0.010 inch (0.12-0.25 mm) after both screws are tightened.

Special socket head screws used with late connecting rods should be lubricated with engine oil before installing. Special screws require a male 3.8-inch hex

wrench. The following procedure should be used when tightening. Tighten both screws evenly to 8-12 ft.-lbs. (10.8-16.2 N•m) torque, then tap lower balance pad with a plastic hammer toward front then toward rear to align cap. Tighten both screws to 40-42 ft.-lbs. (54.2-56.9 N•m), then evenly to final torque of 80-85 ft.-lbs. Side play on crankpin should be 0.005-0.010 inch (0.13-0.25 mm) after both screws are tightened.

PISTONS, RINGS AND SLEEVES

31. The cam ground aluminum alloy pistons are fitted with three compression rings and one oil control ring. All rings are located above the piston pin. Pistons and rings are available in standard size only.

Install compression rings with manufacturers marking facing toward top of piston. Top compression ring is chrome plated and barrel faced and can be installed with either side up if not marked. Second compression ring is bevel faced and should be installed with inside chamfer facing top of piston. Third compression ring is bevel faced and should be installed with inside chamfer facing piston skirt. If oil control ring is three-piece type, expander should be installed in groove, then top and bottom rails should be installed over the expander with ends 180 degrees apart from each other and 90 degrees away from ends of expander. Some two-piece oil control rings have slots toward bottom of piston while on others the oil return slot is between rails. Install expander in groove, then install the oil control ring with ends 180 degrees from ends of expander.

With piston and connecting rod removed, use suitable puller to remove the wet type sleeve. Using suitable micrometers check cylinder sleeves for wear, out-of-round and taper. Specified

cylinder sleeve ID for Model 7010 and 7020 tractors is 3.8755-3.8770 inches (98.437-98.475 mm). ID of cylinder sleeves for all other models should be 4.2495-4.2510 inches (107.937-107.975 mm). For all models maximum sleeve wear should not exceed 0.0015 inch (0.038 mm), out-of-round should not exceed 0.0015 inch (0.038 mm) and taper should not exceed 0.008 inch (0.20 mm). If total sleeve wear is not excessive and if no deterioration of the flange has occurred to reduce protrusion below specified minimum, sleeves may be reinstalled with a life expectancy of approximately one-half to three-fourths of new sleeves.

Cylinder sleeve standout must be maintained at 0.002-0.005 inch (0.05-0.13 mm) above cylinder block surface. To check standout, measure thickness of sleeve flange with micrometer (do not include the 0.0445-0.0475 inch (1.130-1.206 mm) "fire wall" on top end of sleeve) as shown in Fig. 26 and depth of counterbore in cylinder block with a micrometer depth gage as shown in Fig. 27. Then use special shims as necessary to provide the proper 0.002-0.005 inch (0.05-0.13 mm) standout. Shims are available in thicknesses of 0.005, 0.011, 0.015 and 0.020 inch.

NOTE: Do not attempt to measure standout with straightedge and feeler gage after installing sleeve as rough edges on shims may give incorrect standout measurement.

Relationship of cylinder sleeve counterbore to cylinder centerline should also be checked as follows: Place a new cylinder sleeve, flange end, in the chuck of a lathe. True centerline of both ends of sleeve as closely as possible to lathe center using a dial indicator. Machine bottom surface of sleeve flange to a true 90 degree angle with sleeve centerline. Cut a V-section out of sleeve flange and mount a dial indicator to sleeve as shown in Fig. 28. Install test sleeve in cylinder to be checked so bottom of flange fully contacts bottom of cylinder counterbore. Position dial indicator so that spindle is contacting bottom of cylinder counterbore as shown in Fig. 28. Slowly rotate test sleeve assembly and check that sleeve counterbore depth does not vary more than 0.002 inch (0.05 mm). If variation of counterbore depth is excessive or minimum cylinder sleeve standout cannot be achieved it will be necessary to have cylinder counterbore seat recut by a qualified machine shop.

Thoroughly clean all sealing and mating surfaces of block and cylinder sleeve. Insert sleeve in bore without sealing rings; sleeves should be free enough to be pushed into place and then be rotated by hand pressure. If sleeve cannot be inserted and turned by hand, more thorough cleaning is necessary.

When sleeve and bore in cylinder block are properly cleaned, install new sealing rings on dry cylinder sleeve; do not use a lubricant of any kind.

For cylinder liners with three sealing ring grooves install ethylene propylene (black) sealings rings in top two grooves and a silicone (red) sealing ring in bottom groove. For cylinder liners with only two sealing ring grooves install a "Buna N" (black) sealing ring in upper groove and a silicone (red) sealing ring in lower groove. Brush a light coat of **vegetable oil only** in lower sleeve bore of cylinder block. Be extremely careful when installing cylinder sleeve in cylinder block so that sealings rings are not cut on sharp edges of cylinder bore.

NOTE: Sealing rings are easily damaged and extreme care must be used in handling and installation. Sealing ring material expands rapidly when it comes in contact with lubricating oils or permanent anti-freeze.

Check pistons, rings and sleeves against specifications listed in the following table:

Top Compression Ring Side Clearance
　Models 7010 & 7020 ...0.012-0.018 inch
　　　　　　　　　　　(0.30-0.45 mm)
　Model 7045 ....0.0040-0.0065 inch
　　　　　　　　　　　(0.101-0.165 mm)
　All Other Models ...0.0040-0.0055 inch
　　　　　　　　　　　(0.101-0.139 mm)
Second Compression Ring Side Clearance
　Models 7010 & 7020 ...0.020-0.030 inch
　　　　　　　　　　　(0.51-0.76 mm)
　Model 7045 ....0.002-0.004 inch
　　　　　　　　　　　(0.05-0.10 mm)
　All Other Models ....0.003-0.005 inch
　　　　　　　　　　　(0.08-0.13 mm)
Third Compression Ring Side Clearance
　Models 7010 & 7020 ...0.007-0.017 inch
　　　　　　　　　　　(0.18-0.43 mm)
　Model 7045 ....0.002-0.004 inch
　　　　　　　　　　　(0.05-0.10 mm)

All Other Models ....0.003-0.005 inch
　　　　　　　　　　　(0.07-0.13 mm)
Oil Control Ring Side Clearance
　Models 7010 & 7020...0.007-0.017 inch
　　　　　　　　　　　(0.18-0.43 mm)
　Model 7045 ....0.0015-0.0030 inch
　　　　　　　　　　　(0.038-0.076 mm)
　All Other Models ....0.001-0.003 inch
　　　　　　　　　　　(0.03-0.08 mm)
Top Compression Ring End Gap
　Models 7010 & 7020...0.019-0.029 inch
　　　　　　　　　　　(0.48-0.74 mm)
　Model 7045 ....0.015-0.025 inch
　　　　　　　　　　　(0.38-0.64 mm)
　All Other Models ....0.013-0.028 inch
　　　　　　　　　　　(0.33-0.71 mm)
Second Compression Ring End Gap
　Models 7010 & 7020...0.015-0.025 inch
　　　　　　　　　　　(0.38-0.64 mm)
　Model 7045 ....0.025-0.035 inch
　　　　　　　　　　　(0.64-0.89 mm)
　All Other Models ....0.009-0.024 inch
　　　　　　　　　　　(0.23-0.61 mm)
Third Compression Ring End Gap
　Models 7010 & 7020...0.015-0.025 inch
　　　　　　　　　　　(0.38-0.64 mm)
　Model 7045 ....0.007-0.017 inch
　　　　　　　　　　　(0.18-0.43 mm)
　All Other Models ....0.009-0.024 inch
　　　　　　　　　　　(0.23-0.61 mm)
Oil Control Ring End Gap
　Models 7010 & 7020...0.010-0.020 inch
　　　　　　　　　　　(0.258-0.50 mm)
　Model 7045 ....0.007-0.019 inch
　　　　　　　　　　　(0.18-0.48 mm)
　All Other Models ....0.009-0.024 inch
　　　　　　　　　　　(0.23-0.61 mm)
Piston Skirt-to-Sleeve Clearance
　Models 7010
　　& 7020.................0.0045-0.0070 inch
　　　　　　　　　　　(0.114-0.177 mm)
　All Other Models ...0.0025-0.0050 inch
　　　　　　　　　　　(0.063-0.127 mm)

PISTON PINS

32. Piston pins are of the full floating type and are retained in place by snap rings. Remove and install piston pin using a suitable press or by heating piston in 180°F (82°C) water for five minutes. Running clearance of piston pin in piston should be 0.0001-0.0006 inch (0.002-0.015 mm) for all models. Running clearance of piston pin in small end bushing should be 0.0010-0.0017 inch (0.025-0.043 mm) for Models 7010 and 7020, and 0.0014-0.0021 inch (0.035-0.053 mm) for all other models. Renew piston pin and/or connecting rod bushing if clearance between pin and bushing exceeds 0.003 inch (0.07 mm). If necessary to renew small end bushing be sure oil hole in new bushing is properly aligned with oil hole in top of connecting rod. After installation it will be necessary to finish ream or hone new bushing to appropriate ID for proper piston pin clearance.

CONNECTING RODS AND BEARINGS

33. Connecting rod bearings are precision, slip-in type. The bearing inserts can be renewed after removing oil pan and bearing caps. When installing new bearing inserts, be sure that linear projections (tangs) engage the milled slot in connecting rod and bearing cap and that cylinder numbers stamped on rod and cap are in register and face towards camshaft side of engine. Bearing inserts are available in standard size and oversizes of 0.010, 0.020 and 0.040 inch. Check crankshaft connecting rod bearing journals and connecting rod bearings against the following values:

Rod Bearing Journal OD
　Model 7010 & 7020 ...2.3720-2.3735 in.
　　　　　　　　　　　(60.248-60.286 mm)
　All Other Models ......2.7470-2.7485 in.
　　　　　　　　　　　(69.773-69.811 mm)
Rod Bearing Clearance
　Model 7010 and 7020
　　Standard ....0.0009-0.0034 in.
　　　　　　　　　　　(0.022-0.086 mm)
　　Maximum ....0.008 in.
　　　　　　　　　　　(0.20 mm)
　All Other Models
　　Standard ....0.001-0.004 in.
　　　　　　　　　　　(0.02-0.10 mm)
　　Maximum ....0.006 in.
　　　　　　　　　　　(0.15 mm)
Connecting Rod Side Clearance
　All Models ....0.005-0.010 in.
　　　　　　　　　　　(0.13-0.25 mm)

Refer to paragraph 30 for recommended tightening torques, installation procedures and cautions regarding different types of connecting rods.

CRANKSHAFT AND MAIN BEARINGS

34. Crankshaft is supported in seven precision, slip-in type main bearings. If desired main bearing inserts can be renewed with crankshaft installed in engine after removing oil pan, oil pump, oil tube and main bearing caps. If removal of crankshaft is necessary, first remove engine from tractor as outlined in paragraph 14. Then remove clutch, flywheel, engine rear adapter plate, crankshaft pulley, timing gear cover, injection pump drive gear and shaft, idler gear, oil pan, oil pump, engine front plate and camshaft. Remove connecting rod and main bearing caps and lift crankshaft from engine.

Check crankshaft and main bearing inserts against the following specifications:

Main Bearing Journal OD
　Models 7010 & 7020...2.747-2.748 inch
　　　　　　　　　　　(69.76-69.79 mm)
　All Other Models ...3.2465-3.2480 inch
　　　　　　　　　　　(82.41-82.499 mm)

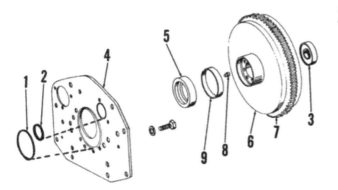

Fig. 29 — Exploded view of engine plate, crankshaft rear seal and flywheel.

1. "O" ring
2. "O" rings
3. Pilot bearing
4. Rear plate
5. Crankshaft rear seal
6. Flywheel
7. Starter ring gear
8. Flywheel dowels
9. Sleeve

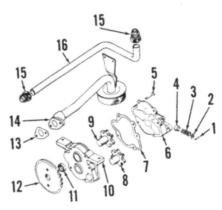

Fig. 31 — Exploded view of oil pump used on Model 7010, 7020 and 7045 tractors.

1. Roll pin	9. Drive shaft assy.
2. Washer	10. Housing
3. Spring (relief valve)	11. Bushing
4. Piston (relief valve)	12. Gear (drive)
5. Place bolt	13. Gasket
6. Cover	14. Suction tube assy.
7. Gasket	15. Tube nut
8. Driven shaft assy.	16. Pressure tube

Main Bearing Clearance
    Models 7010 and 7020
        Standard ....0.0016-0.0043 inch
                            (0.040-0.109 mm)
        Maximum ....0.008 inch (0.20 mm)
    All Other Models
        Standard ....0.0019-0.0051 inch
                            (0.048-0.129 mm)
        Maximum ....0.009 inch (0.23 mm)
Crankshaft End Play
    Models 7010 and 7020
        Standard ....0.004-0.009 inch
                           (0.10-0.23 mm)
        Maximum ....0.015 inch (0.38 mm)
    All Other Models
        Standard ....0.007-0.013 inch
                           (0.18-0.33 mm)
        Maximum ....0.021 inch (0.53 mm)
Main Bearing Cap Bolt Torque
    Models 7010 & 7020 .....130-140 ft.-lbs.
                        (176.1-189.7 N·m)
    All Other Models ....170-190 ft.-lbs.
                        (230.3-257.4 N·m)

Main bearing inserts are available in standard size and 0.010, 0.020 and 0.040 inch oversizes. All main bearing inserts are available as separate upper or lower half. Bearing lower halves do not have oil holes. Crankshaft end play is controlled by thrust flanges at each side of center main bearing. Semi-finished main bearing caps are available for service and must be finished on cylinder block using a line boring bar. Diameter of bearing bore without inserts is 2.9368-2.9375 inches (74.594-74.612 mm) for Models 7010 and 7020, and 3.5607-3.5614 inches (90.441-90.459 mm) for all other models.

CRANKSHAFT OIL SEALS

35. **FRONT SEAL.** The crankshaft front oil seal is located in the timing gear cover and can be renewed as outlined in paragraph 22.

36. **REAR SEAL.** The crankshaft rear oil seal (5–Fig. 29) is installed in the adapter plate (4) at rear of engine. To renew the seal, first remove engine from tractor as outlined in paragraph 14, then proceed as follows: Remove clutch, flywheel and engine rear adapter plate. Remove oil seal from the adapter plate and install new seal with lip of seal to front. Oil seal sleeve (9) on hub of flywheel should be renewed if nicked or worn. Beveled outside diameter of wear sleeve should be toward front (inside) of engine with flywheel installed. New sleeve should be pressed on flywheel flush with edge of flywheel hub.

NOTE: Do not heat or lubricate wear sleeve when installing. Renew the two "O" rings (1 and 2) located between adapter plate and engine, lubricate the oil seal and reinstall adapter plate.

Tighten adapter plate to cylinder block screws to 68-73 ft.-lbs. (92.1-98.9 N·m) and oil pan to adapter plate screws to 18-20 ft.-lbs. (24.3-27.1 N·m). Reinstall flywheel and tighten flywheel retaining screws to 135 ft.-lbs. (182.9 N·m). Complete reassembly by reversing disassembly procedure.

FLYWHEEL

37. **REMOVE AND REINSTALL.** To remove flywheel, first remove engine as outlined in paragraph 14 and unbolt clutch assembly from flywheel. Flywheel is retained to crankshaft with six bolts, one bolt hole is offset so that installation of flywheel is possible in one position only. To renew flywheel ring gear, flywheel must be removed from crankshaft. Inspect clutch friction surface of flywheel and crankshaft rear oil seal wear sleeve (9–Fig. 29) on flywheel hub. Heat new ring gear evenly to 300°-325°F (148°-152°C) (dull red in dark) and install it over flywheel with tooth bevel to front.

When installing flywheel, tighten retaining bolts to 135 ft.-lbs. (182.9 N·m). Complete reassembly in opposite order of removal.

OIL PAN

38. **REMOVE AND REPLACE.** Remove oil drain plug from oil pan and allow oil to drain completely. Remove cap screws and lockwashers securing oil pan to adapter plate, timing gear cover and cylinder block. Jar the pan loose while dropping rear of oil pan first.

The oil pan is sealed at the rear by a large square-cut sealing ring (1–Fig. 29) that is located between adapter plate (4) and rear of engine. Care should be taken not to damage this sealing ring when removing and reinstalling the oil pan as removal of clutch, flywheel and rear adapter plate is required to renew the sealing ring. Place a small quantity of non-hardening sealing compound in the corners formed by adapter plate and cylinder block rails and where front support plate contacts cylinder block rails. Make up two guide studs and install them in diagonal corners of cylinder block. Also have available a greased sheet of shim stock that is larger than the rear seal area of the oil pan. Place shim stock against "O" ring in lower portion of flywheel housing. Position oil pan with front end up and position it over guide studs. Install a cap screw and lockwasher in each corner finger tight. Carefully remove shim stock protecting "O" ring in adapter plate. Remove guide studs and install cap screws and lockwashers which hold rear flange of oil pan to the adapter plate. Tighten rear corner cap screws and adapter plate cap screws alternately. Install remaining cap screws and lockwashers. Tighten all cap screws to 18-20 ft.-lbs. (24.3-27.1 N·m). Install and tighten oil pan drain plug.

OIL PUMP AND RELIEF VALVE

Models 7010, 7020 and 7045

39. **R&R AND OVERHAUL PUMP.** The oil pump can be removed after first removing the oil pan as outlined in para-

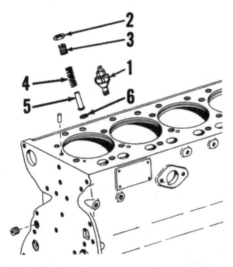

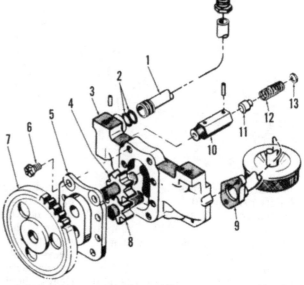

Fig. 33—Exploded view of oil pump used on all models except 7010, 7020 and 7045 tractors.

1. Pressure tube
2. "O" rings
3. Pump body
4. Drive gear & shaft
5. Cover
6. Cap screw
7. Drive gear
8. Idler gear & shaft
9. Inlet tube & screen
10. Relief valve assy.
11. Piston
12. Spring
13. Spring retainer

Fig. 32—The oil pressure sending switch and pressure relief valve on Model 7010, 7020 and 7045 tractors are located on right side of engine.

1. Oil pressure sending switch
2. Locknut
3. Adjusting screw
4. Spring
5. Oil pressure relief valve
6. Relief valve seat

graph 38. The pump is located on lower surface of block and is attached with two cap screws. Remove the suction tube assembly by removing two screws at oil pump and two screws at center main bearing cap. Refer to Fig. 31 for exploded view of pump.

To disassemble, first pull the pump drive gear using a screw type puller and two 3/8-16 bolts. Remove the pump cover and the gear assemblies. Remove roll pin (1 – Fig. 31) and relief valve washer, spring and piston.

Check pump body, gears, shafts and renew any parts excessively worn, scored or damaged. Pump gears and shafts (8 and 9) are available only with shafts.

If pump housing and cover bushings are being renewed press new bushings in from chamfered side of bushing bores. Pump housing (10 – Fig. 31) bushings should be pressed flush for all models except 7045 which should be pressed 0.004-0.007 inch (0.10-0.18 mm) below housing surface. Pump cover (6) bushings should be pressed in until they are 0.03-0.05 inch (0.76-1.27 mm) below surface of cover. Install gear and shaft assemblies (8 and 9). Clearance between gear teeth and housing should be 0.002-0.0045 inch (0.050-0.114 mm). Install pump cover (6) using a new gasket (2). Tighten cover retaining bolts to 18-20 ft.-lbs. (24.3-27.1 N·m) on all models except 7045 which should be tightened to 72-77 ft.-lbs. (97.5-104.3 N·m). Pump gears must turn freely after pump is assembled and should have 0.004-0.009 inch (0.10-0.23 mm) end play. Heat drive gear (12) to 350°-400°F (176°-204°C) and install on

shaft. Drive gear must have 0.040-0.060 inch (1.02-1.52 mm) clearance between gear and front of pump housing. Install assembled pump on engine and tighten retaining bolts to 30-35 ft.-lbs. (40.6-47.4 N·m) on all models except 7045 which should be tightened to 95-105 ft.-lbs. (128.7-142.2 N·m). Backlash between oil pump drive gear and crankshaft gear should be 0.007-0.016 inch (0.18-0.41 mm). Complete installation of oil pump in opposite order of removal.

40. **RELIEF VALVES.** Two oil pressure relief valves are used. The relief valve (3 – Fig. 31) in oil pump body is non-adjustable and should by-pass oil at 75-85 psi (517-586 kPa). This is well above the normal oil pressure and opens only due to surge pressure when engine oil is cold.

The relief valve (Fig. 32) located on the right side of the engine just behind the diesel fuel injection pump or non-diesel governor, is adjustable to control normal engine oil pressure. To adjust the relief valve, loosen locknut (2) and turn adjusting screw (3) in or out until oil pressure is 45 psi (310 kPa) with engine running at 2200 rpm and with oil at normal operating temperature of 180°-200°F (82°-93°C). Tractor is equipped with an oil pressure warning light instead of a pressure gage; therefore, a master pressure gage must be installed in place of oil pressure sending switch (1) when checking or adjusting system pressure. Adjust relief valve only when engine is running at normal operating temperature and speed. Do not attempt to adjust relief valve to compensate for worn oil pump or engine bearings.

All Other Models

41. **R&R AND OVERHAUL PUMP.** Oil pump can be removed after first

removing oil pan as outlined in paragraph 38. Pump is located on lower surface of block with dowel pins and is attached with two cap screws. Oil line bracket must be detached from block before removing pump. Refer to Fig. 33 for exploded view of pump. Drive gear (7) is pressed onto drive shaft. Pump gears and shafts (4 and 8) are available only with shafts. Relief valve (10 through 13) is for surge pressure only. Relief valve is non-adjustable and is preset at 75-85 psi (517-586 kPa). Since this is well above normal oil pressure of 40-45 psi (275-310 kPa), valve opens only due to surge pressure when engine oil is cold.

42. **RELIEF VALVE.** Oil pressure relief valve Fig. 34 located on right side of engine just in front of engine oil filters, is adjustable to control normal engine oil pressure of 40-45 psi (270-310 kPa). To adjust relief valve, turn ad-

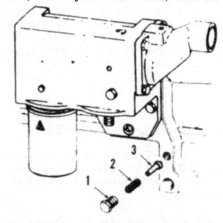

Fig. 34—View showing location of oil pressure adjusting screw (1) on all models except 7010, 7020 and 7045 tractors. Adjusting screw has a Lok-thread and therefore requires no jam nut.

1. Adjusting screw
2. Spring
3. Piston

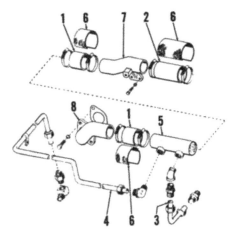

Fig. 35 – Exploded view of engine oil cooler used on Model 7010 and 7020 tractors.

1. Hose	5. Oil cooler
2. Hose	6. Hose heat shield
3. Return tube	7. Water inlet
4. Inlet tube	8. Water outlet

Fig. 35A – Exploded view of engine oil cooler.

1. Gasket
2. Elbow
3. Gasket
4. "O" rings (lube oil side)
5. "O" rings (water side)
6. Housing
7. Insert (relief valve)
8. Plunger
9. Spring
10. Washer (copper)
11. Cap screw
12. Gasket
13. Cooler (lube oil)
14. Tube
15. Insert
16. Oil filters

justing screw (1) in or out until oil pressure is 45 psi (310 kPa) with engine running at 2300 rpm and with oil at normal operating temperature of 180°-200°F (82°-93°C); oil pressure at slow idle speed of 700-750 rpm should then be approximately 20 psi (137 kPa). Tractor is equipped with an oil pressure warning light instead of a pressure gage; therefore, a master pressure gage must be installed in place of oil pressure sending switch when checking or adjusting system pressure. Adjust relief valve only when engine is running at normal operating temperature and speed. Do not attempt to adjust relief valve to compensate for worn oil pump or engine bearings.

NOTE: Adjusting screw has a Lokthread; therefore a gasket nylon pellet or jam nut is not required to prevent oil leakage or loosening.

OIL COOLER

All engines are equipped with an engine oil cooler. Engine oil is also used to cool and lubricate turbocharger. It is

important for cooler to operate correctly for proper cooling. Overheating of oil will usually be indicated by low oil pressure.

Models 7010-7020

43. **R&R AND OVERHAUL.** Refer to Fig. 35 for an exploded view of oil cooler used on Models 7010 and 7020. The oil cooler is located on right side of engine. Normal maintenance consists of renewing hoses (1 and 2) and cleaning or renewal of cooler (5). Make certain heat shields are in place. It may be necessary to use sealer on hoses.

All Other Models

43A. **R&R AND OVERHAUL.** Drain cooling system and remove cap screw from fuel line clamp on top of oil cooler. Remove both oil filter elements. Then, refer to Fig. 35A and remove cap screws securing oil cooler housing (6) and water outlet elbow (2) to cylinder block. Pull cooler out and to rear of engine to remove. Disassembly can be accomplished by removing cap screws securing elbow (2) and tube (14) to housing (6) then remove oil cooler core (13) and relief valve (7 through 11).

Oil passages can be soaked in cleaning solvent; however, it is **difficult** to determine when passages are clean. If badly clogged, oil cooler core should be renewed. If cooler is cleaned with solvent, be sure to flush thoroughly before installing.

Two steel plates can be locally manufactured for pressure testing oil cooler. One of the plates must be connected to air hose for pressurizing the oil passage of cooler. Bolt test plates using heavy gasket material or sheet metal rubber to seal ends of cooler core. Attach air hose and check for leaks with approximately 200 psi (1380 kPa) air pressure. Make certain that gaskets under test plates and air connections are not leaking. Continue to test with unit submerged in hot water until cooler is heated to approximately 150°F (69°C). Cooler should be renewed if core is leaking.

When reassembling, place cooler housing in a vertical position with cast arrow end to top. Install one blue "O" ring and one yellow "O" ring in lower end of housing.

NOTE: Blue "O" ring (5 – Fig. 35A) is the water seal and yellow "O" ring (4) is the oil seal. Use only a vegetable base oil for lubrication of the "O" rings.

Install cooler assembly (13) and align name plate arrow as locating boss with cast arrow on housing (6) within plus or minus ¼-inch (6.3 mm). Install yellow "O" ring (4) in upper section of housing (6), then install blue "O" ring (5) in water inlet (14) and assemble water inlet to housing (6) and tighten cap screws securely. Assemble water outlet (2) and gasket (3) to housing (6) and tighten cap screws. Install oil pressure relief valve (7 through 11) and torque to 30-40 ft.-lbs. (40.6-54.2 N·m).

DIESEL FUEL SYSTEM

The diesel fuel system consists of three basic units: the fuel filter, injection pump and injection nozzles. When servicing any unit associated with the fuel system, the maintenance of absolute cleanliness is of utmost importance.

Probably the most important precaution that servicing personnel can impart to owners of diesel powered tractors is to urge them to use an approved fuel that is

absolutely clean and free from foreign material. Extra precaution should be taken to make certain that no water enters fuel storage tanks. This last precaution is based on the fact that all diesel fuels contain some sulphur. When water is mixed with sulphur, an acid is formed which will quickly erode the closely fitting parts of the injection pump and fuel injection nozzles.

FILTERS AND BLEEDING

44. **FILTERS.** A single, disposable type fuel filter is located on right side of engine. Tractor may be equipped with either a spin-on type or snap-on type filter. Regardless of type of fuel filter incorporated, fuel filter should be examined daily for water and sediment, and drained whenever water is visible. To

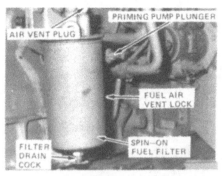

Fig. 36 — View of spin-on type fuel filter used on some model tractors. Refer to paragraph 45 for outline of bleeding procedure.

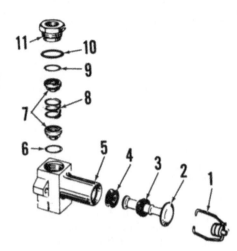

Fig. 38 — Exploded view of fuel primer pump used. Pump is located just ahead of the fuel filter on right side of tractor.

1. Clamp		7. Valves	
2. Plunger		8. Spring	
3. Guide		9. Seal ring	
4. Piston		10. Seal ring	
5. Body		11. Retainer nut	
6. Gasket			

drain filter, open drain cock or remove drain plug at bottom of filter and open vent plug on upper front corner of filter base.

Fuel filter should be renewed after 500 hours of operation. Poor fuel handling and storage facilities will decrease effective life of filter. Filter should never remain in fuel system until a decrease in engine power or speed is noticed as some dirt may enter pump and/or injector nozzles and result in severe damage.

45. **BLEEDING.** Each time filter element is renewed, water is drained or fuel lines are disconnected for any reason it will be necessary to bleed air from system. Exact bleeding procedure will vary depending upon type of fuel filter.

To bleed system equipped with a spin-on type filter, refer to Fig. 36 and proceed as follows: Close filter drain cock and filter air vent plug. Open fuel air vent cock. Loosen knurled nut on bail that holds priming pump plunger in and operate plunger until air-free fuel flows from fuel air vent cock, close vent cock and secure priming pump plunger in its stowed position. Loosen high pressure lines at injector nozzles and crank engine until fuel appears at fittings.

Tighten high pressure line fittings and start engine. Allow engine to run at idle and carefully check fuel system for leaks.

To bleed system equipped with a snap-on type fuel filter, refer to Fig. 37 and proceed as follows: Install drain plug and vent plug. Loosen injection pump fuel inlet line fitting (located in center of injection pump head). Loosen knurled nut on bail that secures priming pump plunger and operate pump until air-free fuel flows from injection pump fuel inlet line fitting. Tighten line fitting and secure pump plunger in its stowed position. Loosen high pressure line fittings at injector nozzles and crank engine until fuel appears at fittings. Tighten high pressure fittings and start engine. Allow engine to run at idle and check fuel system for leaks.

INJECTION NOZZLES

Engines are equipped with a direct type injection system and fuel leaves injection nozzles with enough force to penetrate the skin. When testing nozzles, keep clear of nozzle spray.

46. **LOCATING A FAULTY NOZZLE.** If engine misfires or is rough running, check for defective nozzles as follows: With engine running at low idle speed, loosen high pressure connection at each injector in turn. As in checking spark plugs, faulty injector is the one which least affects running of the engine, when its line is loosened.

If a faulty nozzle is found and considerable time has elapsed since injectors have been serviced, it is recommended that all injectors be removed and serviced or that new or reconditioned units be installed.

47. **REMOVE AND REINSTALL INJECTORS.** Before loosening any lines, wash nozzle holder and connections with clean diesel fuel or kerosene. After disconnecting high pressure and leak-off lines, cover open ends of connections to prevent entrance of dirt or other foreign material. Remove nozzle holder stud nuts and carefully withdraw nozzles from cylinder head, being careful not to strike tip end of nozzle against any hard surface.

Thoroughly clean nozzle recess in cylinder head before reinstalling nozzle and holder assembly. Use only wood or brass cleaning tools to avoid damage to seating surfaces and make sure that recess is free of all dirt and carbon. Even a small particle could cause unit to be cocked and result in a compression loss and improper cooling of fuel injector.

When reinstalling injectors on all models except 7020, always renew dust shield (7 – Fig. 39) and gasket (6). Be sure to install gasket (6) with concave face down.

Injectors used on Model 7020 tractors have a tapered sealing surface machined into nut (5 – Fig. 39) and do not require the use of gasket (6). All other installation steps are the same for all model tractors.

Install nozzle holder washers and nuts and torque in 2 ft.-lbs. (3.7 N·m) increments until each reaches a final torque of 9-12 ft.-lbs. (12.1-16.2 N·m). This method of tightening will help prevent injector from being cocked in cylinder head bore.

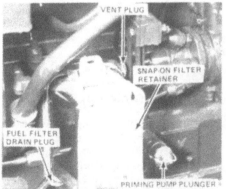

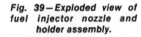

Fig. 37 — View of snap-on type fuel filter used on some model tractors. Refer to paragraph 45 for outline of bleeding procedures.

Fig. 39 — Exploded view of fuel injector nozzle and holder assembly.

1. Holder body
2. Dowel pins
3. Nozzle valve
4. Nozzle
5. Nut
6. Gasket
7. Dust shield
8. Screw (pressure adjust)
9. Washer
10. Locknut
11. "O" ring
12. Spring
13. Spindle

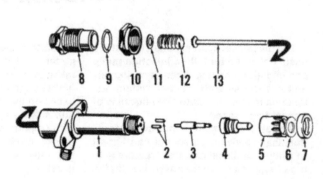

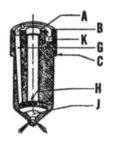
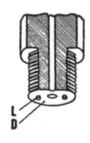

Fig. 40 — Drawing showing text reference points on injector nozzle body and holder body.

A. Pressure surface
B. Pressure surface
C. Shoulder
D. Pressure surface
G. Fuel passage
H. Fuel gallery
J. Needle seat
K. Hole for dowel
L. Dowel

Fig. 41 — Cleaning fuel passage in nozzle body.

Fig. 42 — Cleaning fuel gallery with hooked scraper.

48. NOZZLE TESTING. A complete job of testing and adjusting nozzle requires use of a special tester. Use only clean, approved testing oil in tester tank. Operate tester lever until oil flows; then, attach injector to tester and make following test:

49. OPENING PRESSURE. Close gage valve and operate tester lever several times to clear all air from injector. Then, open gage valve and while slowly operating tester lever, observe pressure at which spray occurs. This gage pressure should be 3100-3150 psi (21.4-21.7 MPa) for Models 7030 with serial numbers below 21089 or 3800-3850 psi (26.2-26.5 MPa) for all other models. If pressure is incorrect, loosen adjusting screw locknut (10 – Fig. 39) and turn adjusting screw (8) in or out as required to increase or decrease opening pressure. If opening pressure cannot be adjusted to correct pressure, overhaul nozzle as outlined in paragraph 52.

50. SPRAY PATTERN. Operate tester handle slowly and observe spray pattern. Nozzle tips have four equally spaced holes. All of the sprays must be similar and spaced equidistantly in a nearly horizontal plane. Each spray must be well atomized. If spray pattern is not as described, overhaul nozzle as outlined in paragraph 52.

NOTE: Rapid operation of tester lever will frequently produce a spray pattern as described even if injector is faulty. Be sure to operate tester lever as slowly as possible and still cause nozzle to open.

51. SEAT LEAKAGE. Wipe nozzle tip dry with clean blotting paper; then, operate tester handle to bring gage pressure to 200 psi (1380 kPa) below opening pressure and hold this pressure for ten seconds. If any fuel appears on nozzle tip, overhaul injector as outlined in paragraph 52.

52. OVERHAUL. Hard or sharp tools, emery cloth, crocus cloth, grinding compounds or abrasives of any kind should NEVER be used in cleaning of nozzles.

Wipe all dirt and loose carbon from injector assembly with a clean, lint free cloth. Carefully clamp injector assembly in a soft jawed vise or injector fixture and loosen locknut (10 – Fig. 39) and back off adjusting screw (8) enough to relieve load from spring (12). Remove nozzle cap nut (5) and nozzle assembly (3 and 4). Normally, nozzle valve needle can easily be withdrawn from nozzle body. If it cannot, soak assembly in solvent such as "Allis Chalmers" carbon and rust remover.

CAUTION: Do not allow solvent to get on the hands or body; use tweezers or basket method to handle parts.

After removing parts from solution, place them in clean diesel fuel for neutralizing. Be careful not to permit valve or body to come in contact with any hard surfaces.

If more than one injector is being serviced, keep component parts of each injector separate from the others by placing in a compartmented pan covered with clean diesel fuel. Examine nozzle body and remove any carbon deposits from exterior surfaces using a brass wire brush. Nozzle body must be in good condition and not blued due to overheating.

All polished surfaces should be relatively bright without scratches or dull patches. Pressure surfaces (A, B and D – Fig. 40) must be absolutely clean and free of nicks, scratches or foreign material as these surfaces must register together to form a high pressure joint.

Using a pin vise with correct size cleaning wire, thoroughly clean the four spray holes in nozzle body end. Cleaning wire should be slightly smaller than spray holes.

Clean out the small fuel feed channels, using a small diameter wire as shown in

Fig. 43 — Scraping carbon from needle valve seat.

Fig. 41. Insert the special groove scraper into nozzle body until nose of scraper locates in the fuel gallery. Press nose of scraper hard against side of cavity and rotate scraper to clean all carbon deposits from the gallery as shown in Fig. 42. Using seat scraper, clean all carbon from valve seat (J – Fig. 40) by rotating and pressing on the scraper as shown in Fig. 43.

Examine stem and seat end of nozzle valve and remove any carbon deposit using a clean, lint free cloth. Use extreme care, however, as any burr or small scratch may cause valve leakage or spray pattern distortion. If valve seat has a dull circumferential ring indicating wear or pitting, or if body is blued, valve and body should be turned over to an authorized diesel service station for possible overhaul.

Before reassembling, thoroughly rinse all parts in clean diesel fuel and make certain that all carbon is removed from nozzle holder nut. Assemble parts while still wet with diesel fuel. Install nozzle assembly and cap nut (5 – Fig. 39) making certain that the valve stem is located in the hole of the holder body and the two dowel pins (2) enter holes in nozzle body. Tighten holder nut (5) to a torque of 40-60 ft.-lbs. (54.2-81.3 N·m).

Install spindle (13), spring (12), adjusting screw (8) and locknut (10) using a new "O" ring (11) and washer (9). Connect injector to a nozzle tester and ad-

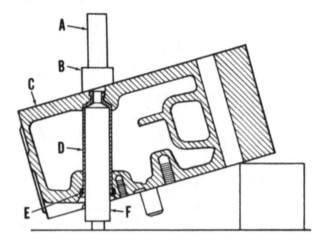

Fig. 44— View showing tool arrangement for injector holder sleeve installation on 7020 models. See text.

A. Swage-ACTP 2059
B. Swage stop-ACTP 2060
C. Cylinder head
D. Sleeve
E. "O" rings (2)
F. Swage adaptor-ACTP 2058

ment on engine that is in operative condition; if engine will not start and improper installation of fuel injection pump is suspected, refer to paragraph 26 for proper timing of fuel injection pump drive gear.

55. R&R INJECTION PUMP. Remove hood panel from right side of tractor. Before loosening any connections, thoroughly clean pump fuel lines and fittings with diesel fuel or petroleum solvent; then, proceed as follows: Shut off fuel supply, remove timing window cover from fuel injection pump and turn engine so that correct timing mark on crankshaft pulley is aligned with pointer on engine front cover as follows:

Model	Injection Timing
7010	18°BTDC
7020	18°BTDC
7030	24°BTDC
7040	16°BTDC
7045	16°BTDC
7050	26°BTDC
7060	18°BTDC
7080	22°BTDC

Injection pump timing marks should be aligned in timing window as shown in Fig. 21. Disconnect fuel supply line, throttle, shut off linkage and fuel return line to pump. Remove clamps from high pressure (nozzle) lines; then, remove lines one at a time from pump and nozzles in order that they were retained in the clamps.

NOTE: When removing injection tubing nut from pump, hold discharge fitting with a wrench to prevent loosening from pump head. All pump nozzles and line openings should be capped or plugged immediately after lines are removed.

Remove inspection cover from front of timing gear cover. Then, remove four cap screws securing pump drive gear to pump drive hub, and remove pump drive

just opening pressure as outlined in paragraph 49. Retest injector as previously outlined. If injector fails to pass above tests, renew nozzle and needle assembly (3 and 4 – Fig. 39).

INJECTOR SLEEVE

Model 7020

53. REMOVE AND REINSTALL. Injection nozzle is positioned in cylinder head by a copper sleeve. Upper end of sleeve is flanged and sealed to cylinder head by two "O" rings. Nozzle end of sleeve is sealed by swaging lower end into cylinder head. Old sleeve must be split and driven out of head using special tool ACTP 2057. After removal of old sleeve be sure to remove sheared tip of sleeve from cylinder head. To install new sleeve, lubricate two "O" rings (E – Fig. 44) with vegetable oil, install them on sleeve (D) and insert sleeve forceably into head. Do not use any sealant on nozzle end of sleeve. Lubricate end of swage tool ACTP 2059 and check fit in nozzle end of sleeve. If swage does not slide easily into end of sleeve, install another sleeve. Assemble tools as shown in Fig. 44 and using a suitable press, force swage against a sleeve until stop is contacted. Remove tools and dress nozzle end of sleeve but remove as little metal as possible.

INJECTION PUMP

54. PUMP TIMING. To check or adjust pump timing proceed as follows: Shut off fuel supply at tank and remove timing window cover (TW – Fig. 20) from injection pump. Turn engine slowly in normal direction of travel until timing marks appear at bottom of timing window in pump housing. Then, continue turning engine slowly until correct timing mark on crankshaft pulley is aligned with pointer on engine front cover as

shown in Fig. 22. Correct timing is as follows:

Model	Injection Timing
7010	18°BTDC
7020	18°BTDC
7030	24°BTDC
7040	16°BTDC
7045	16°BTDC
7050	26°BTDC
7060	18°BTDC
7080	22°BTDC

Timing marks in injection pump window should be aligned as shown in Fig. 21. If not, remove cover from front of timing gear cover and loosen cap screws retaining drive gear to pump hub. Align timing marks by turning pump shaft with a socket wrench. Tighten gear retaining cap screws to a torque of 27-29 ft.-lbs. (36.5-39.2 N·m). Recheck timing by backing the engine up ½-turn, then turning engine slowly to correct timing mark. Check to see that timing marks are aligned in pump timing window and readjust timing if they are not aligned.

NOTE: Timing procedure as outlined assumes normal timing check and adjust-

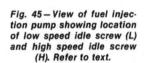

Fig. 45— View of fuel injection pump showing location of low speed idle screw (L) and high speed idle screw (H). Refer to text.

C. Timing cover
F. Fuel filter
H. High speed screw
L. Low speed screw
P. Primer pump

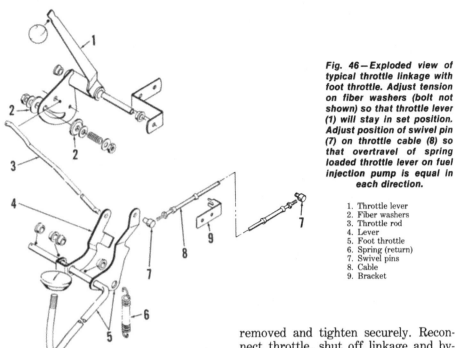

Fig. 46 — Exploded view of typical throttle linkage with foot throttle. Adjust tension on fiber washers (bolt not shown) so that throttle lever (1) will stay in set position. Adjust position of swivel pin (7) on throttle cable (8) so that overtravel of spring loaded throttle lever on fuel injection pump is equal in each direction.

1. Throttle lever
2. Fiber washers
3. Throttle rod
4. Lever
5. Foot throttle
6. Spring (return)
7. Swivel pins
8. Cable
9. Bracket

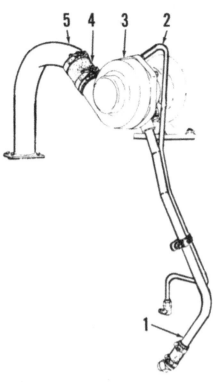

Fig. 47 — View of typical turbocharger unit showing position of pressure line (2) and return line (1).

1. Return line	4. Hose
2. Pressure line	5. Tube to inlet
3. Turbocharger	manifold

gear through opening in timing gear cover. Remove six pump mounting cap screws through opening in timing gear cover leaving one of the top cap screws till last. Support pump and remove remaining cap screw. Pump drive shaft and drive hub will be removed with pump.

Before reinstalling pump, check to be sure that engine has not been turned while pump was removed; correct timing mark on crankshaft pulley should be aligned with pointer on engine front cover. If engine was rotated, make certain that number one piston is on its compression stroke when aligning crankshaft pulley with pointer on engine. Position pump to its mounting plate and start the two upper cap screws to hold pump in place then, start remainder of cap screws and tighten securely.

Remove timing window from pump and rotate pump drive hub clockwise until timing lines in pump are in register. Then, install drive gear to hub and align slots in gear as close to center as possible over threaded holes in hub. Tighten gear retaining cap screws to a torque of 27-29 ft.-lbs. (36.5-39.2 N·m).

Reconnect high pressure lines in reverse order from which they were removed and tighten securely. Reconnect throttle, shut off linkage and by-pass line. Operate hand primer pump until fuel flows "air free" from pump inlet hoses then, connect hose to pump and continue to operate hand primer until a resistance is felt.

56. SPEED ADJUSTMENTS. Actuate throttle hand lever and observe overtravel of the spring loaded throttle arm after both low idle and adjustment and high speed adjustment screws contact stops. Over-travel in each direction should be approximately equal; if not, adjust pivot pin on ends of throttle control cable to obtain equal over-travel. Then, correct speeds are as follows.

Model	Low Idle Rpm	High Idle Rmp	Rated Load Rpm
7010	750-800	2480-2500	2300
7020	750-800	2480-2500	2300
7030	700-750	2500-2550	2300
7040	700-750	2550-2645	2300
7045	700-750	2500-2550	2300
7050	700-750	2500-2550	2300
7060	700-750	2500-2645	2300
7080	725-775	2800-2850	2550

To check and adjust speed, start engine and move hand lever to low idle position and turn screw (L–Fig. 45) in to increase or out to decrease. Move hand throttle lever to high idle speed position and adjust high idle speed screw (H) to obtain correct speed.

After each 1000 hours of operation, turbocharger should be inspected for cleanliness, freedom of rotation and bearing clearance. After each 2000 hours of operation, turbocharger should be disassembled, new seals, gaskets and any other parts that are worn and all bearing tolerances readjusted. When servicing turbocharger, extreme care must be taken to avoid damaging any moving parts.

TURBOCHARGER UNIT

All Models

57. REMOVE AND REINSTALL. Remove right hood panel and disconnect air inlet hose from turbocharger. Remove turbocharger air outlet tube (5–Fig. 47) and exhaust outlet elbow. Disconnect oil inlet and return lines (1 and 2) from turbocharger, then unbolt and remove unit from exhaust manifold.

To reinstall, reverse removal procedure. Leave oil return (1) disconnected and crank engine with starter until oil begins to flow from return port. Connect oil return.

NOTE: Do not start engine until it is certain that turbocharger is receiving lubricating oil.

Tighten stud nuts attaching turbocharger to exhaust manifold to 18-21

TURBOCHARGER

All models are equipped with an exhaust driven turbocharger. Lubrication and cooling is provided by engine oil. After engine is operated under load, turbocharger should be allowed to cool by idling engine at 1000 rpm for 2-5 minutes. Engine should be immediately restarted if it is stalled while operating at high rpm.

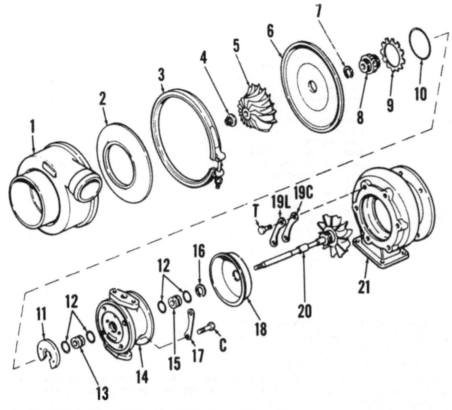

Fig. 48 — Exploded view of AiResearch turbocharger. Use extreme care to prevent damage to parts.

Fig. 49 — View showing method of checking turbine shaft end play. Shaft end play should be checked after unit is cleaned to prevent false reading caused by carbon build-up.

C. Backplate screws	4. Locknut	10. Seal ring	17. Lock plates
T. Turbine housing screws	5. Compressor impeller	11. Thrust bearing	18. Turbine shroud
1. Compressor housing	6. Backplate	12. Bearing retainers	19C. Clamp plates
2. Diffuser	7. Seal ring	13. Bearing	19L. Lock plates
3. Clamp	8. Thrust collar	14. Center housing	20. Turbine wheel and shaft
	9. Spring	15. Bearing	21. Turbine housing
		16. Seal ring	

ft.-lbs. (24.3-28.4 N·m) torque. Exhaust elbow mounting screws should be tightened to 28-33 ft.-lbs. (37.9-44.7 N·m) torque.

AiResearch Turbocharger

58. OVERHAUL Remove turbocharger unit as outlined in paragraph 57. Mark across compressor housing (1 – Fig. 48), center housing (14) and turbine housing (21) to aid alignment when assembling.

CAUTION: Do not rest weight of any parts of impeller on turbine blades. Weight of only the turbocharger unit is enough to damage the blades.

Remove clamp (3), compressor housing (1) and diffuser (2). Remove screws (T), lock plates (19L) and clamp plates (19C); then, remove turbine housing (21). Hold turbine shaft from turning by using the appropriate type of wrench at center of turbine wheel (20) and remove locknut (4).

NOTE: Use a "T" handle to remove locknut in order to prevent bending turbine shaft.

On some turbine shafts, an Allen wrench must be used at turbine end while others are equipped with a hex end and can be held with a standard socket. Lift compressor impeller (5) off, then remove center housing from turbine shaft while holding shroud (18) onto center housing. Remove backplate retaining screw (C), then remove back plate (6), thrust bearing (11), thrust collar (8) and spring (9). Carefully remove bearing retainers (12) from ends and withdraw bearings (13 and 15).

CAUTION: Be careful not to damage bearings or surface of center housing when removing retainers. The center two retainers do not have to be removed unless damaged or unseated. Always renew bearing retainers if removed from grooves in housing.

Clean all parts in a cleaning solution which is not harmful to aluminum. A stiff brush and plastic or wood scraper should be used after deposits have softened. When cleaning, use extreme caution to prevent parts from being nicked, scratched or bent.

Inspect bearing bores in center housing (14 – Fig. 48) for scored surface, out-of-round or excessive wear. Bearing bore diameter must not exceed 0.6228 inch (15.819 mm) and maximum permissible out-of-round is 0.0003 inch (0.007

mm). Make certain bore in center housing is not grooved in area where seal (16) rides. Inside diameter of bearing (13 and 15) must not be more than 0.4019 inch (10.208 mm) and outside diameter must not be less than 0.6182 inch (15.702 mm). Thrust bearing (11) should be measured at three locations around collar bore. Thickness should be not more than 0.1720 inch (4.368 mm) or less than 0.1711 inch (4.345 mm). Inside diameter of bore in backplate (6) must not exceed 0.5015 inch and seal contact area must be clean and smooth. Compressor impeller (5) must not show signs of rubbing with either the compressor housing (1) or the back plate (6). Impeller should have 0.0002 inch (0.005 mm) tight to 0.0004 inch (0.010 mm) loose fit on turbine shaft. Make certain that impeller blades are not bent, chipped, cracked or eroded. Oil passage in thrust faces must not be warped or scored. Ring groove shoulders must not have step wear. Bearing area width must not exceed 0.1758 inch (4.465 mm) and width of groove for seal ring (7) must not exceed 0.0065 inch (0.165 mm). Clearance between thrust bearing (11) and groove in collar (8) must be 0.001-0.004 inch (0.03-0.010 mm), when checked at three locations. Inspect turbine shroud (18) for evidence of turbine wheel rubbing. Turbine wheel (20) should not show evidence of rubbing and vanes must not be bent, cracked, nicked, or eroded. Turbine

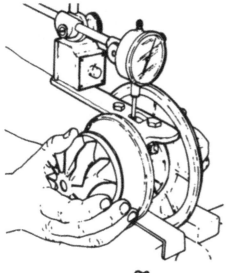

Fig. 50 — Turbine shaft radial play is checked with dial indicator through the oil outlet hole and touching shaft at (D — Fig. 51). Two methods of attaching dial indicator to center housing are shown above.

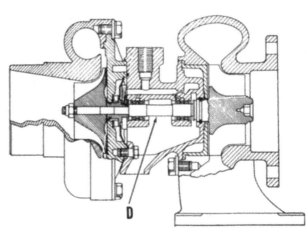

Fig. 51 — Cross-sectional drawing of AiReseach turbocharger. Shaft radial play is checked through oil outlet hole as shown at (D).

wheel shaft must not show signs of scoring, scratching or overheating. Diameter of shaft journals must not be less than 0.3992 inch (10.139 mm) and out-of-round must not exceed 0.0003 inch (0.007 mm). Groove in shaft for seal ring (16) must not be stepped and diameter of hub near seal ring should be 0.682-0.683 inch (17.32-17.35 mm). Check shaft end play and radial clearance when assembling.

If bearing inner retainers (12) were removed, install new retainers using special Kent-Moore tools (JD-274).

CAUTION: Bore in housing may be damaged if special retainer installing tool is not used.

Oil bearings (13 and 15) and install outer retainers using the special tool. Position the shroud (18) on turbine shaft (20) and install seal ring (16) in groove. Apply a light, even coat of engine oil to shaft journals, compress seal ring (16) and install center housing (14). Install new seal ring (7) in groove of thrust collars (8), then install thrust bearing so that smooth side of bearing (11) is toward seal ring (7) end of collar. Install thrust bearing and collar assembly over shaft, making certain that pins in center

housing engage holes in thrust bearing. Install new rubber seal ring (10), make certain that spring (9) is positioned in backplate (6), then install backplate making certain that seal ring (7) is not damaged. Install lock plates (17) and screws (C), then tighten screws to 40-60 in.-lbs. (4.4-6.7 N·m) torque. Install compressor impeller (5) and make certain that impeller is completely seated against thrust collars (8). Install lock nut (4) to 18-20 in.-lbs. (2.0-2.2 N·m), then use a "T" handle to turn locknut an additional 90 degrees.

CAUTION: If "T" handle is not used, shaft may be bent when tightening nut (4).

Install turbine housing (21) with clamp plates (19C) next to housing, tighten screws (T) to 100-130 ft.-lbs. (135.5-176.1 N·m), then bend lock plates (19L) up around screw heads.

Check shaft end play and radial play at this point of assembly. If shaft end play (Fig. 49) exceeds 0.0042 inch (0.106 mm), thrust collars (8 – Fig. 48) and/or thrust bearing (11) is worn excessively.

End play of less than 0.001 inch (0.03 mm) indicates incomplete cleaning (carbon not all removed) or dirty assembly and unit should be disassembled and cleaned. Refer to Figs. 50 and 51. If turbine shaft radial play exceeds 0.007 inch (0.18 mm), unit should be disassembled and bearings, shaft and/or center housing should be renewed. Maximum permissible limits of all of these parts may result in radial play which is not acceptable.

Make certain that legs on diffuser (2 – Fig. 48) are aligned with spot faces on backplate (6) and install diffuser. Install compressor housing (1) and tighten nut of clamp (3) to 40-80 in.-lbs. (4.4-8.9 N·m) torque. Fill reservoir with engine oil and protect all openings of turbocharger until unit is installed on tractor.

Rajay Turbocharger

59. **OVERHAUL.** Remove turbocharger as outlined in paragraph 57. Scribe a mark across turbine housing

Fig. 52 — Exploded view of Rajay turbocharger.

1. Compressor housing
2. Gasket
3. Nut
4. Washer
5. Impeller
8. Shims
9. "O" ring
10. Mating ring
11. Sleeve
11A. Spiral retaining ring
12. Bearing
12A. Wear ring
13. Bearing housing
13A. Bearing housing flange
15. Spring ring
16. Turbine wheel shield
17. Seal ring
18. Turbine wheel & shaft
19. Nut
20. Clamp
21. Gasket
22. Turbine housing
23. Carbon insert
24. Wave spring
25. Washer
26. "O" ring

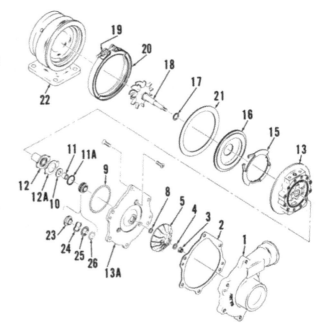

(22—Fig. 52), center housing (13), housing flange (13A) and compressor housing (1) to aid reassembly and proceed as follows: Unscrew nut (19) and remove clamp (20) and turbine housing (22). Unscrew six countersunk screws retaining compressor housing (1) and separate housing from center housing flange (13A). Hold turbine wheel blades with a shop towel and remove nut (3) and washer (4); nut (3) has left hand threads. Remove impeller (5) and shims (8) which may stick to impeller.

CAUTION: Do not allow turbine wheel and shaft to drop when impeller is removed.

Separate turbine wheel and shaft (18), turbine shield (16) and spring ring (15) from center housing (13). Be careful not to damage ring (17) ends when removing ring from turbine shaft. Remove center housing flange (13A) screws and separate flange from center housing (13). Remainder of disassembly is evident after inspection of unit and referral to Fig. 52.

Clean all parts except carbon insert (23) by washing in kerosene or diesel fuel. A nylon bristle brush may be used to clean carbon from parts.

CAUTION: Do not use wire brush, caustic cleaners, etc., on turbocharger parts. Inspect all parts for burring, eroding, nicks, breaks, scoring, excessive carbon build-up or other defects and renew all questionable parts.

Turbine end seal ring (17), "O" ring (9), mating ring (10) and carbon seal components (23, 24, 25 and 26) should be renewed whenever unit is disassembled. Carbon seal components (23, 24, 25 and 26) are available as a unit assembly only. Examine sleeve (11) for burrs, scoring and wear. Sleeve is precision ground and any defect will distort mating ring (10) and cause seal leakage. Note condition of turbine wheel blades, compressor impeller blades and bearing (12). If bearing (12) and/or turbine shaft is worn ex-

cessively, turbine wheel and compressor impeller blades may have rubbed against housings. Shaft clearance in bushing should be measured at compressor impeller end of shaft using a dial indicator with unit assembled. If indicated clearance exceeds 0.022 inch (0.56 mm), bushing (12) and/or shaft (18) should be renewed. Inspect bore in bearing housing (13) for evidence of stuck seal ring (17). If seal ring was sticking in shaft groove, the bore in housing will be grooved and should be renewed. Make certain that seal bore chamfer is clean and smooth to ease installation of shaft and seal ring.

To reassemble turbocharger, proceed as follows: Assemble bearing (12), mating ring (10) and sleeve (11) on turbine shaft (18). Hold sleeve and mating ring tight against shoulder of turbine shaft and check end play of bearing with feeler gage as shown in Fig. 53. End play should be 0.004-0.006 inch (0.10-0.15 mm). If clearance exceeds 0.006 inch (0.15 mm), bearing (12—Fig. 52) should be renewed. **Record amount of end play for use in later step in reassembly.** Remove parts from turbine shaft.

Assemble carbon seal components (23, 24, 25 and 26) and install assembly in center housing flange (13A) so "O" ring (26) is against flange. Install spiral retaining ring (11A) in groove of flange lugs. Press down evenly on carbon seal to be sure seal moves freely and is not binding.

Be sure wear ring (12A) is attached to bearing (12). Lubricate bearing (12) with motor oil and position in bearing housing bore with bearing tabs between center housing lugs. Lightly lubricate bearing (12) face, install mating ring (10) on bearing and lightly lubricate outside face of mating ring.

"O" ring (9) groove may be located in center housing (13) or flange (13A). Ear-

ly models have "O" ring groove located in flange while later models have "O" ring groove located in bearing housing. Do not mix early and late style center housings (13) and flanges (13A). Lubricate "O" ring (9) with silicone lubricant (or light grease) and position "O" ring groove. Refer to previously scribed marks and install flange (13A) on center housing (13). Tighten screws to 80-100 in.-lbs. (8.9-11.2 N·m).

To check bearing clearance proceed as follows: Install turbine wheel and shaft (18) without seal ring (17). Install sleeve (11) on turbine shaft. Attach a dial indicator as shown in Fig. 69. Measure and record total amount of sleeve (11—Fig. 54) end play. If total end play is not within 0.005-0.009 inch (0.13-0.23 mm), check for incorrect assembly. Turbine shaft end play in bearing (previously measured in Fig. 53) should be 0.004-0.006 inch (0.10-0.15 mm) and bearing (12—Fig. 52) end play between bearing housing (13) and flange (13A) should be 0.001-0.003 inch (0.03-0.08 mm). Record total shaft and bearing end play for use when selecting shims (8). Remove turbine shaft and proceed as follows:

Position impeller (5) over sleeve (11). Install compressor housing (1) using a new gasket and three of the attaching screws tightened to 80 inch-pounds (8.9 N·m). Check end play of impeller in housing as shown in Fig. 55. Subtract total shaft and bearing end play (found in preceding paragraph) from impeller end play. In final assembly, add shims (8—Fig. 52) of total thickness of 0.015-0.020 inch (0.38-0.51 mm) less than this determined value. As an example, if impeller-to-housing clearance is 0.049 inch (1.24 mm) and total shaft and

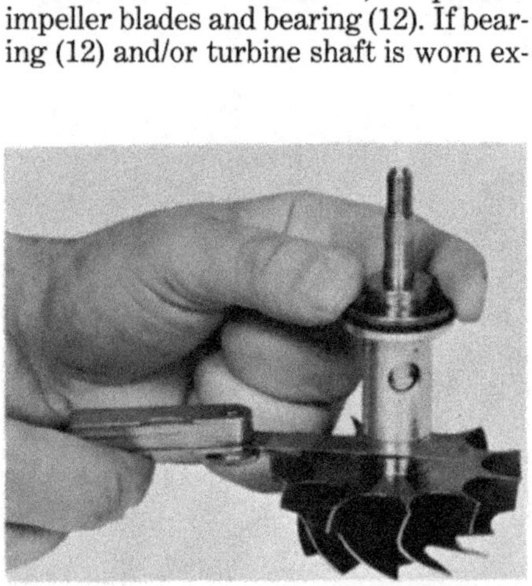

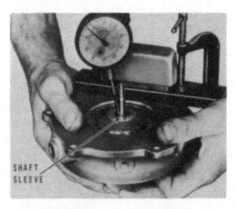

Fig. 53—Measure turbine shaft end play with feeler gage. Early type is shown.

Fig. 54—Total end play of turbine shaft in bearing and bearing housing can be measured as shown. Late type turbocharger is shown but early type is similarly measured. It is usually necessary to push sleeve or spacer down, then push up on turbine shaft when measuring.

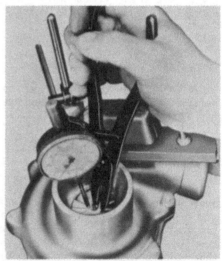

Fig. 55—Measure impeller clearance in housing as shown. Turbine shaft is not installed when checking this clearance.

bearing end play is 0.006 inch (0.15 mm); subtract as follows (following dimensions are in inches):

Impeller end clearance		0.049
Shaft and bearing end play		-0.006
		0.043
Determined value	0.043	0.043
Desired impeller clearance		-0.015 -0.020
Shim thickness required		0.028 0.023

Shims (8) are available in thicknesses of 0.010 and 0.015 inch. The addition of one 0.010 inch and one 0.015 inch thick shim will be within the range of the shim thickness required in the preceding example.

Remove compressor housing (1) and impeller (5). Lubricate groove in turbine shaft (18) with motor oil and install new seal ring (17) in groove. Place plastic seal compressor (furnished with seal ring) over seal ring. Install spring ring (15) in bearing housing and place shield (16) over spring with projections on shield against the flat sections of the spring. Use two small "C" clamps to compress spring and hold shield in place. Insert turbine shaft (18) through hole in shield and bore of bearing. Plastic band used to compress seal ring (17) will be pushed off of seal ring as shaft is inserted in bearing and plastic will disintegrate from heat as soon as engine is started. Rotate turbine shaft to be sure it is a free fit in bearing. Place shim pack (8) (of thickness determined in previous step) on turbine shaft; then, install impeller onto turbine shaft. Install washer (4) and nut (3) on turbine shaft (L.H. threads) and tighten nut to 80-100 in.-lbs. (8.9-11.2 N·m) while holding turbine with shop towel. Check turbine shaft assembly for free rotation.

Reinstall compressor housing (1) and gasket (2) to bearing housing and tighten cap screws to 80-100 in.-lbs. (8.9-11.2 N·m). Install turbine housing and gasket (21). Tighten clamp screw nut to a torque of 15-20 in.-lbs. (1.6-2.2 N·m). Remove plug from bearing housing and fill reservoir with same type oil as used in engine. Protect all openings of turbocharger until unit is installed on tractor.

INTERCOOLER

Models 7045-7050-7060-7080

61. REMOVE AND REINSTALL. Models 7045, 7050, 7060 and 7080 are equipped with an intercooler between turbocharger and engine, which lowers temperature and increases density of air before it enters the engine. Intercooler is mounted inside of intake manifold, Fig. 57.

To remove intercooler, drain cooling system and remove all hood panels. Remove upper radiator hose, by-pass hose, tube from thermostat housing to water pump and tube from thermostat housing to coolant conditioner if so equipped. Disconnect temperature sending unit. Remove thermostat housing, remove engine breather tube and disconnect brace and turbocharger air tube from top of intake manifold. Then, unbolt and remove intake manifold.

To disassemble, remove end cap (7 – Fig. 57) and "O" rings (6) from rear of intake manifold then, wedge header away from front of intake manifold to break gasket seal.

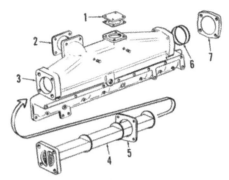

Fig. 57 – Exploded view of intake manifold and intercooler used on 7045, 7050, 7060 and 7080 tractors.

1. Cover	
2. Gasket	
3. Intake manifold	5. Gasket
4. Intercooler core	6. "O" rings
	7. Cap

CAUTION: Do not drive on rear of intercooler core when removing from manifold.

Check intercooler core for leaks or damage. Clean all mounting and mating surfaces and renew all gaskets and "O" rings.

When reassembling, cement new gasket (5) to back side of intercooler core header flange. Then, slide the intercooler core into front of manifold making sure core separation at centerline is in a vertical position. Temporarily install a couple of cap screws to hold core in position while installing rear cap (7). Lubricate and install one "O" ring (6) over end intercooler core making sure it is seated in counterbore of manifold. Install other "O" ring in counterbore of end cap and install end cap, torquing cap screws to 28-33 ft.-lbs. (37.9-44.7 N·m). Install manifold and torque cap screws to 28-33 ft.-lbs. (37.9-44.7 N·m). Finish reassembling in reverse of disassembly.

COOLING SYSTEM

RADIATOR

62. REMOVE AND REINSTALL. To remove radiator, proceed as follows: Remove all hood panels and drain cooling system. Disconnect air cleaner hose, transmission cooler lines and disconnect radiator hoses; then, unbolt and remove radiator from tractor.

Reinstall radiator by reversing removal procedure. Proper coolant level is 2 inches (50.8 mm) below radiator neck.

WATER PUMP

63. REMOVE AND REINSTALL. Remove radiator, loosen alternator mounting screws, then remove drive belts, fan, spacer and pulley. Remove cap screws securing water inlet pipe to water pump and unbolt and remove water pump. If outlet housing is removed from engine block, renew "O" ring to block and "O" ring on tube to oil cooler. All cap screws are torqued to 28-33 ft.-lbs. (37.9-44.7 N·m) except cap screws securing fan and pulley which should be torqued to 30-35 ft.-lbs. (40.6-47.4 N·m).

64. OVERHAUL. Refer to Fig. 58 for cross-sectional drawing of water pump. Using a suitable puller, remove pulley hub (9). Remove snap ring retaining bearing assembly (8) from hub end of pump, then press bearing and shaft

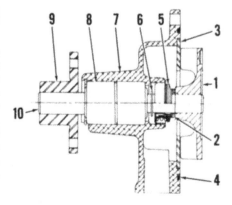

Fig. 58 – Cross-sectional view of water pump used on all models.

1. Impeller	6. Slinger
2. Ceramic seal	7. Body
3. Plate	8. Bearing
4. "O" ring	9. Hub
5. Seal assy.	10. Shaft

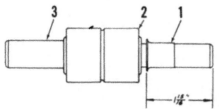

Fig. 59 — View showing measurement for slinger on water pump shaft.

1. Slinger
2. Bearing
3. Shaft

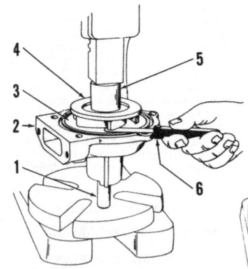

Fig. 60 — View showing installation of impeller and checking clearance. Refer to text.

1. Shaft	4. Impeller
2. Body	5. Collar
3. Plate	6. Feeler gage

Fig. 61 — Measuring correct distance for installing pulley hub. Refer to text.

1. Collar	
2. Pulley hub	4. Impeller
3. Body	5. Base plate

assembly forward out of pump body (7) and impeller (1). Remove impeller and body plate (3) from body. Press seal assembly (5) out towards rear of body.

Renew seal (5) and other parts as necessary. Apply gasket sealer to outer rim of seal and press seal into pump body. On models so equipped, press slinger (1 – Fig. 59) onto rear of shaft to distance shown.

Face of seal assembly must be free of grease or oil before seal assembly is installed. Press shaft and bearing assembly (8 and 10 – Fig. 58) into front of pump body so snap ring retainer can be installed. Position the plate (3) into recess in pump body and press impeller

onto rear end of shaft so there is a maximum clearance of 0.015, measured as shown in Fig. 60, between impeller vanes and body plate. Position water pump with impeller end of shaft firmly supported and press pulley hub onto

front end of shaft with flange side toward pump body. Front edge of hub flange should be four inches from rear surface of pump body as shown in Fig. 61. Install new "O" ring (4 – Fig. 58) into groove in pump body.

ELECTRICAL SYSTEM

ALTERNATOR AND REGULATOR

Delco-Remy Models

65. These alternators are equipped with solid-state regulators which are internally mounted and have no provision for adjustment.

CAUTION: Because certain components of alternator can be damaged by procedures that will not affect a D.C. generator, the following precautions MUST be observed.

a. When installing batteries or connecting a booster battery, the negative post of battery must be grounded.

b. Never short across any terminal of alternator or regulator unless specifically recommended.

c. Do not attempt to polarize alternator.

d. Disconnect all battery ground straps before removing or installing any electrical unit.

e. Do not operate alternator on an open circuit and be sure all leads are properly connected before starting engine.

66. TESTING AND OVERHAUL. The only test which can be made without

removal and disassembly of alternator is the regulator check. If there is a problem with battery not being charged, and battery and cable connector have been checked and are good, check regulator as follows: Make sure alter-

nator has a good, clean ground to engine, start engine, run at a moderate speed and turn on all accessories to check ammeter. If ammeter reading is within 10 amperes of rated output as stamped on alternator frame, alternator

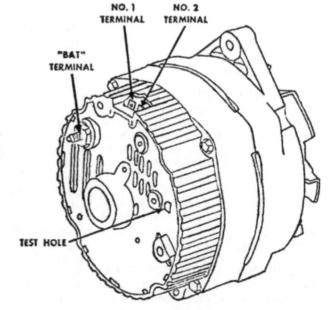

Fig. 62 — View of a Delco Remy alternator showing terminals and test hole. Refer to text.

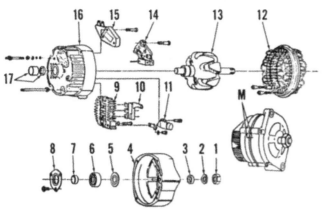

Fig. 63 – Exploded view of Delco-Remy alternator with internal mounted solid-state regulator. Note match marks (M) on end frames.

1. Pulley nut
2. Washer
3. Spacer (outside drive end)
4. Drive end frame
5. Grease slinger
6. Ball bearing
7. Spacer (inside drive end)
8. Bearing retainer
9. Bridge rectifier
10. Diode trio
11. Capacitor
12. Stator
13. Rotor
14. Brush holder
15. Solid state regulator
16. Slip ring end frame
17. Bearing & seal assy

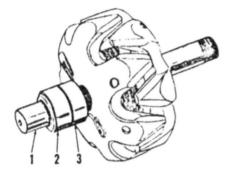

Fig. 64 – Removed rotor assembly showing test points when checking for grounds, shorts and opens.

is not defective. If ampere output is not within 10 amperes of rated output, ground field winding by inserting a screwdriver into test hole, Fig. 62. If output is then within 10 amperes of rated output, renew regulator.

CAUTION: When inserting screwdriver in test hole, tab is within ¾-inch of casting surface. Do not force screwdriver deeper than one inch into end frame.

If output is still not within 10 amperes of rated output, alternator will have to be disassembled. Check field windings, diode trio, rectifier bridge and stator as follows:

To disassemble, first scribe matching marks (M – Fig. 63) on the two frame halves (4 and 16), then remove through-bolts. Pry frame apart with a screwdriver between stator frame (12) and drive end frame (4). Stator assembly (12) must remain with slip ring end frame (16) when unit is separated.

NOTE: When frames are separated brushed will contact rotor shaft at bearing area. Brushes MUST be cleaned of lubricant if they are to be re-used.

Clamp iron rotor (13) in a protected vise, only tight enough to permit loosening of pulley nut (1). Rotor end frame can be separated after pulley and fan are removed. Check bearing surface of rotor shaft for visible wear or scoring. Examine slip ring surface for scoring or wear, and rotor winding for overheating or other damage. Check rotor for grounded, shorted or open circuits using an ohmmeter as follows:

Refer to Fig. 64 and touch ohmmeter probes to points (1-2) and (1-3); a reading near zero will indicate a short circuit to ground. Touch ohmmeter probes to slip rings (2-3); reading should be 5.3-5.9 ohms. A higher reading will indicate an open circuit and a lower reading will indicate an internal short. If windings are satisfactory, mount rotor in a lathe and check runout at slip rings using a dial indicator. Runout should not exceed 0.002

inch (0.05 mm). Slip ring surfaces can be trued if runout is excessive or if surfaces are scored. Finish with 400 grit or finer polishing cloth until scratches or machine marks are removed.

Before removing stator, brushes or diode trio refer to Fig. 65 and check for grounds between points A to C and B to C with an ohmmeter, using lowest range scale. Then reverse lead connections. If both A to C readings or B to C readings are the same, brushes may be grounded because of defective insulating washer and sleeve at the two screws. If screw assembly is not damaged or grounded, regulator is defective.

To test diode trio, first remove stator. Then remove diode trio, noting insulator positions. With an ohmmeter, check between points A and D (Fig. 66) and then reverse ohmmeter lead connections. If diode trio is good it will give one high and one low reading. If both readings are the same, diode trio is defective. Repeat this test at points B and D and at C and D.

Rectifier bridge (Fig. 67) has a grounded heat sink (A) and an insulated heat sink (E) that is connected to output terminal. Connect ohmmeter to grounded heat sink (A) and to flat metal strip (B). Then, reverse ohmmeter lead connections. If both readings are the same, rectifier bridge is defective. Repeat this test between points A and C, A and D, B and E, C and E and D and E. Capacitor (11 – Fig. 63) connects to rectifier bridge and grounds to end frame, and protects diodes from voltage surges.

Test stator windings (Fig. 68) for grounded or open circuits as follows:

Fig. 67 – Bridge rectifier test points. Refer to text.

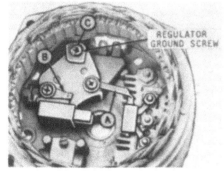

Fig. 65 – Test points for brush holder. Refer to text.

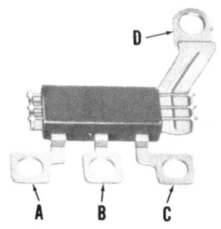

Fig. 66 – Diode trio test points. Refer to text.

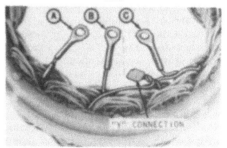

Fig. 68 – View of disassembled stator terminals and "Y" connection.

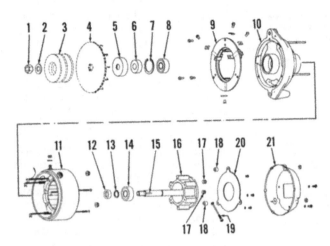

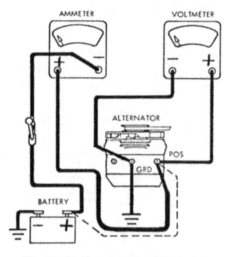

Fig. 69—Exploded view of Niehoff alternator used on some models.

1. Locknut
2. Washer
3. Pulley
4. Fan
5. Bearing cover
6. Bushing
7. Retaining ring
8. Bearing (front)
9. Heat sink
10. Housing & field winding assy.
11. Stator & shell
12. Cup
13. Ring
14. Bearing (rear)
15. Key
16. Rotor
17. Rubber plug
18. Bushing
19. Ground lead
20. Regulator
21. End cover

Fig. 70—Testing alternator. Refer to text.

Connect ohmmeter leads successively between each pair of leads. A high reading would indicate an open circuit.

The three stator leads have a common connection in center of windings. Connect ohmmeter leads between each stator lead and stator frame. A very low reading would indicate a grounded circuit. A short circuit within stator windings cannot be readily determined by test because of the low resistance of windings.

Brushes and springs are available only as an assembly which includes brush holder (14 – Fig. 63). If brushes are reused, make sure all grease is removed from surface of brushes before unit is reassembled. When reassembling, first install regulator and then brush holder, springs and brushes. Push brushes up against spring pressure and insert a short piece of straight wire through hole and through end frame to outside. Be sure the two screws at points (A and B – Fig. 65) have insulating washers and sleeves.

NOTE: A ground at these points will cause no output, or controlled output.

Withdraw wire from under brushes after alternator is assembled.

Remove and inspect ball bearing (6 – Fig. 63). If bearing is in satisfactory condition, fill bearing ¼ full with Delco-Remy lubricant number 1948791 and reinstall. Inspect needle bearing (17) in slip ring end frame. This bearing should be renewed if its lubricant supply is exhausted; no attempt should be made to relubricate and re-use bearing. Press old bearing out towards inside and press new bearing in from outside until bearing is flush with outside of end frame. Saturate felt seal with SAE 20W oil and install seal.

Reassemble alternator by reversing disassembly procedure. Tighten pulley nut to a torque of 40-60 ft.-lbs. (54.2-81.3 N·m).

Niehoff Models

67. **ALTERNATOR AND TRANSISTOR REGULATOR OPERATION.** Alternator consists of an assembly containing the core, bearing and field coil, a stator, rotor and shaft, diode rectifiers, regulator and end cover (Fig. 69). When a current path to field coil is established by the regulator, a magnetic flux is created which flows through a small air gap to rotor, and alternately from pole to pole of stator, inducing alternating three-phase voltage and current in stator coil windings. Alternating current output from stator is applied to a three-phase full wave bridge rectifier which provides direct current to battery.

Six rectifier diodes of the full wave bridge rectifier circuit are imbedded in two circular mounting plates which form a heat sink assembly at front of alternator.

Voltage regulator (20) is an electronic switching device, that senses voltage of charging system and supplies necessary field current to the alternator to accurately control alternator voltage.

At the start of alternator operation, residual magnetism induces sufficient current to turn on the power transistor. The power transistor replaces vibrating points used in mechanical regulators. With power transistor on, field current is permitted to flow through alternator field winding. Current flow causes alternator to provide an output to charging system. When alternator load demand is high, power transistor becomes conductive for longer periods of time and supplies a greater amount of current to field winding. An increase in current flow through the field winding causes an increase in alternator output voltage. As charging system builds up to predetermined setting of regulator, the zener diode in regulator causes the power transistor to turn off. With power transistor turned off, little or no current is applied to field winding, thereby

decreasing alternator output voltage. This on-off field switching action is repeated as often as necessary so average voltage output of alternator is maintained at regulator setting.

68. **TESTING.** Before making any tests be sure belt is tight, cables are clean and making good connections and that battery is good. To make test, turn on all electrical circuits and crank engine (without starting) for a short time to partially discharge battery. Connect test meters as shown in Fig. 70, then start and run tractor at high idle. Observe meter readings; ammeter should read approximately 50-60 amps. As battery is being recharged, voltage should return to regulator setting as follows:

Volts...12
Regulator Setting........................14.2V
Meter Reading (VDC)..............13.7-14.7

If a low or zero output is observed, connect one side of an insulated jumper wire Fig. 71 to ground stud and insert

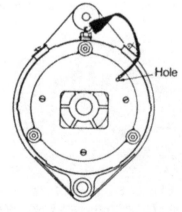

Fig. 71—By-passing regulator test. Refer to text.

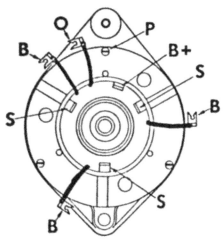

Fig. 72—Testing stator windings, field coil and diode heat sink. Refer to text for procedure.

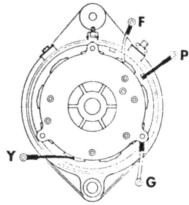

Fig. 73—Testing field coil. Refer to text for procedure.

other end of wire into hole in end cover. Jumper will by-pass regulator portion of charging system and apply full field current. If output is obtained, regulator is bad and should be renewed. If no output is obtained refer to static tests.

69. STATIC TESTS. To perform these tests it is necessary to remove alternator from tractor. After alternator is removed, remove end cover (21–Fig. 69) and pulley and fan (3 and 4). Disconnect three-phase leads (black wires B–Fig. 72) and output lead (red wire O) from heat sink. Then, disconnect field lead (white wire F–Fig. 73), "Y" lead (brown wire Y), ground wire (white wire G) and battery positive lead (brown wire P) from regulator.

To make the different tests, proceed as in following paragraphs.

70. STATOR WINDING. To check for open stator windings, connect an ohmmeter between each successive pair of stator phase leads (B–Fig. 72). Meter should read less than one ohm between each pair of leads. Then, connect ohmmeter between each (B) lead and frame of alternator. No meter needle movement should be seen. If other than zero (0) ohm reading is obtained, stator is grounded and must be renewed.

71. FIELD COIL. Make sure that ohmmeter is corrected to zero reading. Then, connect ohmmeter to field leads (O–Fig. 72) and (F–Fig. 73). It should read 5 ohms plus or minus ½ ohm. If readings are other than stated, field coil should be renewed. If readings are okay connect ohmmeter to (F–Fig. 73) and other ohmmeter lead to alternator housing. There should be no needle movement. If there is, housing and field coil assembly should be renewed.

72. DIODE HEAT SINK. Make sure all leads are disconnected (Fig. 72) from

heat sink before testing diodes. Touch one ohmmeter lead to output terminal (B+) of heat sink and other lead to each of the terminals (S). Observe and record in succession the ohmmeter readings. All readings should be nearly alike (either all high or all low). Then, reverse leads, proceed as before and record readings. All readings should be opposite from first readings. If any readings are the same in both tests, one or more diodes are defective and heat sink should be renewed.

73. OVERHAUL. To disassemble alternator, clamp plulley in padded vise jaws and remove nut (1–Fig. 69). Using a puller, remove pulley and then line up slot in fan with key in shaft and lift fan off shaft. Disconnect wires to heat sink (9), remove three screws securing heat sink to housing and lift heat sink from alternator. Position alternator with shaft down and remove rear cover (21). Remove all leads attached to regulator and remove regulator (20), spacer bushings (18) and ground wire (19). Remove Woodruff key (15), cover (5), spacer bushing (6) and support alternator on two blocks in a press and remove front bearing (8).

NOTE: Position blocks under alternator so stator windings are not damaged.

Remove retainer cup (12), retainer ring (13), rubber plug (17) and press rear bearing (14) from rotor shaft. Remove retaining ring (7) and insert length of ¾-inch (19.0 mm) rod through rear bearing hole and press front bearing out of housing (10). To separate stator assembly (11) from housing (10) remove flanged nuts. Disconnect three black wires and one red wire from field windings, accessible through three openings in housing (10). Separate stator assembly (11) from housing (10) about 3 inches (76.2 mm) then slip sleeves off field winding terminals and unsolder

wires from field terminals. Lift stator assembly off housing.

Clean inside of core to remove any old grease. Inspect all parts and renew any parts that are needed.

When reassembling, place stator assembly (11) on studs of housing and field assembly (10) with wires pointing toward housing and ground stud on stator aligned with threaded hole on housing. Install sleeves on both field leads and bend bare ends of field leads through field coil terminals (white wire and red wire) and solder. Slide sleeves down until field coil terminals are covered. Thread the three black wires and one red wire through opening in housing and slide stator assembly (11) down against housing (10). Connect three black wires and output red wire onto field winding and install nuts on studs and tighten finger tight. Install front bearing (8) using Grade H "Loctite" on outer surface of bearing and press bearing in bore of housing. Install retainer ring (7), flat side against bearing. Install rear bearing (14), retainer ring (13) and retainer cup (12). Apply a thin coat of Grade H "Loctite" to outer surface of rear bearing and press into housing (10). Insert six 1x3 inch (25.4x76.2 mm) pieces of 0.008 inch (0.20 mm) shim stock between each pole of rotor and adjacent stator pole. Be sure shims do not overlap each other. Extra length of shims should protrude from back of alternator. Move stator and shell for best freedom of movement of shims. Tighten the three nuts that were finger tight to a torque of 35 in.-lbs. (3.9 N·m). Pull out shims after nuts are tightened. If shims resist, turn rotor a slight amount. After shims are removed, check with feeler gage for minimum of 0.005 inch (0.13 mm) air gap between rotor poles and adjacent stator poles.

The balance of reassembly is reverse of disassembly.

NOTE: After alternator is assembled, re-energize magnetic circuit as follows:

Connect alternator to battery (positive to positive and negative to negative). Attach a jumper wire to alternator ground stud and insert other end into hole in end cover (Fig. 71). This is a momentary connection only and should not be held for more than a few seconds. This will energize field coil and re-establish magnetic field properties of alternator.

STARTING MOTOR

74. Three different makes of starting motors are used depending upon tractor model and engine application. When servicing starter motor refer to the following specification data:

Allis-Chalmers Starting Motor

Brush Spring Tension.................1-6 oz.
(0.278-1.66 N)
No-Load Test:
  Volts...........................11.7
  Amps (maximum)........................130
  Rpm (minimum).......................3100
Load Test:
  Volts............................8.4
  Amps............................900
  Rpm (minimum)...................900
  Torque.............................18 ft.-lbs.
(24.3 N·m)

Lock Test:
  Volts 5
  Amps (maximum)....................2300
  Torque (minimum)..............45 ft.-lbs.
(60.9 N·m)

Delco Starting Motor
No-Load Test:
  Volts 9
  Amps....................130-160
  Rpm....................5000-7000

Lucas Starting Motor
Brush Spring Tension.................42 oz.
(11.6 N)

No-Load Test:

Volts............................12
Amps..............................100
Rpm....................5500-7500
Load Test:
  Volts.............................8.4
  Amps............................590
  Rpm.............................1000
  Torque.............................16 ft.-lbs.
(11.6 N·m)

Lock Test:
  Volts.............................8.4
  Amps............................980
  Torque............................34 ft-lbs
(46.0 N·m)

TORQUE LIMITER

75. R&R AND OVERHAUL. To gain access to torque limiter for inspection or removal, first split tractor as outlined in paragraph 86, then refer to Fig. 74 and proceed as follows: Remove retaining bolts, retainers, Belleville spring washer, end plate and driven disc from flywheel.

Belleville washer thickness should be 0.156-0.158 (3.96-4.01 mm) for Models 7010, 7020 and 7045, or 0.146-0.150 inch (3.71-3.81 mm) for Model 7080 and 0.140-0.144 inch (3.56-3.66 mm) for all other models. Check free height of Belleville washer as shown in Fig. 75; height should be 0.359-0.379 inch (9.12-9.63 mm) for Models 7010, 7020 and 7045, and 0.359-0.363 inch (9.12-9.22 mm) for all other models. Belleville washer should be renewed if it does not meet both of the preceding specifications or if it is cracked, warped or blued from wear or overheating.

Inspect end plate for wear or warpage. If warpage exceeds 0.014 inch (0.36 mm) end plate should be renewed.

Thickness of end plate should be 0.500-0.510 inch (12.70-12.95 mm) for Models 7010, 7020 and 7045. For all other models thickness of end plate should be 0.337-0.343 inch (8.56-8.71

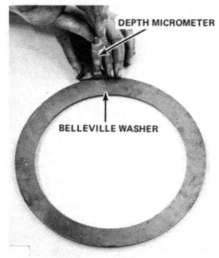

Fig. 75 — Free height of Belleville washer should be checked as shown.

Fig. 74 — View showing installed torque limiter assembly. Note retaining bolt tightening sequence.

mm). Renew end plate if it does not meet any of the above specifications or if it is cracked or blued from wear or overheating.

Check flywheel for wear or damage. If necessary flywheel may be resurfaced, however not more than 0.062 inch (1.57 mm) can be removed before flywheel must be renewed. Outer surface and disc surface of flywheel must be machined the same amount to maintain correct counterbore depth (D–Fig. 76) of 0.173-0.183 inch (4.39-4.65 mm) for all models.

Inspect retainers for wear or cracking. Length (L–Fig. 77) should be 0.892-0.902 inch (22.66-22.91 mm) for

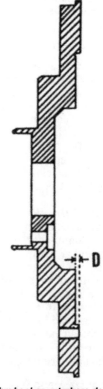

Fig. 76 — Flywheel counterbore depth (D) should be 0.173-0.183 inch (4.39-4.65 mm).

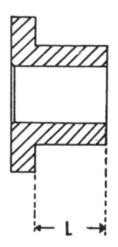

Fig. 77 — Length (L) of Belleville washer retainers should be carefully checked prior to their reuse as outlined in text.

Models 7010, 7020 and 7045 and 0.677-0.687 inch (17.19-17.45 mm) for all other models. If retainer is too long it may be ground off to specification. If retainer is too short it should be discarded and a new one of proper length installed.

Check driven disc for wear, damage, oil on linings and dampener spring failure. Note that partial failure of dampener springs or their mountings can cause a rhythmic gear train rattle at lower engine speeds.

Diameter of driven disc on Model 7010, 7020 and 7045 tractors is 11.875 inches (301.63 mm). Disc thickness should be 0.370-0.390 inch (9.40-9.90 mm). Lining material is "Cerra Metallic".

All other model tractors use a "Raybestos" lined driven disc that is 12.00 inches (304.8 mm) in diameter and 0.297-0.323 inch (7.45-8.89 mm) thick.

Note that some Model 7040, 7060 and 7080 tractors were equipped with a metallic pad lined driven disc. This type disc is 12.00 inches (304.8 mm) in diameter and 0.372-0.388 inch (9.45-9.85 mm) thick. Because this type driven disc is thicker than the standard disc, 0.040 inch (1.02 mm) thick shims were installed under the retainers to maintain proper pressure on friction surfaces. If metallic driven disc is to be placed back into service be sure to reinstall the retainer shims. Because only the "Raybestos" lined disc is available as a service part it will be necessary to discard the retainer shims when assembling the torque limiter.

Thoroughly clean all surfaces with brake cleaner making sure that any oil film is removed. If there was an oil leak make certain that it has been corrected. Note that an oil leak will cause a torque limiter failure, conversely, the heat of a torque limiter failure will cause secondary oil leaks. Make sure that all oil leaks have been found and corrected. Install driven disc with long hub toward flywheel. Install end plate, Belleville spring washer, retainers and retaining bolts. Tighten retaining bolts finger tight then carefully center driven disc in flywheel. Tighten retaining bolts in sequence shown in Fig. 74 to 90-100 ft.-lbs. (121.9-135 N·m). Check compressed height of Belleville washer as shown in Fig. 78. Height should be 0.200-0.220 inch (5.08-5.59 mm) for Models 7010, 7020 and 7045. For all other models height should be 0.190-0.230 inch (4.83-5.841 mm). If compressed height of Belleville washer is not correct alter length of retainers and/or renew Belleville washer as required.

Fig. 78 — Compressed height of Belleville washer should be checked as shown.

76. **TORQUE LIMITER BREAK-IN.** After installation of a new driven disc or end plate, or resurfacing of flywheel, torque limiter must be broken-in as follows: Place tractor in first, second or third gear and high range. Drive tractor forward at approximately 200 rpm and aggressively apply brakes until engine is killed or torque limiter breaks free. Immediately release clutch, restart engine if necessary, and allow engine to run at idle for three minutes to cool driven disc. If engine was killed no further break-in will be required. However if torque limiter broke loose it will be necessary to repeat the above procedure (possibly as many as 10 times) until torque limiter can no longer be broken loose.

NOTE: Prolonged slipping of torque limiter will rapidly damage driven disc and related parts to the point where tractor will again have to be split.

"POWER-DIRECTOR"

The "Power-Director" clutches are actuated by hydraulic pressure. To shift from one speed to another on the go, simply step on HI or LO shift button on floor of operator's platform. When HI shift button is depressed, oil is directed to rear clutch piston, clutch is engaged and transmission input shaft is driven at engine speed. With LO shift button depressed, front clutch is engaged and transmission input shaft is driven at reduced speed.

Depressing tractor clutch pedal automatically shifts "Power-Director" clutch into LO speed, so that all starts are made in low speed.

TESTS AND ADJUSTMENTS

Models So Equipped

77. **CLUTCH ADJUSTMENT.** Adjustment of "Power-Director" clutch is made in different places depending on whether clutch has rod or cable type valve control linkage.

If rod type linkage is used, refer to Fig. 79 and adjust cable retaining "U" rod until there is 0.040 to 0.060 inch (1.02-1.52 mm) of clearance between yoke that straddles spool and front spring plate when clutch pedal is in UP position.

Fig. 79 — View showing rod type linkage adjustment. Refer to text.

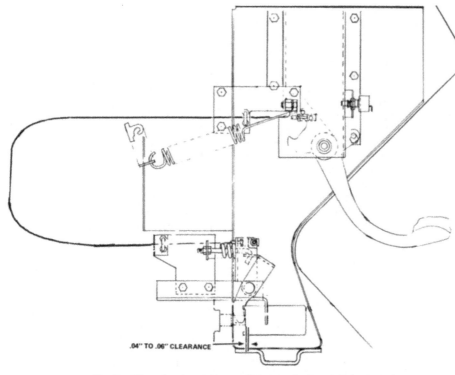

Fig. 80 — View showing cable type linkage adjustment. Refer to text.

clutch or transmission brake. Start engine and run until transmission oil reaches operating temperature. Depress clutch until transmission brake is engaged then throttle engine down until transmission oil pressure warning light begins to flicker on. Note engine rpm. Engage Low range clutch and again throttle engine down until transmission warning light begins to flicker on. This should be 50-100 rpm higher than when just the transmission brake was applied. Repeat this test a third time with High range clutch engaged. Transmission oil warning light should come on at nearly the same rpm as for Low clutch test. If transmission oil warning light comes on at a higher pressure in any one range, that is the clutch with the leak.

80. REGULATING VALVE PRESSURE. To test regulating valve pressure install a 0-600 psi (0-4100 kPa) gage in top test port of valve as shown in Fig. 82. Start engine and allow transmission oil to reach a temperature of 130°-140°F (54°-60°C). Pressure reading on gage should be 195-205 psi (1346-1414 kPa) at 2300 rpm. Pressure will fall off to approximately 150 psi (1035 kPa) at idle. This pressure should remain the same regardless of position of clutch pedal or whether transmission is in High or Low range.

If regulating valve pressure is incorrect refer to paragraph 85 for outline of shim installation.

81. HIGH AND LOW CLUTCH PRESSURE. To test high and low clutch pressure attach 0-600 psi (0-4100 kPa) gages to test ports, located on rear of pressure regulating valve, as shown in Fig. 83. Start engine and allow transmission oil to reach a temperature of 130°-140°F (54°-60°C). Engage either the high or low clutch and run engine at 2300 rpm. Gage reading for engaged clutch should be 195-205 psi (1346-1414 kPa). Gage reading for disen-

If cable type linkage is used, refer to Fig. 80 and adjust cable retaining "U" bolt in slots until there is 0.040-0.060 inch (1.02-1.52 mm) clearance between yoke and front spring plate.

78. GEROTOR PUMP. To test gerotor pump a Flo-Rater will be needed. Refer to Fig. 81 and proceed as follows:

Remove pump line running from rear right-hand side of transmission housing to clutch valve assembly as shown in Fig. 81. Note that on some models it will be necessary to remove the auxiliary fuel tank to gain access to pump fittings. Connect inlet hose of Flo-Rater to fitting at transmission housing. Connect outlet hose of Flo-Rater to clutch valve. Open Flo-Rater restrictor valve completely. Then, start tractor and operate hydraulic system until oil temperature is 130° to 140°F (54°-60°C). With engine speed set a 2300 rpm, gerotor pump should produce 14-17 gpm (52.9-64.3 Lpm) flow at 200 psi (1380 kPa) pressure. If flow is less than 13 gpm (49.2 Lpm), pump should be removed and repaired. Refer to paragraph 151 for pump overhaul procedure.

79. CLUTCH LEAKAGE. It is not necessary to install any test equipment to check for leaks in either High or Low

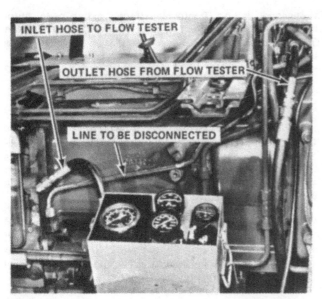

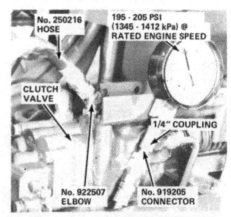

Fig. 81 — View showing Flo-Rater being used to check gerotor pump.

Fig. 82 — View showing proper installation of pressure gage to test regulating valve pressure.

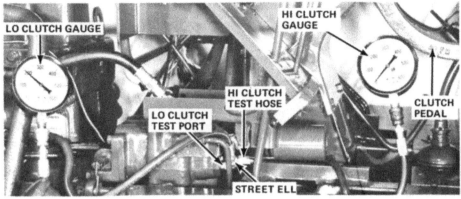

Fig. 83 — View showing proper installation of pressure gages to test High and Low clutch pressures.

gaged clutch should be zero. If gage readings are not correct the problem is probably in the valve or pump. If gage readings are correct the problem is probably in the clutch.

82. CLUTCH CROSSOVER PRESSURES. With gages connected as outlined in paragraph 81, crossover pressure during high/low or low/high shift may be checked.

When shifting from low to high, pressure in high range gage should rise to 195-205 psi (1346-1414 kPa) quickly, while pressure in low range gage should slowly fall. There should also be a momentary load placed on the engine while both sets of clutches are partially engaged.

When shifting from high to low, pressure in low range gage should rise at the same rate that pressure falls in high range gage.

If pressures do not react correctly, problem is most likely in the shifting spool portion of the pressure regulating valve or in the modulating or selector spools. If reaction of pressure gages is correct but the shift is still bad, problem is probably in the clutch.

83. INCHING PRESSURE. To check inching pressure refer to Fig. 83 and proceed as follows: Install a 0-600

psi (0-4100 kPa) gage in low clutch test port. With engine running at 2300 rpm and clutch pedal fully depressed gage should read zero pressure. By gradually letting out on clutch pedal it should be possible to gradually, in 50 psi (344 kPa) increments, increase gage pressure until full operating pressure of 195-205 psi (1346-1414 kPa) has been reached. Working pressure should be reached just before clutch pedal is fully released. If gage pressure cannot be regulated, check clutch cable or linkage, inching spool, and inching springs.

84. TRANSMISSION BRAKE. Attach a 0-600 psi (0-4100 kPa) gage to

transmission brake outlet on clutch valve as shown in Fig. 84. This may be done by removing existing elbow from valve and replacing it with a suitable tee.

With engine running and transmission oil at normal operating temperature push clutch pedal completely down. With clutch pedal fully depressed and engine running at 2300 rpm gage pressure should be 195-205 psi (1346-1414 kPa).

Test for leakage in brake circuit by holding clutch pedal down and shutting off engine. Gage pressure should not drop immediately.

If transmission brake pressure cannot be obtained, check clutch cable or linkage adjustment. If transmission brake pressure is normal, but shifting cannot be performed without gear clash, brake is probably being overcome by friction from warped clutch plates.

CONTROL AND SHIFT VALVES

All Models So Equipped

85. R&R AND OVERHAUL. To remove control and shift valve, first

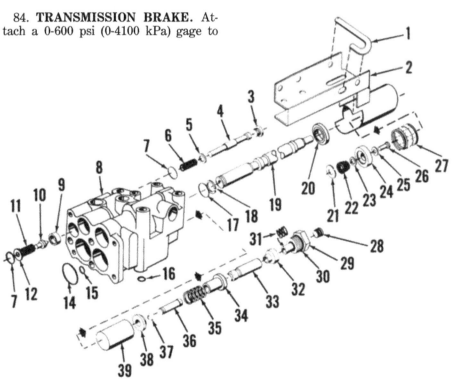

Fig. 85 — Exploded view of Power-Director control valve used on all models so equipped. There are two of each item 28 through 39 used in the control valve.

1. Pin	10. Valve (cooler relief)	21. Washer
2. Bracket	11. Spring	22. Spring
3. Plug	12. Washer	23. Washer
4. Spool (pressure regulating)	14. "O" ring	24. Stop washer
5. Washer (shim)	15. "O" ring	25. Washer
6. Spring	16. "O" ring	26. Cap screw
7. "O" ring	17. "O" ring	27. Spring
8. Body	18. Retainer	28. Plug
9. Seat	19. Spool (inching)	29. Plug
	20. Seal	30. "O" ring

31. Spring	
32. Seat	
33. Selector spool	
34. Seat	
35. Spring	
36. Pin	
37. Ring	
38. Seat	
39. Piston (accumulator)	

Fig. 84 — View showing proper installation of pressure gage to test transmission brake function.

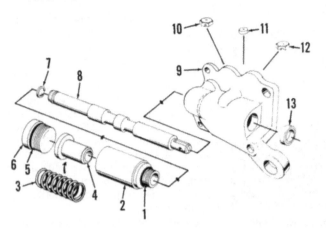

1. "O" ring
2. Retaining plug
3. Spring
4. Retainer
5. "O" ring
6. Plug
7. Retainer
8. Spool (shifting)
9. Valve
10. Orifice (two way)
11. Plug
12. Orifice (one way)
13. Seal

20-24 ft.-lbs (27.1-32.5 N·m). Then, adjust linkage as outlined in paragraph 77.

"POWER-DIRECTOR" CLUTCH SPLIT

All Models

86. Remove front end weights from tractor to prevent front end assembly from tipping forward when disconnected from transmission housing. Remove right and left hood assemblies. Remove panel at front of console and remove top hood assembly. Disconnect battery cable. Remove left platform panel, and remove step and brace from left side of tractor. Disconnect vent line and fuel return lines from top of fuel tank. Shut off fuel from main fuel tank (large tank). Remove cross-over fuel line between main tank (left side) and auxiliary tank (right side) if so equipped. Plug auxiliary fuel line with plastic cap if tractor is so equipped. Loosen main fuel tank support straps at trunnion on inner side of tank. Support tank on floor jack. Disconnect battery wire from fuel level sending unit and disconnect ground wire from bolt on platform support. Remove fuel tank support straps, lower fuel tank to floor and move from under tractor. Disconnect speed control cable, fuel shut-off cable and engine hour-meter cable. Drain oil from power director clutch compartment (drain plug in bottom cover). Disconnect wiring harness to engine at right side of console by

remove left side of panel on steering column support. Clean valve and area around it. Disconnect all tubes and control linkage. Disengage shifting linkage from slot in spool. Remove cap screws securing valve to transmission housing. Slide valve forward to clear shifting linkage and remove valve.

To disassemble valve, insert an Allen wrench through center of bracket (2 – Fig. 85) to remove socket head cap screw (26), then remove bracket. Remove shifting spool assembly from front of valve. When shifting valve (Fig. 86) has been removed, cooler by-pass valve (10 – Fig. 85) and pressure regulating valve (4) can be removed.

NOTE: If it is only necessary to adjust regulating valve or cooler by-pass pressure, remove only the shifting valve assembly.

Pressure regulating valve (4 – Fig. 85) pressure is adjusted by adding or removing shims (5) which are 0.022 inch (0.55 mm) thick. Each additional shim will increase pressure approximately 11.6 psi (79.9 kPa).

Cooler by-pass valve (10 – Fig. 85) is spring-loaded and shim adjusted. During cold weather operation, if more than 220-240 psi (1516-1654 kPa) is required to force oil through oil cooler, by-pass valve will open and allow oil to flow directly to the sump. Each shim (12) for cooler by-pass valve is 0.035 inch (0.88 mm) thick and will alter by-pass pressure by approximately 20.5 psi (141.3 kPa).

To remove accumulator piston assemblies (39) and selector spools, remove plugs (29) at rear of valve. Selector spools (33) can be removed through these bores. Then, using a soft rod, push accumulator pistons out front of valve. Remove inching spool (19) through front of valve. The shifting spool (8 – Fig. 86) can be removed after removing plug (6). Withdraw spool (8), spool retainer (4) and return spring (3). On face of shifting valve (9) that fits to inching portion of valve, there are two restrictor plugs which have an interference fit in body.

To remove plugs, tap threads in plugs, insert a cap screw and pull plugs with cap screw. This operation will destroy plugs, and new ones will have to be installed on reassembly. Located behind each plug is a flow restrictor. Restrictor (12) serving high range accumulator piston is a one-way and flat side of this restrictor should be installed toward shifting spool (8). Restrictor (10) for low range accumulator piston is the two-way type. Small orifice, 0.028 inch (0.71 mm) diameter, of this restrictor should be installed toward shifting spool (8). New retaining plugs should be pressed into valve body until they contact limiting shoulder.

When reassembling, lubricate all seals and "O" rings and reverse disassembly procedure. Tighten bolts attaching shifting valve to inching valve to a torque of

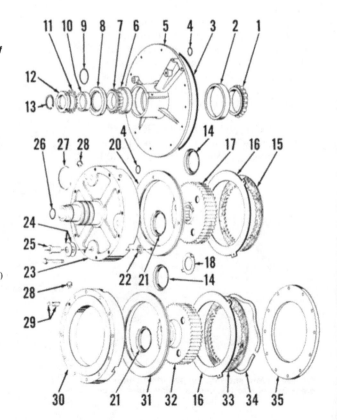

Fig. 87 — Exploded view of Power-Director clutch.

1. Cone
2. Cup
3. "O" ring
4. "O" ring
5. Cover
6. Cup
7. Cone
8. Seal
9. "O" ring
10. Sleeve
11. Lock washer
12. Adjustment nut
13. Snap ring
14. Seal
15. Driving plate (4 used)
16. Separator plate (6 used)
17. Hub (front)
18. Thrust washer
19. Piston
20. Piston
21. Seal
22. "U" cup seal (4 used)
23. Housing
24. Belleville washer (24 used)
25. Retainer
26. "O" ring
27. Seal ring (3 used)
28. Plug
29. Spring & ball
30. Housing
31. Piston
32. Hub (rear)
33. Driving plate (2 used)
34. Wave spring
35. End plate

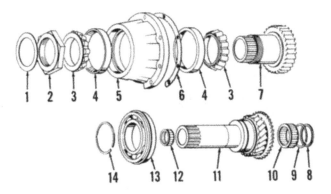

Fig. 88 — Exploded view of transmission input shaft.

1. Shim
2. Locknut
3. Cone
4. Cup
5. Retainer
6. Shim
7. Shaft & gear
8. Sleeve
9. Snap ring
10. Bearing
11. Input shaft
12. Needle bearing
13. Bearing & snap ring
14. Snap ring

releasing catches and pulling connector apart. Disconnect starting motor wires (orange, green and brown) from front of console. Disconnect fuel inlet and return lines at right rear of engine. Use floor jack to assist in separating tractor, then block rear section securely and safely (block wheels and under drawbar and transmission housing). Place wedge shaped wood block between front axle and side frames at both sides to prevent tipping sideways. Attach a hoist to engine lift eyes and take up slack in hoist. Remove the four bolts attaching side frames to front of transmission housing. Install guide studs in the two bottom holes. Remove four bolts attaching engine mounting plate to transmission housing. Roll engine and front end of tractor forward far enough to provide work clearance, and block securely and safely.

Reassembly is reverse of disassembly procedure.

"POWER-DIRECTOR" CLUTCH

87. REMOVE AND REINSTALL. To remove Power-Director clutch assembly, it is necessary to split engine from transmission as outlined in paragraph 86. Remove cap screws attaching power director cover to front of transmission. Install two 8-inch (203.2 mm) long guide studs and two ½x4 inch cap screws in bosses on cover to use as handles. Then, slide clutch assembly from transmission.

When reinstalling, shim pack (1 – Fig. 88) is used between rear clutch hub and bearing adjusting nut on high range hollow drive shaft. If no new parts were used in clutch pack, same shim pack may be used. To select correct shim pack proceed as follows:

Using a one-inch square bar 15 inches (381 mm) long as a straightedge, turn clutch assembly upward and place straightedge across clutch and plate. Measure between mounting cover surface and straightedge and record the measurement. Then, measure between straightedge and rear (high range) clutch hub. Subtract second measure-

ment from first and record this reading for later use. Now place straightedge across clutch mounting surface at front of transmission housing. Then, measure distance between straightedge and bearing adjusting nut (2 – Fig. 88). Subtract this dimension plus 0.020 inch from the last reading. This will give shim pack thickness required. Clearance permissible is 0.013-0.027 inch (0.33-0.69 mm). Shims are available in thicknesses of 0.010, 0.020 and 0.030 inch.

Using new "O" ring, reassembly is reverse of disassembly. Torque all cap screws to 28-33 ft.-lbs. (37.9-44.7 N·m).

88. OVERHAUL. Remove snap ring at front of clutch and drive main pto drive shaft rearward using a soft hammer. Refer to Fig. 87 and proceed as follows:

Position clutch assembly so that support cover is supported on bench and with rear clutch housing facing upward. Remove the 12 bolts attaching end plate and clutch housings together. Mark front and rear clutch housings so that they are reassembled in same positions. Remove rear clutch hub (32) and end plate (35). Remove clutch plates, wave springs and separator plates. Remove rear clutch housing (high range) from front clutch housing (low range). Remove front clutch hub (17) and clutch plates. Remove piston from rear clutch housing. Separate front support cover from front (low range) clutch housing and remove piston from clutch housing as follows: Bend up locking tabs (11) and remove spanner nut (12) at front of support cover with a spanner wrench. Remove locking washer. Place front support cover in a press with cover on top and low range housing downward. Support front cover and press clutch housing from cover. Remove the four cap screws attaching Belleville spring washers (24) to front of clutch housing. Remove four spacers (25). Clutch piston may now be removed from low range clutch housing. Check piston seals and renew if necessary. Use a hollow tube type driver (3-3/16 inch OD and 3 inch ID) and drive front bearing cone and oil

seal from support cover. Drive from rear side of cover. Remove front bearing cup by driving it forward, using a tube type driver 3.235 inches (82.17 mm) in diameter. Remove rear bearing cup, using an inside puller. Remove rear bearing cone from front clutch housing, using a knife edge pulling attachment. Check and clean front bearing hub orifice plug check valve. All parts must be thoroughly cleaned and inspected before assembling. Renew all parts that show excessive wear or damage. New parts may be damaged when improperly installed. Use tubular type driver tools for installing bearing cups and cones. Press front bearing cup (6) into place with cup back face seated against shoulder in cover. Press rear bearing cup (2) into place with cup back face seated against shoulder in cover. Press rear bearing cone (1) onto front of low range clutch housing (23) with cone back face seated against shoulder on housing. Install inner and outer clutch piston seals (14 and 21) in piston grooves with seal lips facing toward oil pressure when assembled. Install the four "U" cup seals (22) in place in front clutch housing with seal lips facing toward oil pressure. Lubricate inner area of seals and also piston seals and housing surfaces. Carefully install piston (20) into front housing, making sure seals are not damaged. Use a feeler gage blade or similar tools on lips of seals. Place five Belleville spring washers (24) over each spring retainer (25). Lubricate retainers and insert them through housing and seals. Install cap screws through spring retainers and thread into clutch piston. Torque the four cap screws to 20-25 ft.-lbs. (27.1-33.8 N·m). Position front clutch housing on bench with front side (bearing side) upward. Install three sealing rings (27) in grooves of housing. Apply lubriplate on sealing rings and sealing surfaces of cover. Place support cover over front housing and carefully assemble over sealing rings. Press front bearing cone in place with cone back face upward. Install "O" ring seal (9) in groove next to bearing cone. Place seal sleeve (10) over "O" ring seal. Install oil seal (8) in hub of cover, using sealer on outer surface of seal. Seal lip (garter spring side) must face toward front bearing. Drive oil seal into cover until it is 0.060-0.090 inch (1.52-2.28 mm) past flush with front edge of seal sleeve. Install nut locking washer (11) with bent tabs facing upward. Install spanner nut (12) with chamfered side next to locking washer. Clamp support cover in a vise, and attach a dial indicator. Using a spanner wrench, tighten nut to provide 0.001-0.006 inch (0.03-0.15 mm) end play in bearings. Lock nut securely with locking washer. Position clutch support

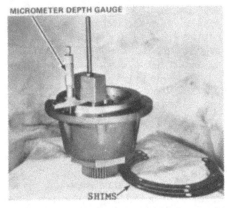

Fig. 89—View showing measuring for correct shims. Refer to text.

Fig. 90—View showing measuring from shim surface of housing to outer race of bearing. Refer to text.

cover on bench with front clutch housing facing upward.

NOTE: Front (low range) and rear (high range) clutch friction plates are different on some models. Do not mix these friction plates.

All separator plates are alike on both models. Install a separator plate next to clutch piston. Next install a friction plate (w/internal teeth). Install thrust washer (18) and front clutch hub in housing. Use grease to hold thrust washer in place. Continue to install clutch plates alternately in above manner until three separator plates and three friction plates are installed. Some models have four friction plates. Install piston seals on rear clutch piston in same manner as for front clutch piston. Lubricate seals and inside surface of housing. Install piston in housing, being careful not to damage seals. Use a feeler gage blade or similar tool to start lip of seals into housing.

"POWER-DIRECTOR"HIGH RANGE OUTER (HOLLOW) SHAFT AND COUNTERSHAFT GEAR

89. **R&R AND OVERHAUL.** To remove outer (hollow) high range shaft and front countershaft gear, it is necessary to split engine from front of transmission housing as outlined in paragraph 86. Then, remove Power-Director clutch as outlined in paragraph 87 and proceed as follows: Remove platform section necessary for access to top of transmission. Remove two cap screws securing Power-Director shifter bracket to cover. With shift lever in neutral position, remove cap screws securing shift cover to transmission and lift cover off.

NOTE: Front and rear cap screws at center of cover are a special dowel cap screw for positioning cover to transmission.

Lock transmission from rotating by sliding front shift collar forward and engage center shift collar. This will permit removal of countershaft nut.

NOTE: High range hollow shaft and its bearing retainer, and front countershaft gear must be removed together with gear teeth in mesh.

This nut will require a 2½ inch socket. If high range hollow shaft and its bearing retainer is to be disassembled, large nut must be removed prior to removal. This nut will require a 4-inch socket similar to the OTC 1914-4 socket wrench. Remove the six cap screws that attach bearing retainer to transmission housing. Slide hollow shaft with bearing retainer and front countershaft gear off together with gear teeth in mesh. Remove countershaft gear from countershaft and hollow shaft with bearing retainer from low range hollow shaft. If low range shaft is to be removed refer to paragraph 90. To separate high range hollow shaft (2 through 7–Fig. 88) from bearing retainer, support bearing retainer in a press. Use a step plate adapter at end of hollow shaft and press shaft from retainer. Remove bearing cone from hollow shaft, and bearing cups from retainer if renewal is necessary.

When reassembling, refer to following paragraph: Assemble Power-Director high range hollow shaft to bearing retainer. Press rear bearing cone on shaft with cone back face seated against shoulder. Install bearing cups into bearing retainer with cup back face seated

against shoulder. Lubricate bearings and assemble shaft into retainer. Install front bearing cone and bearing adjusting nut. Torque nut to give 0.001-0.004 inch (0.03-0.10 mm) end play in shaft, using a dial indicator. Stake nut securely to shaft. Stake nut in four places, 90 degrees apart. This nut requires a wrench similar to OTC 1914-4. Select a shim stack to give 0.0005-0.005 inch (0.013-0.13 mm) clearance between bearing retainer and outer race of ball bearing as follows:

Use a micrometer depth gage and measure from shim surface of bearing retainer to ball bearing contact surface as shown in Fig. 89. Now measure from shim surface of housing to outer race of ball bearing as shown in Fig. 90. Subtract Step 2 from Step 1. The dimension obtained in Step 3 will be the space between bearing retainer and housing. Install a shim stack that is 0.001-0.003 inch (0.03-0.08 mm) thicker than the Step 3 dimension. Shims are available in thicknesses of 0.010, 0.012 and 0.014 inch.

Pilot the high range hollow shaft and bearing retainer assembly (with shim stack in place) over low range shaft. Mesh teeth of front countershaft gear with high range drive gear, and slide them back into position on their shafts. Install cap screws in bearing retainer and torque to 28-33 ft.-lbs. (37.9-44.7 N·m). Install a new nut on countershaft and torque to 290-320 ft.-lbs. (392.9-433.6 N·m), and stake securely to shaft.

Balance of reassembly is reverse of disassembly.

TRANSMISSION (STANDARD)

Some standard transmissions are fitted with main shaft gear bushings and lock pins, while in all other units the main shaft gears ride directly on the shaft. Disassembly and overhaul of both units is identical except where noted.

90. **R&R AND OVERHAUL.** To remove transmission, first split tractor as outlined in paragraph 86. Drain oil from transmission and differential. Remove shift cover as outlined in paragraph 91. Support front of platform and

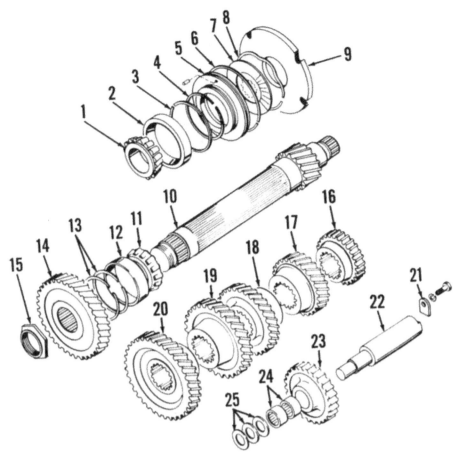

Fig. 91 — Exploded view of countershaft, transmission brake and reverse idler gear and shaft. Item 17 is used with 20 speed. If tractor has 16 speeds, a spacer replaces item 17.

1. Cone
2. Cup
3. Snap ring
4. Sealing ring
5. Piston
6. Sealing ring
7. Friction plate
8. Wave spring
9. Brake plate
10. Countershaft
11. Cone
12. Cup
13. Snap ring
14. Gear
15. Nut
16. Reverse gear
17. Gear (2nd)
18. Gear (3rd)
19. Gear (4th)
20. Driven gear
21. Lock
22. Shaft (reverse idler)
23. Gear
24. Bearing
25. Thrust washers

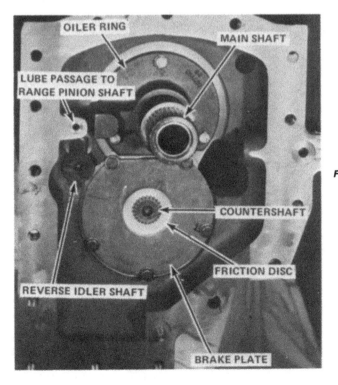

Fig. 92 — View of rear of transmission.

remove right fuel tank, if so equipped. Remove protective pan from below hydraulic pump assemblies. Remove hydraulic pumps and lines. Place a support stand or jack under transmission and unbolt and roll transmission away from differential housing.

With transmission out of tractor remove high range (hollow) shaft as outlined in paragraph 89. Before low range hollow shaft can be removed from housing, it will be necessary to move countershaft rearward far enough to drop countershaft driven gear to bottom of housing. Proceed as follows:

Remove transmission brake parts from rear of countershaft (Fig. 92). Remove brake plate (9 – Fig. 91), friction plate (7) and brake piston (5). Remove snap ring (3) retaining countershaft rear bearing cup. Remove lock clip (21) from reverse idler shaft (22) and slide shaft rearward. Remove idler gear (23) and thrust washers (25). Press countershaft rearward until driven gear (20) can drop down to bottom of housing. Pull low range hollow shaft with gear from bearing bore. Refer to Fig. 93 and Fig. 87 for exploded view. Remove ball bearing retaining snap ring (14 – Fig. 88) and press shaft from bearing if renewal is necessary. Remove sleeve (8) and snap ring (9) at rear of hollow shaft to renew roller bearing. Remove needle bearing (12) at front end of shaft if renewal is necessary. Remove snap ring (36 – Fig. 94) from rear of main shaft. Remove oil seal ring (37) at rear of main shaft. Remove the two seal rings (20) from grooves of shaft. Remove snap ring (35) from rear of transmission housing. Remove snap ring (1) at front end of shaft that retains the fourth and fifth gear splined collar.

At this point it will be necessary to identify which type of transmission is being overhauled, i.e., with or without main shaft gear bushings. If transmission is 16-speed and incorporates a spacer instead of second gear (18 – Fig. 94) it was produced with main shaft bushings. Likewise, if a spacer is used in place of third gear (12) transmission does not incorporate main shaft gear bushings.

On 20-speed transmissions, after removing snap ring (1), slide splined collar (2), shift collar (3) and fourth gear (6) forward on main shaft far enough to allow visual inspection for thrust washers (7 and 9) and snap ring (8). If snap ring and both thrust washers are present transmission was produced without main shaft bushings. If only one thrust washer (9) is present transmission incorporates main shaft bushings.

After identification of transmission type has been made proceed with overhaul.

On transmission with main shaft bushings, press main shaft rearward until rear bearing cup is free of housing bore. Remove fourth and fifth gear shifter coupling (3) and splined collar (2) through opening in top of housing.

On transmissions without main shaft bushings, slide fourth and fifth gear shifter coupling (3) and fourth gear (6) forward on main shaft until third gear snap ring (8) can be removed from its groove. Slide third gear (12) and second and third gear shifter coupling (14) forward until second gear locking thrust washer (15) can be unlocked from mainshaft. Slide second gear (18) and first and reverse sliding gear (25) forward until the first gear locking thrust washer (26) can be unlocked from the main shaft.

On all models, move main shaft rearward by hand and remove fourth gear (6) over front end of shaft. Be careful not to drop or lose bushing lock pin (4) if so equipped. Remove thrust washers as necessary to allow removal of third gear (12). Be careful not to drop or lose bushing lock pin (10) if so equipped. Remove second and third gear shift coupling (14) and splined collar (13). Remove second gear (18) and associated thrust washer (15) or bushing (17); be careful not to drop or lose bushing lock pin)16) if applicable.

NOTE: On Model 7020 16-speed tractors, a spacer is used in place of third gear. Also snap ring (8) and thrust washers (7 and 9) in front of third gear have been eliminated. On all other model 16-speed tractors, a spacer is used in place of second gear (18).

Remove first and reverse sliding gear (25) and splined collar (24). Remove first gear (29) and associated thrust washer (26) or bushing (28); be careful not to

drop or lose bushing lock pin (27) if applicable. Remove main shaft from housing bore and remove thrust washer (30) from shaft. Remove bearing cones (32 and 33) from main shaft and front cup from housing if renewal is necessary.

Countershaft and gears were partly removed prior to removing main shaft and gears. Front countershaft gear is removed with Power-Director high range hollow shaft. Countershaft driven gear was dropped down to bottom of housing so that Power-Director low range hollow shaft could be removed. Countershaft brake was removed and countershaft moved rearward to allow driven gear to drop down to bottom of housing. Continue to pull countershaft rearward and remove driven gear (20–Fig. 91) and front bearing cone. Remove fourth, third, second and reverse gears from housing. Remove rear bearing cone from shaft, and front bearing cup from housing bore, if renewal is necessary. Be careful not to damage rear countershaft bore where transmission brake sealing rings ride.

NOTE: On 7020 16-speed tractors a 2.581-2.583 inch (65.56-65.61 mm) wide spacer is used in place of third countershaft gear.

Thoroughly clean and inspect all gears, bearings and component parts and renew as necessary before reassembling. Refer to Fig. 91 and reassemble countershaft as follows:

Install rear bearing cone (1–Fig. 91) on countershaft with cone back face seated against shoulder next to the 16 tooth gear. Front bearing cup (12) and two retaining snap rings (13) are installed in housing bore **after** installation of low range input shaft. Enter front end of countershaft through rear bearing bore and install reverse gear (16) over

end of shaft with longer protruding hub facing forward. Install second gear (17) over end of shaft with longer protruding hub facing forward.

NOTE: On Model 7020 16-speed tractors a 2.581-2.583 inch (65.56-65.61 mm) wide spacer is used in place of third gear. On all other model 16-speed tractors a 2.581-2.583 inch (65.56-65.61 mm) wide spacer is used in place of second gear.

Install third gear (18) over end of shaft with longer protruding hub facing rearward. Install fourth gear (19) over shaft with longer protruding hub facing forward. Place front bearing cone (11) in place in front bearing cup with cone back facing rearward. Install countershaft driven gear (20) in place with protruding hub side rearward. Insert countershaft through driven gear and let it rest in bearing bore with end of shaft just flush with front of housing. Install a strap under countershaft so that you will be able to lift it later. Countershaft is to be left in this stage of assembly until after installation of reverse idler, main shaft and gears, and low range input shaft.

Place reverse idler shaft (22) into rear housing bore with lock tang notch upward. Place reverse idler gear (23) over shaft with longest protruding hub side forward, or with rounded tooth side rearward. Place three thrust washers 0.123-0.126 inch (3.12-3.20 mm) thick over end of idler shaft and push shaft into place in housing. Install lock clip, lockwasher and cap screw. Torque cap screw to 35-45 ft.-lbs. (47.4-60.9 N·m). Use Loctite 277 on threads of cap screw.

Now install main shaft as follows: It is suggested that gears, gear bushings, splined collars and couplings be stacked on main shaft as shown in Fig. 96. Then select a snap ring (1–Fig. 94) that will allow 0.006-0.012 inch (0.15-0.30 mm) end play for transmissions without main shaft bushings. For transmissions with main shaft bushings the thickest snap ring (1) that will seat in shaft groove should be selected. Set aside snap ring (1) for later installation. Install rear bearing cone (32) on rear of main shaft with cone back facing rear of shaft (taper towards front of shaft), then install bearing cone (33) with cone back against cone back of bearing cone (32)–bearing cone tapers should point away from each other. Install snap ring (36), press bearing cones against snap ring. Install snap ring (31) in transmission housing then install and bottom bearing cup (32) against snap ring.

On transmissions with main shaft gear bushings place 0.180-0.182 inch (4.57-4.62 mm) wide thrust washer (30) against front bearing cone (32).

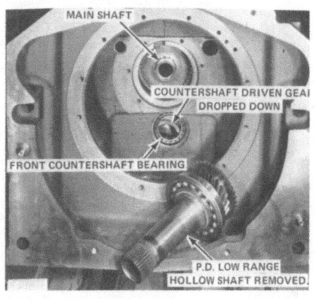

Fig. 93 — Front of transmission with power-director low range shaft removed and countershaft driven gear dropped down.

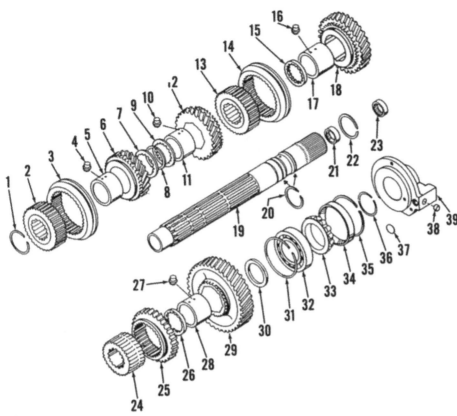

Fig. 94 — Exploded view of standard transmission main shaft assembly. Splined washer (7) and snap ring (8) are not used on models equipped with gear bushings.

1. Snap ring	11. Bushing	21. Bearing	30. Thrust washer
2. Splined collar	12. Gear (3rd)	22. Snap ring	31. Snap ring
3. Shift coupling	13. Splined collar	23. Seal	32. Taper roller bearing
4. Pin	14. Shift coupling	24. Splined collar	33. Bearing cone
5. Bushing	15. Splined thrust washer	25. Sliding gear	34. Bearing cup
6. Gear (4th)	16. Pin	(1st & reverse)	35. Snap ring
7. Splined thrust washer	17. Bushing	26. Splined thrust washer	36. Snap ring
8. Snap ring	18. Gear (2nd)	27. Pin	37. Seal
9. Splined thrust washer	19. Mainshaft	28. Bushing	38. Plug
10. Pin	20. Sealing rings	29. Gear (1st)	39. Housing

On all models, install main shaft through rear bearing bore of housing.

On transmissions with main shaft gear bushings check width of gear bushings with width of gear hubs. Bushings should be 0.006-0.010 inch (0.15-0.25 mm) wider than gear hubs. During installation use grease (Lubriplate) to retain lock pins in bushings. Install first gear bushing (28) on main shaft with lock pins (27) forward.

Use grease (Lubriplate) on all gear bushings and gear hubs. Place first gear (29) on main shaft with shifter teeth forward and slide to rear of housing.

On transmission without main shaft gear bushings install first and reverse gear splined thrust washer (26), 0.3005-0.3015 inch (7.632-7.658 mm) wide, on shaft. Install first and reverse splined collar (24), 1.850-1.852 inch (46.99-47.04 mm) wide, onto shaft with end void of splines forward.

On transmissions with main shaft gear bushings install first and reverse splined collar (24), 2.158-2.160 inch (54.81-54.86 mm) wide, onto shaft with chamfered end of spline lock groove rearward.

NOTE: To prevent jumping out of reverse, 7030 tractors after serial number 3650 and 7050 tractors after serial number 2585 and all 7040, 7060 and 7080 tractors, incorporate a new design splined collar. This new design splined collar may be installed in older tractors if jumping out of reverse is a problem. This collar should be installed with end void of splines forward.

Install first and reverse sliding gear (25 – Fig. 94) over splined collar with shifter fork groove forward.

On transmission with main shaft gear bushings install second gear bushings (17) with lock pins (16) in place over shaft with lock pins forward.

Install second gear (18) with shifter teeth forward.

NOTE: On all model tractors except 7020 a spacer is used in place of second speed gear and bushing.

On transmissions without main shaft gear bushings install splined thrust washer (15), 0.3005-0.3015 inch (7.632-7.658 mm) wide, next to second gear.

Install second and third gear splined collar (13) over shaft with short splines forward.

NOTE: If jumping out of second gear is a problem, a new design splined collar is available as a service replacement part. This new design splined collar was installed in production effective with serial number 2048 on Model 7040, serial number 1963 on Model 7060 and serial number 1139 on Model 7080.

Install second and third shift coupling (14) with chamfered end forward. On models so equipped, install third gear bushing (11) and lock pins (10) on shaft with lock pins rearward. Install third gear (12) with shifter teeth rearward.

NOTE: On Model 7020 16-speed tractor a 1.879-1.884 inch (47.73-47.85 mm) wide spacer is used in place of third gear (12), thrust washers (7 and 9) and snap ring (8).

Install thrust washer (9), 0.100-0.102 inch (2.54-2.59 mm) wide, on shaft. On transmissions without main shaft gear bushings also install snap ring (8) and thrust washer (7), 0.100-0.102 inch (2.54-2.59 mm) wide. On models so equipped, install fourth gear bushing (5) lock pins (4) on shaft with lock pins forward. Install fourth gear (6) with shifter teeth forward.

NOTE: On Model 7030 after serial number 1101 using main shaft gear bushings and later Model 7050 using main shaft gear bushings, components (5, 6, 9, 11 and 12 – Fig. 94) were changed and

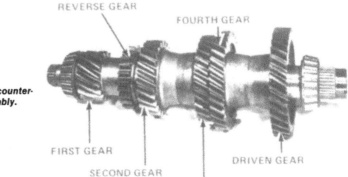

Fig. 95 — View of counter-shaft after assembly.

must be installed as a set in early Model 7030 and 7050 transmissions to maintain correct gear location. Early components are not available; a complete set of components (5, 6, 9, 11 and 12) must be installed if any one component is faulty on early models.

Install fourth and fifth splined collar (2) on front of shaft. Install shift coupling (3) on splined collar with chamfered side forward.

NOTE: On all model tractors with main shaft gear bushings, except 7040, 7060 and 7080, splined collars and shift couplings are alike. On Model 7040, 7060 and 7080 tractors the fourth and fifth gear splined coupling and shift collar allow for deeper engagement of fifth gear and are

identified by a machined groove around chamfered side of shift collar.

NOTE: In late production Model 7040, 7060 and 7080 tractors with mainshaft gear bushings, second, third and fourth gears have been changed. Shifter teeth on second and third gears of later models are closer to shifter coupling for deeper engagement. Fourth gear has had shifter teeth moved and number of teeth changed from 23 teeth to 25 teeth for better field working speeds. The new second and third gears are interchangeable with older model tractors. To install the new design fourth gear in an older model tractor the countershaft gear and shifter coupling must also be changed.

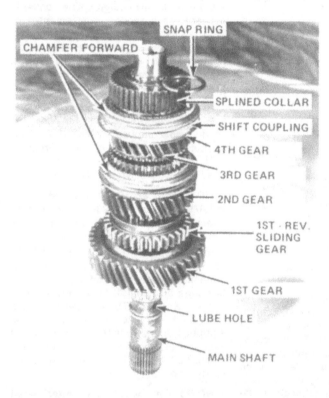

Fig. 96—Assemble main-shaft on bench to properly choose correct select-fit snap ring.

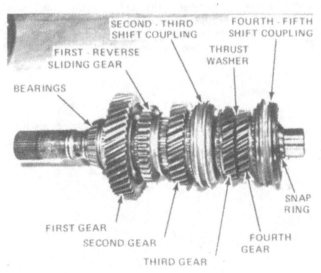

Fig. 97—View showing arrangement of gears and components on main shaft.

Temporarily support front end of main shaft by installing Power-Director low range hollow shaft. This will aid in obtaining correct bearing adjustment, and aid with installing correct thickness snap ring (1—Fig. 94) that was preselected during bench assembly of main shaft and gears. Push shaft forward until front bearing cone is seated in bearing cup. Install rear bearing cup (34) with cup back face rearward. Select a snap ring (35) that will allow 0.001-0.006 inch (0.03-0.15 mm) shaft end play. Snap ring (35) is available in thicknesses of 0.103 to 0.124 inch in steps of 0.003 inch. Remove Power-Director low range hollow shaft from bore of housing. Make sure assembled parts of main shaft are all tight and fit properly. Install snap ring (1) that was preselected during bench assembly of main shaft. Snap ring (1) is available in thicknesses from 0.078 to 0.126 inch. After installation of snap ring (1) check side clearance of all main shaft gears. For transmission without main shaft gear bushings, gear side clearance should be 0.004 inch (0.10 mm) or greater. For transmissions with mainshaft gear bushings, gear side clearance should be 0.006-0.010 inch (0.15-0.25 mm).

Before installing Power-Director low range hollow shaft (11—Fig. 88) inspect bearings (10, 12 and 13) for wear or damage and renew as necessary. Press against numbered side of bearing (12) during installation. Position bearing (12) so it is 0.090-0.120 inch (2.29-3.05 mm) below front edge of shaft. If a new roller bearing (10) is to be installed first remove pressed in sleeve (8), bearing retaining snap ring (9) then bearing (10). Install new bearing and snap ring. Press sleeve (8) into rear of shaft until it is 0.125-0.155 inch (3.18-3.94 mm) past flush with face of gear. Press bearing (13) onto shaft and position bearing so outer snap ring is away from gear. Install bearing retaining snap ring (14) in groove of shaft (11). Push shaft and bearing assembly into bore of housing until bearing outer snap ring is seated against housing.

Raise countershaft with previously installed strap and reinstall rear bearing cup (2—Fig. 91) part way into bore. Install front bearing cone (11) with back face rearward. Install front bearing cup (12) into bore and install snap rings (13). Press countershaft forward until front bearing cup is seated against snap rings. Complete installation of rear bearing cup (2) and install snap ring (3). Check countershaft end play, which should be 0.001-0.006 inch (0.02-0.15 mm). Countershaft end play is controlled by snap ring (3) which is available in thicknesses of 0.030-0.146 inch in steps of 0.003 inch. Rear countershaft adjusting

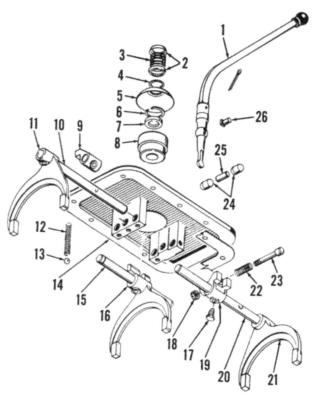

Fig. 98 — Exploded view of transmission shift cover assembly.

1. Shift lever
2. Washers
3. Spring
4. Sealing ring
5. Dust cover
6. Snap ring
7. Pivot washer
8. Cover insert
9. Shift lug (4th & 5th)
10. Shift rail (4th & 5th)
11. Fork (4th & 5th)
12. Detent spring
13. Detent ball
14. Cover
15. Shift rail (2nd & 3rd)
16. Fork (2nd & 3rd)
17. Lock screw
18. Nut
19. Shift lug (1st & reverse)
20. Shift rail (1st & reverse)
21. Fork (1st & reverse)
22. Plunger spring
23. Plunger latch
24. Interlock plunger
25. Interlock pin
26. Pivot pin (2 used)

snap ring (3) must be taken in and out with care to avoid damaging brake piston seal surface.

Install new seal rings (4 and 6) in grooves of piston (5). Lubricate seal rings and housing seal surface and install piston in counterbore of housing. Install friction disc (7) over splines of countershaft. Install wave spring (8), using a suitable lubricant to hold into place if necessary. Install brake plate (9) and torque cap screws to 28-33 ft.-lbs. (37.9-44.7 N·m).

Install Power-Director high range hollow shaft as outlined in paragraph 89 and Power-Director clutch as outlined in paragraph 87. Install transmission into tractor in reverse order of removal. When transmission is installed, install shift cover, using a new gasket, and with all forks in neutral position. Install cover straight down making sure that shift forks fit properly into grooves of shift collars. Center cover using two dowel cap screws. Install remaining cap screws.

SHIFTER ASSEMBLY

91. **R&R AND OVERHAUL.** To remove shifter assembly, remove platform sections necessary for access to top of transmission. Shift transmission to neutral and remove cap screws securing cover and lift cover from transmission.

Front and rear cap screws at center of cover are special dowel bolts.

To remove gear shift lever (1 – Fig. 98), remove pin from shift lever. Then, remove washers and spring (2 and 3), sealing ring (4) and dust cover (5). Remove snap ring (6), pivot washer (7) and lift shift lever and two pivot pins from cover. Cover insert can be pressed or driven from cover if renewal is necessary. To reinstall lever, reverse removal procedure.

92. **SHIFT RAILS.** Mount shift cover in a vise with first and reverse shift rail (20 – Fig. 98) upward. Remove lock wires from all three shift lugs and remove lock screws (17). Rotate first and reverse shift fork (21) upward to release detent ball from notches and slide shift rail rearward slowly. Use caution by placing hand over holes to catch detent ball when end of rod releases it. Remove shift rail with shift fork from cover, and remove shift lug from shaft. Remove detent spring. Remove cover from vise and dump out one interlock plunger (24) and interlock pin (25). Place cover back into vise in same manner, and remove center (second and third) shift rail (15). Catch detent ball in same manner by placing hand over holes in cover. Slide shift rail rearward and remove second and third shift fork and shift rail from cover. Remove detent

spring. Remove cover from vise and dump out second interlock plunger (24). Place cover back in vise in same manner as before. Remove fourth and fifth shift rail (10) by sliding it forward. Detent ball and spring for fourth and fifth shift rail is located at bottom of interlock drilled passage. Remove shift forks from shift rail if renewal is necessary.

NOTE: Transmissions with 16 speeds have the second gear omitted, and use a snap ring on front end of second and third shift rail (15) to prevent shifting into second gear position.

Shift rails must be removed and installed as follows:
   A. First & Reverse shift rail
   B. Second & Third shift rail
   C. Fourth & Fifth shift rail
When installing start with step C and proceed to steps B and A in that order.

When reassembling proceed as follows: Place shift cover in vise with interlock plunger passage upward. Drop a detent spring and ball to bottom of this passage. Fourth and fifth shift fork (11 – Fig. 98) and first and reverse shift fork (21) can be attached to shift rails. Enter fourth and fifth shift rail (10) at front cover and place shift lug over rail. Depress detent ball against spring and slide rail over ball. Place in neutral position. Drop an interlock plunger down passage on top of rail. Install second detent spring and ball at front of cover. Enter center shift rail (15) through front of cover, depress detent ball against spring and slide rail over ball. Install second and third shift fork (16) over rail. Move rail rearward until interlock pin hole is near rear support. Insert interlock pin through shift rail and push rail rearward to neutral position. Drop in second interlock plunger next to center second and third shift rail. Install third detent spring and ball. Depress ball against spring and enter shift rail through rear of cover and slide over ball. Move shift rail rearward and install first and reverse shift lug (19).

NOTE: If latch plunger (23) has been removed from shift lug for any reason, it must be adjusted by nut until end of plunger is flush with face of shift lugs.

Move shift rail forward to neutral detent position. Install and tighten lock screws in shift lugs and the (center) second and third shift fork. Install lock wires to prevent screws from becoming loose.

To install, place shift lever and transmission gears in neutral position. Install cover gasket and cover assembly with two special dowel cap screws located in center front and center rear holes to align cover. Install and tighten remaining cap screws.

TRANSMISSION (POWER SHIFT)

Some models are optionally available with a Power Shift transmission which has three planetary gear sets and a single countershaft. Shifting is accomplished by six hydraulically actuated wet clutches used in combination to provide six forward speeds and one reverse. Wet clutches are controlled by a cam in the Power Shift valve, located on right side of transmission, which is connected by mechanical linkage to operator console shift lever.

In each speed, two of the six clutches are engaged while the other four are disengaged. Power Shift valve has a rod linkage to inching valve; this provides a mechanical connection to operator inching clutch pedal allowing operator to start up and shift gradually as with a standard transmission. It is not necessary to use inching valve to start or shift the Power Shift, except for range transmission shifting. When clutch pedal is depressed all the way, transmission brake is applied.

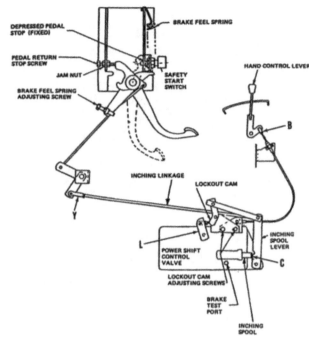

Fig. 99 — View of inching pedal linkage and adjustment points.

ADJUSTMENTS

93. **SHIFT LEVER.** MODELS 7010, 7020 AND 7045. Disconnect Power Shift cable at shift lever (L–Fig. 99) and set shift lever in fifth speed position (second detent from rear). Set hand control in fifth speed position. Adjust cable end at lever (L) until hand control lever is visually centered in gate on console or by determining equal distances when lever is moved forward and rearward until detent forces are felt. Tighten jam nut.

94. ALL OTHER MODELS. Move Power Shift lever (L–Fig. 99) on valve to first speed which is third detent position from front. Then, set hand control lever in first speed which is third position from rear on quadrant. Now adjust upper ball joint (B) until hand control lever is centered in quadrant and moves an equal distance in both directions.

95. **INCHING (CLUTCH) PEDAL.** MODELS 7010, 7020 AND 7045. Refer to Fig. 99 and depress inching pedal completely. Adjust yoke (Y) until a clearance of 0.001-0.020 inch (0.03-0.51 mm) is obtained between inching spool lever and inching spool at point (C). Release inching pedal and adjust pedal return stop screw until a clearance of 0.10-0.30 inch (2.5-7.6 mm) is obtained between inching spool lever and inching spool at point (C) with inching spool held

in against its internal stop. Pedal return stop screw may be reached by removing engine side cowling cover and insulation blocks, then reaching in from top.

96. ALL OTHER MODELS. Refer to Fig. 99 and turn return stop completely in, then hold inching clutch pedal in depressed position against fixed stop. Now adjust lower inching rod yoke (Y) until inching spool lever clears end of inching spool by 0.001 to 0.020 inch (0.02-0.50 mm). Release pedal and tighten jam nut on lower inching rod. Adjust pedal return stop screw until it just contacts pedal and tighten jam nut.

97. **FIFTH AND SIXTH SPEED LOCKOUT.** Before making this adjustment, adjust shift lever as outlined in paragraph 93 or 94.

With transmission operating in fifth or sixth speed, it should automatically shift to fourth speed when inching pedal is depressed. To adjust lockout linkage refer to Fig. 99 and place hand control lever in sixth speed position. Adjust lockout cam by loosening adjusting screws and sliding mounting bracket in slotted holes until lockout cam just contacts shift lever (L). Tighten mounting screws and check adjustment by depressing inching pedal. Hand control

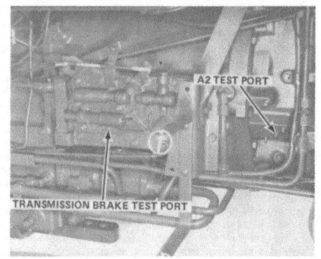

Fig. 100 — View of test port for transmission brake.

lever must move to the fourth speed position. Be careful not to move lockout cam too far forward or tractor will not stop when operating in fourth speed. Recheck shift cable adjustment.

98. BREAK STOP FEEL SPRING. ALL MODELS EXCEPT 7010, 7020 AND 7045. Before making this adjustment, shift lever and inching pedal linkage adjustment must first be completed as previously outlined.

Refer to Fig. 100 and attach 0-300 psi (0-2000 kPa) gages in A2 port and transmission brake test port. Then, place range transmission shift lever in park, start engine and place power shift lever in 3rd speed position. Observe pressure on gages as inching pedal is moved. There should be no reading on either gage when inching pedal is just starting to contact brake stop feel spring (Fig. 99). If there is pressure, loosen the two cap screws holding brake stop feel spring and adjust plate up or down until there is no pressure reading on either gage when pedal is just contacting spring arm. If A2 pressure is obtained when contact is made, move plate down until there is no reading.

99. MODELS 7010, 7020 AND 7045. Attach two 0-400 psi (0-2700 kPa) pressure gages to transmission break pressure test ports shown in Fig. 103. Place range transmission shift control lever in "PARK" position, start tractor and place power shift transmission hand control lever in third speed position. Monitor pressure on gages as inching pedal is moved. There should be no pressure reading on either gage until inching pedal is just starting to contact break feel spring arm. If adjustment is necessary, turn break feel spring adjusting screw in or out until no pressure reading is obtained on either gage when

inching pedal is just touching break feel spring arm.

100. SAFETY START SWITCH. Refer to Fig. 99 and adjust safety start switch until switch button moves 0.050-0.180 inch (1.27-4.57 mm) before inching pedal contacts switch and begins to move switch mounting arm.

PRESSURE TESTS

There are six test ports for checking clutches on right side of transmission as shown in Figs. 101 and 102. C1, C2 and "A" regulating valve pressure test ports are located behind auxiliary fuel tank. If tractor is so equipped, auxiliary fuel tank will have to be removed. There are also test ports in control valve cover (Fig. 103) for testing transmission brake, inching pressure and modulating pressure. Pressure testing can be accomplished with a single 0-600 psi (0-4000 kPa) pressure gage, however, a bank of six gages with long enough connecting hoses so that gages could be read from operators seat while testing tractor would be preferable.

101. REGULATOR VALVE PRESSURE. Refer to Fig. 102 and install test gage in port "A". Pressure at regulator is the oil used for the six clutches. Pressure should be 210-230 psi (1440-1580 kPa) with oil at working temperature, transmission in neutral and engine running at rated speed. If pressure is not as stated add or remove shims between spring and spool end (Fig. 107) as required. Each 0.030 inch thick shim will increase pressure approximately 10 psi (69 kPa).

102. CLUTCH PRESSURE. Power shift clutches are tested at ports shown in Figs. 101 and 102. Tests can be made with one 0-600 psi (0-4000 kPa) gage, but six gages with pressure hoses save a lot of changing if all clutches are to be checked. Test ports are 1/8-inch or ¼-inch pipe thread. Following data will tell which clutches are applied in each gear range.

Gear	Clutches Applied
Reverse	A2 & C2
Neutral	None
1st	A2 & B1
2nd	C1 & B1
3rd	C1 & A2
4th	C1 & B2
5th	C1 & A1
6th	A2 & A1

The clutches that are applied should test 190-210 psi (1310-1440 kPa). If a clutch or clutches don't check out be sure and make all other tests outlined in paragraphs 101, 103, 104 and 105. Then, if the other tests are OK there is probably a leaking or broken seal in the clutch or clutches that don't check out.

103. MODULATOR VALVE. The modulator valve provides a smooth shift from one gear range to another and purpose of this test is to evaluate the timed pressure changes of the off going and the on coming clutches. To make this test, it is best to use three 0-600 psi (0-4000 kPa) gages. Test will be explained using the 1st and 2nd transition; however, any gear transition can be used by checking the following table to determine which clutch is going off, remaining on and coming on.

Gear	Going off	Remaining on	Coming on
1st to 2nd	A2	B1	C1
2nd to 3rd	B1	C1	A2
3rd to 4th	A2	C1	B2
4th to 5th	B2	C1	A1
5th to 6th	C1	A1	A2

Procedure for making this test is as follows: Shift tractor into park position,

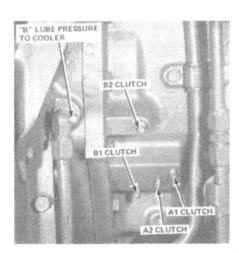

Fig. 101—View showing some of the test ports for power shift clutches. Refer to Fig. 102 for other test ports.

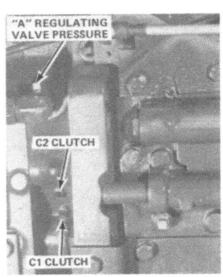

Fig. 102—View showing some of the test ports for power shift clutches. Refer to Fig. 101 for other test ports.

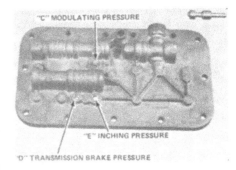

Fig. 103—Power shift cover plate showing test ports. Refer to text.

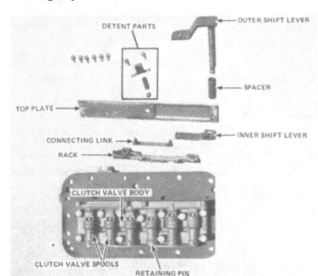

DETENT PARTS — OUTER SHIFT LEVER

SPACER

TOP PLATE —

CONNECTING LINK —

RACK —

INNER SHIFT LEVER

CLUTCH VALVE BODY

CLUTCH VALVE SPOOLS

RETAINING PIN

Fig. 104—Power shift detent, shift lever and components.

the tractor until oil temperature is 150°-180°F (66°-82°C). Then, connect a flow tester to port ("B"–Fig. 101). With engine running at rated speed, relief valve should open at 160-180 psi (1100-1200 kPa). A 0.030 inch shim (Fig. 107) will increase relief valve pressure approximately 9 psi (62 kPa).

CAUTION: Do not operate transmission more than one minute without lube oil flow.

POWER SHIFT VALVE

106. **R&R AND OVERHAUL.** Drain oil from transmission until oil level is below valve cover. Swing out batteries and secure in open position. Remove battery retainer and platform brace assembly. If tractor is equipped with auxiliary fuel tank, unbolt and lay tank under tractor. Disconnect shift cable and remove entire inching clutch linkage from valve cover plate. Remove remaining cap screws from valve cover and remove valve assembly. Place valve assembly on a clean work bench for further disassembly.

To disassemble refer to Fig. 104 and place rack in neutral position (2nd detent from front of rack) and remove inner shift lever and connecting link. Then, remove the ten cap screws securing clutch valve body to cover plate. Being careful not to damage mating surfaces between valves, plates and gaskets, remove top plate and detent mechanism from clutch valve body and remove rack and connecting link. Clutch valve spools may be removed by pulling retaining pins at bottom of valves and removing retaining washer, spring, spool and actuator pin from each bore (Fig. 105). Remove outer shift lever (Fig. 104) and spacer. Remove modulator spool stop clip from 3rd annulus of modulator spool bore (Fig. 106).

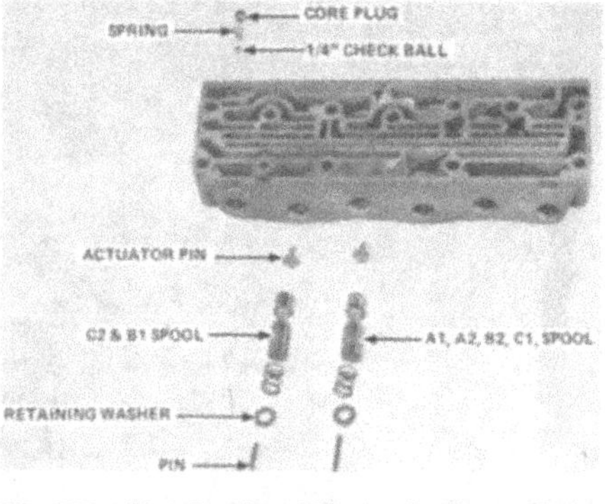

install gages in clutch ports (A2, B1 and C1–Figs. 101 and 102). Start tractor, shift into 1st gear and observe gages A2 and B1. Pressure should be 190-210 psi (1310-1440 kPa). Now shift to 2nd gear and observe gages; pressure at A2 should drop to zero in approximately one second. Pressure at B1 should stay on but pressure will drop some and then build back up to 190-210 psi (1310-1440 kPa) after approximately one second. C1 pressure is coming on and should go from zero to 190-210 psi (1310-1440 kPa) during the first second. If readings are not as stated, trouble is probably in modulator valve; refer to paragraphs 106 through 113 for overhaul of valve.

104. **INCHING VALVE.** This test is performed by installing two pressure gages in ports (B1 and A2–Fig. 101). Inching will be checked on A2 gage with transmission shifted into reverse. Braking will be checked on gage B1 with clutch pedal fully depressed. Engine speed should be set to approximately 1000 rpm. Proceed as follows: As clutch pedal is slowly depressed, pressure on gage A2 will remain at 190-210 psi (1310-1440 kPa) during approximately the first 20 percent of pedal travel.

After 20 percent travel, as pedal is further depressed, pressure should drop rapidly to 80-100 psi (551-689 kPa) and then slowly decrease to zero when pedal is depressed 70 percent of its travel. Between 70 and 80 percent, both gages should be at zero. At approximately 80 percent of pedal travel, additional pedal effort should be required because of brake feel spring and at this time, pressure gage connected to port B1 should rapidly rise to 30 psi (206 kPa) and remain at this pressure during the remaining travel.

NOTE: Between approximately 70 and 80 percent of pedal travel there must be a point where both A2 and B1 gages are at zero. If not and there was always some pressure, a dragging condition would exist.

If there is a problem, refer to overhaul of inching valve, paragraphs 106 through 113.

105. **LUBE PRESSURE.** Relief valve setting may be checked by running

SPRING — CORE PLUG
1/4" CHECK BALL

ACTUATOR PIN

C2 & B1 SPOOL
A1, A2, B2, C1, SPOOL

RETAINING WASHER

PIN

Fig. 105 — Clutch valve shift spools shown. Refer to text.

Fig. 106 — Back side of cover plate showing location of modulator spool stop.

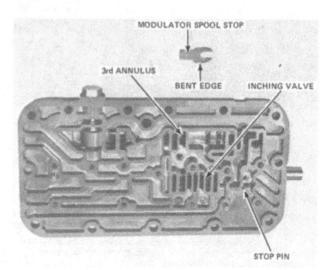

MODULATOR SPOOL STOP

3rd ANNULUS

BENT EDGE   INCHING VALVE

STOP PIN

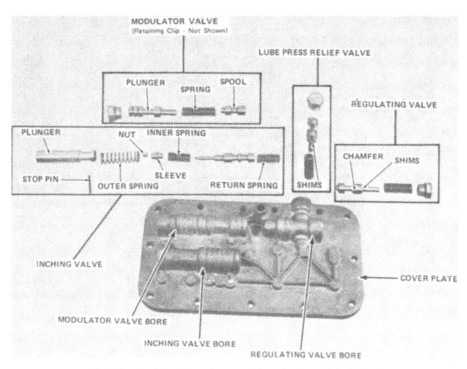

Fig. 107 — Exploded view of cover plate and valves. Refer to text.

Inching, modulator, pressure regulating and lube pressure relief valves can now be removed by removing plugs.

Clean all parts and passages and blow dry with compressed air. Inspect all parts for scoring and damaged seals and renew as necessary.

When reassembling, refer to Fig. 107 for correct installation of valves and proceed as follows:

107. INCHING VALVE. Install inner spring which has a free length of 1.32 inches (33.5 mm) on stem of valve spool, followed by sleeve and nut. Tighten nut until it bottoms on shoulder of spool. Install return spring which has a free length of 0.776 inch (19.71 mm) in front of spool, then install spring and spool into inching valve bore of cover plate.

NOTE: Bore is larger at rear, so make sure spring and spool are installed in smallest bore.

Install outer spring which has a free length of 2.30 inches (58.4 mm) in bore, then install inching plunger. Hold plunger in place and install stop pin in small hole in back of front plate. Drive pin in until it hits against front of cover plate. Do not strike gasket sealing surface of cover plate. Check inching plunger for freedom of movement. Press new seal into place around plunger with lip toward front. Press seal in until rear of seal is flush with front of chamfer.

108. MODULATING VALVE. Install modulator spool, spring and plunger

together and slide in modulator valve bore of cover plate with plunger to rear. Modulator valve spool and plunger must move freely in valve bore. Install "O" ring plug at rear. Install spool stop around spool in 3rd annulus from front with bent edges of stop pointing toward rear (Fig. 106).

109. PRESSURE REGULATOR VALVE. Assemble shims and spring to pressure regulator valve spool and slide in bore (Fig. 107). Spring has a free length of 1.67 inches (42.4 mm) and spool has two lands with a chamfer on rear of front land and a length of 2¾ inches (60.85 mm). Be sure spool slides freely in bore.

110. LUBE PRESSURE VALVE. Assemble shims and spring to valve spool and insert in vertical bore of valve cover. Spring has a free length of 1.54 inches (39.1 mm) and spool is similar to pressure regulator spool, but shorter and has no chamfer on spool land.

111. CLUTCH VALVE. If modulator valve check ball has been removed, reinstall ¼-inch ball and small spring in hole in top of valve body next to bore for C2 clutch valve. Install new core plug to hold check ball and spring in place. Turn clutch valve body assembly over and install clutch valve actuator pins, spools and springs. All springs have an OD of 0.52 inch (13.2 mm) and a free length of 0.68 inch (17.2 mm).

NOTE: Spools for C2 and B1 clutches are different than the other four. They have a longer top land. See Fig. 105.

Make sure all spools slide freely in their respective bores, then install retaining washers and pins. Pins must not project past face of housing. Install pin connecting link to rack with link toward inside and cotter pin to outside (Fig. 108). Lubricate sliding surfaces of rack and install in top of clutch valve body with lever for attaching control link to front of clutch valve side of body. Attach cover plate to top of clutch valve body with raised end toward rear. Place rack in neutral position (2nd detent notch from front) and install detent ball, spring and cover on top of cover plate. Torque cover plate cap screws evenly to 20 in.-lbs. (2.2 N·m).

112. CLUTCH VALVE TO COVER PLATE. Make sure all passages and surfaces of cover plate, clutch valve and separator plate are clean and dry. Assemble new gasket, separator plate, and another new gasket to dowel pins on

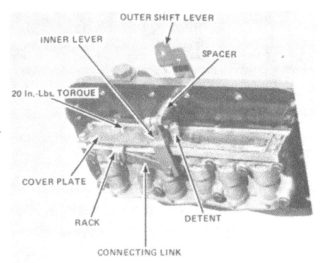

Fig. 108 — Shift linkage for power shift valve.

49

Fig. 109—Sequence for torquing valve body to cover plate. Torque to 11 ft.-lbs. (14.9 N·m).

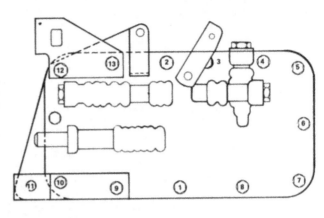

Fig. 110—Sequence for torquing cover plate to transmission. Torque to 30 ft.-lbs. (40.6 N·m).

cover plate, making sure all gasket holes align with holes in separator plate. Separator plate should project beyond cover plate at lower rear corner just below inching plunger. Turn clutch valve assembly so rack is on top and attach to inside of cover plate assembly using dowel pins in cover assembly for alignment. Secure assemblies together with 10 cap screws torqued evenly to 11 ft.-lbs. (14.9 N·m), using the sequence shown in Fig. 109. Place new "O" ring seal on outer shift lever shaft (Fig. 108) and insert shaft through hole in boss on cover plate. Attach spacer and inner lever to other end of shaft. Attach connecting link to inner lever and rack. Operate shift linkage to check assembly.

113. CLUTCH VALVE TO TRANSMISSION. Attach valve assembly loosely to transmission housing making sure all seals are in place and in good condition. Reattach inching and shift linkage to rear upper and lower bolts. Tighten attaching bolts to 30 ft.-lbs. (40.6 N·m) torque using sequence shown in Fig. 110.

Readjust inching and shift linkage as previously outlined. Check regulator valve pressure as outlined in paragraph 101 and lube pressure as outlined in paragraph 105. Adjust as necessary.

POWER SHIFT UNIT

114. **REMOVE AND DISASSEMBLE.** To remove power shift transmission, first split engine from power shift housing as follows: Remove front end weights from tractor to prevent front end assembly from tipping forward when disconnected from transmission housing. Remove right and left hood assemblies. Remove panel at front of console and remove top hood assembly. Remove protective pan from below hydraulic pump assemblies. Disconnect battery cables from battery to prevent short circuits and remove left platform panel, step and brace. Remove vent line and fuel return lines from top of fuel tank at left side of tractor. Shut off fuel

from main fuel tank (large tank). Remove cross-over fuel line between main tank (left side) and auxiliary tank (right side) if so equipped. Plug auxiliary fuel line with plastic cap if tractor is so equipped. Loosen main fuel tank support straps at trunnion on inner side of tank. Support tank on floor jack. Disconnect battery wire from fuel level sending unit and disconnect ground wire from bolt on platform support. Remove fuel tank support straps, lower fuel tank to floor and move from under tractor. Drain oil from power shift transmission and differential. Disconnect wiring harness to engine at right side of console by releasing catches and pulling connector apart. Disconnect starting motor wires (orange, green and brown) from front of console. Disconnect fuel inlet and return lines at right rear of engine. Remove clamps attaching cooler oil lines to front right side of transmission housing. Remove two clamps attaching cooler lines at lower right side of engine. Remove clamp from power steering hoses at rear right side of engine. Disconnect oil cooler lines at right side of transmission housing. Plug oil lines with plastic caps. Disconnect power steering hoses from power steering cylinder and plug hoses and hose connections on steering cylinder. Remove oil line from rear of transmission oil filter to entrance at rear left side of transmission housing. Place floor jack under power shift at front of transmission and place a metal stand under front of transmission housing. Use floor jack to assist in separating tractor, then block rear section securely and safely block wheels and under drawbar and transmission housing. Place wedge shaped wood block between front axle and side frames at both sides to prevent tipping. Attach a hoist to engine lift eyes and take up slack in hoist. Remove four bolts attaching side frames to front of transmission housing and install guide studs in the two bottom holes. Remove four bolts attaching engine mounting plate to transmission housing, roll engine and front end of tractor forward

far enough to provide work clearance and block securely and safely. Then, disconnect external lines and any linkage that interferes with removal of transmission housing. Unbolt hydraulic pumps and valve body on right side of transmission housing. With differential blocked securely, unbolt and remove transmission housing. The transmission housing is in two sections front and rear. Stand housing up on front housing. Remove bearing retaining nut from rear of countershaft. Remove the six cap screws from bolt circle at rear of transmission housing. Remove cap screws attaching front and rear housing together. Attach hoist to rear housing and carefully lift rear housing up until it is clear of transmission assembly. Place rear housing clear of work area. Remove snap rings from around output shaft and remove large stake nut from rear of countershaft. Pull gear and bearing cone from rear of countershaft. Use two 3/8-16x10 inches long bolts as a lifting tool for attaching hoist to rear clutch

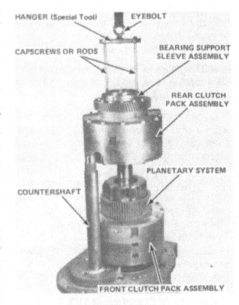

Fig. 111—View showing removal of rear clutch pack assembly.

pack. Attach these cap screws or rods to a length of strap steel properly drilled with an eyebolt at center as shown in Fig. 111. Thread these cap screws or rods into bearing support sleeve assembly. Attach hoist to eyebolt and lift rear clutch pack assembly (C1 and C2 housing) from planetary system. Place clutch pack on bench for disassembly. Remove the twelve tie bolts from front clutch pack. Use ACTP 3055 planetary hanger or devise a lifting tool similar to the one shown in Fig. 112 that will thread onto output shaft for attaching a hoist as shown. Lift planetary system up slowly until it is clear of front clutch pack. The B2 clutch hub will catch on A2 clutch friction plates and will lift them from the A2 clutch housing. Drop these plates off by hand and allow them to remain with A2 clutch housing. Place planetary assembly on a support stand to disassemble. Place front transmission housing in a horizontal position. Remove seal and snap ring from groove of input shaft located at front of ball bearing (Fig. 114) in hub of A1 clutch housing. Pull input shaft (Fig. 113) from A1

clutch housing, being careful not to lose the ball bearing retaining pin. Remove snap rings retaining A1 clutch hub and front sun gear, and remove from input shaft. Remove bearing adjusting nut and lockwasher from front of A1 clutch hub (Fig. 114). USE SPECIAL TOOL ACTP 3052 as a driving plug to fit front surface of clutch hub and pilot in bearing bore of clutch hub. If ACTP 3052 is not available, a plug can be made with dimensions as follows: outside diameter overall 3.031 inches (76.99 mm), step section diameter 2.441 inches (61.60 mm), step section length 0.114 inch (2.89 mm), overall thickness of plug 0.750 inch (19.05 mm). Place driving plug into front end of A1 clutch hub and press it rearward until front bearing cone is free of clutch housing. Clutch housing and countershaft must be removed together with gear teeth in mesh. Turn front housing back up on end and compress the B1 clutch stack as shown in Fig. 115. Remove snap ring retaining screw, and remove snap ring from groove of housing (Fig. 116). Remove the "C" clamps and bar used to compress clutch stack. Remove the B1 clutch end plate, wave springs, friction plates, separator plates and clutch piston from front housing (Fig. 117). Remove piston seals from clutch piston. Remove bearing cups from clutch housing if bearing renewal is necessary. Inspect and clean all lube oil and clutch oil passages (Fig. 117).

115. **OVERHAUL COMPONENTS.** Disassemble the clutches starting with

rear clutches C1 and C2 (first ones that were removed). Proceed as outlined in following paragraphs.

116. REAR CLUTCHES C1 AND C2. Place clutch pack on bench with support sleeve flange downward and end plate upward. Remove the twelve bolts

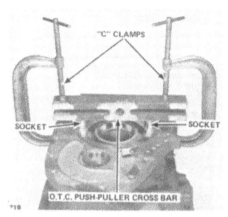

Fig. 115—Compressing B1 clutch stack. Refer to text.

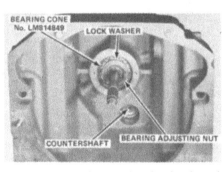

Fig. 114—View of clutch housing bearing adjusting nut.

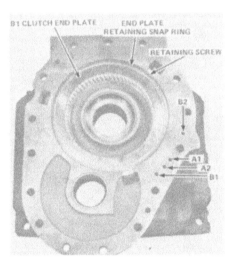

Fig. 116—View showing end plate snap ring and retaining screw.

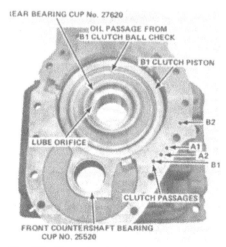

Fig. 117—View showing B1 clutch piston with clutch stack removed.

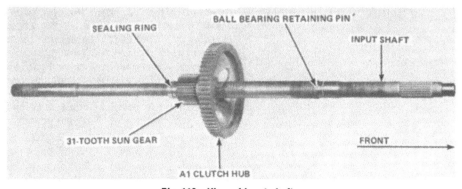

Fig. 112—Planetary unit, output shaft and lift tool.

Fig. 113—View of input shaft.

from clutch pack. Remove end plate from C1 clutch housing (Fig. 121). Remove wave springs, friction plates and separator plates from C1 clutch housing. Remove C1 clutch housing from C2 clutch housing. Remove clutch piston from C1 clutch housing. Remove wave springs, friction plates and separator plates from C2 clutch housing. Remove clutch piston from C2 clutch housing. This clutch piston is different from the other five clutch pistons, which are all alike. Remove rear ring gear and thrust washer from C2 clutch housing. Inspect check ball and spring in both clutch housings and make sure all oil passages are clean. Refer to Fig. 118 and remove bearing adjusting (or retaining) snap ring from groove at front of sleeve. Press rear sun gear and support

sleeve rearward until front bearing cone is free of sleeve. Remove front bearing cone and slide sleeve assembly from C2 clutch housing. Remove bearing cups from C2 clutch housing if bearing renewal is necessary. The 75 tooth gear may be removed from C2 clutch housing if renewal is necessary. To remove rear sun gear (Fig. 119) from support sleeve, press gear into sleeve until gear is seated tightly against shoulder in sleeve. This will free gear retaining pin so that it can be removed from hole in sleeve. During operation, this sun gear attempts to work out from sleeve and will bind retaining pin so that it is impossible to remove pin without pressing gear back into sleeve. Pull sun gear from sleeve and remove the four keys. Check sealing rings and renew if worn or

damaged. Check internal needle bearing and renew if necessary.

Clean and inspect all parts and renew any showing excessive wear or damage. Reassemble as follows: Press needle bearing into support sleeve until it is 1.060-1.160 inches (29.46-26.92 mm) past flush with front face of sleeve (Fig. 119). Install the four keys into slots of sleeve with larger diameter toward inside of sleeve. Press the 41 tooth sun gear into sleeve with keys engaged in slots until sun gear bottoms in counterbore of sleeve. Press the four steel balls, 0.375 inch (9.53 mm) diameter, into oil passages of sleeve until they are flush with surface of sleeve. Press cup plug into lube passage until it is 0.060 inch (1.52 mm) past flush with sleeve surface. Press bearing cone onto bearing support sleeve assembly, with cone back face seated against shoulder on sleeve. Install three sealing rings in grooves of sleeve (Fig. 118). Install 75 tooth gear onto C2 clutch (rear clutch) housing, and retain with twelve cap screws. Torque cap screws to 38-45 ft.-lbs. (51.4-60.9 N·m) for all models except 7010, 7020 and 7045 which should be torqued to 30-37 ft.-lbs. (40.6-50.1 N·m).

NOTE: Later transmissions will have 7/16-inch cap screws retaining 75 tooth gear to C2 clutch (rear clutch) housing. These should be torqued to 65-75 ft.-lbs. (88.0-101.6 N·m).

Make sure there is no thread interference by checking under all cap screw heads with a thin feeler gage. Remove cap screws and clean threads and threaded holes in C2 clutch housing using "Locquic" primer grade T. Apply 3-4 drops of "Loctite" 242 to each cap screw and tighten to recommended torque.

NOTE: Cap screws retaining 75 tooth gear to C2 clutch housing can lose torque. This can be caused by chamfer of cap screw hole being too large, thus reducing bearing area under head of cap screw. If OD of this chamfer measures more than 0.460 inch (11.68 mm) the 75 tooth gear should be renewed. This problem was corrected effective with the following Model/serial numbers: 7020-2489, 7045-1974, 7060-6577. Some 75 tooth gears were remachined to reduce this chamfer. These gears are marked "c/s" on the machined side. When reinstalling one of these gears be sure side marked "c/s" is placed next to heads of cap screws and away from surface of C2 clutch housing.

Press bearing cup into the 75 tooth gear with cup back face seated against C2 clutch (rear clutch) housing. Press bearing cup into front of C2 (rear) clutch housing assembly with cup back face

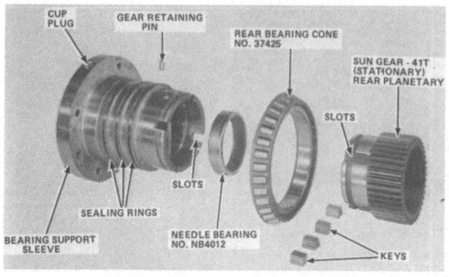

Fig. 118 — C2 clutch housing and support sleeve. Refer to text.

Fig. 119 — Exploded view of bearing support sleeve assembly.

seated against shoulder in housing (Fig. 118). Assemble the C2 (rear) clutch housing assembly over bearing support sleeve assembly so that bearing cup in gear mates with bearing cone on sleeve assembly. Insert sun gear retaining pin into hole in sleeve, then install bearing cone with cone back face outward, and retain with snap ring. Select thickest snap ring that will seat in groove of sleeve. Maximum end play should not exceed 0.005 inch (0.13 mm). Use dial indicator for checking end play (Fig. 120).

When assembling C1 and C2 clutches make sure all oil passages are clean and that check ball (¼-inch dia.) and compression spring are in place. Use an assembly gage rod with a 0.180 inch (4.57 mm) diameter to compress spring and ball as shown in Fig. 120A. Screw in outer plug until it bottoms on rod. The remaining three plugs should be flush to 0.030 inch (0.76 mm) below flush with face of housing. Install inner and outer seals on clutch piston. Lubricate seals and surface of clutch housing with transmission oil to aid in proper installation and avoid damage to sealing rings. Install clutch piston into C2 clutch housing, using a feeler gage blade or similar tool to start lip of seals into housing. C2 clutch piston is different than the others, and uses a smaller diameter inner seal. Install thrust washer into C2 clutch housing. Lubricate thrust washer with transmission oil. Place rear ring gear into C2 clutch housing with turned down section of external spline facing upward. Install a separator plate next to clutch piston. Install a friction plate and wave spring. Note that C2 clutch uses thin separator plates which are 0.084 inch (2.13 mm) thick and friction plates with the rounded type spline teeth as shown in Fig. 122. Friction plates are to be dipped in transmission oil prior to installing. Repeat the above procedure until four separator plates, four friction plates, and four wave springs have been installed. Wave springs should be install-

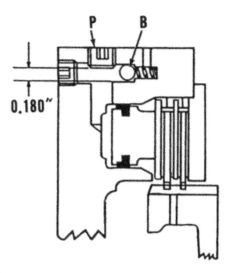

Fig. 120A—When installing check ball (B) and spring in clutch housings insert a 0.180 inch (4.57 mm) gage rod through hole in face of housing and depress ball and spring until retaining plug (P) is installed.

ed with cut ends facing toward clutch piston, and gaps staggered so they are not in line. Compress plate stack flush with end of C2 housing, and retain with assembly clips (Fig. 121). Now assemble the C1 clutch as follows: Clean all oil passages in C1 clutch housing. Insert spring, check ball and plug as previously outlined and shown in Fig. 120A. Install retaining ring in groove on inside bore of C1 clutch housing. Install inner and outer seals on piston. Lubricate seals and seal surface of C1 clutch housing with transmission oil. Use a feeler gage blade or similar tool to start lip of seals into housing. This will aid proper installation of clutch piston and avoid damage to sealing rings. Install a separator plate next to clutch piston. Install a friction plate and wave spring. Note that C1 clutch uses thick separator plates which are 0.120 inch (3.05 mm)

SPLINE TEETH ON FRICTION PLATE FOR A1, A2, B2, C1 CLUTCH

SPLINE TEETH ON FRICTION PLATE FOR B1 AND C2 CLUTCH

Fig. 122—View of two kinds of friction plates and respective clutch assemblies.

thick and friction plates with flat type spline teeth. Friction plates are to be dipped in transmission oil prior to installing. Repeat the above procedure until five separator plates, five friction plates, and five wave springs have been installed. Wave springs should be installed with cut ends facing clutch piston, and gaps staggered so they are not in line. Compress plate stack flush with housing and retain with assembly clamps (Fig. 121). Place one "O" ring into counterbore of oil passage in face of C2 clutch housing. Install C1 clutch housing assembly onto C2 clutch housing, aligning the corresponding oil feed passages of housings. Separator plate tab slots should be in line on both housings when properly positioned. Install end plate and retain the entire stack with twelve cap screws (3/8-16x6½ inches) and locknuts. Torque cap screws or locknuts to 38-45 ft.-lbs. (51.4-60.9 N·m). Remove assembly clamps from C1 and C2 clutch plate stacks. To check final clutch assembly, apply air pressure through the two oil passages in bearing support sleeve, and observe for correct operation of each clutch. Excessive air leakage or improper actuation of clutches indicates an incorrect assembly. Lubricate check ball assemblies with transmission oil just prior to application of air to aid in seating balls. Use 120 psi (827 kPa) air pressure. Lay unit aside until time of installation.

117. **OUTPUT SHAFT AND PLANETARY.** Output shaft should be supported so front planet carrier with B2 clutch hub is held above surface of bench. Remove snap ring retaining rear planet carrier assembly to output shaft, and remove thrust washer from output shaft. Remove rear planetary carrier

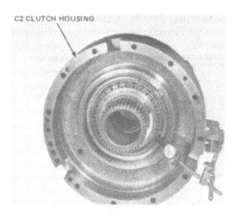

Fig. 120—Checking bearing end play with dial indicator.

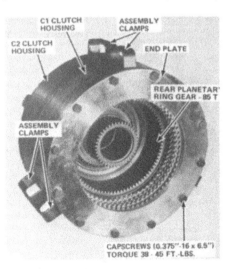

Fig. 121—View of rear clutch pack assembled.

assembly from output shaft (Fig. 123). Remove middle ring gear from middle carrier assembly, being careful to locate the four cut-out spaces in oil baffle over the four planet gears (Fig. 124.) Lift middle planet carrier assembly and front ring gear from front planet carrier assembly. The three cut-out spaces in ring gear oil baffle must be located over the three ring gears of front planet carrier assembly. Remove middle sun gear from middle planet carrier (Fig. 123). Remove retaining snap ring from inside of front planet carrier assembly that retains carrier to output shaft. This snap ring cannot be removed from carrier until one planet gear has been removed from carrier. Remove output shaft from front planet assembly. Front and middle planet gears are individual needle rollers, 28 rollers per gear. Rear planet gears use caged roller bearings, two

bearings per gear. To remove planet gears from planet carriers, remove cross pin from carrier that retains planet pins. Remove planet pins by pressing on solid end of pin (opposite grooved end). Carefully remove gears and thrust washers together. Thrust washers will assist in retaining rollers in hub of gears.

Clean and inspect all parts and renew any showing excessive wear or damage. Then reassemble as follows: Starting with front planet carrier install two rows of rollers, with spacer between roller sets into hub of planet gears. There are 14 rollers per row. Place one thrust washer on each side of gear hub to help retain rollers during installation into planet carrier. Front carrier has three 27 tooth planet gears. Place gear with rollers and thrust washers into planet carrier. Use an assembly pin of

smaller diameter to align gear and thrust washers with holes in carrier. Press planet pin into carrier from rear, through thrust washers and rollers in gear hub. Planet gears must turn freely in carrier. Assembly pins are available under special tool No. ACTP 3053. Press cross pin into carrier from either side to retain planet pins. Press pin flush with either face of carrier. Install second planet gear same as first. Then, before installing the last planet gear, place planet carrier retaining snap ring into planet carrier before installing the third planet gear. Snap ring will lie between planet gears and splined hub of carrier.

NOTE: This snap ring is impossible to get in place inside carrier after the third gear is installed without deforming snap ring.

Install third planet gear. Install oil deflector to rear face of carrier and retain with three cap screws (¼-20x½-inch). Torque cap screws to 9-12 ft.-lbs. (12.1-16.2 N·m). Install a snap ring in rear grooves of front planet ring gear. Install middle planet carrier through front of ring gear and retain with second snap ring. Install oil baffle at front of ring gear and retain with snap ring. Gap of snap ring must straddle dimple in oil baffle. Install two rows of rollers, fourteen rollers per row, into hub of planet gear. Install spacer between the two roller sets. Place one thrust washer on each side of gear to retain rollers during installation into carrier. Place gear with rollers and thrust washers into planet carrier. Use aligning pin of reduced diameter to align gear, rollers and thrust washers with holes in carrier. Press planet pin into carrier from rear face, through thrust washers and roller of gear. Planet gear must turn freely on pin.

NOTE: Press gear against thrust washers with a soft copper bar to free gear, if necessary.

Press retaining cross pin into carrier until flush with either face of carrier. Then, install the two other planet gears. Install two roller bearing assemblies into each hub of 22 tooth planet gears, and place one thrust washer on each side of gear. Place gears with bearings and thrust washers into planet carrier. Press planet pin into carrier from front side through thrust washers and gears with roller bearings. Planet gears must turn freely on pin. Care must be taken so that flat on one pin groove is indexed to permit assembly of retaining cross pin. Press each cross pin into carrier until flush with either side. Now install front planet carrier assembly onto front splines of output (hollow) shaft. Secure with snap ring previously placed in front

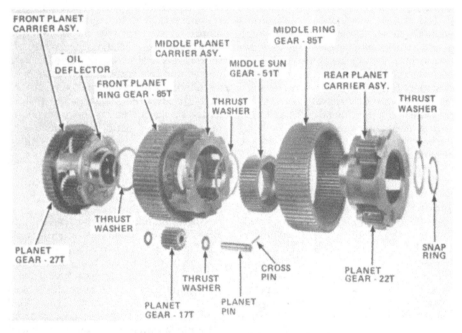

Fig. 123 — Exploded view of planetary system.

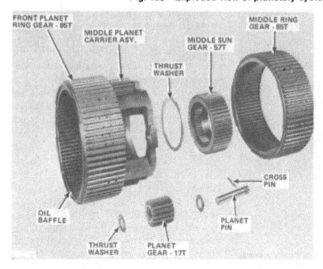

Fig. 124 — Exploded view of middle planetary.

planet carrier assembly. Place output shaft with front carrier vertically on an assembly stand which supports the shaft and allows front carrier to rest on installed snap ring. Place a thrust washer onto front carrier. Lubricate thrust washers with transmission oil. Install middle planetary carrier assembly with front planet ring gear over front carrier assembly on output shaft. Locate the three cut-outs in oil baffle over the three planet gears of front carrier assembly. Install second thrust washer into middle carrier assembly. Lubricate thrust washer with transmission oil. Install the 51 tooth middle sun gear onto output shaft with shorter counterbore toward front planet carrier. Install middle ring gear over middle carrier assembly, being careful to locate the four cut-out spaces in oil baffle over the four planet gears of middle carrier assembly. Install rear planetary carrier assembly onto hollow output shaft with planet gears facing away from middle carrier assembly. Place thrust race over output shaft and secure complete planetary system to output shaft with snap ring. Select thickest snap ring that will seat in groove of shaft. Maximum end play must not exceed 0.005 inch (0.13 mm). Check to ensure that front and middle ring gears rotate freely.

118. A1 CLUTCH HOUSING. Remove the four sealing rings from A1 clutch housing hub (Fig. 125). If necessary to remove B1 clutch hub, 61 tooth gear, or rear bearing cone from A1 clutch housing, remove six retaining cap screws with lock plates and pull gear, clutch, hub and bearing cone from clutch housing (Fig. 126). Remove ball bearing retaining snap ring and remove bearing if renewal is necessary (Fig. 125). When reassembling, press ball bearing into front of A1 clutch housing, with pin hole (notch) in inner race toward inside of housing. Install bearing retainer snap ring (Fig. 125). If remov-

ed, press 61 tooth gear and B1 clutch hub on A1 clutch housing. Use two aligning studs to properly align bolt holes. Retain hub and gear to A1 clutch housing with six ½-13x3 inch cap screws and three lock plates. Torque cap screws to 90-110 ft.-lbs. (121.9-149.0 N·m) and secure by bending a corner of lock plates at each cap screw head (Fig. 126). Press bearing cone on A1 clutch housing, with cone back face seated against face of B1 clutch hub. Install four sealing rings in grooves of A1 housing assembly (Fig. 125).

119. A2-B2-A1 CLUTCHES. Remove all of the friction separator plates and wave washers from all clutches. Remove the three pistons and piston seals. Clean and inspect all parts and renew any showing excessive wear or damage. The three clutches will be assembled later as transmission is reassembled.

120. REASSEMBLE AND REINSTALL POWER SHIFT TRANSMISSION. To reassemble, make certain housing is thoroughly cleaned and proceed as follows: Press bearing cup into center bore from front (bell end) face of housing, with cup back face seated against shoulder in bearing bore. Press bearing cup into center bore from rear of housing, with cup back face seated against shoulder in bearing bore. Press bearing cup into countershaft bore, with cup back face seated against shoulder in bearing bore. Position front housing with front face down and rear face upward. Install inner and outer sealing rings on B1 clutch piston. The A1, A2, B1, B2, and C1 clutch pistons and sealing rings are alike. The C2 clutch piston and inner sealing rings are different. Apply transmission oil on seals and housing bores to aid proper installation and avoid damage to sealing rings. Install piston with seals into front housing assembly. Lip edge of seals always point toward oil side of piston. Use a feeler

gage blade or similar tool on lips of seals to aid in installation. Install a separator plate next to clutch piston, then a friction plate and wave spring. Thin separator plates, 0.084 inches (2.13 mm) thick, are used in B1 clutch as well as A1, B2 and C2. Friction plates with rounded inner teeth (as shown in Fig. 122) are used in B1 clutch. Continue installing plates in order until four separator plates, four friction plates, and four wave springs have been installed. Install end plate. Compress the stack down by placing two sockets between a crossbar and end plate. Use "C" clamps to clamp bar to flange of housing. Use a bar similar to the OTC push-puller crossbar (Fig. 115). Install end plate retaining snap ring, starting ring at retaining screw hole so snap ring retaining screw can be installed at ring end gap (Fig. 127). Torque screw to 9-12 ft.-lbs. (12.1-16.2 N·m). Install check ball, spring and plug in passage of B1 clutch piston housing if removed. Lubricate check ball with transmission oil and apply air pressure to corresponding oil passage and observe correct operation of B1 clutch (Fig. 127). Install the 63 tooth gear, if removed, at short splined end of countershaft. Press bearing cone onto countershaft with cone back face seated against hub of 63 tooth gear. Install a new stake nut to retain gear and bearing cone. Torque nut to 290-320 ft.-lbs. (392.9-433.6 N·m) and stake nut securely to shaft. Install A1 clutch housing assembly, and countershaft with gear, simultaneously into front transmission housing assembly. Care must be taken that successive engagement of B1 clutch hub with B1 clutch friction plates takes place without force. Engagement is complete when bearing cone on A1 clutch housing assembly seats on bearing cup in front transmis-

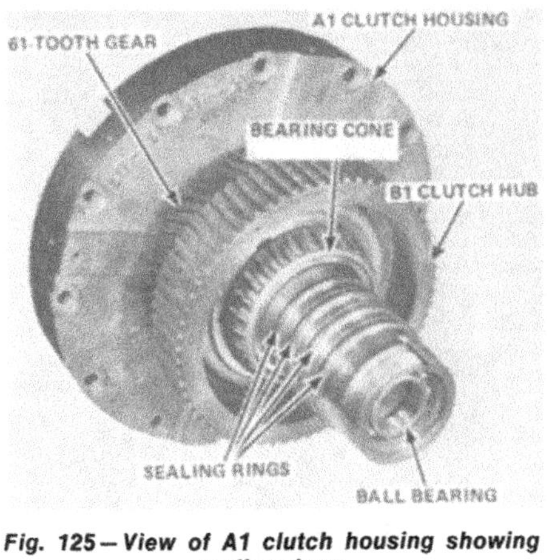

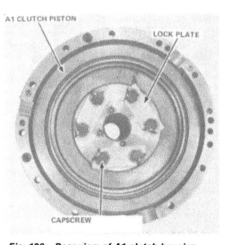

Fig. 125 — *View of A1 clutch housing showing sealing rings.*

Fig. 126 — Rear view of A1 clutch housing.

Fig. 127 — *Installing end plate retaining snap ring.*

sion housing assembly, and bearing cone on countershaft seats on bearing cup in front housing (Fig. 128). Retain A1 clutch housing to front housing as shown. Position front transmission housing assembly (with A1 clutch housing installed) horizontally, while supporting the loosely installed countershaft. Install bearing cone, lockwasher and locknut on front of A1 clutch housing assembly (Fig. 129). Adjust nut to obtain 0.001-0.006 inch (0.03-0.15 mm) end play in bearings; using a dial indicator to measure end play. Bend lockwasher tabs over nut in three places to secure nut. Position front transmission assembly with front face down as before. Install A1 clutch hub on splines of shaft with offset portion that engages clutch plates facing toward front end of shaft. Install the 31 tooth front sun gear on splines of input shaft at rear of A1 clutch hub. Secure hub and gear to shaft with two snap rings. Install cast iron sealing ring in groove of shaft at rear of 31 tooth sun gear. Install ball bearing retaining pin into input shaft with round end protruding. Install input shaft with A1 clutch hub and 31 tooth sun gear through A1 clutch housing, aligning pin on input shaft with notch in ball bearing

inner race located inside A1 clutch housing. When pin is in notch, a dimension of 1 inch can be measured from A1 clutch housing to back side of A1 clutch hub. When pin is not in notch, the dimension will be 1-1/8 inches (28.57 mm). Install inner and outer sealing rings on A1 clutch piston. Apply transmission oil on seals and seal surface of A1 clutch housing. Install piston with seals in clutch housing being careful not to cut seals. Install a separator plate next to clutch piston. Install a friction plate and wave spring. Wave springs must be installed with ends downward and end gaps staggered between slots in clutch housings. Use thin separator plates, which are 0.084 inch (2.13 mm) thick and friction plates with flat type teeth (Fig. 122) for A1 clutch. Apply transmission oil to all friction plates during assembly. Install second separator plate, friction plate and wave spring. The A1 and B2 clutches use two friction plates. The B1 and C2 clutches use four friction plates. The A2 and C1 clutches use five friction plates. Install inner and outer seals on piston of B2 clutch. Lubricate seals and seal surfaces of clutch housing. Install piston into B2 clutch housing, being careful not to cut seals. Install a separator plate next to clutch piston. Lubricate friction plates and install friction plate and wave spring. Install second separator plate, friction plate and wave spring. B2 clutch uses thin separator plates which are 0.084 inch (2.13 mm) thick and friction plates with flat type spline teeth. Install two "O" rings in counterbores of oil passages in face of A1 clutch housing. Place assembled B2 clutch housing onto A1 clutch housing, aligning corresponding oil feed passages. Install inner and outer seals on piston of A2 clutch. Lubricate

seals and seal surfaces of clutch housing. Install piston into A2 clutch housing, being careful not to cut seals. Install a separator plate next to clutch piston. Lubricate and install a friction plate and wave spring. Continue this procedure until five separator plates, five friction plates and five wave springs have been installed. A2 clutch uses thick separator plates which are 0.120 inch (3.05 mm) thick and friction plates with flat type spline teeth. Compress the plate stack and retain with assembly clamps (special tool ACTP 3054). Place one "O" ring into counterbore of oil passage in face of B2 clutch housing. Place A2 clutch housing assembly onto B2 clutch housing aligning corresponding oil feed passages (Fig. 130). Install end plate over A2 clutch housing and install the 12 cap screws and locknuts. Torque cap screws to 38-45 ft.-lbs. (51.4-60.9 N·m). Remove the three assembly clamps from A2 clutch plate stack. Separator plate tab slots should be in line on all three housings when properly positioned. To check complete assembly, apply air pressure to the three oil passages that operate A1, B2 and A2 clutches and observe operation of each clutch. These oil passages are located in rear machined flange of front transmission housing (Fig. 130). Lubricate check ball assembly in each clutch housing prior to air checking. Use 120 psi (827 kPa) air pressure for checking. Install planetary that was assembled in paragraph 117. Attach a hanger that will thread onto end of input shaft for attaching hoist and install planetary system (Fig. 131). Slowly lower planetary system down through friction plates of A2 and B2 clutches, and engage front sun gear with front

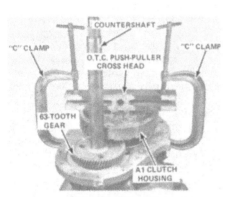

Fig. 128 — Retaining A1 clutch housing to install front bearing cone.

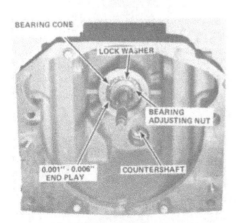

Fig. 129 — View showing the A1 clutch housing bearing adjusting nut.

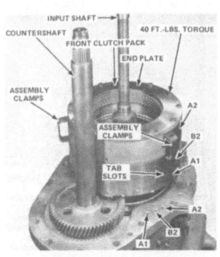

Fig. 130 — Front housing with front clutch pack installed.

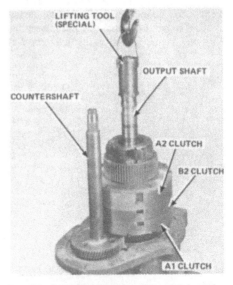

Fig. 131 — Installing planetary into clutch pack. Refer to text.

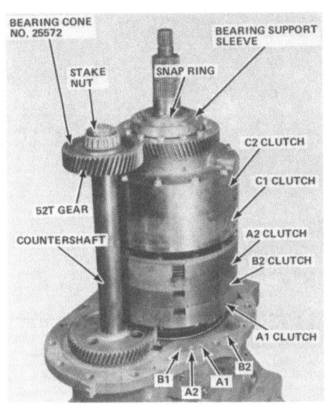

Fig. 132 — View of clutch packs assembled ready to install housing.

rear transmission housing down over bearing support sleeve, and allow to rest on gasket of front housing face. Align six cap screws in bearing support sleeve, and torque to 38-45 ft.-lbs. (51.4-60.9 N·m) (Fig. 134). Install fifteen cap screws in front transmission housing, and torque to 180-220 ft.-lbs. (243.9-298.1 N·m). Install ball bearing over output shaft into bore of transmission housing. Install stake nut and torque to 20-25 ft.-lbs. (27.1-33.8 N·m). Stake nut securely to shaft in three places. Install snap ring in housing bore to retain ball bearing. Install snap ring in groove on output shaft. Install bearing cup in countershaft bore in transmission housing. Torque the previously installed stake nut on countershaft to 290-320 ft.-lbs. (392.9-433.6 N·m).

planetary carrier assembly. Rotating output shaft and front ring gear while gently lowering planetary assembly will ensure complete engagement of friction plates. Planetary system is completely lowered into place when it is in position shown in Fig. 131. Using rear clutch pack hanger that was made for disassembling, attach hanger to bearing support sleeve flange for attaching hoist (Fig. 111). Attach hoist and slowly lower rear clutch stack down gently over output shaft, engaging the five friction plates in C1 clutch with middle ring gear, and engaging rear ring gear and rear sun gear with planetary gears of rear carrier assembly. Rotating middle ring gear while gently lowering rear clutch stack will ensure complete engagement of friction plates. Install ball bearing retaining snap ring in groove of output shaft. Install 52 tooth gear and bearing cone on countershaft. Install bearing cone with cone back face seated against gear. Install stake nut loosely at this time (Fig. 132). Install two "O" rings in clutch oil passages, and one "O" ring in lube oil passage counterbore on rear flange of bearing support sleeve. Install four "O" rings and a new gasket (dry) on front transmission housing flange face. Attach a lifting chain to rear transmission housing flange for attaching a hoist. Use second threaded hole from top at right side of housing, and third threaded hole from bottom at left side of housing (Fig. 133). Lower

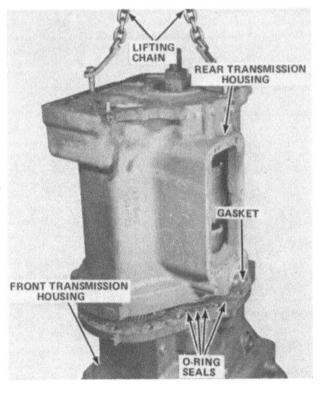

Fig. 133 — Installing rear transmission housing. Make sure "O" rings are in place.

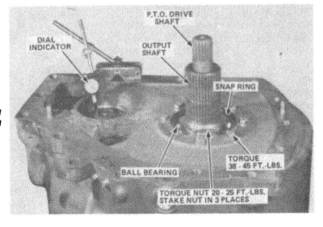

Fig.134 — View showing countershaft end play being checked.

Select a snap ring which gives 0.001 to 0.006 inch (0.03-0.15 mm) end play in countershaft and install in groove of housing. Position assembled transmission horizontally and install core hole plug with "Loctite" in front countershaft bore. Install snap ring over front of input shaft and into groove of shaft to retain ball bearing. Install oil seal in cover, using "Loctite" on outer diameter of seal. Place gasket (dry) on cover and install on transmission housing. Install the four cap screws and torque to 38-45 ft.-lbs. (51.4-60.9 N·m). As a final check of transmission assembly, apply air pressure at valve mounting flange of rear transmission housing to the six oil passages and observe operation of clutches (Fig. 135). Balance of reassembly is reverse of disassembly. Adjust as outlined in paragraphs 93 through 100.

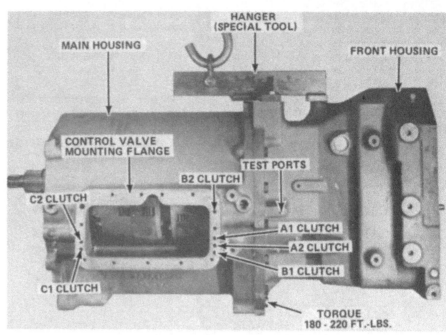

Fig. 135 — View of transmission ready for air check of clutches before installing in tractor.

RANGE TRANSMISSION

The range transmission is located in front portion of tractor rear housing. This transmission provides a Hi, Lo, neutral and a parking lock position.

To remove range transmission gears and shafts, tractor rear main housing must be separated from transmission housing and differential assembly removed. See Fig. 140 for an exploded view of range transmission shafts and gears.

R&R AND OVERHAUL

121. To remove rear main housing assembly, first drain all oil compartments. Support tractor under transmission housing and block front axle to prevent tipping. Loosen and raise rear of platform or cab and support securely. Disconnect necessary hydraulic lines and electrical connections. Support front and rear of rear main housing with split stands and/or rolling floor jacks. Remove cap screws attaching rear main housing to back of transmission. Roll rear housing rearward until clear of tractor. With housing clear of tractor remove rear wheels and hydraulic lift housing.

122. To disassemble transmission, proceed as follows: Remove final drives from main housing. Remove oil splash plate from top of gear train. Remove tube for three point hitch and its support bracket from rear of housing. Remove

plug from rear bearing bore. Remove rear bearing cup retaining snap ring from bearing bore. Remove bearing retaining nut from rear of shaft. Remove front bearing retaining snap ring from groove of shaft. Attach a slide hammer and ½-13 thread adapter to rear of input shaft. Pull shaft until rear bearing cup is free of housing bore. Continue to pull shaft until front bearing cone is free of shaft. Remove slide hammer and adapter from shaft. Move shaft rearward by hand and remove gear, front bearing cone and snap ring through top of housing. Remove spacer from shaft.

Remove front bearing cup from housing bore and rear bearing cone from shaft if renewal is necessary.

Remove all lube lines and lube line fittings from brake housing. Attach a sling to differential assembly and support with hoist. Then, remove cap screws securing brake back-up plate and remove plate and brake discs. Remove both brake housing assemblies from side of main housing and lift out differential. Keep shims with brake housing as they are removed for aid in reassembling. Shift range transmission into park and loosen nut on pinion shaft (Fig. 136).

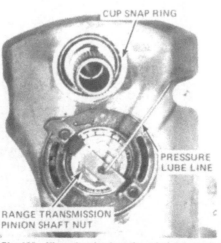

Fig. 136 — View showing location of pinion shaft nut located behind lube line support bracket.

Fig. 137 — View of shift fork attaching cap screws and High-Lo shift fork.

Remove top cover of range transmission and remove two cap screws retaining Hi-Lo shift fork to shift rail (Fig. 137). Remove cap screws retaining side shift cover to main housing and lift upward on Hi-Lo shift fork until it clears shift rail and remove cover. Remove pto shaft oil seal at rear bearing bore of range compartment. Remove snap ring retaining front bearing cup in housing bore. Remove snap ring retaining high range gear to hollow shaft. There are some special tools that will be needed for removing the hollow shaft. These tools can be made from scrap material as shown in Fig. 138. The 2¼-inch (63.5 mm) spacer is used between the two gears (Fig. 139), and the 2-inch (50.8 mm) spacer is used between the front gear and housing. Step plug is inserted at rear of hollow shaft and is used to press shaft forward. Press hollow shaft forward until rear bearing cone is free of shaft and front bearing cup is free of housing bore. As shaft is moved for-

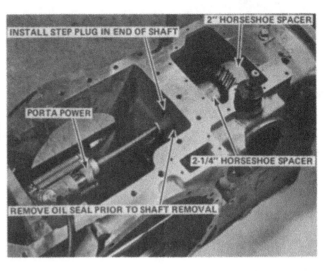

Fig. 139 — Use spacers as shown when removing hollow shaft (3 — Fig. 140).

ward, remove low and high range drive gears. Remove front bearing cone from shaft and rear bearing cup from housing bore if bearing renewal is necessary. Remove bearing adjusting nut (20 — Fig. 140) from front of pinion shaft. Remove snap ring (31) from groove of shaft that retains splined shift collar (30).

NOTE: Do not scratch shaft surface while removing snap ring.

Press shaft rearward until front bearing cone is free of shaft. Move pinion shaft rearward and remove components. Remove rear bearing cup from housing bore, and rear bearing cone from shaft if renewal is necessary. Remove rear bearing cup retaining snap ring (26) and oil seal (27). Remove front bearing adapter (18) from housing bore and remove bearing cup from adapter.

123. Clean and inspect all parts and renew any showing excessive wear or damage. Any disassembly or service required on range transmission shift cover will be obvious after an examination of unit and reference to Fig. 141.

To correctly locate pinion shaft, variable thickness snap rings (26 — Fig.

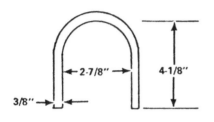

3/8" THICK BY 2-1/4" WIDE STEEL STRAP

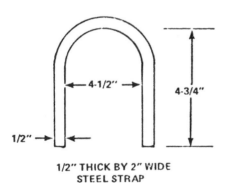

1/2" THICK BY 2" WIDE STEEL STRAP

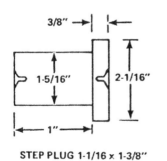

STEP PLUG 1-1/16 x 1-3/8" BAR STOCK

Fig. 138 — View showing dimensions of special tools for removing hollow shaft.

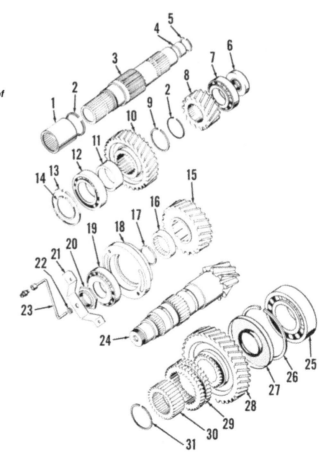

Fig. 140 — Exploded view of range transmission.

1. Coupling
2. Snap ring
3. Hollow shaft
4. Bushing
5. Snap ring
6. Seal
7. Bearing
8. Gear (lo-range)
9. Snap ring
10. Gear (hi-range)
11. Spacer
12. Bearing
13. Snap ring
14. Snap ring
15. Gear (hi-range)
16. Spacer
17. Snap ring
18. Bearing adapter
19. Bearing
20. Nut
21. Support
22. Grommet
23. Tube
24. Pinion shaft
25. Bearing
26. Snap ring
27. Seal
28. Gear (lo-range)
29. Coupling (hi-lo range)
30. Collar
31. Snap ring

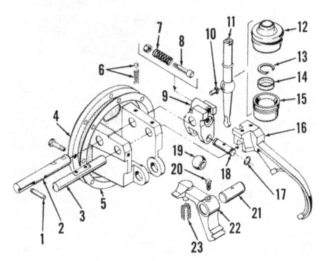

Fig. 141—Exploded view of range transmission cover and components.

1. Plunger
2. Rod
3. Rod
4. Cover
5. Gasket
6. Spring & ball detent
7. Spring
8. Plunger
9. Carrier
10. Pivot pin
11. Shift lever
12. Cover
13. Snap ring
14. Washer
15. Insert
16. Fork
17. Snap ring
18. Shaft
19. Roller
20. Set screw
21. Shaft
22. Parking pawl
23. Spring

140) are installed in rear bearing bore at back face of rear bearing cup. These snap rings are furnished in thicknesses of 0.177-0.192 inch in steps of 0.003 inch. To select correct snap ring, the following method is used: Measure width of rear bearing from surface of cone back face to surface of cup back face. Add bearing measurement to dimension etched on end of pinion shaft. Then, subtract this sum from dimension stamped on rear of housing. Select a snap ring with thickness as close as possible to the last result. Install selected snap ring in housing. Install seal (27) with lip facing forward, and press in flush with chamfer of housing bore. Do not press seal against snap ring. Install rear bearing cup into housing with cup back face seated lightly against snap ring. Then, install rear bearing cone on pinion shaft with cone back seated against shoulder of pinion. Install front bearing cup in bore of bearing adapter (18) with back face of cup seated against shoulder. Place pinion shaft through rear bore of housing and install low range gear (28) with shifter

teeth forward, collar (30), shift coupling (29) over collar and retaining ring (31).

NOTE: Lubricant should be applied to shafts and gears to provide lubrication until pressure from tractor lubrication system reaches them.

Install high range gear (15) with shifter teeth rearward, spacer (16) and snap ring (17) in groove of shaft. Install and position adapter so pressure lube tube bracket will fit in notch in front face of housing. Install front bearing cone with cone back facing forward and install bearing adjusting nut (20) but do not preload bearings. For proper bearing adjustment, use following procedure: Measure torque required to rotate shaft with gears and oil seal in place. Tighten bearing adjustment nut until rotating torque increases 10-20 in.-lbs. (1.12-2.24 N·m), then stake nut securely to shaft. Reinstall oil tube and tube support bracket at front of pinion shaft and housing. Then, install range shift fork in groove of shift coupler.

Install hollow shaft as follows: Install rear bearing cup in housing with cup

back face seated against shoulder. Press front bearing cone on shaft with cone back face rearward and slightly past flush with edge of snap ring groove. Do not install snap ring at this time. Place spacer (11) over shaft next to bearing cone. Enter shaft through front bearing bore and install high range drive gear (10), snap ring (9), snap ring (2) and low range drive gear (8). Lubricate rear bearing cone and place in bearing cup. Press hollow shaft (3) into bearing cone until cone back face is seated against hub of gear (8) and gear is seated tight against snap ring (2). Install snap ring (9) and high range drive gear (10) on shaft. Install front bearing cup in housing with cup back face forward. Press front bearing cone on until it is seated against spacer, spacer is seated against the high range drive gear and gear is seated tight against snap ring. Select thickest snap ring (14) that will seat in groove at front of front bearing cone. Snap rings are furnished in thicknesses of 0.078 to 0.108 inch in steps of 0.003 inch. Select a bearing cup snap ring (13) that will give 0.001-0.006 inch (0.03-0.15 mm) end play of hollow shaft. Bearing cup must be seated tight against snap ring while checking end play with a dial indicator. Snap rings are furnished in thicknesses of 0.150 to 0.189 inch in steps of 0.003 inch. Install oil seal (6) for power take-off drive shaft in rear bore of range transmission. Install seal with lip facing rearward and press into bore until flush with rear surface of housing. Reinstall shaft by reversing disassembly procedure.

Reinstall differential assembly in rear main frame and check carrier bearing preload as outlined in paragraph 124 and backlash as outlined in paragraph 125.

Complete reassembly of tractor by reversing disassembly procedure.

MAIN DRIVE BEVEL GEARS AND DIFFERENTIAL

The differential is carried on tapered roller bearings. Models 7010, 7020, 7030 and 7040 are equipped with a two-pinion differential while all other models have four-pinion differential. Differential locking assembly is built into differential.

ADJUSTMENT

124. CARRIER BEARING PRE-LOAD. To adjust carrier bearing pre-

load, first remove bevel pinion shaft as outlined in paragraph 122. Then, reinstall differential assembly in rear main housing with ring gear on right side (of normal drive pinion position). Install right-side brake housing assembly (with bearing cup in place) without shim stack and torque four equally spaced cap screws to 70-80 ft.-lbs. (94.8-108.4 N·m) for all models except 7010, 7020 and 7045 which should be torqued to 55-65 ft.-lbs. (74.5-88.0 N·m). Install left-side

brake housing assembly with enough shims so that bearings will not be preloaded. Be careful not to damage rings while installing brake housing over oil sealing rings in differential cover. Torque four equally spaced cap screws to 70-80 ft.-lbs. (94.8-108.4 N·m) for all models except 7010, 7020 and 7045 which should be torqued to 55-65 ft.-lbs. (74.5-88.0 N·m). Adjust differential bearings by adding or removing shims behind left brake housing until rolling

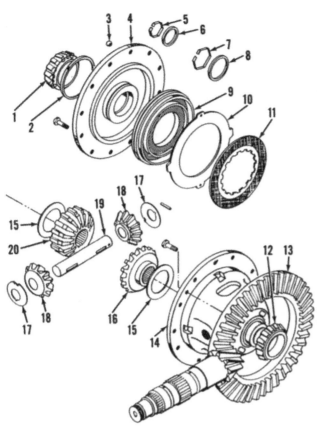

Fig. 142—Exploded view of two-pinion differential assembly.

1. Bearing cone
2. Seal ring
3. Ball
4. Cover
5. Expander ring
6. Sealing ring
7. Expander ring
8. Sealing ring
9. Piston
10. Separator plate
11. Friction plate
12. Bearing cone
13. Ring gear
14. Housing
15. Thrust washer
16. Left side gear
17. Thrust washer
18. Gear
19. Pinion shaft
20. Right side gear

torque of differential is 10-25 in.-lbs. (1.12-2.8 N·m) for all models except 7010, 7020 and 7045 which should be 10-40 in.-lbs. (1.12-4.48 N·m). Shims are furnished in 0.010, 0.012 and 0.015 inch thicknesses. The rolling torque can be measured by wrapping a cord around ring gear mounting flange and connecting a spring scale to cord. A pull of 2-8 pounds (8.9-35.6 N) should be registered on scale. Remove differential assembly from housing, keeping shims intact. With bearing preload determined, refer to paragraph 123 for procedure for reinstalling pinion shaft and to paragraph 125 to set backlash between bevel gear and pinion.

125. BACKLASH ADJUSTMENT. Adjust pinion and ring gear backlash in following manner: Set backlash on ring gear to 0.010-0.013 inch (0.25-0.33 mm) by moving shims from one side to the other as required. Use a dial indicator to check backlash. Backlash should be checked in several places around circumference of ring gear. Torque the two brake housing cap screws (3/8-16x7/8-inch) to 28-33 ft.-lbs. (37.9-44.7 N·m) and the twelve attaching cap screws to 70-80 ft.-lbs. (94.8-108.4 N·m) for all models except 7010, 7020 and 7045 which should be torqued to 55-65 ft.-lbs. (74.5-88.0 N·m) on each side of housing.

R&R BEVEL GEARS

126. The main drive bevel pinion is also the range transmission lower shaft. Procedure for removing, reinstalling and adjusting pinion is outlined in paragraphs 123 through 125.

To remove bevel ring gear, follow procedure outlined in paragraph 127 for R&R of differential. The ring gear is secured to differential cover by twelve cap screws which should be torqued to 105-125 ft.-lbs. (142.2-169.3 N·m) on two-pinion differentials or 150-170 ft.-lbs. (203.2-230.3 N·m) on four-pinion differentials. The two 3/8-16x3/4-inch cap screws should be torqued to 28-33 ft.-lbs. (37.9-44.7 N·m).

R&R DIFFERENTIAL

127. Models 7010, 7020, 7030 and 7040 are equipped with a two-pinion differential. Models 7045, 7050, 7060 and 7080 are equipped with a four-pinion differential. Removal procedure is essentially the same for both types of differentials.

To remove differential, remove hydraulic lift as outlined in paragraph 155. Refer to paragraph 128 for removal procedure of final drive and planetary drive.

With final drives and hydraulic lift assembly removed, attach a hoist to dif-

ferential assembly, remove both brake housings and carefully lift differential from rear housing.

To disassemble, remove the twelve bolts that attach cover and ring gear to differential housing. Remove cover assembly. Ring gear is attached to differential housing with two additional cap screws (3/8-16x3/4-inch). Remove these two cap screws if removal of ring gear is necessary. Remove differential lock piston (9—Fig. 142 or 144) from cover. Remove bearing cone from cover and differential housing, if renewal is necessary. Remove differential lock plates and side gear from differential housing.

On models with two pinions, remove lock pin retaining pinion shaft (Fig. 143) in differential housing. Remove pinion shaft, pinions and thrust washers and left side gear. On models with four pinions, refer to Fig. 144 and remove the short pinion shafts (20), then long pinion shaft (19). Remove pinion gears and thrust washers and the left side gear and thrust washer.

When reassembling two-pinion differential refer to Fig. 142 and proceed as follows: Place differential housing on bench with flange side up. Install bearing cone (1) on hub of differential cover (4) with cone back face seated against shoulder. Install oil sealing rings (2) in grooves of cover hub for differential lock. Install bearing cone (12). Install left side gear (16) and thrust washer (15). Install pinion gears (18) and thrust washers (17). Install pinion shaft (19) through thrust washers and pinions with lock pin hole in alignment. Install lock pin (Fig. 143) and drive in slightly past flush with outside of housing. Attach ring gear to housing flange with two

Fig. 143—View showing pinion gear shaft lock pin used on two-pinion differential.

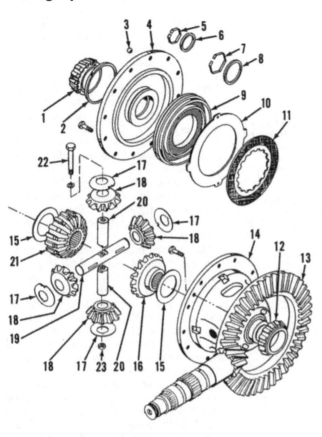

Fig. 144—Exploded view of four-pinion differential assembly.

1. Bearing cone
2. Sealing ring
3. Ball
4. Cover
5. Expander ring
6. Sealing ring
7. Expander ring
8. Sealing ring
9. Piston
10. Separator plate
11. Friction plate
12. Bearing
13. Ring gear
14. Housing
15. Thrust washer
16. Left side gear
17. Thrust washer
18. Pinion gear
19. Long pinion shaft
20. Short pinion shaft
21. Right side gear
22. Retaining bolt
23. Nut

tial refer to Fig. 144 and proceed as follows: Install bearing cone (1) of cover (4) with cone back seated against shoulder. Install sealing rings (2) in grooves of cover hub. Install bearing cone (12) to hub of housing (14) with cone back seated against its shoulder. Install side gear (16) and thrust washer (15) into housing. Install long pinion shaft (19), two pinions (18) and thrust washers (17) into housing (14). Install remaining two pinion gears (18) and thrust washers (17), and short pinion shafts (20) into housing. Make sure that the stepped ends of the short pinion shafts are seated into the flats of the long pinion shaft. Install pinion shaft retaining bolt (22) and nut (23), and torque to 38-45 ft.-lbs. (51.4-60.9 N·m). Short pinion shafts should be seated into flats of long pinion shaft so a 0.005 inch (0.13 mm) feeler gage will not fit between either short shaft and long shaft. Attach ring gear to housing flange using two 3/8-16x3/4-inch cap screws and tighten to 28-33 ft.-lbs. (37.9-44.7 N·m). Install side gear (21) and thrust washer (15) into housing (14). Install a friction plate (11) next to housing then a separator plate (10). Install plates alternately until four of each have been installed. Install a new expander ring (5) and sealing ring (6) in cover (4). Install a new expander ring (7) and sealing ring (8) onto outer diameter of piston (9). Lubricate sealing rings and carefully install piston into cover making sure sealing rings are not damaged. Attach cover and piston assembly to housing using twelve 9/16-18x1½-inch cap screws tightened alternately across diameter of cover to a torque of 150-170 ft.-lbs. (203.2-230.3 N·m). Refer to paragraphs 124 and 125 for preload and backlash adjustment.

3/8-16x3/4-inch cap screws torqued to 28-33 ft.-lbs. (37.9-44.7 N·m). Install right side gear (20) and thrust washer (15) with gear teeth meshing with pinion gear teeth. Install a friction plate (11) next to differential housing. Next install a separator plate (10) on top of friction plate. Install these plates alternately until three of each has been installed. Install a new expander ring (5) and sealing ring (6) in groove of differential cover. Install expander ring (7) and sealing ring (8) on outer diameter of piston (9). Lubricate all sealing rings and carefully install piston into cover, making sure sealing rings are not damaged. Install piston with center cavity towards plates. Attach cover and piston assembly to differential housing using twelve ½-13x1½-inch cap screws. Apply "Loctite" to cap screws and tighten alternately across diameter of cover to a torque of 105-125 ft.-lbs. (142.2-169.3 N·m).

When assembling four-pinion differen-

FINAL DRIVE

The final drive assemblies consist of rear axle and planetary reduction unit.

The use of a stand to support final drive during disassembly and overhaul is recommended. Allis-Chalmers special tool ACTP 3003 is available for this purpose. If necessary a final drive support stand to hold axle in vertical position can be made from a piece of tubular steel having an inside diameter of 3-21/32 inches (92.8 mm) x 13½ inches (342.9 mm) long. Close bottom end of tube by welding a ¼-inch (6.35 mm) thick plate of the appropriate size. Weld three legs, 1½ inch (38.1 mm) x 15 inch (381.0 mm) pipe, to bottom end of tube to form a base for stand.

128. REMOVE AND REINSTALL. To remove final drive, first drain all oil

compartments. Remove any bracket that is attached to final drive housing. Remove right side fuel tank. Support tractor under transmission housing using a suitable stand or the transmission section of Allis-Chalmers splitting stand special tool ACTP 3014-2. Loosen fasteners and raise rear of cab or platform and support it using suitable jack stands. Disconnect all necessary electrical wires and hydraulic lines. Support final drive housing using a suitable stand or the rear section of Allis-Chalmers splitting stand special tool ACTP 3014-3. Remove cap screws attaching final drive housing to rear of transmission. Remove rear wheel and tire assemblies. Roll final drive housing out from under cab or platform assembly. All work can be performed with housing

mounted on stand.

Reinstall the final drive housing by reversing removal procedure.

129. OVERHAUL PLANETARY AND REAR AXLE. Remove final drive as outlined in paragraph 128, refer to Fig. 145 and remove sun gear (11). Remove cap screw (12), retainer plate (13), and shim stack (14) that retains planetary carrier to rear axle and remove carrier assembly (7). Remove ring gear (9) from sleeve housing (held by two dowel pins). Remove cap screw and lock clips (6), retaining planet pins to carrier, and remove pins (5). Carefully remove planet gears (4), side thrust washers (1), bearing rollers (2) and spacer (3). Remove the two bearing adjusting snap rings (23 and 24) from

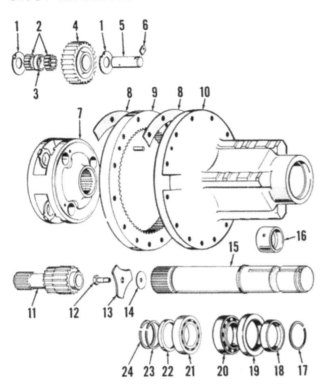

Fig. 145 — Exploded view of rear axle and planetary unit.

1. Thrust washer
2. Needle rollers
3. Spacer
4. Planet gear
5. Planet pin
6. Lock
7. Carrier
8. Gasket
9. Ring gear
10. Housing
11. Sun gear
12. Cap screw
13. Plate
14. Shim
15. Axle
16. Bushing
17. Snap ring
18. Wear sleeve
19. Seal
20. Bearing (outer)
21. Bearing (inner)
22. Retainer
23. Snap ring
24. Snap ring

groove of axle. Press axle outward until inner bearing retainer (22) and bearing cone are free of axle. Remove axle shaft from sleeve housing. Remove wear sleeve (18) and bearing cone from axle if axle or bearing renewal is necessary. Remove inner and outer bearing cups from sleeve housing if renewal is necessary.

NOTE: On 7020 tractors with short axles, a wear sleeve (18) is not used. The short axle uses a bearing retainer ring instead of a wear sleeve and the axle seal sealing surface is the axle.

When reassembling proceed as follows: Press wear sleeve (18) on axle with chamfered side seated against shoulder on axle. Install outer bearing cone on axle with cone back face seated against wear sleeve. Install outer bearing cup and inner bearing cup with cup back faces seated against their backing shoulders in sleeve housing. Install axle into sleeve housing and stand upright on outer end of axle. Axle should be placed in a proper stand when positioned upright on outer end of axle. Install inner bearing cone with cone back face upward. Drive cone down until seated in bearing cup. Install inner bearing retainer (22) and press down until seated against bearing cone. An assembly tool (spacer) is necessary to properly preload axle bearings. This tool or spacer can be made from a piece of steel tubing having an inside diameter of 3-9/16 inches (90.49 mm), an outside diameter of 4-1/16 inches (103.19 mm) and a length of 3-7/16 inches. (87.31 mm). This tool or

spacer will fit between bearing retainer (22) and retainer for planet carrier, using axle cap screw (12) to preload bearings. Install assembly tool (spacer sleeve) against bearing retainer (22). Place planet carrier retainer (13) over assembly tool and insert axle cap screw. Tighten cap screw until 60-85 in.-lbs. (6.7-9.2 N·m) of rolling torque is required to rotate sleeve housing. See Fig. 146. Remove assembly tool and install two snap rings (23 and 24) as required to retain rolling torque of 60-85 in.-lbs. (6.7-9.2 N·m). The thickest snap rings that will fit in groove should be used. Retainer will hold preload until snap rings are installed. Snap rings are furnished in thicknesses of 0.101 to 0.158 inch in steps of 0.003 inch. Install oil seal (19) at outer end of sleeve housing with lip of seal facing toward outer bearing. Use final drive support stand to hold axle in vertical position.

To reassemble planet carrier, proceed as follows: Assemble planet gears (4 – Fig. 145) in planet carrier. Apply Lubriplate to inside diameter of planet gear hubs. Apply Lubriplate to needle rollers and stack in place in hub of gear. Install rollers (2) until 26 rollers have been placed in gear. Install spacer (3) with Lubriplate on top of first set of rollers. Continue to install second set of rollers until 26 more have been put in place (a total of 52 rollers for each planet gear).

NOTE: The Model 7010 and 7020 tractors only have 20 rollers (2 – Fig. 145) per row or 40 rollers per gear.

Place planet gears with rollers, spacer and thrust washers in position in planet carrier. Insert lock end of planet pin through carrier from side opposite lock clip side. Pin should be hand slide fit in this bore, but becomes a press fit in bore on lock clip side. Pin must be aligned so notch for lock clip will be in register with lock clip (6).

NOTE: A dummy shaft having a diameter of 2.275 inches (57.78 mm) and a length of 2.575 inches (65.41 mm) can be inserted into hub of planet gear after rollers have been installed. Shaft will hold rollers in place during installation of gear to carrier. Planet pin will push shaft out of gear as it is being inserted. Dummy shaft can be made from steel, plastic or hard wood. Dummy shaft can also be used when installing roller bearings in planet gears.

With planet pins in place, install lock clips, lockwashers and cap screws. Torque cap screws to 28-33 ft.-lbs. (37.9-44.7 N·m). Install planet carrier assembly on axle. Shims (14) must be installed between end of axle and axle retaining plate (13) to provide 0.060-0.080 inch (1.52-2.03 mm) end play for planet carrier. After carrier has been placed on axle, add shims to end of axle until flush with planet carrier hub. Then, add an additional 0.070 inch (1.78 mm) shim stack and install retaining plate with cap screw. Shims are furnished in thicknesses of 0.010, 0.015 and 0.090 inch. Install ring gear (9) to axle sleeve housing flange by fitting it over the two locating dowel pins and the three planet gears. Install sun gear (11) through brake splined hub into differential side gears. Install final drive assembly to rear main housing. Make sure alignment is maintained during assembly and gear teeth are in mesh. Do not use bolts to force housings together as this could cause damage. Torsion bar should be installed and aligned as second final drive is installed. Torque cap screws to 180-220 ft.-lbs. (243.9-298.1 N·m).

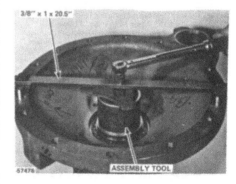

Fig. 146 — View showing checking rotating torque of axle bearings. Refer to text.

BRAKES AND CONTROL VALVE

Brakes on all models are actuated hydraulically and are self-adjusting. Tractors are equipped with wet type disc brakes which are accessible after removing rear axle and planetary drive assemblies. Refer to paragraph 135 for overhaul.

Brake operation can be accomplished with engine inoperative because of a one-way check valve located on inlet side of brake valve. This check valve closes when hydraulic pressure ceases and thus provides a closed circuit which permits operating brake control valve with oil trapped within the circuit.

Service (foot) brakes MUST NOT be used for parking or any other stationary job which requires tractor to be held in position. Even a small amount of fluid seepage would result in brakes loosening and severe damage to equipment or injury to personnel could result. USE PARK LOCK when parking tractor.

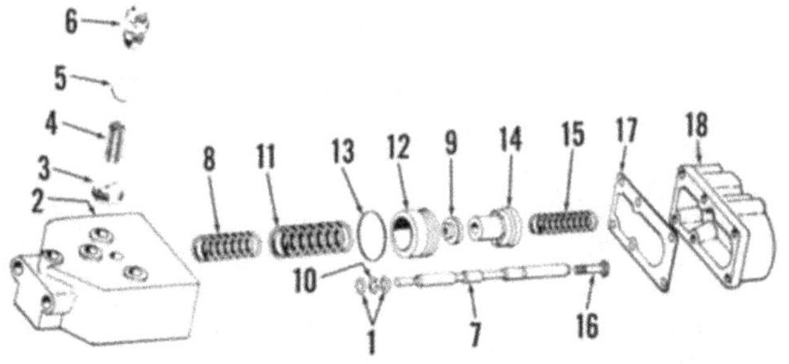

TESTING BRAKES

130. Oil for brake operation is supplied from piston pump. Operating

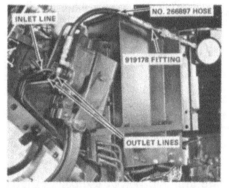

Fig. 147—Testing operating pressure of brakes.

Fig. 148—View showing brake pedal adjustment. Refer to text.

Fig. 149—Exploded view of brake control valve.

1. "O" ring
2. Body
3. Orifice
4. Spring
5. Ball
6. Inlet fitting
7. Spool
8. Spring
9. Washer
10. Snap ring
11. Spring
12. Piston (large)
13. "O" ring
14. Piston (small)
15. Spring
16. Screw
17. Gasket
18. Cap
19. Cap screw

pressure to brake valve is controlled by the pressure reducing valve. To test brakes, start engine and apply brakes. Brakes should feel solid at approximately one inch (25.4 mm) pedal movement. If brakes feel spongy, actuate several times at high pedal pressure. If sponginess persists, loosen brake line fittings at brake valve and bleed all air. If this does not correct sponginess, remove brake outlet line at brake valve and install a 0-600 psi (0-4000 kPa) gage (Fig. 147). Depress brake pedal gradually. Pressure should increase gradually to a maximum pressure of 280-410 psi (1900-2800 kPa). Brake pedal should feel solid after about one inch (25.4 mm) of travel. If pressure does not reach maximum, trouble will probably be found in pressure compensator valve or incorrect stand-by pressure. See paragraph 150 for pressure compensator valve and paragraph 136 for incorrect stand-by pressure. If you suspect an internal leak in brake valve, stop tractor engine and depress brake pedal. Pedal should become solid after about 1½ inches (38.1 mm) of travel. If pedal continues downward movement, trouble is probably caused by internal brake valve leakage. Remove inlet line to brake valve. Again pump brakes with engine shut off. If oil is pumped out of brake inlet fitting, check ball in fitting is not seating properly. If no oil is pumped out of inlet fitting, leakage is probably due to damaged seals on actuator piston inside valve. If brake valve, stand-by pressure and reducing valve pressure all check okay, problem is most likely in brake area.

BRAKE ADJUSTMENT

131. The only external adjustment that can be made on brakes are the brake pedal maximum travel and brake control valve adjustment.

132. To make brake pedal adjustment proceed as follows: Adjust left pedal stop bolt until there is 5/8-inch (15.87 mm) between pedal stop boss on pedal and inside front of valve mounting bracket. See Fig. 148. Right pedal should be adjusted even with left pedal at this time. After pedals are adjusted, valve spools should be turned in or out of trunnion, until trunnion is loose in its pedal mounting. If this adjustment is not made correctly, brakes could be energized all the time. After adjustments have been made, install cotter pin in hole in brake valve spool and bend over trunnion to retain spool to pedal adjustment.

BRAKE CONTROL VALVE

133. R&R AND OVERHAUL. To remove the individual wheel brake valve, remove right steering support side sheet. Clean valve and area around it. Remove cotter pin out of end of brake valve spools. Disconnect all lines. Insert a small punch in the cotter pin hole and unscrew spool out of trunnion at brake pedal. See Fig. 148. Remove cap screws attaching valve to its mounting bracket.

134. With brake control valve removed, refer to Fig. 149 and disassemble valve as follows: Remove cap (18 – Fig. 149) from valve body (2), then identify each spool with its bore.

NOTE: Care must be taken that pistons and spools are reassembled into their respective bores. Pistons are hand-lapped and should remain together.

Remove large piston (12) and springs (8 and 11) from body (2). Remove screw (16) from spool (7). This will allow removal of spring (15), small piston (14), snap ring (10) and washer (9). Remove inlet fitting (6), spring (4) and steel ball (5) from inlet port. Remove orifice (3).

Clean and inspect all parts. Pay particular attention to pistons (12 and 14) and their bores. Be sure orifice (3) is clean as well as all other oil passages. Renew springs (8 and 11) if any doubt exists as to their condition. Use all new "O" rings and gaskets and reassemble by reversing disassembly procedure.

NOTE: Retaining cap screw (16) must be installed with Loctite and torqued to 25-27 ft.-lbs. (33.8-36.5 N·m).

After valve is installed, adjust brake pedals as outlined in paragraph 132.

BRAKE ASSEMBLIES

135. R&R AND OVERHAUL. Removal of either brake is accomplished by removing final drive as outlined in paragraph 128. Refer to Fig. 150 for exploded view of brake. Remove cap screws attaching brake back-up plate (8) to brake housing (3). If not necessary to remove brake housing (3), leave it in place and reinstall two 3/8-inch cap screws 180

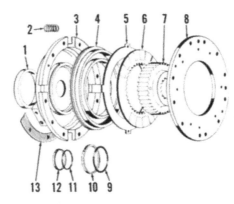

Fig. 150—Exploded view of hydraulic brakes.

1. Cup	
2. Spring	8. Back-up plate
3. Housing	9. "O" ring
4. Piston	10. Sealing ring
5. Separator plate	11. "O" ring
6. Friction plate	12. Sealing ring
7. Brake hub	13. Shim

degrees apart to retain brake housing, shim stack and differential assembly. Remove brake hub (7) and brake plates

(5 and 6). Remove brake piston assembly (4) from brake housing. Remove the three compression springs (2). Inspect brake plates and piston sealing rings and renew any parts showing excessive wear or other damage.

NOTE: If necessary to remove brake housing for any reason, it will also be necessary to remove lift shaft housing and support differential assembly. Remove oil lines and fittings, prior to removing brake housing.

It is not necessary to remove rear main housing from tractor to repair brakes. Brakes must be removed if internal repair of rear main housing is necessary.

Install the sealing rings and brake piston into housing. Remove cap screws used to retain brake housing, shim stacks and differential assembly. Install brake plates, separator plates and back up plate. Torque the twelve ½-inch cap screws to 55-65 ft.-lbs. (74.5-88.0 N·m) and the two 3/8-inch cap screws to 28-33 ft.-lbs. (37.9-44.7 N·m).

POWER TAKE-OFF

OPERATING PRESSURE

136. The pto clutch is supplied with oil from piston pump circuit. First step in testing pto clutch is to check the low or stand-by pressure, as follows: Install a 0-3000 psi (0-20 MPa) gage in the test port at the tee on pto valve as shown in Fig. 151. Place remote control levers in "Hold" position and lower three-point hitch. Start tractor engine and run at low idle rpm. Pressure reading should be 350-600 psi (2400-4100 kPa). Increase engine speed to 2300 rpm and again check pressure. It should remain at 350-600 psi (2400-4100 kPa). If stand-by pressure is incorrect problem is in low pressure valve located in piston pump compensator mechanism. This compensator is located on front of piston pump. The spool which maintains the constant 350-600 psi (2400-4100 kPa) could be sticking and giving a high or low reading. Also, spring used in conjunction with this spool could need more or less shims installed. For adjustment procedures refer to paragraph 150. When stand-by pressure has been found to be correct or has been corrected, refer to Fig. 152 and proceed as follows: Remove plugs from top of pto cover at rear of tractor and connect two 0-600 psi (0-4100 kPa) gages at these outlets. Start tractor engine and check pressure on right hand, or pto brake test port.

Gage should read 280-410 psi (1900-2800 kPa). Engage pto clutch and read pressure on left hand, or pto clutch test port. This gage should also read 280-410 psi (1900-2800 kPa). If both gages read low it indicates a problem in the pressure reducing valve area or a leak further downstream. If only one gage reads low it indicates excessive leakage in that compartment only. It should take one second for pressure in pto clutch test port to reach its full pressure. If

pressure rises too slow or too fast it indicates a problem in the modulator valve area.

A flow tester may be used to check each component for leakage if desired. Install flow tester as shown in Fig. 153. Also install a 0-600 psi (0-4100 kPa) gage in pto clutch test port and pto brake test port as shown in Fig. 153.

Start and run tractor engine until oil reaches a temperature of 130°-140°F (54°-60°C). Open flow tester restrictor

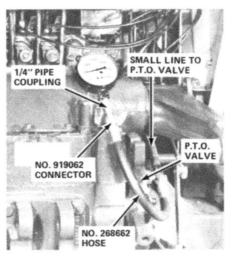

Fig. 151—View showing proper installation of pressure gage when testing pto clutch low or stand-by pressure.

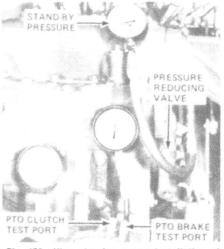

Fig. 152—View showing proper installation of pressure gages for testing pto clutch and pto brake operating pressures.

valve and observe pto brake pressure gage. Pressure should be 280-410 psi (1900-2800 kPa) with zero flow registered on flow tester. If flow is registered it indicates a leakage in pto brake internal line or piston seals. Engage pto clutch and observe flow tester flow. There should be zero flow. If flow is registered it indicates a leakage in pto internal line or piston seals.

To remove pto main shaft and pto idler shaft, first remove hydraulic lift housing. If just the output shaft and clutch shaft are to be removed, remove bottom plate on pto housing.

R&R AND OVERHAUL
OUTPUT SHAFTS & CLUTCH

137. **OUTPUT SHAFT 540-1000 RPM.** To remove the dual 540-1000 rpm shaft proceed as follows: Remove cover at bottom and pto shield from rear of housing. Remove gear retaining nut (4–Fig. 154) from shaft through opening at bottom of housing. This nut is 2-3/8 inches (60.3 mm) across the flats and will require a crowfoot, double hex open type wrench. Some type of slide hammer adapter will be necessary to pull pto shaft. Either of two types may be used. An adapter made from an old universal joint yoke can be adapted directly to splined shaft. Splined shaft must be bolted to center shaft flange. An adapter can also be made that will adapt directly to center shaft flange, instead of to splined shaft. Pull shaft rearward until front bearing cone contacts hub of gear and its retaining nut. Hold bearing cone tight against nut and gear and pull shaft with slide hammer until

front bearing cone is removed from shaft. Catch bearing cone, nut, gear and thrust bearing as shaft is moved rearward. Loosen bolts in rear bearing retainer, pry flange of hollow shaft (2) rearward until hollow shaft gear (17) can be removed, and remove gear. Continue to pry hollow shaft rearward until rear bearing cup is free of housing bore. Remove snap ring and bearing cone from hollow shaft if renewal is necessary. Remove oil seal (12) from bearing retainer (10). Remove internal snap ring (13) from front of hollow shaft, and remove two roller bearings (14) and bearing spacer (15). Remove oil seal (3) from inside diameter of hollow shaft. Remove front bearing cup if renewal is necessary. Inspect all parts and renew as necessary.

Reassembly is reverse of disassembly procedure.

138. **1000 RPM OUTPUT SHAFT.** Refer to Fig. 155 and proceed as follows: Remove gear retaining snap ring (3) from groove in output shaft. Remove bearing and oil seal retainer (11) at rear of housing, and remove from shaft. Pull shaft rearward until gear (4) and snap ring are against front bearing cone, and gear is against rear wall of housing. Pull on shaft while driving against gear hub to remove front bearing cone from shaft. A slide hammer with adapter can be attached to shaft to aid in bearing removal. Hold front bearing cone, snap ring and gear to prevent them from falling to the floor, or temporarily install bottom cover to catch them. With shaft removed retrieve front bearing cone, snap ring and gear. Remove rear bearing cone from shaft if renewal is necessary. Remove front bearing cup from housing and rear bear-

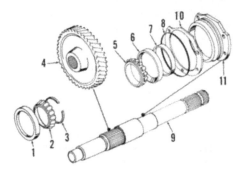

Fig. 155—Exploded view of 1000 rpm output shaft.

1. Cup		
2. Cone		7. Seal
3. Snap ring		8. "O" ring
4. Gear (1000 rpm)		9. Shaft (output)
5. Cone		10. Shim
6. Cup		11. Retainer

ing cup from retainer if renewal is necessary. Remove oil seal from retainer.

Inspect and renew any parts that are damaged or worn. Reassembly is reverse of disassembly.

139. **CLUTCH ASSEMBLY.** Remove pto brake cover (1–Fig. 156) at rear of housing. Brake piston (8) is sealed to this cover and will be removed with it. Separate brake piston from cover and inspect sealing rings (2 and 3). Remove brake disc (7) from end of shaft. Remove pto brake adapter (5) from rear bore of housing. Remove pto clutch, shaft and gear as an assembly through opening at rear of housing and place on bench for disassembly.

NOTE: Pto output shaft must be removed before pto clutch, shaft and gears can be removed.

To disassemble, clamp clutch hub (19) in a soft-jawed vise and remove bearing

Fig. 153—View showing proper installation of flow tester for checking pto clutch hydraulic circuit.

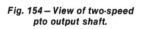

Fig. 154—View of two-speed pto output shaft.

1. Shaft (output)
2. 540 rpm shaft
3. Seal
4. Nut
5. 1000 rpm shaft
6. Cone
7. Cup
8. Cone
9. Cup
10. Retainer
11. "O" ring
12. Seal
13. Internal ring
14. Roller bearing
15. Spacer
16. Gear (1000 rpm)
17. Gear (540 rpm)
18. Thrust bearing
19. Thrust race
20. Ring
21. Pin
22. Lockwire

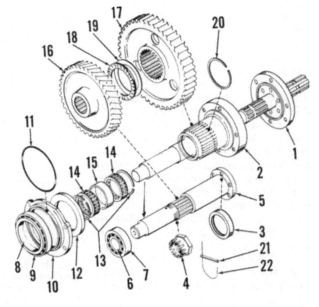

adjusting nut (21) from front end of shaft. Support clutch housing and press front end of shaft through bearing cones on shaft and in clutch hub. Remove shaft from clutch assembly. Remove snap ring retaining rear bearing cone on shaft. Support gears remaining on shaft in press bed and press shaft through gears and rear bearing cone. Remove snap ring (25), end plate (26), friction plates (27), wave washers (28) and separator plates (29). Then, remove clutch piston (30).

With clutch disassembled, clean and inspect all parts. Renew all sealing rings and any other parts showing excessive wear or other damage.

NOTE: On 1000 rpm clutch and shaft assemblies, a spacer is used instead of gear (14 – Fig. 156).

Reassemble as follows: Install steel expander and sealing ring in groove of clutch housing and clutch piston. Install clutch piston (30) into clutch housing, being careful not to damage sealing rings. Use lubricant on sealing surfaces of housing and piston. Install a separator plate (29) next to piston and a friction plate (27) next to separator plate. Install a wave spring (28) between each separator plate. Continue to install plates alternately until they all have been installed. Install end plate (26) and its retaining snap ring (25) in appropriate groove. Install the two internal snap rings inside clutch hub. Press clutch hub bearing cups in clutch hub with back face of cups toward snap rings. Lubricate and install rear bearing cone in clutch hub, with back face toward rear of hub. Install clutch hub in pto clutch. Install "O" ring type sealing ring (23) on clutch shaft. Lubricate clutch hub and sealing rings and install clutch assembly on shaft. This can be done by supporting clutch hub and inserting clutch shaft through pto clutch and pressing from non-threaded end. Extreme care should be used while pressing shaft into clutch to avoid damage to sealing "O" rings. The "O" rings seal shaft to clutch. Install front clutch hub bearing cone on shaft. Back face of cone should be installed to threaded end of shaft. Install front shaft bearing cone with back face to rear of shaft. Front bearing cone should not be pressed tight against bearings in clutch hub gear. Install front bearing adjustment nut (21). Clamp clutch housing in a vise. Measure in.-lbs. of torque required to rotate clutch hub. This can be accomplished by wrapping a cord around gear and pulling on it with a spring scale. Clutch plates and bearings should be lubricated prior to taking reading. After initial reading is obtained, tighten bearing adjusting nut

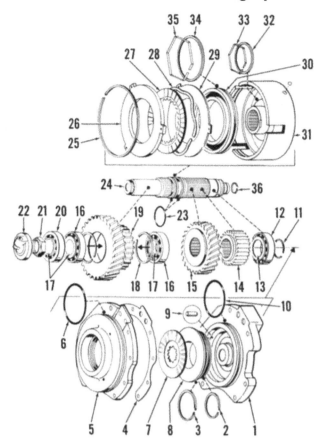

Fig. 156—Exploded view of pto clutch and brake assembly.

1. Cover
2. Seal
3. Seal
4. Shim
5. Adapter
6. "O" ring
7. Friction plate
8. Brake piston
9. Pin
10. "O" ring
11. Snap ring
12. Cup
13. Cone
14. Gear (540 rpm)
15. Gear (1000 rpm)
16. Cup
17. Cone
18. Snap ring
19. Hub
20. Cup
21. Nut
22. Plug
23. "O" ring
24. Clutch shaft
25. Snap ring
26. End plate
27. Friction plate
28. Wave washer
29. Separator plate
30. Piston
31. Hub
32. Seal
33. Expander
34. Seal
35. Expander
36. Seal

until an additional 5-10 in.-lbs. (0.5-1.1 N·m) of torque is required to rotate gear. This would be an additional pull of 3-4 pounds (13.4-17.8 N) on spring scale. Stake nut securely to shaft. Install the 26 tooth gear on rear of pto clutch shaft. Install the 18 tooth gear on rear of shaft. Longer hub of gear should be to front of shaft. Install rear bearing cone on pto clutch shaft. Install front bearing cup in bore of housing with cup back face seated against shoulder. Install rear bearing cup in pto brake adapter (5) with cup back face seated against shoulder. Install a new "O" ring seal (6) in groove of adapter. Install clutch and shaft assembly through opening at rear of housing and hold in position. Select shims (4) to obtain 0.001-0.006 inch (0.03-0.15 mm) end play in shaft and install brake adapter. Shims are furnished in thicknesses of 0.010, 0.012 and 0.015 inch. Use a dial indicator to check shaft end play. Adapter is attached with two cap screws 180 degrees apart. Torque cap screws to 28-33 ft.-lbs. (37.9-44.7 N·m). Shims, adapter and brake cover have two holes that fit over oil tube connectors. Install sealing rings (2 and 3) on brake piston. Lubricate sealing rings and inside surface of brake cover and assemble piston into cover. Dowel pin in cover must align with hole in piston. Install "O" ring seal (10) in outer groove of brake cover. Lubricate seal and sealing

surface of adapter. Install brake friction plate (7) on end of shaft splines. Make sure sealing ring in end of shaft is not damaged. Install brake cover (1) at rear of adapter, using two "O" ring seals over oil tube connectors. Insert the six cap screws and torque evenly to 28-33 ft.-lbs. (37.9-44.7 N·m).

140. REINSTALL 540-1000 RPM OUTPUT SHAFT AND GEARS. Install front bearing cup (7 – Fig. 154) with cup back face seated against shoulder in housing bore. Install one internal snap ring (13) in rear groove of hollow shaft. Install one roller bearing (14) next to snap ring, then install bearing spacer (15) and second roller bearing (14). Install second snap ring (13) to retain roller bearings. Install oil seal at rear of hollow shaft with lip of seal facing toward roller bearings. Install seal flush to 0.030 inch (0.76 mm) past flush of shaft surface. Install oil seal in rear bearing retainer with lip of seal facing toward inside of retainer. Press seal flush to 0.030 inch (0.76 mm) below flush with surface of retainer. Lubricate lip of seal and surface of hollow shaft, and install bearing retainer over hollow shaft. Bearing retainer cap screws must be placed in holes at this time. Install "O" ring seal (11) in groove of retainer. Place hollow shaft in an upright position on bench and place bearing cup on top of bearing retainer with cup back face next to retainer. Place bearing cone over hollow shaft with cone back face upward. Press or drive bearing cone on shaft until it is just past snap ring groove, and install snap ring. Place the 52 tooth gear over hollow shaft and press gear tight against snap ring. Bearing cone, at this time, must be tight against hub of gear.

NOTE: If hub of gear is not seated against snap ring, and bearing cone is not seated against hub of gear, correct bearing adjustment cannot be maintained.

Install hollow shaft assembly into rear bore of housing. Bearing cup must be started into bearing bore first, then follow up with bearing retainer. Use caution while starting retainer so that "O" ring seal is not damaged. Press or drive bearing retainer in only far enough to start cap screws. Place the 52 tooth gear through opening at bottom of housing and over splines of hollow shaft with recessed hub side rearward. Tighten bearing retaining cap screws evenly to a torque of 20-24 ft.-lbs. (27.1-32.5 N·m). Lubricate roller bearing surface of center pto shaft (5) and insert it through hollow shaft (2). Place thrust bearing (18) and bearing race (19) over end of hollow shaft with race next to hub of the 52 tooth gear. Place the 41 tooth gear in position at front of hollow shaft with grooved hub side rearward. Hold gear in position and slide center pto shaft through hub of gear. Place bearing adjusting nut (4) over front end of shaft. Place front bearing cone in bearing cup, and press or drive the shaft in until cone back face is seated against shoulder on shaft. Thread bearing adjusting nut on shaft against hub of gear, leaving some end play in shafts. Using an inch-pound torque wrench and adapter, measure rolling torque required to turn center shaft. Record this torque reading. Now tighten bearing adjusting until rolling torque increases 20 in.-lbs. (2.2 N·m) more than previous torque reading. Stake nut or install pin (21) and lockwire it to shaft. Reinstall bottom cover using a new gasket.

141. REINSTALL 1000 RPM OUTPUT SHAFT AND GEAR. Install front bearing cup (1 – Fig. 155) in bore of housing. Install oil seal (7) in retainer (11) with closed side of seal 0.030 inch (0.76 mm) past flush with rear surface of retainer. Install rear bearing cup (6) in retainer with cup back face seated against its shoulder. Install rear bearing cone (5) with cone back face seated against shoulder on shaft. Enter shaft through rear bore of housing and install gear (4) over shaft with long hub side

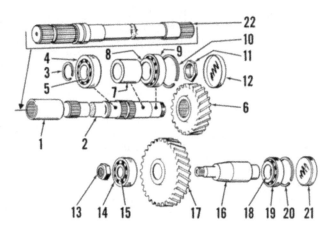

Fig. 157 – Exploded view of pto main shaft and idler shaft.

1. Coupling
2. Main shaft (rear)
3. Snap ring
4. Cup
5. Cone
6. Main gear
7. Spacer
8. Cone
9. Cup
10. Snap ring
11. Nut
12. Core plug
13. Nut
14. Cup
15. Cone
16. Idler shaft
17. Idler gear
18. Cone
19. Cup
20. Snap ring
21. Core plug
22. Main shaft (front)

forward. Install snap ring (3) over shaft next to gear hub. Place front bearing cone (2) into front bearing cup and press shaft into front bearing cone until back face of cone is seated against shoulder on shaft. Install a new "O" ring seal (8) in groove of bearing retainer. Lubricate "O" ring seal, lip of oil seal, bearing bore and surface on shaft. Select shims (10) to obtain 0.001-0.006 inch (0.03-0.15 mm) end play of shaft and install bearing retainer. Shims are furnished in thicknesses of 0.010, 0.012 and 0.015 inch. Torque retainer cap screws to 28-33 ft.-lbs. (37.9-44.7 N·m). Make sure bearing cones and cups are seated while checking shaft end play with a dial indicator.

Install pto bottom cover and torque cap screws to 45 ft.-lbs. (60.9 N·m).

R&R AND OVERHAUL MAIN & IDLER SHAFTS

142. To remove the main shaft (2 – Fig. 157), first remove hydraulic lift housing and proceed as follows: Remove oil splash plate from top of gear train. Remove plug (12) from rear bearing bore. Remove rear bearing cup retaining snap ring (10) from bearing bore. Remove bearing retaining nut (11) from rear of shaft. Remove front bearing retaining snap ring (3) from groove of shaft. Attach a slide hammer and ½-inch thread adapter to rear of input shaft. Pull shaft until rear bearing cup is free of housing bore. Continue to pull shaft until front bearing cone is free of shaft. Remove slide hammer and adapter from shaft. Move shaft rearward by hand and remove gear (6), front bearing cone and snap ring through top of housing. Remove spacer (7) from shaft. Remove front bearing cup from housing bore and rear bearing cone from shaft if renewal

is necessary. Assembly of main shaft is reverse of disassembly.

To remove idler shaft proceed as follows: Remove plug (21 – Fig. 157) at rear bearing bore of main housing. Remove rear bearing retaining snap ring (20) at rear bearing cup. Remove bearing retaining nut (13) at front of shaft. Block between gears to hold idler shaft from turning as nut is loosened. Use a ½-inch slide hammer adapter at rear of idler shaft (16) and pull until front bearing cone is free of shaft. A block should be used between back of idler gear and rear of housing while pulling shaft to prevent damage to gear and housing. Gear and shaft are a tapered fit. Remove rear bearing cone from shaft, and front bearing cup from housing if bearing renewal is necessary.

When reassembling, install front bearing cup (14) in bore of housing with cup back face seated against shoulder. Install rear bearing cone (18) on shaft with cone back face seated against shoulder. Enter shaft through rear housing bore and place idler gear (17) into position in housing, with large hole diameter rearward. Place front bearing cone (15) in bearing cup and retain with gear. Enter shaft through gear and into bearing cone. Use slide hammer and 1/2-inch thread adapter and drive shaft into bearing cone. Install a new retaining nut (13) at front end of shaft. Install rear bearing cup (19) with cup back face rearward. Torque nut at front end of shaft to 140-175 ft.-lbs. (189.7-237.1 N·m) and stake nut securely to shaft. Select a snap ring (20) to obtain 0.001-0.006 inch (0.03-0.15 mm) end play in shaft. Snap rings are furnished in thicknesses of 0.078 to 0.110 inch in steps of 0.004 inch. Install a new plug (21) at rear bearing bore with cupped side inward (flat side outward). "Loctite" sealant should be applied to sealing surface of cup.

HYDRAULIC LIFT SYSTEM

TEST AND ADJUSTMENTS

143. TORSION BAR PRELOAD ADJUSTMENT. Always adjust torsion bar preload before making other linkage adjustments if any looseness is felt in draft arms. Remove all implements and loads from draft arms and make sure torsion bar pivot bushings are well lubricated. Refer to Fig. 158 and proceed as follows: Loosen jam nuts on preload adjusting screws and turn screws in approximately two turns tight

against tractor frame. Back screws out then retighten until they are just finger tight and tighten jam nuts while holding adjusting screws in this position.

144. TRACTION BOOSTER LINKAGE ADJUSTMENT. Before making any adjustments on Traction Booster linkage, remove all implements and loads from draft arms and make sure that torsion bar preload adjustment is correct. Refer to Fig. 159 and proceed as follows: Adjust over-travel spring to a

length of six inches (152 mm) by turning locknut on lower end of spring rod. Swing draw bar completely to one side. Place range shift lever in "PARK" position and start engine. Run engine at 1000 rpm, place "Position Control" lever fully forward and "Traction Booster" lever fully to rear. Stop engine and measure distance from lower side of drawbar to lift arm pin at outer end of lift arm. If distance is greater than 37 inches (939.8 mm) adjust lift arm turnbuckles out. If distance is less than 35 inches (889.0 mm) adjust lift arm turnbuckles in. Repeat above procedure until distance is consistently between 35 and 36 inches (889.0-914.4 mm).

145. SENSITIVITY ADJUSTMENT. The sensing arm that is attached to right torsion bar arm incorporates three rod guide locating holes. Fig. 159 shows rod guide located in center position which is used for average draft loads. For heavy draft loads place rod guide in hole closest to torsion bar. This reduces amount of lift arm motion induced by "Traction Booster" system as draft load fluctuates. For light draft loads place rod guide in outermost hole to increase lift arm motion with less load fluctuation.

146. HAND LEVER FRICTION ADJUSTMENT. "Traction Booster" lever friction should be adjusted until lever will stay where it is placed while tractor is in operation. To adjust lever friction turn adjustment nut (A – Fig. 160) located on top left-hand side of console. "Position Control" lever friction should be adjusted tightly enough to prevent lever from moving forward from

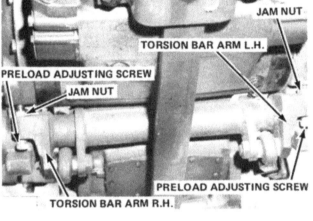

Fig. 158 — View showing adjustment points for torsion bar preload.

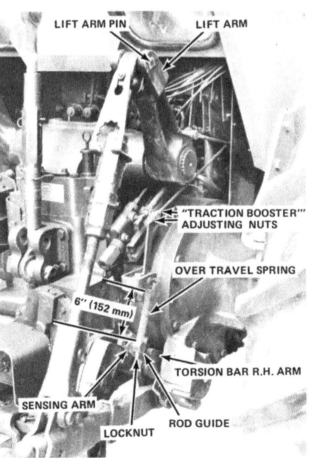

Fig. 159 — View showing adjustment points for traction booster linkage.

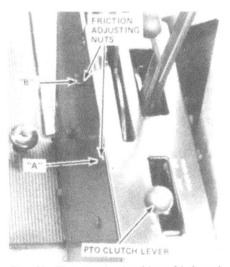

Fig. 160 — View showing hand lever friction adjustment points.

position where it is placed while tractor is in operation. To adjust lever friction, turn adjusting nut (B). If this fails to provide ample friction loosen jam nut on opposite end of shaft and tighten shaft ¼ turn. Retighten adjustment nut (B).

147. **PRESSURE RELEASE ADJUSTMENT.** If remote control valve lever does not return to the "HOLD" position when remote cylinder reaches end of its stroke, refer to Fig. 161 and adjust remote valve spring as follows: Remove rubber cap of remote valve spring retainer from remote valve to be adjusted. With engine running at 2300 rpm, use an appropriate Allen wrench to turn adjusting screw out just enough to allow valve to return to "HOLD" position. Reinstall protective rubber cap.

148. **HIGH PRESSURE CHECK.** Connect a flow tester to remote couplers as follows: Connect inlet hose of flow-tester to upper remote outlet. Connect outlet hose of flow-tester to lower remote coupler. Open restrictor valve on flow-tester to full open position. Refer to Fig. 162. Open flow control valves on remote spools to maximum flow position. Position remote valve hand lever in raise position. With engine running at 1000-1500 rpm, turn flow-tester restrictor valve until a pressure reading of approximately 1500 psi (10.3 MPa) is obtained. Let tractor operate at this setting until temperature of hydraulic oil reaches 130°-140°F (54°-60°C). After oil reaches checking temperature, increase engine speed to 2300 rpm. Turn flow-tester restrictor valve inward until a pressure of 2000 psi (13.7 MPa) is obtained. At this time, pump volume should read a minimum of 17 gpm (64.3 L/min). Turn restrictor valve of flow-tester inward until all flow has been stopped. At this time, a pressure

reading of 2300-2500 psi (15.8-17.2 MPa) should be registered. While making these tests, remote valve hand lever will have to be tied or held in raise position to keep it from cancelling and returning to hold. Another way of keeping lever in position, is to remove rubber plug from cap on back of remote spool valve and turn in on self-cancelling adjusting screw until lever will not self-cancel. If this method is used, it will be necessary to readjust self-cancelling pressure of valve. This can be done by backing out on screw. Check setting by opening and closing flow-tester restricting valve while observing pressure at which valve returns to its hold position. Remote valves should self-cancel at 1700-2000 psi (11.7-13.7 MPa). If 17-22 gpm (64.3-83.2 L/min) flow reading is not obtained, rotate flow control valve while observing flow rate to make sure it is set at maximum opening. Check flow at another spool valve to eliminate any possible malfunction in valve as being a cause of the problem. If flow is still not correct, disconnect line from three-point hitch control valve to pto valve and cap it. This will eliminate any leakage which may have been taking place at low pressure circuit. Recheck flow at remote control valve. If flow is still incorrect, problem is probably in piston pump or compensator. If it is decided that problem is at the pump or compensator, perform test again, starting with flow-tester valve open. If, as you increase pressure on pump, by restricting flow with flow-tester, pump volume sharply decreases, problem is most likely in high pressure valve. High pressure valve is located in compensator located on front of pump. Refer to paragraph 150 for pump overhaul.

PUMPS

Three hydraulic pumps supply oil under pressure to power the various hydraulic functions of tractor. These three pumps are all located together under the range transmission where they are driven at engine speed by a shaft from the pto gear train. Front (piston) pump supplies hydraulic oil for three-point hitch, power take-off control valve, differential lock control valve and control valve for brakes. Second pump or gear pump is equipped with a flow divider which splits its output into priority and secondary flows. This pump priority flow supplies oil for power steering, cooling and filtering oil in rear axle reservoir. Secondary flow from gear pump supplies lubricating and cooling oil for brakes and pto clutch. Rear pump is a gerotor type. It is recessed into range transmission housing and is not visible from outside of tractor. This

pump supplies oil for Power Director and/or Power Shift, lubrication and cooling for Power Director and/or Power Shift and lubrication, filtering and cooling in range and main transmissions. Flow rating for the piston pump is 20 gpm (75.7 L/min) for all models. Flow rating for gear pump is 9.5 gpm (35.9 L/min) for all models except 7010, 7020 and 7045 tractors which have a pump flow rating of 11.0 gpm (41.6 L/min). Gerotor pump is flow rated at 15.0 gpm (56.7 L/min) for Model 7010, 7020 and 7045 tractors and 13.5 gpm (51.0 L/min) for all other models.

All models except 7010, 7020 and 7045 are also equipped with a fourth scavenger pump which is located in the Power Director clutch housing and is driven by the front gear of the transmission countershaft. This pump transfers sump oil from the Power Director housing to the range transmission housing.

149. **R&R HYDRAULIC PUMPS.** Before removing hydraulic pumps be sure to clean pumps and connections thoroughly. When removing pumps, it will be necessary to drain lubricant from transmission, range transmission and differential housings. Remove intake line running from right filter to intake of piston or front pump. Remove intake line running from surge chamber to intake of front pump. Remove line running from left-hand filter to intake of gear pump. Remove cap screws attaching plate to bottom of range transmission. This plate is located directly to rear of hydraulic pump, and 7.4 gallons (28 liters) of oil will be drained when this plate is removed. Remove intake screen, attached to rear of gerotor pump.

NOTE: This intake screen is accessible after removal of plate on bottom of range transmission.

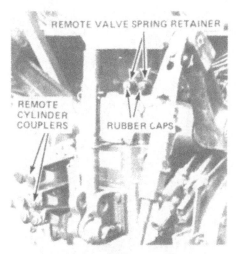

Fig. 161 — View showing location of remote valve spring retainers. Adjust remote valve lever pressure release as outlined in text.

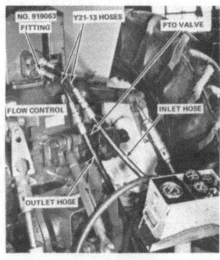

Fig. 162 — View showing checking high pressure circuit. Refer to text.

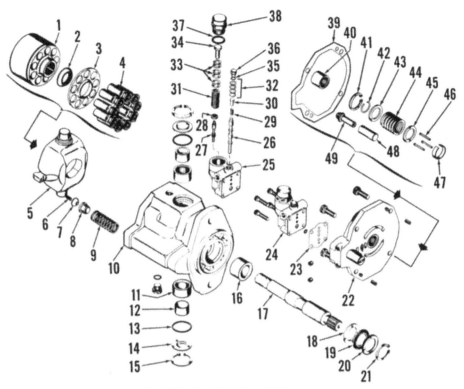

Fig. 163 — Exploded view of piston pump.

1. Block	14. Cover	26. Spool (flow)
2. Pivot	15. Snap ring	27. Spool (pressure)
3. Spider	16. Needle bearing	28. Retainer
4. Pistons	17. Shaft	29. Spring
5. Swashplate	18. Race	30. Stop
6. Roll pin	19. Thrust bearing	31. Spring
7. Button	20. Race	32. Shims
8. Guide	21. Retainer	33. Shims
9. Spring	22. End plate	34. Stop
10. Housing	23. Gasket	35. "O" ring
11. Needle bearing	24. Compensator assy.	36. Plug
12. Race	25. Housing	37. "O" ring
13. "O" ring	(compensator)	38. Plug

39. Gasket
40. Needle bearing
41. Snap ring
42. Retainer-early model pumps
43. Washer (special)
44. Spring
45. Washer
46. Pins
47. Keeper
48. Piston (control)
49. Pin

ed properly overheating of transmission fluid will result.

150. OVERHAUL PISTON PUMP. Thoroughly clean outside of pump. Clamp pump drive shaft in a soft-jawed vise with body of pump up. Refer to Fig. 163 and proceed as follows: Remove Allen head retaining screws holding compensator assembly (24) to end plate (22). Remove compensator assembly, "O" ring and gasket (23). Remove four cap screws from end plate (22) then use a plastic mallet to tap the end plate loose. Pull end plate straight out, remove gasket (39) and control piston (48). Invert pump assembly and let piston block (1) fall out into your hand.

On early model pumps it will be necessary to construct the special tool shown in Fig. 163B and proceed as follows to remove piston block from pump body. Set special tool on piston block with two fingers centered on outside land of piston block. Insert two 3/8-16x4½-inch cap screws through holes in special tool and screw them into tapped holes in aluminum case. Tighten cap screws evenly until piston block is depressed approximately 1/16 inch (1.6 mm). Use a pair of sharp tools to remove retaining ring (42 – Fig. 163) from drive shaft. After removing retaining ring remove special tool and lift piston block from pump body. Note that retaining ring (42) is used only in early model pumps.

If pistons (4) did not come out with piston block, lift them out along with spider (3) and pivot (2). Piston block assembly need not be disassembled unless internal pins (46) or spring (44) is damaged.

CAUTION: Spring (44) is highly compressed and snap ring (41) should not be removed without compressing the spring. The following procedure should be used if spring is to be removed from piston block.

Place a 3/8 IDx1⅛-inch flat washer over the end of a 3/8-16x3¼-inch cap screw and insert cap screw through center of piston block. Place another flat washer over threaded end of cap screw and center it in washer (45). Install an appropriate nut on cap screw and use it to compress spring (44) inside piston block. Remove snap ring (41). Remove cap screw, nut and washers. Withdraw washer (43), spring (44), washer (45), pins (46) and keeper (47).

On early model pumps withdraw shaft (17) from housing. Remove snap ring (21) from shaft and lift off thrust washers (18 and 20) and bearing (19).

On later model pumps remove snap ring (48 – Fig. 163A) from housing (10) and withdraw shaft (17) along with

Disconnect all other external lines attached to hydraulic pump. Remove all line clamps in immediate pump area. Move lines to provide room for pump to be removed from tractor. Remove the two ½-inch cap screws attaching piston pump to gear and gerotor pump. By moving piston pump ahead approximately two inches (50.8 mm), pump can be removed. After piston pump has been removed, the three cap screws attaching gear and gerotor pump assembly to range transmission housing can be removed. Move gear and gerotor pump assembly forward until it clears housing enough for removal.

Install externally mounted hydraulic pumps in opposite order of removal.

The scavenger pump which is used in all model tractors except 7010, 7020 and 7045 can be removed from Power Director compartment without separating engine from transmission. Drain oil from Power Director clutch compartment and remove cover from bottom of Power Director clutch housing. Remove cap screws holding oil baffle to scavenger pump. Remove cap screws

that secure pump to transmission housing. Pull down pump to separate pump tube from tube in transmission housing. This is a slip joint that is "O" ring sealed.

To install scavenger pump in transmission first install a new "O" ring in groove of pump tube and lubricate it with transmission fluid. Place oil pump baffles (9 and 10 – Fig. 164A) in housing. Insert pump assembly through hole in bottom of clutch housing and guide pump tube of transmission housing. Install pump mounting cap screws. Check pump drive to driven gear backlash. Gear backlash should be 0.005-0.030 inch (0.12-0.76 mm). If backlash is less than 0.005 inch (0.12 mm) shim pump away from transmission housing. Install oil baffle (9 – Fig. 164A) across front face of oil pump so it fits close to pump and seal all gaps using a suitable silastic sealer. Make sure rubber portion of second baffle (10) contacts side of transmission housing and side and bottom of clutch housing cover. Tighten all baffle retaining cap screws securely. Install clutch housing bottom cover.

NOTE: If oil pump baffles are not install-

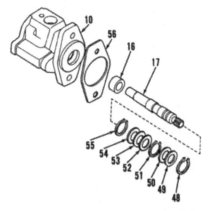

Fig. 163A — Exploded view of late model piston pump housing and shaft assembly.

10. Housing	51. Snap ring
16. Bushing	52. Thrust washer
17. Shaft	53. Bearing
48. Snap ring	54. Thrust washer
49. Spacer	55. Snap ring
50. Spacer	56. Gasket

Fig. 163B — To remove piston block from pump body of early model pumps it will be necessary to fabricate a special tool to the dimensions shown.

A. 13/32 inch (10.3 mm)
B. 6-5/8 inch (168.2 mm)
C. 5-19/32 inch (142.0 mm)
D. 3/8 inch (9.5 mm)
E. 3/4 inch (19.0 mm)
F. 3-3/8 inch (85.7 mm)
G. 3-1/4 inch (82.5 mm)
H. 1-5/16 inch (33.3 mm)

spacers (49 and 50) and associated snap rings, thrust washers and bearing. Remove snap ring (51) from shaft and withdraw thrust washers (52 and 54) and bearing (53).

Remove two covers (14 – Fig. 163), two "O" rings (13), two inner races (12) and two bearings (11). The swashplate (5) can now be lifted out of housing (10). Remove spring guide (8) and spring (9). Do not remove button (7) or roll pin (6) unless they are loose or worn. Remove plug (38), stop (34), shims (33), spring (31), retainer (28) and spool (27). Remove plug (36), stop (30), shims (32), spring (29) and spool (26).

Clean all parts in a suitable solvent and carefully inspect for wear, damage, metal flaking and metal build-up. Piston shoes should not be lapped. Renew any worn or damaged parts and all "O" rings and gaskets. Prior to assembly lubricate all parts in a clean, high viscosity lubricating oil.

The swashplate pivot bearings (11) are a slip fit. Shaft bearing (16) is a press fit. Install all bearings with numbered end out. Install spring (9) and guide (8) in housing (10). Install swashplate (5) in housing then insert bearings (11) and inner races (12) over swashplate shaft and into housing. Numbered end of race should face outward and chamfered ID should be inward. Install "O" rings (13) around OD of bearings. Install covers (14) and snap rings (15). Install seals and spacers on shaft (17) in opposite order of removal. Install shaft assembly in housing (10) and on later models, secure in place with snap ring (48 – Fig. 163A). On all models, install pin keeper (47 – Fig. 163) in piston block (1). Install pins (46) in splined grooves of piston block with heads inward. Install washer (45),

spring (44), washer 43) with chamfered ID up. As in removal, use cap screw, nut and flat washers to compress spring and install snap ring (41). Install pivot (2), spider (3) and pistons (4) in piston block. Install piston block assembly in housing; piston shoes must be in contact with swashplate. Be sure all parts are in their proper positions and move freely. Install gasket (29). Install piston (48) on end plate (22). Install end plate and torque retaining screws to 27-31 ft.-lbs. (37-42 N·m). Install spool (26), spring (29), shims (32), stop (30), "O" ring (35) and plug (36). Install spool (27), retainer (28), spring (31), shims (33), stop (34), "O" ring (37) and plug (38). Install gasket (23) and related "O" ring, then install compensator assembly (24) and tighten

retaining screws to 8-10 ft.-lbs. (11-14 N·m).

Lay pump aside for later installation.

151. OVERHAUL GEAR AND GEROTOR PUMP. Thoroughly clean outside of pump, refer to Fig. 164 and proceed as follows: Remove the four 12-point cap screws from end of housing (33). Remove "O" rings (34). Withdraw housing from pump drive shaft. Remove thrust plate (29). Remove outer gear (30) and inner gear (31) from housing (33). Remove key (32) from drive shaft and gear (21). Remove the eight 12-point cap screws from gear pump housing (20), then use a plastic mallet and tap the end of drive shaft (21) to separate adapter

Fig. 164 — Exploded view of gerotor and gear pump.

1. Plug
2. "O" ring
3. Cap
4. Spool
5. Plug
6. "O" ring
7. Orifice
8. Spring
9. Shims
10. Cap
11. "O" ring
12. Plug
13. Wear plate
14. Gasket
15. Seal
16. Seal
17. Seal
18. Gasket
19. "O" ring
20. Housing (gear pump)
21. Drive gear
22. Idler gear
23. Star washer
24. Shim
25. Spring
26. Ball
27. Adapter
28. Seal
29. Plate
30. Outer gear (gerotor)
31. Inner gear (gerotor)
32. Key
33. Housing (gerotor pump)
34. "O" ring

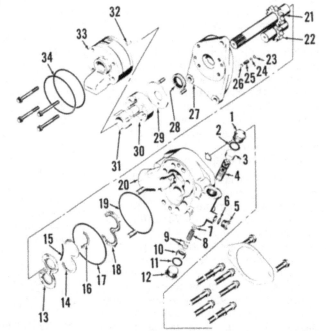

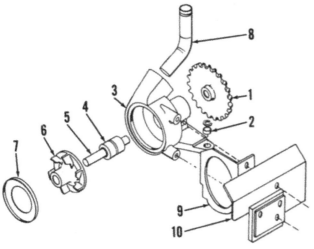

Fig. 164A — Exploded view of scavenger pump used on all models except 7010, 7020 and 7045. Bearing (4) and shaft (5) are not available separately.

1. Gear
2. Bushing
3. Housing
4. Bearing
5. Shaft
6. Impeller
7. Plug
8. Tube
9. Front oil baffle
10. Side oil baffle

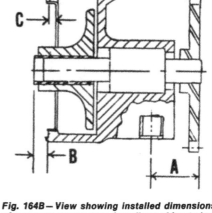

Fig. 164B — View showing installed dimensions of scavenger pump gear, impeller and front plug. See text for outline of pump assembly procedure.

A. 1.250 inches (31.75 mm)
B. 0.170-0.180 inch (4.31-4.57 mm)
C. 0.291-0.301 inch (7.39-7.64 mm)

(27) from pump housing (20). Withdraw drive shaft and gear (21) and idler gear (22) from housing (20). Remove "O" ring (19). Remove wear plate (13), seal (15), gasket (14), seal (16), "O" ring (17) and gasket (18) from housing (20). Remove shaft seal (28) from adapter (27). It is not necessary to remove star washer (23), washer (24), spring (25) or steel ball (26) unless there appears to be damage to these parts. Place gear pump housing in a soft-jawed vise and remove plugs (1 and 12), caps (3 and 10), shims (9), spring (8), orifice (7) and spool (4) from flow divider.

Clean all parts in a suitable solvent and dry with compressed air. Renew all gaskets and "O" rings. Inspect all parts for wear, damage, metal flaking and metal build-up.

Minimum idler gear or drive gear shaft diameter is 1.1220 inch (28.498 mm), renew shaft and gear as an assembly if necessary. If gear width is less than 0.4970 inch (12.623 mm) gear and shaft assembly should be renewed. Break edge of sharp gear teeth with fine emery cloth.

If the ID of any bushing exceeds 1.130 inch (28.70 mm) the housing in which it is located should be renewed as the bushings are not available separately. Associated bearings should show no signs of scoring or galling. ID of gear housing (33) should not exceed 3.004 inch (76.30 mm) and should be free of any scoring. Gerotor side of adapter (27) should be free of any scoring and wear should not exceed 0.0015 inch (0.038 mm). Inspect gear pocket of housing (20) for scoring and wear. Clearance between edge of gear teeth and inside surface of gear pocket should not exceed 0.010 inch (0.25 mm). Inspect flow divider spool bore which should be free from any deep scratches.

Prior to assembly lubricate all parts in a clean high viscosity lubricating oil.

Items (13, 14, 15, 16, 17 and 18) should be assembled into a subassembly then installed into the gear pocket of pump housing (20). Bronze side of wear plate (13) should be facing pump gears. Be sure cut out contours of wear plate match contours of gear pocket. Install "O" ring (19) in housing (20). Install idler gear assembly (22) in pump housing with long shaft end through wear plate. Install drive gear assembly (21) in pump housing with short end of shaft through wear plate. Install steel ball (26), spring (25), shim (24) and retainer (23) if they were removed. Install adapter (27). Install adapter retaining cap screws and tighten them to 80-100 ft.-lbs. (108-136 N·m). Install shaft seal (28) into adapter. Use No. 601 Loctite on OD of seal. Install seal with lip facing towards rear. Install key (32) into drive shaft. Install inner gear (31) and outer gear (30) into housing (33). Install thrust plate (29) over gears with bronze face toward gears. Coat both sides of gasket between gerotor pump housing and adapter with No. 3 Permatex. Install gerotor housing assembly onto adapter and tighten retaining cap screws to 27-30 ft.-lbs. (37-41 N·m). Install "O" rings (34) onto housing (33). Before installing flow divider spool (4), examine spool carefully. Distance from end of spool to the four scroll drillings is further from one end than the other. Install long end of spool towards flow divider test port. Install orifice (7) and spring (8) in long end of spool then install spool in housing as previously outlined. Install caps (3 and 10) and plugs (1 and 12). Tighten plugs to 15-20 ft.-lbs. (20-27 N·m).

Install flange gasket on end plate of piston pump and coat with No. 3 Permatex. Engage splines of piston pump drive shaft with internal splines of gear pump shaft and secure pumps together using appropriate retaining screws. Tighten retaining cap screws to 70-80 ft.-lbs. (95-108 N·m). Install pumps on tractor and connect all lines. Then, fill piston pump half full of system oil. Fill oil reservoir and start engine and run at low idle.

CAUTION: Pump should immediately pick up oil and fill system; if there is no indication of oil pick up in 60 seconds, stop engine and determine cause.

After system starts to show signs of oil pickup, operate control valve to purge system. Continue to operate system slowly with no load until system responds fully. Check lines for leaks and reservoir fluid level.

151A. OVERHAUL SCAVENGER PUMP. Refer to Fig. 164A and proceed as follows: Using a suitable press and press plate remove drive gear (1) from pump shaft. Be sure to support drive gear as close to hub as possible. Note that it may be necessary to heat shaft to release grip of previously applied Loctite. Remove plug (7) then press impeller (6) and shaft (5) out of housing (3). Plug (7) is disposable and will have to be renewed upon reassembly of pump. Inspect all parts for wear or damage and renew as necessary.

Assemble pump by pressing shaft assembly into housing using a press sleeve that will contact the outer race of bearing (4). Be sure long end of shaft (5) is facing impeller side of pump and bearing (4) is pressed in until it contacts shoulder in housing (3). Apply a thin coat of Loctite (AC part 927938) to gear end of shaft. Press gear (1) onto shaft until distance (A – Fig. 164B), measured from center of pump mounting hole to face of gear, is 1.250 inches (31.75 mm). Apply

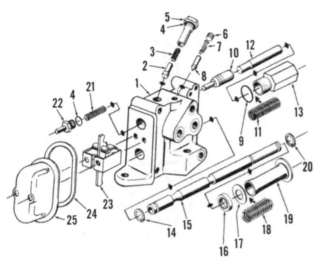

Fig. 165 — Exploded view of three-point hitch control valve.

1. Body
2. Poppet (lift check)
3. Spring
4. "O" ring
5. Plug
6. Plug
7. Spring
8. Poppet (pilot check)
9. "O" ring
10. Spool (priority)
11. Spring
12. Pin
13. Cap
14. Snap ring
15. Spool (control)
16. Seal
17. Washer
18. Spring
19. Sleeve
20. Retainer
21. Spring
22. Spool (feedback)
23. Bracket
24. Gasket
25. Cover

screw retaining feedback bracket to valve body should be torqued to 20-25 ft.-lbs. (27-34 N·m). To avoid distorting rear cover of valve, screws retaining cover should be torqued to 18-30 in.-lbs. (2.0-3.4 N·m). Tie bolts securing remote valves or plate to side of three-point control valve should be evenly torqued to 12-15 ft.-lbs. (16-20 N·m). Mounting nuts securing valve to lift arm housing should be torqued to 70-80 ft.-lbs. (95-108 N·m).

153. R&R AND OVERHAUL REMOTE VALVE. Remove operators seat and seat mounting bracket from tractor platform. Thoroughly clean area around valves to prevent dirt from entering hydraulic system. Because of close fits involved with hydraulics, extreme cleanliness must be observed. Remove "E" rings and pins attaching spools to control linkage. Remove pivot shaft support and remote valve outlet(s). Remove tie bolts retaining remote valve to three-point control valve, and lift valve from tractor.

Remove cap screw retaining cover (15 – Fig. 166) at rear of actuator spool. Slide actuator spool (4) out rear of valve body. Clamp linkage end of spool in vise.

CAUTION: Avoid scratching or nicking spool.

Remove screw retaining the self-cancelling mechanism to spool. Remove

a thin coat of Loctite (AC part 927938) to impeller end of shaft. Press impeller (6 – Fig. 164A) onto shaft until distance (B – Fig. 164B), measured from end of impeller bushing to machined face of housing, is 0.170-0.180 inch (4.31-4.57 mm). Press plug (7 – Fig. 164A) into housing until distance (C – Fig. 164B), measured from front face of housing to front face of plug, is 0.291-0.301 inch (7.39-7.64 mm). Check pump for freedom of movement, impeller blades must not touch front plug. If tube (8 – Fig. 164A) was removed from housing (3) it must be sealed to housing during reassembly using Loctite (AC part 930235).

CONTROL VALVES

152. R&R AND OVERHAUL HITCH VALVE. The three-point hitch control valve is mounted on top of lift arm housing. Tractor seat and seat bracket must be removed to gain access. Disconnect all lines secured to right side of valve stack. Remove pivot shaft supports. Remove nuts retaining valve to lift arm housing. Clean exterior of valve and remove tie bolts. Separate remote valves from three-point control valve if tractor is equipped with remote valves. Remove cover (25 – Fig. 165) at rear of valve. Remove socket head cap screws retaining feedback bracket (23) to valve body. Remove feedback spool assembly (22) and spring (21). Remove snap ring (14) from draft control spool (15). Withdraw draft control spool through front of valve body. Pressure control spring (18) and sleeve (19) can be removed, if necessary, by removing snap ring (20) at front end of spool. Remove cap (13) covering priority valve mechanism. Remove priority spring (11), pin (12) and valve (10). The lift check poppet (2) is contained under hex plug (5) located on

top of valve body. Pilot check poppet (8) is located under the 1/8-inch N.P.T. socket type plug (6). Front portion of draft control spool is sealed to valve body with a lip type seal (16). All other seals used in valve are of the "O" ring type, except the gasket used at rear cover.

All internal parts should be inspected prior to assembly. They should contain no deep nicks or scratches.

A suitable lubricant should be used as valve is reassembled. Assembly procedure is reverse of disassembly. Cap

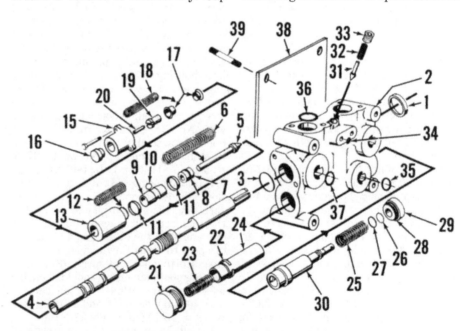

Fig. 166 — Exploded view of remote control valve.

1. Seal	11. Collar	21. Plug	31. Poppet (pilot check)
2. Body	12. Spring	22. Snap ring	32. Spring
3. "O" ring	13. Barrel	23. Spring	33. Plug
4. Spool (control)	14. Cover	24. Spool (load check)	34. "O" ring
5. Poppet (kickout)	15. Cover	25. Spring	35. "O" ring
6. Spring	16. Plug	26. "O" ring	36. "O" ring
7. Seat	17. Seat	27. Back-up ring	37. "O" ring
8. "O" ring	18. Spring	28. "O" ring	38. End plate
9. Sleeve	19. Screw	29. Bushing	39. Tie bolt
10. Ball	20. Setscrew	30. Spool (flow control)	

self-cancelling mechanism making sure none of the small parts are lost. Parts making up spool and self-cancelling mechanism are shown in Fig. 167. All parts should fit closely to spool, but should slide freely. Screw that retains this mechanism to the spool contains another screw that is used to adjust self-cancelling pressure. This screw has a nylon insert and should be renewed if there is not some resistance to turning. Spool is sealed to valve body at front end, by a lip type seal (1–Fig. 166). Opposite end of spool and self-cancelling mechanism are sealed to valve body with "O" rings. Flow control spool (30) can be removed out front of valve body after removing bushing type nut (29). Load check spool (24) can be removed after removing plug (21) at rear of valve. Pilot check poppets (31) can be removed after removing the 1/8-inch N.P.T. socket type plug (33) located at upper portion of valve body.

Check all valve parts for scoring, etc. Any damaged parts should be renewed.

As valve is reassembled, cleanliness of all parts should be maintained. Lubricate mating parts with a suitable lubricant prior to assembly. A lubricant should also be used on all seals. Install remote outlet on valve and install valve on tractor in reverse of removal. Torque tie bolts holding valve sections together evenly to 12-15 ft.-lbs. (16-20 N·m). Cap screws attaching remote outlets to remote valve should be torqued to 28-33 ft.-lbs. (38-45 N·m).

Check valve spool for freedom of operation after valve is assembled. Set self-cancelling pressure to 1800-2100 psi (12.4-14.4 MPa). If spool will not self-cancel, loosen and retighten tie bolts, as uneven tightening of these bolts can distort valve body enough to interfere with free movement of spool.

154. R&R AND OVERHAUL PTO CLUTCH VALVE. Thoroughly clean

Fig. 168 — Exploded view of pto clutch valve.

1. Plug
2. "O" ring
3. Shim
4. Spring
5. Reducing valve
6. Plug
7. "O" ring
8. Spool retainer
9. Spring
10. Retainer
11. "O" ring
12. Plug
13. "O" ring
14. Body
15. Snap ring
16. Spool
17. Seal
18. Accumulator plunger
19. Spring
20. Spring
21. Accumulator piston
22. Cap
23. Snap ring

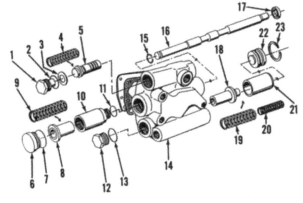

area around valve before beginning disassembly. Disconnect all hydraulic lines from valve and disconnect mechanical linkage. Remove attaching cap screws and pull valve to right until it clears lines running to pto clutch and brake. Place valve on workbench and remove top plug (1–Fig. 168). Withdraw shim (3), spring (4) and pressure reducing valve (5). Remove plug (6) and withdraw spool retainer (8), spring (9) and spool (16). Remove snap ring (23). Remove plug (12) and use a soft rod to push accumulator plunger (18), springs (19 and 20), accumulator piston (21) and cap (22) out of body (14).

Clean all parts in a suitable solvent and inspect for nicks, scores and wear. Renew all seals and "O" rings. Front of spool (16) is sealed to body with a lip type seal (17). All other seals are "O" rings.

Assemble valve in reverse order of disassembly. Install valve on tractor in reverse order of removal. After valve has been mounted on tractor tighten attaching cap screws evenly to 28-33 ft.-lbs. (38-45 N·m).

ROCKSHAFT HOUSING AND LIFT CYLINDER

155. REMOVE AND REINSTALL HOUSING. To remove lift arm housing, move hitch to fully lowered position and disconnect three point lift links from rockshaft arms. Remove seat, mounting bracket and platform section to which front of seat bracket is attached. Disconnect all lines and links that might interfere. Remove cap screws attaching lift arm housing to rear main housing. Attach a sling to top of lift arm housing and lift housing off tractor with a hoist.

Reinstall lift arm housing on tractor in reverse order of removal. The 3/4-inch cap screws should be torqued to 245-275 ft.-lbs. (332-373 N·m) and 5/8-inch cap

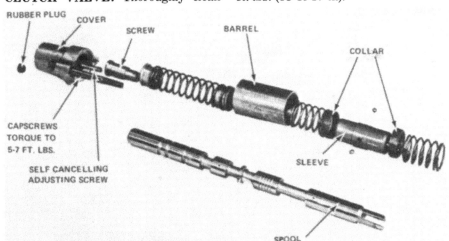

Fig. 167 — View of spool and self-cancelling mechanism.

Fig. 169 — View showing lift arm housing positioned on work bench for disassembly.

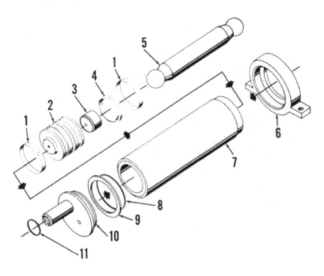

Fig. 170 — Exploded view of lift cylinder.

1. Wear rings
2. Piston
3. Insert
4. Seal
5. Rod
6. Brace
7. Tube
8. "O" ring
9. Back-up ring
10. Cap
11. "O" ring

screws should be torqued to 140-160 ft.-lbs. (190-217 N·m).

156. R&R AND OVERHAUL LIFT CYLINDER. Remove lift arm housing as outlined in paragraph 155 and place lift arm housing on workbench with bottom side up. Refer to Fig. 169 and proceed as follows: Remove piston rod from crank arm. Remove lockwashers, cap screws and washers that attach cylinder support brace to lift housing then remove brace from end of cylinder. Remove cylinder from end cap assembly. Remove end cap from lift housing. Remove "O" ring seal from housing bore. Remove "O" ring seal and back-up ring from end cap. Remove seal and wear rings from piston.

Clean all parts in a suitable solvent and inspect for wear or damage. Renew as necessary. Renew all "O" rings and seals.

Install "O" ring (11 – Fig. 170) in end cap (10). Install back-up ring (9) and "O" ring (8) on pressure side of end cap, and insert end cap into bore of tube (7). Install piston seal (4) in center groove of piston (2) with seal lip facing pressure side of piston. Install wear rings (1) in grooves in both ends of piston. Insert piston into tube with piston rod side facing toward end of tube that has been machined for support brace (6). Install support brace over tube and attach to lift arm housing using appropriate cap screws and hardened washers. Torque cap screws to 47-54 ft.-lbs. (64-73 N·m) and secure with lockwires.

Install piston rod (5) in piston and crank arm. Tighten cap screws in crank arm finger tight against piston rod, then back off cap screws ¼ to ½ turn and secure in this position with lockwires.

157. R&R AND OVERHAUL ROCK-SHAFT. To remove rockshaft, first re-

move lift housing as outlined in paragraph 155. Remove snap rings (5 – Fig. 171) retaining right and left lift arms, and remove lift arms (6) and thrust washers (10). Remove snap rings (8) from grooves at each side of the crank arm (13) at center of shaft. Pull shaft from left side of housing and remove crank arm and retaining snap rings. Remove seal rings (7 and 9) from grooves of housing. Remove bushings (2 and 3) if renewal is necessary.

If new bushings are being installed, press the two outer bushings 3/4-inch (19 mm) past flush with outer surface of

housing. Press center bushing to a dimension of 12-7/16 inches (316 mm) from left surface of housing. Bushings are prefit and require no reaming after installation. Install "O" ring seal (9) in groove of housing at left side. Install "O" ring seal (7) in groove at right side of housing. Lubricate "O" ring seals and bushings prior to installing lift shaft. Enter shaft from left side of housing and install a snap ring for retaining crank arm. One shaft spline and crank arm have a punch mark for indexing arm to shaft. Place crank arm in position in housing and enter end of shaft through crank arm. Place second crank retaining snap ring over shaft. Push shaft through housing until first snap ring contacts end of splines; then place snap ring over splines. Align punch mark on shaft splines with punch mark on crank arm, and push shaft splines through crank arm. Install snap rings in grooves at each side of crank arm. Install same number of thrust washers (10) at each side of housing that were removed from shaft. Add washers as required to limit side movement of shaft to less than two washers. An even amount of washers must be used on both sides of housing. Install and align lift arms with punch marks on ends of shaft. Install lift arm retaining snap rings (5). When attaching piston rod to crank arm, adjust cap screws finger-tight against piston rod; then back off each cap screw (14) ¼ to ½ turn and secure cap screws with lock wire.

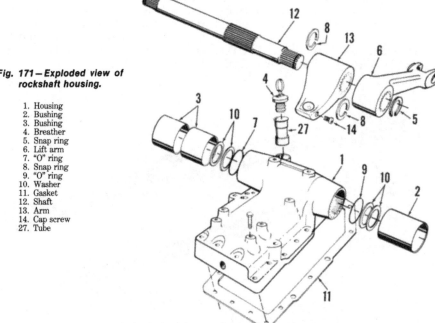

Fig. 171 — Exploded view of rockshaft housing.

1. Housing
2. Bushing
3. Bushing
4. Breather
5. Snap ring
6. Lift arm
7. "O" ring
8. Snap ring
9. "O" ring
10. Washer
11. Gasket
12. Shaft
13. Arm
14. Cap screw
27. Tube

CAB AND PROTECTIVE FRAME

CAB, PLATFORM FRAME AND CONTROL MODULE

All Models So Equipped

158. **REMOVE AND REINSTALL CAB.** Set rear wheel tread to provide a minimum of 55 inches (1397 mm) between inside edges of tires with wheels equally spaced from center of tractor. Refer to Fig. 172 and proceed as follows: Remove "U" bolts that attach cab support brackets to rear axle housings. Heat nuts to 200°F (93°C) to break Loctite seal. Remove spacer nut from "U" bolt at front of axle. It is not necessary to remove cab support, isolator or cab mounting bracket from cab.

Refer to Fig. 173 and proceed as follows: Remove cover from left side of console. Disconnect control cables from control levers and remove tie wraps from cable anchors. Pull control cables through holes in rear of cab. Remove floor mat moulding, retainers and floor mat from cab. Remove center floor section from cab. Disconnect side shaft cable from range shift and park lock lever at the rear of main housing. Remove entrance handle from side of instrument panel cowl. Disconnect cab wiring harness from tractor wiring harness. Remove carriage bolts and cap screws attaching cab to platform frame assembly. Devise a sling made from cables and hooks that will attach to cab so a hoist can be used to lift cab from tractor. Lift cab slowly and carefully to make sure that all bolts, cables and electrical wiring are removed or disconnected. Place cab in a suitable place where it can be covered and protected from damage.

If for any reason the right and left cab mounting brackets, isolators and supports have been removed they must be reinstalled using the correct size and grade of cap screws, and torqued as follows: Install 5/8-16x3¼-inch, grade 5 cap screws in upper front holes of both brackets. Install 5/8-16x1¾-inch grade 5 cap screws in remaining three holes of each bracket. Torque the nuts of these eight grade 5 cap screws to 140-160 ft.-lbs. (190-217 N·m). Install a 5/8-18x3½-inch grade 8 cap screw through washer, isolator, cab support and mounting bracket on each side of tractor, then torque nut to 180-220 ft.-lbs. (244-298 N·m).

NOTE: Do not substitute cap screws of a lesser size or grade in case original bolts become lost or damaged. Use only cap screws of the size and grade previously specified.

Use a sling made from cables and hooks to attach cab to a suitable hoist. Lift cab high enough that it can be lowered onto tractor with all mounting holes in alignment. Place two shims between cab supports and rear axle housings. Install "U" bolts in center position around axle housing as shown in Fig. 172. Use Loctite TL-222 on threads of both "U" bolts. Install spacer nut on front (short end) of "U" bolt and tighten nut until "U" bolt contacts bottom of axle housing. Install spacer, washer and nut on rear (long end) of "U" bolt. Torque nuts evenly to 70-80 ft.-lbs. (95-108 N·m).

NOTE: "U" bolts are made from special material and should not be substituted with any other type of "U" bolt. Use only "U" bolts provided by Allis-Chalmers for this purpose.

Install two ½-13x¾-inch, grade 2 cap screws and lockwashers through holes on left side of cab and one ½-13x¾-inch, grade 2 cap screw and lockwasher through mounting hole on right side of cab. Install two ¾-16x1-inch carriage bolts through holes in front of cab. Use 3/8-16x¾-inch grade 2 cap screws in remaining mounting holes. Attach cab grounding cable to right front carriage bolt. Attach cab wiring harness to tractor wiring harness. Reconnect park lock

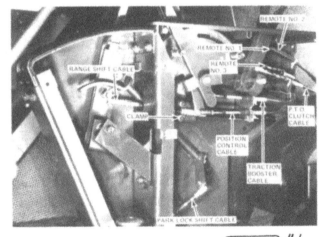

Fig. 173 — View showing console with left cover removed.

Fig. 172 — View showing correct location of shim and long and short spacer nuts of cab mounting bracket.

Fig. 174 — Illustration showing location of cable routing holes in rear of cab.

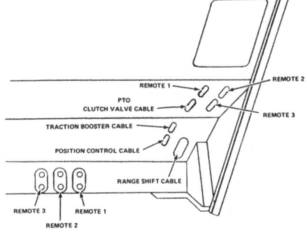

cable to park lock lever. Refer to Fig. 174 and feed control cables through their appropriate holes in rear of cab and attach cables to control levers. Install cab floor and attach it to platform frame using 3/8-16x3/4-inch cap screws. Install floor mat, moulding and retainers.

159. R&R PLATFORM FRAME AND CONTROL MODULE. Remove cab from tractor as outlined in paragraph 158. Remove step and brace from left side of tractor. Remove main and auxiliary fuel tanks. Remove batteries and battery box. Disconnect throttle and fuel shut-off cables from fuel injection pump. Disconnect and plug power steering hoses at steering cylinder. Disconnect brake oil line from brake valve and plug all open line ends. Disconnect oil lines to power director valve and remove valve from tractor. Remove 5/8-11x1¼-inch, grade 8 cap screws attaching front platform frame support to transmission housing. Attach a suitable hoist to control module and platform frame assembly, using a sling made from cables and hooks. Lift control module and frame assembly carefully to make sure wiring harness, oil lines and control linkage have been disconnected. Remove rear platform support channel from rear main housing.

Install rear platform support to rear main housing top angle facing forward, using two ½-13x1¼-inch, grade 2 cap screws and lockwashers. Attach rear platform support channel to support using two ½-13x1¼-inch, grade 2 cap screws with lockwashers and nuts. Attach a sling made from cables and hooks to control module and platform frame assembly. Using a suitable hoist lift assembly into place on tractor. Attach platform frame assembly to transmission housing using two 5/8-11x1¼ inch, grade 8 cap screws with lockwashers and nuts. Reconnect brake valve oil lines and hoses. Connect throttle and fuel

shut off cables to fuel injection pump. Reinstall fuel tanks and reconnect fuel lines. Reinstall battery box and batteries.

PROTECTIVE FRAME

All Models So Equipped

160. REMOVE AND REPLACE. Set tractor rear wheel tread to provide a minimum of 55 inches (1397.0 mm) between inside edge of tires. Rear tires should be equally spaced from center of tractor. Remove "U" bolts that attach protective frame support brackets to rear axle housings. Refer to Fig. 173 and remove cover from left hand side of console. Disconnect control cables from control levers and remove tie wraps from cable anchors. Pull cables through holes in rear of protective frame. Remove retainer strips that attach floor mat to platform and remove mat. Remove center platform section and disconnect shift shaft cable from range shift and park lock lever in rear main housing. Remove cable from cable anchor. Disconnect tractor wiring harness from protective frame wiring harness and remove clamp securing harness to tractor. Using cables and hooks attach a suitable hoist under roof section of protective frame being careful not to damage plastic roof section. Support weight of frame and remove the 5/8-18x1¼ inch, grade 5 cap screws that attach frame to each side of platform assembly. Move protective frame rearward until it is clear of rear edge of platform, then lift slowly until it is clear of tractor. As frame is being lifted clear of tractor check to be sure that all wiring, cables and hoses have been disconnected.

Attach a suitable hoist under roof section of protective frame and lift frame into position on tractor, being careful to position front edge of seat support under rear edge of left and right hand

platform sections. If protective frame support brackets have been removed they must be reinstalled as follows: Install a 5/8-18x3½ inch, grade 5 cap screw with nut and lockwasher in top forward hole of each bracket. Install 5/8-18x1¾ inch, grade 5 cap screws with lockwashers and nuts in the remaining holes of each support bracket. Torque all eight cap screws to 140-160 ft.-lbs. (190-217 N·m).

CAUTION: Do not substitute cap screws of a lesser size or grade in case original ones become lost or damaged. Use only cap screws of size and grade specified in text.

Use a 5/8-18x1¼ inch cap screw to attach each side of protective frame to platform assembly. Slack off hoist and allow protective frame support brackets to rest on axle housings. Install "U" bolts around axle housings and through holes in support brackets, and install lockwashers and nuts. Torque "U" bolt nuts evenly to 70-80 ft.-lbs. (95-108 N·m) and cap screws to 100-115 ft.-lbs. (136-156 N·m).

CAUTION: "U" bolts are made of special material and should not be substituted by any other type of "U" bolt. Use only "U" bolts supplied by Allis-Chalmers for this purpose.

Connect protective frame wiring harness to tractor wiring harness and install wiring harness clamp. Reconnect park lock cable to park lock lever using a plastic tie wrap to attach cable to its anchor. Feed control cables through holes in rear of protective frame as shown in Fig. 174. Install center platform section using 3/8-16x3/4 inch, grade 2 cap screws with flat washers, lockwashers and nuts. Torque these cap screws to 20-24 ft.-lbs. (27-33 N·m). Install floor mat and retainer clips. Install left hand panel on console.

General Torque Recommendations

Use the following torque *recommendations* as a guideline when a specification for a particular fastener is not available. In many cases manufacturers do not provide torque specifications, especially on older models.

Consider fastener condition carefully when referring to either a recommendation or a specification. If fastener reuse is appropriate, select the minimum value to account for fastener stretch. Softer fasteners or those securing softer materials, such as aluminum or cast iron, typically require less torque. In addition, lubricated or unusually long fasteners typically require less torque.

Determine fastener strength by referring to the grade mark on the bolt head. The higher the grade is, the stronger the fastener.

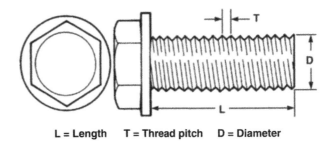

L = Length   T = Thread pitch   D = Diameter

Determine fastener size by measuring the thread diameter (D), fastener length (L) and thread pitch (T).

Size and Pitch	SAE grade 1 or 2 bolts	SAE grade 5 bolts	SAE grade 8 bolts
1/4—20	4-6 ft.-lbs.	6-10 ft.-lbs.	9-14 ft.-lbs.
1/4—28	5-7 ft.-lbs.	7-11 ft.-lbs.	10-16 ft.-lbs.
5/16—18	6-12 ft.-lbs.	8-19 ft.-lbs.	15-29 ft.-lbs.
5/16—24	9-13 ft.-lbs.	15-21 ft.-lbs.	19-33 ft.-lbs.
3/8—16	12-20 ft.-lbs.	19-33 ft.-lbs.	28-47 ft.-lbs.
3/8—24	17-25 ft.-lbs.	26-37 ft.-lbs.	36-53 ft.-lbs.
7/16—14	22-32 ft.-lbs.	31-54 ft.-lbs.	51-78 ft.-lbs.
7/16—20	27-36 ft.-lbs.	40-60 ft.-lbs.	58-84 ft.-lbs.
1/2—13	34-47 ft.-lbs.	56-78 ft.-lbs.	80-119 ft.-lbs.
1/2—20	41-59 ft.-lbs.	64-87 ft.-lbs.	89-129 ft.-lbs.
9/16—12	53-69 ft.-lbs.	69-114 ft.-lbs.	102-169 ft.-lbs.
9/16—18	60-79 ft.-lbs.	78-127 ft.-lbs.	115-185 ft.-lbs.
5/8—11	74-96 ft.-lbs.	112-154 ft.-lbs.	156-230 ft.-lbs.
5/8—18	82-110 ft.-lbs.	127-175 ft.-lbs.	178-287 ft.-lbs.
3/4—10	105-155 ft.-lbs.	165-257 ft.-lbs.	263-380 ft.-lbs.
3/4—16	130-180 ft.-lbs.	196-317 ft.-lbs.	309-448 ft.-lbs.
7/8—9	165-206 ft.-lbs.	290-382 ft.-lbs.	426-600 ft.-lbs.
7/8—14	185-230 ft.-lbs.	342-451 ft.-lbs.	492-665 ft.-lbs.
1—8	225-310 ft.-lbs.	441-587 ft.-lbs.	650-879 ft.-lbs.
1—14	252-345 ft.-lbs.	508-675 ft.-lbs.	742-1032 ft.-lbs.
1 1/8—7	330-480 ft.-lbs.	609-794 ft.-lbs.	860-1430 ft.-lbs.

NOTES